C0-ATX-005

Orchid Biology—Reviews and Perspectives, II

Other Books and Reviews by Joseph Arditti

Orchid Biology: Reviews and Perspectives, I

Factors Affecting the Germination of Orchid Seeds (*Botanical Review 33* [*1*])

Experimental Plant Physiology (with A. S. Dunn)

Experimental Physiology: Experiments in Cellular, General, and Plant Physiology (with A. S. Dunn, first author)

Experimental Animal Physiology (with A. S. Dunn, first author)

Orchid Biology

REVIEWS AND PERSPECTIVES, II

EDITED BY

JOSEPH ARDITTI

Department of Developmental and Cell Biology
University of California, Irvine

COMSTOCK PUBLISHING ASSOCIATES a division of
CORNELL UNIVERSITY PRESS | Ithaca and London

First published 1982 by Cornell University Press.
Published in the United Kingdom by Cornell University Press Ltd.,
Ely House, 37 Dover Street, London WIX 4HQ.

International Standard Book Number 0-8014-1276-5
Library of Congress Catalog Card Number 76-25648
Printed in the United States of America
Librarians: Library of Congress cataloging information appears on the last page of the book.

The paper in this book is acid-free and meets the guidelines for permanence and durability of the Committee on Production Guidelines for Book Longevity of the Council on Library Resources.

CONTENTS

ILLUSTRATIONS

Plates

Figures

BOARD OF EDITORS

The following individuals have acted as a board of editors by providing comments, ideas, advice, reviews, and suggestions.

AUTHORS

JOSEPH ARDITTI is Professor in the Department of Developmental and Cell Biology, University of California, Irvine. He joined the department in 1966, one year after receiving his Ph.D. from the University of Southern California. Dr. Arditti became interested in orchids as an undergraduate while working part-time as a grower for the late Roy J. Scott, who at the time was a well-known orchid amateur in California. His research centers on orchid development, physiology, and phytochemistry.

POPURI NAGESWARA "DANNY" AVADHANI received his M.Sc. from Andhra University, India, and his Ph.D. from Durham University. He is presently Associate Professor of Botany at the National University of Singapore. Dr. Avadhani has worked on problems related to carbon fixation since his postgraduate years and developed a special interest in orchid physiology after joining the National University of Singapore in 1960.

CHEN SING-CHI (S. C. Chen or Xin-qi Chen) graduated from the Agricultural College of Fukien in 1953 and was a postgraduate student in the Department of Plant Taxonomy and Phytogeography, Institute of Botany, Academia Sinica, Peking, for four years. He specializes in Orchidaceae and Liliaceae and is one of the authors of several volumes of *The Flora of China.* At present he is Associate Professor of Botany at the Institute of Botany, Academia Sinica, and a member of the standing committee for the compilation of *The Flora of China.*

MARK A. CLEMENTS is a staff member of the National Botanic Gardens, Canberra, Australia. He was educated at Canberra College of Advanced Education, taking a degree course in applied science with a major in ecology. A special interest in orchids, particularly the southern Australian terrestrial species, led to his present area of research, the symbiotic germination of the terrestrial orchids with particular emphasis on the specific nature of the relationship between fungi and orchid seeds. He is interested in all aspects of this group of orchids.

GALFRID CLEMENT KEYWORTH "STALKY" DUNSTERVILLE was born in England, the son of Stalky in Rudyard Kipling's *Stalky & Co.,* and spent his childhood in India. On the advice of his father, he majored in the then new science of petroleum engineering at University of Birmingham and, after holding many positions with Shell Oil Company throughout the world, eventually became president of its Venezuelan subsidiary. His interest in nature led him to orchids, and his many talents made him one of the most productive and eminent orchidologists of our time. In addition to numerous articles published in orchid publications around the world, Mr. Dunsterville has written several books—the best known of which is the six-volume *Venezuelan Orchids Illustrated,* coauthored with Leslie Garay of Harvard University and produced in collaboration with his wife, E. "Nora" Dunsterville.

E. "NORA" DUNSTERVILLE has collaborated with her husband in the production of *Venezuelan Orchids Illustrated.* In fact, he freely admits that without her help "these books might never have seen the light of day."

ROBERT ERNST was trained as a chemical engineer and first became interested in orchids as a hobby. Subsequently he combined his interest in orchids and scientific training and became an Adjunct Assistant Professor in the Department of Developmental and Cell Biology at the University of California, Irvine, where he earned his Ph.D. in 1979. His research interests are centered on the biological effects of surfactants and the physiology and development of orchids.

GERTRUD FAST received her doctorate from the University of Munich in 1948 and joined the staff of the Institute for Soil Science and Plant Nutrition (headed by F. Penningsfeld) in Freising-Weihenstephan, West Germany—a center for agricultural and horticultural research in Bavaria. When comprehensive studies on nutrient requirements of ornamental plants were initiated there, she undertook to study the orchids and quickly became an internationally known authority on the subject.

CHONG JIN GOH is Senior Lecturer in the Botany Department, National University of Singapore. After receiving his Ph.D. from the University of Newcastle-upon-Tyne, England, in 1967, he joined the staff of the National University of Singapore and is currently working on various aspects of orchid physiology.

GEOFFREY HADLEY, Senior Lecturer in the Botany Department, University of Aberdeen, Scotland, holds a doctorate in fungal physiology from the University of Birmingham. After working on the physiology of sporulation in fungi, he turned to orchid mycorrhizal research as successor to Dorothy G. Downie at Aberdeen. He has investigated aspects of host-fungus relations, mainly in north temperate orchids, and worked for a year with tropical species in Malaya.

ROBERT M. HAMILTON, formerly Associate Professor at the University of British Columbia School of Librarianship, later the Assistant Librarian for Collections at U.B.C., has compiled several reference books on Canada and recently has produced a number of indexes to orchid illustrations and guides to the care of orchids. He is now retired and lives in Richmond, British Columbia.

GORO NISHIMURA has taught floriculture at the Osaka Horticultural High School since 1978. He was educated at the Shimane and Chiba universities and holds B.Sc. (1969) and M.Sc. (1972) degrees in agriculture. Between 1972 and 1978 he studied orchid morphology at the University of Osaka Prefecture. During this period he spent two years (1974–1976) doing research work at the University of California, Irvine. At present his research centers on the morphology of orchid seedlings.

HUGH A. POOLE is Laboratory Manager and Chief Consultant with A & L Southern Agricultural Laboratories, Ft. Lauderdale, Florida. He received his B.S. (1970) and M.S. (1971) degrees in ornamental horticulture from the University of Florida, Gainesville, and his Ph.D. in floriculture at Cornell University in 1974. From 1974 to 1980 Dr. Poole was Assistant Professor in the Department of Horticulture, Ohio State University, and an extension research specialist at the Ohio Agricultural Research and Development Center, Wooster. His research has centered on floricultural production problems including nutrition, media, growth regulators, irrigation, and energy management for commercial greenhouse operators. His present consulting covers production and management problems for commercial growers.

ADISHESHAPPA NAGARAJA RAO received his B.Sc. (Hons.) and M.Sc. degrees from Mysore University, India, and Ph.D. from the University of Iowa and is Professor of Botany at the National University of Singapore. He is the author of many research papers on the anatomy, embryology, morphogenesis, tissue culture, and conservation of tropical orchid species.

THOMAS J. SHEEHAN is Professor of Ornamental Horticulture, Department of Ornamental Horticulture, Institute of Food and Agricultural Sciences (IFAS), University of Florida, Gainesville. He received his A.B. in botany in 1948 at Dartmouth College and his M.S. (1950) and Ph.D. (1952) degrees in floriculture from Cornell University. Dr. Sheehan has been a member of the IFAS faculty since 1954. His teaching and research have centered around the physiological problems involved in the production of floricultural crops. His speciality is orchids, and he is the coauthor of *Orchid Genera Illustrated* with his wife, Marion R. Sheehan, and the author of the chapter "Orchids" in *Introduction to Floriculture,* edited by Roy A. Larson.

MICHAEL S. STRAUSS earned his doctorate at the University of California, Irvine, with a dissertation on flower physiology of orchids. He is now Assistant Professor of Biology at Northeastern University in Boston and authors a regular column on orchid tissue culture for the *Orchid Review.*

TANG TSIN (T. Tang or Jin Tang) received his B.Sc. at the University of Agriculture, Peking, in 1926. He was Associate Professor and Associate Curator of the Fan Memorial Institute of Biology until 1949. From 1935 to 1938 he worked as a visiting botanist at the Royal Botanical Garden, Kew, England, and at herbaria in Europe. Since 1949 he has been a Professor in the Department of Plant Taxonomy and Phytogeography, Institute of Botany, Academia Sinica, Peking. His publications include several volumes of *The Flora of China* in collaboration with F. T. Wang.

NORRIS H. WILLIAMS is Associate Curator in the Department of Natural Sciences, Florida State Museum, University of Florida, Gainesville. He received his Ph.D. in 1971 from the University of Miami. His research centers on the systematics and evolution of the Orchidaceae, pollination biology of species of the Orchidaceae, floral fragrances, and the biology of euglossine bees.

PREFACE

From the outset my concept of *Orchid Biology* was that it would develop into a series of volumes presenting a balanced scientific view of orchids through exhaustive and critical literature reviews by international experts. I edited the first volume with this goal in mind and am continuing along the same lines here.

The prefatory chapters in these books are patterned in format after those in the *Annual Review of Plant Physiology*, which are written by influential figures in plant physiology and provide personal insights into the work and minds of people who have made major impact on their fields of research. My policy is to invite eminent orchidologists who are past retirement age and who have been associated with scientific institutions to contribute prefatory chapters and write on any subject they wish.

The prefatory chapter by G. C. K. "Stalky" and E. "Nora" Dunsterville in this volume meets my requirements admirably. In it they describe an expedition to Auyán-tepui, a remote table mountain in Venezuela, where they collected and studied orchids. I was lucky to accompany Stalky and Nora on a similar trip to Angel Falls in 1966 and saw them at work. Those not fortunate enough to have done so can at least read about it, and, I hope, sense some of their kindness, generosity, humor, intelligence, and many talents.

Stalky is well known, of course, as author with L. A. Garay of Harvard University of the six volumes of *Venezuelan Orchids Illustrated.* While still in the field, Stalky often photographs or draws many of the orchids he studies; several of his photos and drawings appear in this book. I have watched him draw a tiny orchid amid clouds of hungry mosquitoes (with socks on his hands and a beekeeper's hat on his head as protection against bites) in the Venezuelan jungle. He and Nora take many plants from the jungle to their home atop a hill in the village of El Hatillo near Caracas, where Nora grows them. Altogether this account of a collecting expedition provides a real and rare glimpse into the creative process of a team of most productive, influential, and clever orchidologists.

Orchids were grown in China for thousands of years before they were fully appreciated in the West. This interest has continued into the present. It was natural, therefore, for me to invite Professor Tang Tsin, one of the most famous orchidologists in China, and Dr. Chen Sing-chi to contribute an article. Happily they agreed. I was given the first draft of the chapter on my visit to Peking in September 1979 and edited it there on trains to Tsingtao and Shanghai and in Canton before returning the manuscript for revisions and retyping. This chapter may well be the first review on the orchids of China to be published in the United States (or the West) in many years.

Mycorrhiza plays a very important role in the life cycle of orchids, since in nature their seeds cannot germinate without fungal infection, and the association between the two organisms continues for the entire life span of orchid plants. Orchid mycorrhizae have been the subject of many studies during the last eighty years, and a considerable amount of information has accumulated. Geoffrey Hadley, a leader in the field, reviews this subject in Chapter 3 and places it in a modern context.

I doubt if the pollination mechanisms of any other group of plants are as interesting as those of the orchids. Charles Darwin's book *The Various Contrivances by Which Orchids Are Fertilized by Insects* is certainly a classic, and so is *Orchid Flowers: Their Pollination and Evolution* by L. van der Pijl and Calaway H. Dodson. In Chapter 4 Norris H. Williams, a student of Dodson, reviews the role played by one group of bees in orchid pollination.

Although carbon fixation by orchids has interested scientists for approximately a century, most of the important research has been carried out during the last thirty years (the papers by Erich L. Nuernberg and his students are the first of the modern era). I wish we knew more about the subject, but enough information is available at present to justify a review. Having visited the Botany Department of the National University of Singapore almost annually during the last decade, I consider it in some ways to be my university away from home. For this reason I am especially pleased to coauthor Chapter 5 with my friends Danny, Chong Jin, and Rao.

The mineral nutrition of orchids is of considerable interest, not only to physiologists and ecologists but also to growers who would like to learn how to fertilize their orchids properly. Chapter 6 by Hugh A. Poole and Thomas J. Sheehan provides this information.

Most orchids are grown for their blossoms (the so-called jewel orchids are an exception), so it is not surprising that interest in the physiology of flowers and their induction is considerable. The available information is widely scattered, and a review was therefore long overdue and very necessary. Our hope is that Chapter 7 will spur additional research on the subject. I am very glad to have as coauthors Chong Jin Goh, a leading authority on the subject, and my former student Michael S. Strauss.

Preparation of indexes is a duty authors and editors do not relish and an undertaking in which they have limited expertise. Fortunately, expert indexers do exist. Robert M. Hamilton, who also knows orchids, is one of them and has indexed this volume.

My mail often brings practical questions about orchids. To answer at least some of them I decided to include practical manuals as appendixes to these volumes. The positive reaction to the tissue-culture propagation manual in the first volume indicates that they serve a useful purpose. The appendix in this volume deals with orchid seed germination. Unlike the appendix in volume I, the one here was written by several experts: Mark A. Clements, Gertrud Fast, Geoffrey Hadley, Goro Nishimura, Robert Ernst, and myself.

As in the first volume, each chapter (except the first) was subjected to prepublication review by at least two experts from the *Orchid Biology* Board of Editors to ensure accuracy and accepted scientific standards. In addition I have edited each chapter to achieve a degree of uniformity in style. Since orchid nomenclature is in a constant state of flux, I have not insisted on any "correct" name, spelling, or system of classification.

For simplicity or because of their use in the original literature it is sometimes necessary to employ trade names of chemicals, equipment, apparatus, products, and so forth. Such use does not imply endorsement of the named products, and no such endorsement is intended. Nor is any criticism or lack of endorsement implied or intended for similar items that are not mentioned.

I thank the Board of Editors for their help and advice, Dee Ostlin and Kay Franklin for typing, and the staff of Cornell University Press for maintaining the high standards they established in the first volume.

JOSEPH ARDITTI

Irvine, California

1

Auyán-tepui: Reminiscences of an Orchid Search

G. C. K. DUNSTERVILLE and E. DUNSTERVILLE

Somewhat late in lives previously devoted to business, we developed an urge to study tropical orchids as a serious hobby and were lucky to find ourselves in Venezuela, an ideal locality for enjoying this new interest. We soon lost our fears that the material for our studies would prove too limited to last out the coming years: previous investigators, we found, had by no means swept the field clean. More important still, we soon realized that in the orchid family the flowers themselves are such an essential element in botanical classification that the professional field botanist must collect with this in mind: the taxonomists are not going to thank him for a fine collection of sterile material. So it is inevitable that such collectors are going to pass over a very considerable number of plants that do not happen to be in flower at the time and this practice would leave a gap that we could help to fill not only by collecting flowered plants for others to study but also by collecting live but unflowered plants of possible interest so that when they did flower we could feed new material to the taxonomists and, for that matter, study them ourselves. If eventually, by illustrating all our finds,[1] describing them, and then picking the brains of experts we could help to spread knowledge of them, any efforts we might make in exploring for orchids would be amply justified. This idea of field research into Venezuela's orchids took hold and absorbed us, eventually leading us by foot, water, air, or road to a hundred out-of-the-way corners of this beautiful country, on short trips and long trips, cheap ones and costly ones, dull ones and highly memorable ones—such as when, in 1964, we explored for orchids on the top of Auyán-tepui, one of the largest of Venezuela's enormous "Lost World" sandstone table mountains, known as *tepuis* or *cerros,* depending on their locality. At that early stage in our orchid-hunting career, very few people had set foot on this summit, and even fewer had gone beyond the transverse cliff, several hundred feet high, that separates the northern half from the southern and which we learned to refer to as the Wall.

To find a new orchid species, or even perhaps a new genus, was the special hope that led us on, and with this in mind the expedition we were now making had both the northern and the southern halves of this great *tepui* as its objective. Five days out from our starting point at Guayaraca at the foot of the *tepui,* heavy clouds were gathering as we halted for the day not far from the base of the Wall. Distant thunder could be heard and some heavy raindrops were spattering down. A big storm was clearly brewing up trouble for us. Rudy, our guide and one of the very few people who had ever set foot on the northern half of the *tepui,* glanced at our self-supporting "igloo" tent, comfortably set up on the clean stone bed of the almost dry river, and advised a move to higher ground in the forest lining the river banks: flash-flooding rivers such as this one are a very real menace not only to comfort but on occasion also to life and limb. We made rapid calculations on a piece of paper. A fall of two inches of rain in two hours on the twenty square miles of almost bare rock that formed the upstream gathering area of the river was well within the bounds of possibility. If the immediate drain-off of this water was 80 percent, then 400 million cubic feet of water would be passing through our twenty-yard-

[1]See Plates 1-3, 1-4, 1-5, 1-6, 7-1-**A** and 7-1-**B** for examples.—Ed.

Plate 1-1. E. "Nora" and G. C. K. "Stalky" Dunsterville observing an orchid (photo courtesy of the Dunstervilles).

wide streambed over a period of, say, four hours, allowing for some hold-back in the riverbed itself. The result would be a rise of water anywhere from two feet flowing at two hundred miles per hour to two hundred feet flowing at two miles per hour. It could be, of course, that somewhere we had lost a decimal place or two, but a glance at vegetation on the banks showed that rises of at least ten feet had been common. There was no doubt that Rudy's advice should be followed.

By good fortune the forested banks of the river at this point were filled with sand from previous floods. A liberal use of the machete quickly cleared a small piece of relatively flat ground large enough for our tent, but by the time the tent had been set up dusk was on us. Supper was cooked and eaten in a hurry and we turned in. During the night rain poured down while from our safely elevated site we listened with unanxious interest to the increasing sound of the river, congratulating ourselves on our (or Rudy's) foresightedness. Around two in the morning the weather cleared and we crept out to see how close to the top of the bank the river had come. The bare rock was still there—the total rise had been a scant three inches. Only later, examining the river, did we realize that at this point, as indeed in several other places, the riverbed was deeply cracked and most of its water here flowed underground for quite a long stretch.

But we were not shamefaced over our unnecessary caution. On any trip into remote and basically inaccessible places such as where we were, caution had to be the keynote and a dozen worn-out platitudes take on real meaning. "Better safe than sorry" is as valid

Plate 1-2. The south face of Auyán-tepui from the base at Guayaraca (photo courtesy of the Dunstervilles).

on a floodable riverbed as "Stand not upon thy dignity but slide upon thy butt" on a slippery rock slope. "Make haste slowly," "Look before you leap," "All that glitters is not a safe handhold"—the list of such wise tags is endless. It was costing us a lot of money and muscle to get where we were, and it would be silly to waste it in an early demise just because the dictionary defines a platitude as "an empty remark made as if it were important."

In this same spirit of sympathy with the platitudes we were (as always) carrying anti-snake-bite serum and large pieces of yellow cloth to be spread on the ground as a signal to passing aircraft (if any?) that help was needed. Neither then nor later was either of these items needed, but their weight was negligible and their safety potential great.

Our expedition was the direct outcome of a visit to Auyán-tepui the previous year when, having reached the top, we had been turned back by bad weather. We had seen enough to realize that only the overused word "fantastic" could begin to describe it and knew that we would not rest happy until we had visited it again with time to explore it in some detail and in weather to match the needs of exploration. The intervening months had given us an opportunity to make a map of the summit from old aerial photos (from which it seemed no proper map had yet been made) and then to work out plans for our present three-to-four-week expedition. Rudy knew the way to get up and hoped he could remember the way across to where the famous airplane of Jimmie Angel still lay abandoned ever since 1937 when Angel, hoping to find gold, landed it in boggy ground. He failed to find his gold but found his falls instead.

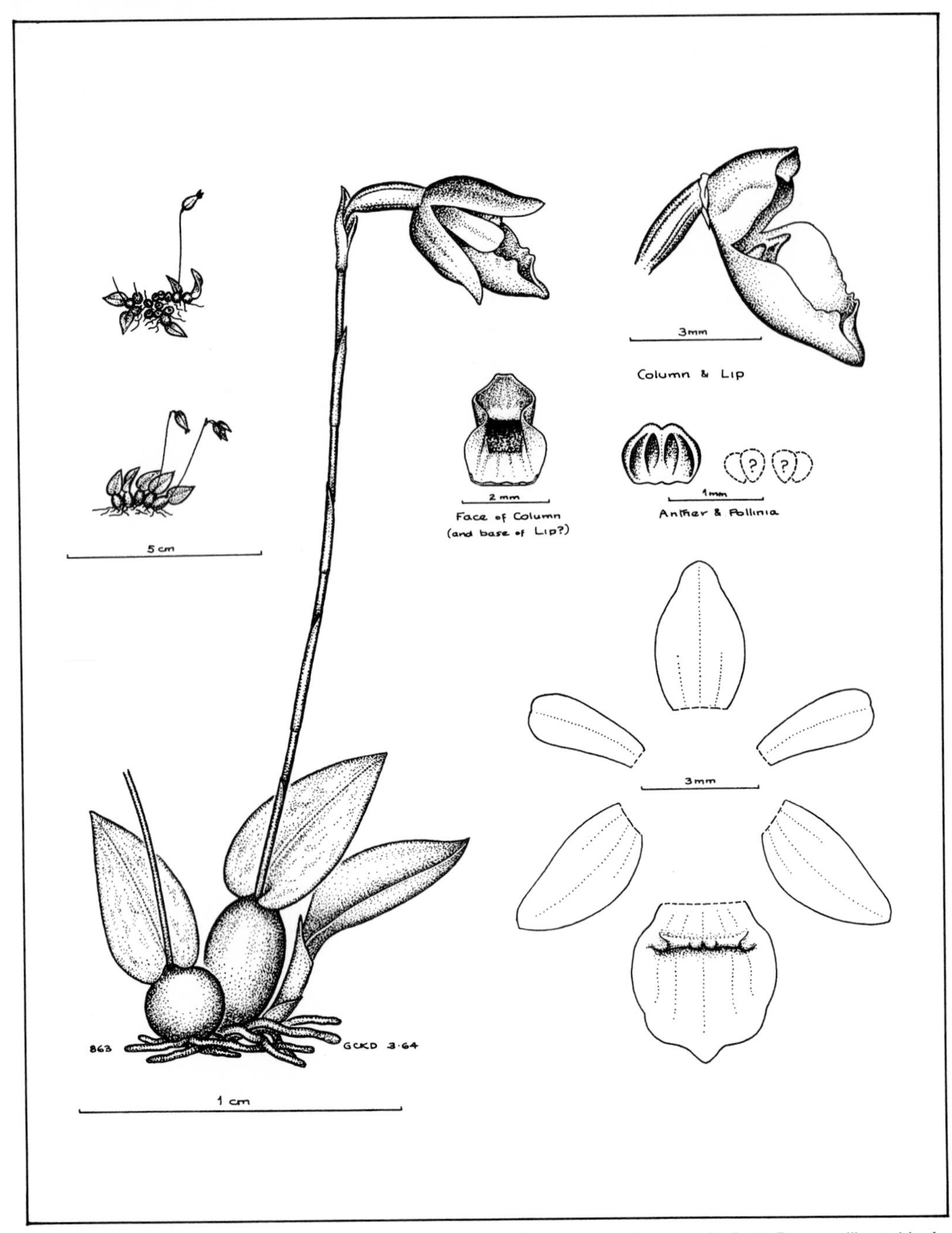

Plate 1-3. Pinelia alticola Garay & Dunsterv. (drawing by G. C. K. Dunsterville, from G. C. K. Dunsterville and L. A. Garay, *Orchids of Venezuela,* Botanical Museum of Harvard University, 1979).

Making the map had given us a fair idea of what we were facing. Roughly, this not-so-flat-topped mass of rock is in the form of a thick-limbed letter V. The base is about 7 miles wide, the northwest limb some 25 miles long, the northeast limb shorter, and the

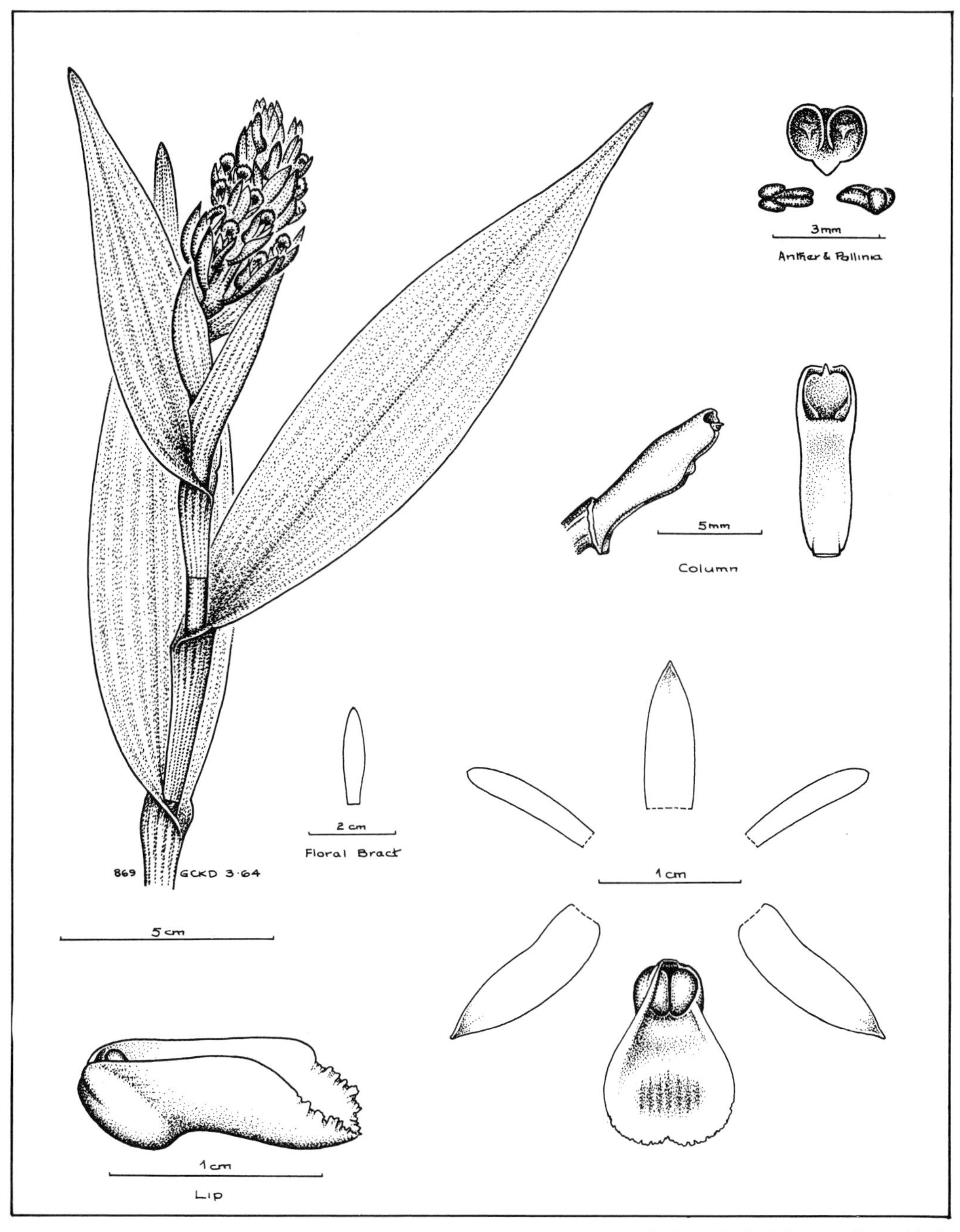

Plate 1-4. Elleanthus norae Garay & Dunsterv. (drawing by G. C. K. Dunsterville, from G. C. K. Dunsterville and L. A. Garay, *Orchids of Venezuela,* Botanical Museum of Harvard University, 1979).

two are spread apart at the northern end by about 20 miles. The heart of the V is a deep and relatively narrow gorge with giant cliffs, Angel Falls spouting from one of them. The whole of this enormous V, 300 square miles or more in area, is raised 5,000 to 6,000 feet above the surrounding country on great cliffs that form a coastline so far broached in only one spot, where years ago the topographer-explorer Cardona found where a diagonal cleft in a 600-foot vertical cliff at the south end provides an access. Elsewhere the cliffs run unbroken, a sheer drop of 600 to 3,000 feet, from the base of which steep slopes run down to the plains below.

From this 8,000-foot southern tip of the V, the sandstone mass dips gently to the north while most of the surrounding walls remain high, thus concentrating the drainage toward the central part of the V, and where the streams of this portion gather to make the drop into the central gorge of the Churún River the altitude is but little over 5,000 feet. Here, close to this point of confluence, was the riverbed campsite that we had so rapidly albeit unnecessarily, abandoned.

As we flew over the *tepui* before landing in the savanna near the southern tip, the sight of an almost waterless Angel Falls had confirmed expectations of a very dry mountain top and a trip severely handicapped by having to carry our water with us, a "portage" problem augmented by the fact that, anticipating many treeless campsites on the bare-rock summit, we were taking a bulky tent instead of our usual light jungle hammocks and plastic tarpaulins. Cooking and eating breakfast, washing up, breaking camp, and packing for the road all take time, and with a fairly large party of Indian porters like ours, the first step along the trail is seldom made before 8 A.M. This first morning at our camp at Guayaraca below the southern tip of the *tepui* was no exception. Our Indians tied seemingly unmanageable conglomerations of bundles onto their homemade *guayare* backpacks, hitched the loads onto their backs, and at the scheduled hour we were on our way. Normally about 25 kg is a fair load for an Indian porter on a trip where much climbing is in view, but at the start there is always more food-weight than at the end, and as the porters had demanded twice the daily wages they had received the year before, we felt that a call for some extra effort was called for, and they started with about 35 kg per load. Surprisingly, they cheerfully accepted this. Later in the excursion we noted that the parts they most seemed to enjoy were the toughest, any trip or stagger in places where even an unloaded person might be excused some instability bringing much kidding of the offender and peals of high-pitched laughter.

Starting from Guayaraca, our first day's trek took us through very steep forest to the flat "step" of Danta at about 5,000 feet, and then on up to camp for the night at 6,000 feet in the shelter of a great isolated boulder called "El Peñon" on the steep open slopes between Danta and the foot of the final barrier-cliff. Bellbirds had been clanging their anvils at us from near and far, and despite the energy needed for the steep ascent we felt very much at peace and "at home", a feeling that in later years we found to be characteristic of all our visits to the Venezuelan Gran Sabana region of the *tepuis*.

We had found the Danta plateau just as full of ground orchids as on our first trip, but "keep moving" was the present need, and all we could do was to promise ourselves a closer look at them on the way back. From El Peñon the climb to the base of the 600-foot vertical cliff that bars the top was dry where the year before it had been soggy-wet, and the general aspect of the surroundings were almost as different as desert from swamp. This time, in the dry, the tangled roots of the "Devil's Network" guarding the final cliff had lost the slippery terrors of our previous trip, but was still an uncomfortable struggle that made us very glad to have younger and more agile muscles carrying the loads for us.

Time and again, while we were teetering on some perilous spot, crawling under some restricting overhang, or clambering hand over hand up (or down) a tricky section, that feeling of "What a hero am I!" would be jarred by the thought that the heavily and cumbersomely laden Indians were not only doing the same thing but doing it faster, with much less fuss and, in bare feet, much more securely.

Once more the contrast between "last time" and "this time" struck us as, after a pause for elevenses, we tackled the final stretch to the top, consisting of a diagonal gash cutting steeply across the face of the cliff. Before, slippery rocks had made every upward movement questionable, every downward one a near disaster. This time we had that mountain-goat grip on good dry rock that made us feel we could almost climb straight up a vertical face. And as we finally stepped out on the bare flat rocks of the top at 8,000 feet, (a point known as El Libertador) we were greeted by fair views and fine sunshine, so much more welcome than the swirling mists and rains of our earlier visit even if the surroundings thereby lost some of their dramatic aspect.

On the brink of one of the many chasms splitting the surface at this end of the mountain we stopped for lunch. Despite the general dryness, Ignacio, the Indian in charge of the porters was convinced that there would be places ahead of us with sufficient water for camping so we had decided not to handicap ourselves by carrying water with us. Nevertheless, the weight of our food was too much to be taken to the top in one haul from the bottom, and to allow some of the younger porters to go back the next day to bring up the rest we decided to make the afternoon's trek a short one and to camp at a patch of dwarf forest a little over a mile from the spot where we were lunching—that is to say, a little over a mile for a rifle bullet. For us, with a number of enormous gashes sliced through the rock at right angles to our direct course and stretching far to left and right, the effective distance was something altogether different. Deviating first a long way to the right we managed to reach the bottom of the first chasm and climb its far side, then had to make more such diagonal or even backward tacks to traverse succeeding ones. Once we left our lunch spot we were on ground new to us, and the drama of this fantastic scenery was once more impressing us, to be reinforced as each successive day took us farther into the heart of the mountain. From El Libertador northward to the base of the Wall, everything is mainly a vast expanse of bare rock dipping gently to the north, almost black in color except where new rock-falls show a light pink, unweathered surface, and is slashed with long and sometimes seemingly bottomless chasms, some wide enough to have forested floors, some so narrow they can be jumped by a bold spirit. Far in the distance was the Wall, and although we were now on the "flat" summit great cliffs bounded our horizon on both sides while all around were enormous piles of boulders and weird, deeply weathered rock-shapes. Even the relatively flat rock underfoot was a maze of curious forms cut from the sandstone by wind and rain. In its own right this was a landscape dramatic enough to take our breath away. To realize that it was an isolated piece of Venezuela, twice the size of the Island of Barbados, yet lifted up to 8,000 feet and ringed by hundreds of miles of great vertical cliff walls left us bereft of adjectives adequate to express our feelings.

Fairly early in the afternoon we reached the promised water and set up camp. Thanks to the dryness of the rock, our nylon rope had been needed only at one spot on the way, where a slip while traversing the steep side of a chasm could have led to a fatal drop to

Plate 1-5. Catasetum pileatum (photograph by G. C. K. Dunsterville).

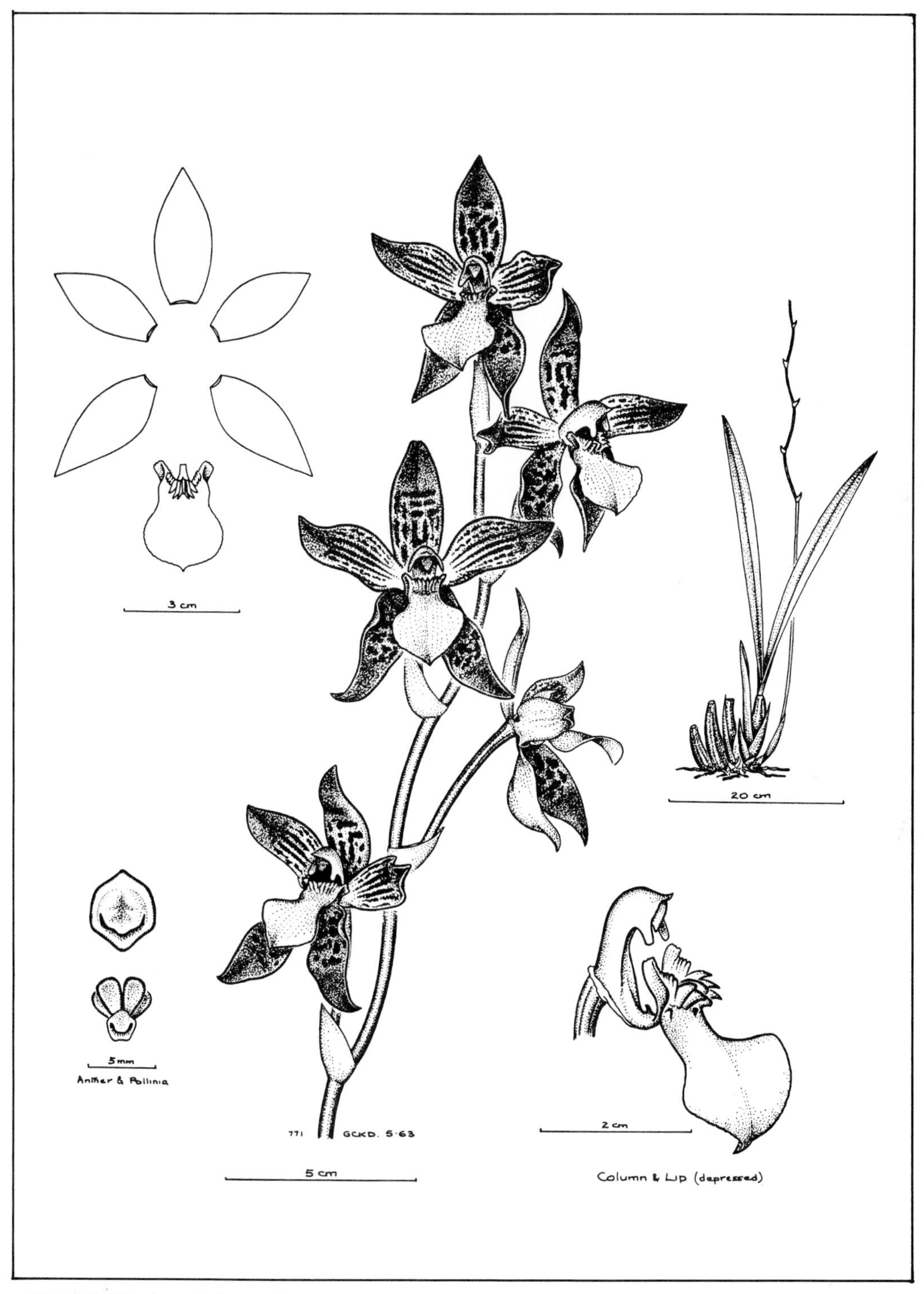

Plate 1-6. Mendoncella burkei (Rchb.f.) Garay (drawing by G. C. K. Dunsterville, from G. C. K. Dunsterville and L. A. Garay, *Orchids of Venezuela,* Botanical Museum of Harvard University, 1979).

the bottom. The new campsite was at about 7,400 feet on the edge of a dense copse of stunted shrubby trees perhaps five acres in extent, growing in what appeared to be almost pure peat. In wet weather this would surely be extremely soft and boggy, but under the dry conditions prevailing its peaty nature was soon proven by its starting to smoulder under our campfire. Hidden in the tangled roots in the heart of this copse lay pools of stagnant brown water, looking distinctly unappetizing in the gloomy light that was all that could filter through from the bright sunshine outside. But water is water, and brown but perfectly palatable water is nothing unusual in these parts. Stagnant though they were, these pools could have little animal and certainly no human contamination, and if the content of insect corpses was high we were not disposed to worry about it. Properly boiled it would, and did, make tea; though the altitude took some of the sting out of the boiling temperature and the tea was most decidedly not up to standard. True Limeys, while we are cosmopolitan enough to feel a meal unfinished without a cup of black coffee at the end of it, we feel tea even more indispensable as a pick-me-up after a day's work.

Immediately after our arrival at this spot, some of the young porters had been sent hurrying back to El Peñon to spend the night and return the next day with the missing loads. We wandered around until supper, orchid hunting, of course, and went early to bed. That night the temperature dropped to 44°F, and we were glad not only of our tent and sleeping bags but of yet more steaming tea.

Before our first trip to the summit, a kind friend had told us that he had reliable information (from whom he did not say) that the top was orchidless and our trip would be a waste of time. And indeed, on the bare rock around El Libertador we had found almost none apart from a small *Habenaria mesodactyla* that grows almost in our backyard outside Caracas. So we had been left wondering what success, if any, we would have on the more extended exploration of the summit we were now starting. We need not have worried. There was evidence here and there from gnarled remains of old fallen tree stumps that at least the southern half of the top had once borne considerable expanses of true forest, not just the scattered patches of dense dwarf growth set in surroundings of basically bare rock that now faced us. But wherever this rock bore even the thinnest sprinkling of soil it was not easy to avoid treading on the many orchid plants taking advantage of it. Most of these were true soil-lovers, but some were evidently epiphytic species that had adapted with considerable success to a treeless existence. Dominant among the former were large and healthy plants of *Eriopsis biloba*, mostly in full bloom and flaunting their highly colored many-flowered racemes high above the other short vegetation. Frequent plants of *Oncidium warmingii* also sent up delicate spikes of small but most attractive flowers, and large plants of *Oncidium nigratum* with extensive suberect inflorescences were also well in evidence, their large pseudobulbous bases looking extremely healthy despite their exposed situation. Principal among the epiphytic orchids adapting to the treeless landscape were very healthy plants of *Acineta alticola* with a luxurious growth of pseudobulb and leaf, but whose normally pendent inflorescences could do no better than to burrow into the soil and there abort. With no chance to advance through seed production, the plants were nevertheless fully enjoying their static existence and looked well set for many centuries yet of life. Only in the boggiest of boggy ground on top, it seemed, were conditions so adverse to orchids that they spurned it, and even here an occasional species such as *Duckeella alticola* showed its head.

We continued orchid hunting around camp the next morning until the porters got back from El Peñon at midday. Leaving behind some supplies for the return trip, we then made a short afternoon move, mostly over bare rock and skirting great chasms here and there, to our next camping halt at about 7,000 feet. Here another small patch of dense dwarf forest (that we named "Oso Woods") gave wind shelter to the tent, and the soil was much less boggy as it drained easily into a wide stretch of bare, smooth, pinkish white sandstone dipping to the north in gentle steps. That this was, in wet season, a streambed was shown by the whiteness of the sandstone: "unwashed" rock would have weathered to the usual very dark grey. At the moment, however, only a small trickle of water dribbled its way down, but was more than enough for tea-making, rice-boiling, and, where some pools had accumulated below, a very good bath in water that reached 70°F where it had absorbed and retained some of the warmth of the sun. From the point of view of orchids, this spot was marked by large plants of *Epidendrum nocturnum* with stems to a meter tall, leaves 17 × 7 cm, and magnificent flowers of more than 15 cm spread.

So far we had seen little animal life on the summit, but at this site we were visited by some delightfully tame birds with red heads and yellow bellies, and aggressively inspected from extremely close range by a pugnacious hummingbird, most resentful of our intrusion. Later we saw more of both and a number of black oriolelike birds flitting around in squeaky groups. An occasional crested sparrow showed itself, and on one trail through close forest we came on a fat *poncha* ambling along, a bird rather like a guinea fowl on longish legs and much prized for eating. Thanks to an ample supply of other types of food we ourselves were under no temptation to harm it, but to the Indians anything that moves is at all times fair game, and we had difficulty in restraining them from stoning the bird. On other occasions they showed great skill in attracting a variety of small birds by making squeaking noises, sucking the back of the wrist, and while we admired their success in this, we had trouble in stopping them from catching the poor things. Apart from birds, animal life seemed limited to a few lizards and a solitary small rat: while we had seen snakes on the way up, on the top we saw none. We woke one morning to a sound like the cough of a jaguar, but where there is "tiger" there is almost always plenty of tapir, or at least some sizable animals for the jaguar to feed on. Although such prey are frequent on the lower slopes around Auyán-tepui we saw no positive traces of them on top, so perhaps there was also no jaguar. Nor, despite Conan Doyle and his Professor Challenger, could we see any iguanodon or pterodactyl—a bitter disappointment. The top was blessedly free of malevolent insects while we were there, though small ticks had been plentiful on the way up. In particular it was nice not to be in danger of wasps, which in the Venezuelan interior, together with an enormous species of ant, the Indians seem to fear more even than jaguar or snake.

Starting off the next morning we left behind not only more food for the return trip but also such spares in the way of boots, flashlight batteries, and the like that experience so far led us to feel were unlikely to be needed for another ten days or so, thus finally reducing our loads to the capacity of the porters without need for further double trips.

Until the very last part of the day's walk we were on rock (or treading on masses of terrestrially or lithophytically growing orchids) and the slope was all downhill. It was not hard, therefore, to cover six crow-flight miles by lunchtime even though the crow's miles proved very different from the ones we had to follow, winding our way hither and yon in a purposeful series of zig-zags to circumnavigate the continuous series of crevasses or other obstacles. Time and again we would wonder why we could not head straight across a large expanse of flat rock only to see the reason later when, from the side, we could peer into the depths of a long slit, invisible until it was almost under our feet.

Some day, someone with money and leisure will no doubt find a more direct trail across this mountain. In the meantime, however twisting the present trail may be, it is a trail that gets you there. We remembered our platitudes once more, and we followed that trail in the knowledge that a shortcut is the longest way round, and that any deviation might lose us a whole day or even more. The trail we were following was basically the one worked out by Angel's party on their escape from the plane, and we were following it more than a quarter-century later. In the woods and thick scrub we were later to encounter, old trails will remain visible for many years, but on bare rock the trail as such is invisible. Here it was marked mainly by small piles of stones (to which we added many more as we went) and by Rudy's previous turning over of loose stones to show their pink underbellies, contrasting strongly with the surrounding darkly weathered rock, a method that is valid only for the few years that it takes the newly exposed surface to weather to a matching darkness. Across boggy stretches the trail became faint in the extreme, which was scarcely surprising, but all in all, though we often strayed and had to search for the route, Rudy's memory of his earlier trip some seven years before was remarkably accurate.

Our north-trending trail first climbed to higher and more complicated ground on our left below a high north-south cliff, the face of which we followed as closely as copious old rock falls would permit. From near the end of this wall the trail tacked eastward until stopped by the vertical edge of a deep valley where an almost dry tea-colored stream was winding lazily from pool to pool on a rocky bed some 150 feet below. With more tacking motions we then did our best to follow this edge until finally the precipice lowered itself into true montane forest almost at river level. Here we were greeted by the towering ten-foot leafy stems of an *Elleanthus* species that seemed new to us and which was later named *E. norae* by Leslie Garay of Harvard, the first of several new orchid species eventually named for "E.D.", coauthor of this chapter.

Walking through the forest parallel to the river for two hours that afternoon we finally found a nice, almost dry patch of flat-rock riverbed and decided to camp. "Walking through the forest" does not, probably, bring up any image in the mind at all comparable to the action involved. Basically, the ground on either side of the river was just an enormous pile of once loose rocks and boulders, some the size of a highway trailer, many the size of three-story buildings. Many of the gaps between these rocks, but by no means all, had over the years accumulated vegetation and soil, and finally the forest had grown to mask it all. But to walk through it was anything but straightforward. To the right and left, between the tree roots, holes appeared, ten, twenty, maybe thirty feet deep. A solid-looking cover of leaf mold a foot to one side or other of the trail (or even sometimes directly on it) would fall away at a touch of the foot to reveal a lethal trap. Nothing was straight, in either the horizontal or vertical sense, and there was as much up-and-downing to the mile as there was twisting and winding. Any looking for orchids, of which there were many, had to be done from a standstill as once on the move eyes and feet remained inseparable. It was slow and fatiguing progress but never tedious and always fascinating, enriching our growing catalogue of orchid species with well-known epiphytic items such as *Mendoncella jorisiana* and rarities such as a mini-miniature *Pinelia* that proved later to be a new species and was named *P. alticola.*

Here and there along its course the river in dry season disappears for long stretches completely underground, and where we selected our campsite there was so little flowing on the surface that it seemed certain that more was below than met the eye on top. Enough remained, however, for our tea-making, and some good pools lower down offered tempting but still very cold bathing despite our drop to about 5,500 feet. This was near the site where later we felt it safer to move from the riverbed to the forest, but at the moment the weather was only beginning to threaten.

We left River Bottom camp on the morning after our arrival there, and a further two-hour struggle through the forest brought us finally to the base of the Wall that hinders access to the northwestern arm of the V that forms Auyán-tepui. The northeastern arm is all bare rock, shown by the air photos to be terribly broken up in chasms and gashes far greater than anything we had yet traversed. Nobody had yet explored it, and its apparent lack of vegetation held little to induce us to try, although it undoubtedly bears millions more orchid plants of the types we were already treading on. By contrast, the northwestern arm, or at least its eastern edge, showed on our air photos a complete change from our present rock-and-chasms to something more gently rolling and with large patches of dark vegetation which we assumed would prove to be forest.

It was on this northwestern part that Jimmie Angel had landed, and his party of four had perforce had to find a way down the Wall before they could reach the southern section and the only known "exit" from the top. Since then only nine "racionales" with accompanying Indians had broached this northwest arm before ourselves, almost half of this number being Rudy with three companions in 1959. Clear sign-posting of the way up the Wall had never been possible in the tumbled rocks at its base and it remained for Rudy to relocate the route. Fortunately some indications turned up to show the start of the climb through and up the rock fall which forms the lower half of the Wall, a steeply rising section covered by the usual thin layer of "soil" and scrubby forest, and filled with the usual booby traps for the unwary foot. Thanks to basically dry weather the final cliff section was not overly difficult, and enough clean handholds were available to lead us to scorn the rope that Rudy was dying for us to use—quite rightly on his part, as a slip or a false handhold would certainly have led to a serious result, not only for the slipper but for everyone else below him. But the porters were going up with no rope, packs and all, and we felt it would be just too sissy to use it ourselves, though we realized that coming down (always more difficult) it would be definitely advisable to be safe rather than sorry.

For a half-mile back from the top edge of the Wall, the ground continues to rise slowly and is covered by piles of giant rocks, tumbled this way and that, making a labyrinth that must have been heartbreaking to Jimmie Angel's party trying to find a way through it.

Never *on* these rocks, but always twisting and winding through gaps between them and often under them, the trail we followed would make the famous Hampton Court Maze in London look simple. Sometimes going forward but more often seeming to go backward, we were still buried in their depths when we saw a pool of water gleaming from a deep hole between rocks and decided to stop for lunch. Too thirsty to wait while one of the porters struggled to find a way to fill our canteens from the pool, we sampled water stored in some of the many bromeliad (brocchinia) plants around us. Each plant produced a cup of cool water liberally loaded with dirt and dead insects, but when the bulk of the impurities had settled, what remained was quite palatable.

After lunch our struggle through the maze continued, but old trail signs were more obvious and our trail-losings were not so frequent. Finally we were through, and topping the gentle rise beyond we looked out over the new part of the mountain. True to the impression of the air photos, the landscape was now entirely different. Before us stretched a gentle downslope of open grass, leading to rolling hills of grass and forest that we hoped would eventually lead us to the Angel plane and to many new orchids. For a change we could march straight forward and would no longer have to walk five miles to progress two. Or so we thought, until we realized that we were about to enter a bog, and even making a great loop to avoid it we still had to trudge tiringly through soft ground about as comfortable to walk over as sacks of loosely packed straw. Finding the start of the trail over this section was not easy, but once found it was generally not too difficult to pick it up again, and eventually when all our feet had tramped it down it looked good for many more years yet. Finally, when our course had at last maintained a northerly direction for long enough, we skirted some more chasmed rocks on our left and reached a nice campsite on a small river with the usual half-dry stream sliding down the usual rock-slab bottom. By now the cloudy weather was threatening to turn into rain, so we set up one of our plastic tarpaulins over a suitable cooking-eating spot on some boulders completely spanning the stream, and pitched our tent on a sandy spot on the bank somewhat lower downstream.

So far we had examined and noted orchids rather than doing any serious collecting, but now that we had reached the final leg of the trip we felt the time had come to start work. Not far away lay the edge of a large expanse of the dwarf forest that covers the eastern side of the northwestern leg to a width in some places of about five miles. Here and there some larger yellow-topped trees showed their heads, holding promise of exciting orchids below. So the next morning we sent two porters off to make a trail toward the yellow trees. We had expected them back by lunch, but they were still on the job at midday, so after lunch we set off along their new trail to see what they had achieved. After an easy stretch across fairly open ground we entered their trail into the dwarf forest. "Entered" is a word carefully chosen in this context for we found the forest so thick a tangle that the trail literally became a tunnel. Mostly composed of trees not over thirty feet in height, it appeared to be a forest growing through and over the dead body of a previous forest, or was perhaps the result of a continuing process of the old slowly falling on its face through lack of soil support and the new coming up concurrently. Whatever the explanation, the result was that one walked through the forest tunnel on an elevated tangle of old fallen trees intertwined with roots and recumbent branches. Seldom were our feet nearer than two feet from the ground; never were they on it, and mostly they were anywhere from three to six feet above it. Progress was slow in the extreme, and the porters must have worked very hard to achieve the mile of trail that they had. For us, of course, it is never a hardship to go slowly, above all when there are orchids all around us, and this mile of trail occupied us for a whole busy afternoon, finding old orchid friends and some "probably new" ones. The next day our course was supposed to go roughly northward, but again it seemed to go in any direction but north. This time we were no longer twisting to avoid boggy ground but instead were skirting around patches and tongues of forest and scrub. Our experience of the forest by the yellow trees was quite enough to convince us that, silly as it might seem on the face of it, a circular mile over open ground was infinitely quicker than cutting a straight hundred yards through this type of forest. By now we were well into the heart of the rolling part of the mountain and away from flat rock slabs, cracks, chasms, and cliffs. Our course led us up and down steep, dry, orchid-and-scrub-covered slopes between 5,700 and 6,000 feet in elevation, broken here and there by patches of black, rough-surfaced crystalline rock weathering off into thin curved plates that were raised loosely above the solid rock below. This weathering gave a broadly spheroidal surface to the patches of bare rock that here and there showed through the thin red soil, quite distinct from the flat surfaces of the sandstone we had traversed further south, and raised our hopes that this abrupt change of rock and topography would produce different vegetation or forest with new and exciting orchids, despite the fact that the principal orchids to catch the eye, growing as freely as ever in the thin soil or on bare rock, were still the same species that grew on the sandstone.

By early afternoon we had reached a second small stream gurgling over the coarse black rock of its bed, and with no knowing when or where we might find another adequate source of water we pitched camp. This was to be our base for trips to Jimmie Angel's plane, and for orchid hunting in the scrubby savanna around us and in the dwarf forest that lay close to the east. The top of Angel Falls lay probably no more than six miles away to the northeast and was a tempting target, but by the time we had dealt with our immediate area, including the plane, the anticipated extra days required for six miles of tunneling through dense dwarf forest was more than our schedule and food supplies could cope with, and the idea was reluctantly ruled out.

The Angel plane now lay a short four miles to the north, and for once the trail next day led fairly straight for its objective. First steeply rolling as before, it then crossed a long stretch of open savanna alternating with wide patches of head-high shrubs, many with flowers. Finally, after a hard struggle through a section of dwarf forest where a tongue from the main forest to our right barred our way, we came out onto further savanna of a softer, almost boggy, nature and in another mile had reached the plane. From a distance, the plane, though obviously lying unusually close to the ground, looked almost new, its aluminum body and wings still gleaming brightly in the sun. Close up, the impression faded. The wheels were deeply sunk in the marshy ground, the external surfaces were bent and torn, and the inside of the plane, stripped of all its instruments and seats, was in a state of ruin. Looking around us we could see how easily the pilot had been deceived into thinking this soft, grassy semibog was hard savanna, and could admire the skill with which he managed to avoid a fatal somersault on landing, when his wheels sank into the ground and his tail went up in the air. But once safely down it must have been immediately apparent that the plane was there for keeps. While we were examining the plane, along with a couple of the porters, the others back in camp had been cutting new trails for us into the forest there, but when we checked them on the next day the forest proved to be the same dense, tangle-rooted mass that we had found at the earlier camp. Here and there whole stretches of small trees seemed to have been flattened, all in one direction, as if blasted by some catastrophic wind, though no such wind could strike one small spot and leave the rest standing. Once more we were struck by the extraordinary way these dense dwarf forests have developed and perhaps some day we shall know the answer. Meanwhile they remain as full of questions as they are full of orchids, the latter mainly small or miniature though here and there a yard-long leaf showed the presence of a big houlletia plant. But we could not stay there forever just wondering: our time was up and we had to start back.

On the twelfth day since we started up the mountain we began the return trip. After a rainy night at an intermediate stop we were once more threading our way through the labyrinth of the Wall and, this time with the aid of the rope, dropping carefully down its face. That night we camped again with our two tents on the flat sandstone bed of the river at River Bottom camp.

Six hours uphill walk the next day lifted us gently from the 5,600-foot River Bottom to the 7,000-foot Oso Woods. Once clear of the initial forest section it was a relief to have clean solid rock again beneath our feet. Even if its surface was often so deeply and intricately cut by wind and rain that it was seldom advisable to take our eyes from our feet for more than a few paces, there were no longer roots to trip us up or leaf-hidden traps to fall into.

Dawn the following day was misty and drizzly, and so it remained for all that morning. Dry rock was now wet and slippery, and progress called for reduced speed and increased caution, above all when we reached the zig-zag section in and out of the great chasms north of El Libertador. Rudy's nylon rope, where the route crossed the steeply inclined section of smooth rock above a vertical drop, was now a very welcome aid, though we felt that a rope of less elasticity would be far more helpful. Nylon no doubt bears fantastic weights, but as it gives no effective support until you have stretched it a yard or two, it becomes a very tricky thing for the unskilled to use as a handhold.

At Libertador we lunched and for a brief spell before the inevitable clouds began to form on the cliff face had a wonderful view to the south and southeast over the woods and savannas 6,000 feet below and across to the distance where other *tepuis* reared their cliff-bound tops. When clouds put an end to this we continued our way, the porters with their bundles popping one by one like rabbits into the narrow secret hole between the boulders that block the top of the diagonal gash in the cliff that gives access to the mountain. From here, all the way down the difficult drop to the base of the cliff 600 feet below, cries of laughter from the porters formed a noisy background, cries partly inspired by a going-home mood but mainly by the narrow escapes from being swung by the heavy bundles on their backs into headlong falls from boulder to boulder. At one point "G.C.K." nearly lost his "E." when the latter, sliding wisely on her backside, nevertheless went a bit too far and disappeared suddenly into a seemingly bottomless three-foot-wide cleft. It turned out to be only about six feet deep so all was well, but many such clefts and holes were twenty feet deep or more, and to be jammed tight into the bottom of such a spot with a broken arm or leg would have been anything but a joke. Luckily the patron saint of foolish amateurs had us under his wing, and by 4 P.M. we had traversed once more the "Devil's Network," successfully found some pieces of a rare and new-to-us violet-flowered *Vargasiella venezuelana* that we had seen on the way up, and were once more camping under the great overhang of El Peñon.

The weather was now clear again, though clouds continued to form as ever along the rim of the great cliffs now towering high above us. While Rudy and some of the porters went the next day right down to Guayaraca with all our equipment and some of our increasing number of orchid bundles, we stayed on for a day and a night at Danta and were able to make a proper collection of the orchids that we had only looked at on the way up. Apart from a few small epiphytes, such as the very pretty *Masdevallia picturata* and the very "miserable" *Epidendrum miserrimum,* the rest were naturally almost all terrestrials. The largest among these were head-high clumps of a most handsome sobralia that later proved to be a new species and was named *Sobralia infundibuligera:* we and others had actually seen plants of this elsewhere in the Guayana before our Auyán-tepui trip, but we were lucky enough to be the first to find it in flower. Smaller clumps of the very rare *Selenepedium steyermarkii* were also around and again were luckily in flower: the species is most difficult to grow in captivity, and although the piece we brought back survived for a while it was never happy and finally faded away. Rather smaller, and tending to be hidden by the knee-high general vegetation, were many plants of *Mendoncella burkei* and occasional plants of *Houlletia odoratissima:* the latter were not in flower but kindly did flower for us later at home, while the mendoncella plants, both here and on the summit, were in full bloom, with sturdy erect spikes of large, white-lipped, browny-green flowers of very solid consistency. It was particularly interesting to note the wide variation in color intensity of flowers of this latter species. About ten terrestrial species of epidendrums were around, but all were common elsewhere, such as the ubiquitous *E. elongatum.* One solitary plant of *Cattleya lawrenceana* caught our eye, growing lithophytically and not too happily so high above its normal range, but at least more happily than another very poor specimen we had seen on the summit itself where a stray seed had somehow managed to take hold: both these plants were left to continue the struggle, with a delicate pat on the pseudobulbs to help them along.

When two of the porters got back from Guayaraca the next day to help with the rest of the baggage, we were ready for the final drop through steep forest, collecting again as we went. Of very few species were we collecting more than a plant or two, but despite this economy we nevertheless had quite an appreciable volume of plants to add to our camping impedimenta and to substitute fairly light but bulky loads for the concentrated heavy ones of food that we had by now used up.

We had been playing safe with our time schedule and were two days ahead of the date set for a small plane to pick us up. The campsite where we waited tended to be hot, with the late afternoon sun beating straight into the forest edge. For relief, we moved our tent farther into the trees and on return from our local orchid hunt crossed the savanna to the only suitable bathing site a quarter of a mile or so away. Suitability in such matters, of course, is a subjective condition depending on personal criteria. In this case the water, a series of stagnant pools, sole relics of a wet-season stream ambling through boggy ground, was still well up to our standards for a good wash. It was deep, it was not too visibly dirty (though not exactly transparent), it was cool without being cold, and it contained (we hoped) nothing to bite us. If its banks were muddy and if a large pile of fresh horselike droppings near one end showed that tapirs also used it as a bathtub, we were not in a mood to complain. It cooled us off, and we came out of it at least feeling cleaner than when we went in. After dark, while we slept, leaf-cutting ants got to work cutting our orchids to bits and transporting their booty to a nest deep in the forest. Only after considerable depradations had been caused did we discover what was going on, whereupon the sacks were quickly suspended free of the ground and the ants smoked out by smouldering sticks waved under the bundles. We might, of course, have foreseen the possibility of such an infliction, but it had never happened to us before and we had taken no precautions. Fortunately, though a number of orchids were stripped almost to the bare stems, the leaves are not always supremely important, and though the plants would suffer a setback we hoped that most would survive.

Flying back below the west wall of Auyán-tepui we could see what seemed like dark and stormy weather ahead. Horrifyingly, this turned out to be not a work of nature but of man. Great sections of forest were burning to both sides of us and straight ahead, burning for no rhyme or reason other than that some of the very few Indians in the area had been careless with matches, doing their unwitting best to turn their native forests in a few years from green and productive coolness to dry eroding desert. At the moment, smoke seeping up between live green trees in a dozen wide-spaced locations was the main evidence of damage being done: the fires would continue creeping along the leafy mold of the forest floor until the rains came. From above, the trees as yet showed little harm, but the roots of many would be mortally struck, and by next year large patches would be barren forest. This was a truly sad way to end such an inspiring trip, but eventually the distraction of unpacking and sorting out our plants in Caracas washed the taste of the smoke away.

Many weeks later, when we had been able to check with our orchid mentor, Leslie Garay, curator of Harvard's Oakes Ames Orchid Herbarium, we were able to sum up our findings and contribute effectively to the general study of the flora of Auyán-tepui being worked up by another botanical friend, Julian Steyermark of the Caracas Botanical Institute, who intensively explored the top not long after ourselves.

In total we had tracked down approximately 140 species of orchids on the flank and summit of the mesa: of these about 100 were identifiable and the rest would have to wait until the live plants we had brought home recovered from the journey, took hold in their new home, and produced flowers. The large *Sobralia infundibuligera,* the ten-foot *Elleanthus norae,* and the very tiny *Pinelia alticola,* when they proved to be new species, were clearly the "tops" of our collection, and a number of other species proved to be new records for the Guayana region of Venezuela. We were satisfied with our findings, but, above all, we had enjoyed a trip into dramatic and almost untouched country and had collected memories that would never fade.

PHYTOGEOGRAPHY

2

A General Review of the Orchid Flora of China*

CHEN SING-CHI and TANG TSIN

*The literature survey pertaining to this chapter was concluded in September 1979; the chapter was submitted in October 1979, and the revised version was received in December 1979.

Brief History

In ancient Chinese literature, from *Shih Ching* (The Book of Songs) to *Li Chi* (The Book of Rites) and other classics, one can find a few names of orchids, such as *ni* (Plate 2-1), the somewhat doubtful *chien,* and *lan* (Plate 2-1). *Ni* is first mentioned, in what is perhaps the earliest record of orchids in the world, in a section of the Book of Songs (about 10th-6th century B.C.), in connection with a description of the customs of the people of Chen. In this section, "Chen Feng," there is a ballad entitled "Fang yu chueh chao" (On magpies' nest on the dike) that contains the following words:

> Chung tang yu pi, chiung yu chih ni.
> Shui chou yu mei? Hsin yen ti ti.
>
> (Saying that there is pottery on the road, and fragrant ribbon grass on the mound,
> Who had deceived my darling? I am worried and on the alert.)

Apparently, *ni* refers to ribbon grass, which is now known as *Spiranthes sinensis* (Pers.) Ames, according to *Mao Chuan* (Mao's Exegesis, probably prepared during the Western Han Dynasty), which gives a clear explanation of the word *ni:* "ni shou tsao yeh" (*ni* is ribbon grass), and to *Mao Shih Tsao Mu Niao Shou Chung Yu Shu* (Exegesis of Herbs, Trees, Birds, Beasts, Insects, and Fishes in Mao Shih) written by Lu Chi of Wu during the Three Kingdoms period in the 3d century, in which the writer states: "ni wu se tso shou wen, ku yue shou tsao" (*ni* is colorful and like a ribon, so it is called ribbon grass). The word *chien,* a synonym of *lan* based on Lu Chi's work just mentioned, is also found in the Book of Songs, but in a different section. A ballad under the heading "Cheng Wei" in the section "Cheng Feng" gives the following description:

> Cheng yu wei fang huan huan hsi,
> shih yu nu fang ping chien hsi.
>
> (As the water in Cheng and Wei begins to rise,
> young men and women are carrying *chien* in their hands.)

It is still uncertain, however, whether *chien* and *lan* are true orchids, for in many ancient writings *lan* seems to refer to several common fragrant plants rather than exclusively to orchids. For example, we find: "Yu chi shu lan chih chiu wan, yu shu hui chih pai mu" (I have grown nine *wan*[1] of *lan,* as well as a hundred *mu*[2] of *hui*) in *Li Sao* (Poem of Greivance), by the 4th century B.C. poet and statesman Chu Yuen, and "tai fu chih hsun, chu hou chih lan" (senior officials carry *hsun,*[3] while dukes carry *lan* in their hands)

[1]A unit of area, about 1.5–2 ha.

[2]A unit of area equal to 0.0667 ha.

[3]*Hui* and *hsun* may be both fragrant plants belonging to Labiatae, according to *An Account of Flora of the Southern Regions* written by Chi Han of the Tsin Dynasty.

Plate 2-1. The Chinese words *ni, chien,* and *lan* (calligraphy by S. C. Chen).

in the Book of Rites, probably published during the Western Han Dynasty. Hsu Shen (30–124 A.D.), in his *Shuo Wen Chieh Tzu* (Analytical Dictionary of Characters), said: "lan hsiang tsao yeh" (*lan* is a fragrant grass). Yen Shih-ku (581–645 A.D.) also did not regard it as an orchid, but as some other fragrant plants in his exegesis of the text of *Han Shu* (A History of the Han Dynasty). He pointed out that *lan* was exactly the same plant which was called *tse lan* during that time ("lan chi chin tse lan yeh"). *Tse lan* (*lan* of the marshes) is now *Eupatorium,* (Compositae). Although these sources differ in opinion about what plant *lan* is, it seems to be correct from the botanical point of view that what was called *lan* (at least as mentioned in these books) must have been a plant cultivated over a large area. At least it was rather common in central and northern China, where the Chung Yuen culture existed during that time. Other authors have stated that what is called *che lan* by Confucius (551–479 B.C.) is a true orchid, or *Cymbidium,* for the habitat of *che lan* he described is no different from that of wild cymbidiums. Confucius wrote: "che lan sen yu shen lin pu yi wu ren erh pu fang" (*che lan* that grows in deep forests never withholds its fragrance even without being appreciated). Obviously, it stands to reason that the far-reaching influence of Confucius' teachings during a long period in Chinese history played an important part in assigning the name *lan* to cymbidiums, as indicated by S. Y. Hu (1971).

Turning to more technical and scientific literature, the earliest book on record about orchids is *Shen Nung Pen Tsao Ching* (The Materia Medica of the Mythical Emperor). This famous book on medicine, probably prepared during the Han Dynasty, mentions several well-known orchid names that are still being used, such as *shih hu* (*Dendrobium*), *chih chien* (*Gastrodia elata*), and *pai chi* (*Bletilla*). On the other hand, one of the earliest purely Chinese botanical treatises, *Nan Fang Tsao Mu Chuang* (An Account of Flora of the Southern Regions), by Chi Han of the Tsin Dynasty, uses two orchid names, *chai tzu ku* (*Luisia*) and *shih hu* (*Dendrobium*), in describing a plant called *chi li tsao,* and indicates that *hui* was also called *hsun tsao* (with its leaves opposite and ramielike). In addition to these, there are several hundred ancient books on medicine and botany in the scientific history of China. Treatises on orchids can be found in most of them.

One of the most important botanical works dealing with the cultivation of orchids is *Chin Chang Lan Pu* (Treatise on Orchids of Chin Chang) by Chao Shih-keng, published in 1233. This work describes 22 kinds of orchids (divided into two groups, purple-flower orchids and white-flower orchids) and cultivation techniques. It seems that in addition to cymbidiums a few genera with broad leaves, such as *Phaius* and *Calanthe,* were also included. Another rather comprehensive work is the *Lan Pu* (Treatise on Orchids), written by Wang Kuei-hsueh of southern Fukien a few years later (1247). There are six chapters in this work, which describes 37 species or varieties and many cultivation techniques. It is evident that the cultivation of orchids became very popular during that time, at least among the superior classes, scholars and the rich. Two famous ancient orchid paintings (both *Cymbidium goeringii*) make the same point. One of them (Plate 2-2), now in the Palace Museum in Peking, was painted by Chao Meng-chien (1199–1264, who also called himself Chao Tsu-ku or Yi Chai) during the Southern Sung Dynasty. The other (Plate 2-3), painted on January 15, 1306 (the lunar calendar), by Cheng Sze-shiao (1239–1316, who also called himself Cheng Suo-nan or Suo-nan Won), is now probably in Taiwan. Many additional treatises on Chinese orchids have appeared, such as *Lan Yi* (Changes of Orchids) by Lu Ting-wong in 1250, *Lan Shih* (A History of Orchids) by Tan Hsi-tzu in 1368, *Lan Hui Ching* (Orchid Mirror) by Tu Yung-ning in 1811, *Lan Hui Tung Hsin Lu* (Reports of Both Lan and Hui) by Hsu Chi-lou in 1865, *Lan Yen Shu Lue* (A Brief Description of Orchids) by Yuen Shih-chun in 1876, and *Lan Hui Hsiao Shih* (A Brief History of Orchids) by Wu En-yuen in 1923. Some existed as manuscripts, which were formally published in 1933 in Chao Yi-chen's work, *Yi Hai Yi So* (A Scoop from the Sea of Arts).

There are also many accounts of Chinese orchids in Japanese and European publications, of which J. Matsuoka's work (1728; see Miyoshi, 1932) is of special interest.

The first scientific name was assigned to Chinese orchids by Linnaeus in *Species Plantarum* (1753). In this work he described a well-known orchid plant, *Epidendrum ensifolium* (now *Cymbidium ensifolium*), which had been collected in China by his student Peter Osbeck who visited Huang Pu, a harbor near Canton, in August of 1752 (Bretschneider, 1880). Many European missionaries and explorers subsequently came to China, and as a result, an increasing number of plant specimens were collected and sent to Europe. Among them, of course, were included a few orchids, which were described as new. The

Plate 2-2. Spring orchid (*Cymbidium goeringii*), an ancient painting by Chao Meng-chien (1199–1264), the eleventh descendant of the first emperor of the Sung Dynasty. This painting is presently preserved in the Palace Museum in Peking. The poem on the left is by the painter:

The summer heat steams along Hen[a] and Hsiang.[b]
The natural fragrance blown out cheers us fresh and cool.
The plant has been introduced from western Chekiang.
It blooms only once a year with one or two scapes.
—Translated by F. H. Wang and S. C. Chen

publications, however, were fragmentary and scattered until J. Lindley published his classical book *Species and Genera of Orchidaceous Plants* (1830–40). In it he listed all 33 species (belonging to 21 genera) of Chinese orchids then known. Lindley's work on the Orchidaceae, as well as that by O. Swartz (1800), C. L. Blume (1858), H. G. Reichenbach (1852–1900), and others, is still very important to the study of Chinese orchids.

The poem on the right is by the collector Ku Tsin(?):

Who would accept that a National Flower is not a common grass;
It is a species of Spring Orchid from Shau Hsi.[c]
Whither roams the descendant of the Prince today?
Seeing the bow Wu Hao[d] we also remember the late minister at Lake Ding.[e]

—Translated by Chun-jo Lui

---

[a]Hen Yang, a city of Hunan Province.
[b]Hsiang Kiang, a river of Hunan Province.
[c]A village of Chekiang Province.
[d]The name of a famous ancient bow of a great general.
[e]There is a legend that the Yellow Emperor, a legendary ruler of Chinese people, had gone to the sky at Lake Ding.

R. A. Rolfe was the first English botanist to make a special study of Chinese orchids. In Forbes and Hemsley's *Index Florae Sinensis* (1903), he presented a complete review of the Orchidaceae of China, Korea, and the Ryukyu Islands, in which 77 genera and 270 species of Chinese orchids were enumerated, including one new genus and many new species. A later treatise on the orchids of these regions is *Orchediologiae Sino-Japonicae*

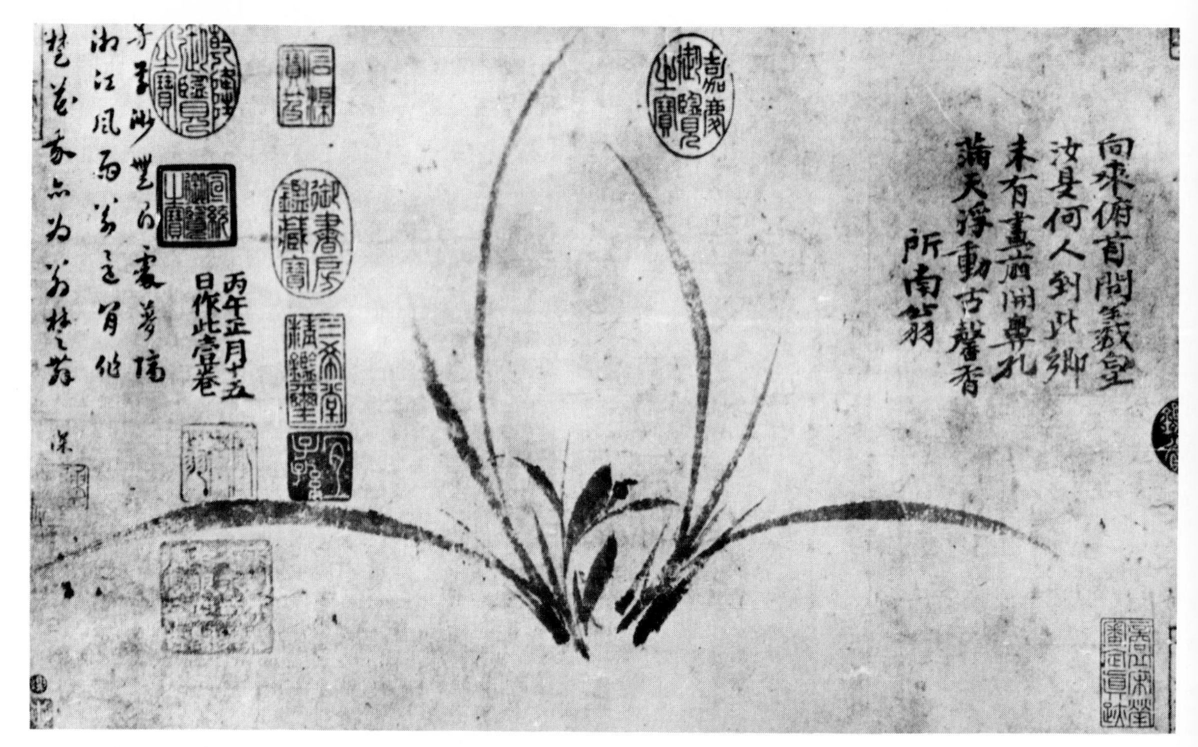

Plate 2-3. Spring orchid (*Cymbidium goeringii*), an ancient painting by Cheng Sze-shiao (1239–1316, Southern Sung Dynasty to Yuan Dynasty), who was a famous patriotic painter and called himself Suo-nan Won ("old man living in the house which faces toward the south," meaning that he cherishes the memory of the Southern Sung Dynasty). He painted orchids without drawing the soil, because his Sung Dynasty's "territory" had been occupied by the Yuan Dynasty from Mongolia. It was painted on January 15, 1306 (the lunar calendar) and is probably preserved in Taiwan now. The poem on the right is by the painter:

Always bowing my head I have been asking the legendary King
  Your identity and your journey.
Though no nostril has ever been tingled by a painting,
  The fragrance of long ago is hovering on the air.
—Translated by Chun-jo Lui

The poem on the left is by the collector Chen Shen(?):

No more is the fragrant grass now.
  Along the Hsiang River in wind and rain a dream resides.
Leisurely you paint the Chu[a] Flower,
  Which I also write a few words to eulogize.
—Translated by S. C. Chen

[a]Chu, name of an ancient kingdom during the Warring Kingdoms (475–221 B.C.), roughly modern Hupeh and Hunan; Chu Flower = orchid.

Prodromus by R. Schlechter (1919). In China, he recorded 102 genera and 582 species, including 67 genera and 214 species from Taiwan and 13 genera and 23 species from Tibet. His list includes one new genus and 75 new species. Rolfe's and Schlechter's works are still considered to be indispensable references, though not a few of Schlechter's new species are invalid. In respect to orchids in Taiwan, B. Hayata (1906–21), N. Fukuyama (1932–44), G. Masamune (1932–78), and others have published many new taxa, but owing to lack of comparisons with plants from the continent the validity of some of these species is problematic and must be clarified during our compilation of *The Orchid Flora of China.*

Since the early 1930's, one of the authors (T. Tang) has studied the Orchidaceae of China and neighboring areas in cooperation with Professor Fa-tsuang Wang. We compared specimens we collected with the types at Kew (including those from Lindley's herbarium), the British Museum (Natural History), the Linnean Herbarium, the Edinburgh Botanical Garden, and other herbaria in Europe. As a result we have written "Materials on the Orchidaceae of Eastern Asia" (in manuscript), thereby laying a foundation for an understanding of the orchid flora of China.

Several more recent publications in Chinese are: "Key to the Subfamilies, Tribes, Subtribes and Genera" (Orchidaceae) by T. Tang and F. T. Wang published in 1954 (revised edition by Z. H. Tsi and K. Y. Lang, 1979); *Iconographia Cormophytorum Sinicorum* (Orchidaceae) by K. Y. Lang and Z. H. Tsi, 1976; *Flora Hainanica* (Orchidaceae) by T. Tang and S. C. Chen, 1977. *The Flora of Taiwan* (Orchidaceae) by T. S. Liu and J. H. Su, published in 1978, is in English. In addition, S. Y. Hu's *The Orchidaceae of China* I–IX (1971–75) and other works (1965–77) contain a wealth of valuable information. Edited by T. Tang, F. T. Wang, and S. C. Chen, *The Orchid Flora of China* in three volumes (part of *Flora Reipublicae Popularis Sinicae,* which will be complete in 80 volumes) is being compiled and will be completed during the next five to seven years.

Genera

China extends from the temperate zone in the north to the northern edge of the tropical zone in the south (Plate 2-8). Within its boundaries are many climates and topographical features; some of them are complex. The precipitation in many parts is moderately heavy, favoring luxuriant orchid growth. According to Tang and Wang's key, 136 genera were known in 1954. This list, however, does not include the genera *Yoania, Disperis, Dicerostylis* (= *Hylophila*), *Ascocentrum,* and *Trichoglottis,* which grow in Taiwan. Published in the same year, *A Dictionary of the Families and Genera of Seed Plants in China* by Foon-chew How includes the same genera, and the author considered that they comprise a total of 1008 species. In *The Orchidaceae of China II* by S. Y. Hu (1971), 151 genera and 1200 species were recognized (excluding two genera of Apostasieae), among which, however, *Ponerorchis* is a congenus of *Orchis, Symphosepalum* of *Neottianthe, Hemihabenaria* of *Habenaria, Tricosma* of *Eria, Ione* of *Sunipia, Arethusantha* of *Cymbidium, Birmannia* (Chinese species) of *Kingidium, Haraella* of *Gastrochilus,* and *Staurochilus* (Chinese species) of *Trichoglottis.* Thus there are actually only 142 genera. In *Iconographia Cormophytorum Sinicorum* V (1976) and "Key to Subfamilies, Tribes, Subtribes and Genera" (1979) mentioned above, *Monomeria* and several additional genera of Vandeae were added, yet some are still missing, especially those from Taiwan.

Altogether, our list of orchids native to China includes 158 genera (including two genera of Apostasieae[4]) and 993 species in eight tribes and five subfamilies (Table 2-1).

Clearly, China has a rich orchid flora. There are representatives of all eight tribes and five subfamilies from the most primitive, the Apostasioideae, to the most advanced, the Epidendroideae. This is in line with the richness of the whole Chinese angiosperm flora.

Among the 158 genera, 79 are terrestrials (excluding "saprophytes"),[5] mostly widely distributed: 65 are epiphytic and are mainly found in Yunnan, Kwangtung, Kwangsi, and Taiwan. Small genera constitute the majority, more than two-thirds with less than five species. Only a few genera have more than 30 species and most of them are cosmopolitan or Asiatic tropicals with wide distribution (*Habenaria, Dendrobium, Liparis, Calanthe, Platanthera, Bulbophyllum, Cirrhopetalum, Eria,* and *Herminium*).

Endemism is notable. There are altogether eight endemic genera: *Porolabium, Smithorchis, Tsaiorchis, Changnienia, Tangtsinia* (Plate 2-4), *Sinorchis* (Plate 2-5), *Diplandrorchis* (Plate 2-6), and *Ischnogyne;* and three subendemic genera: *Bulleyia, Hancockia,* and *Holcoglossum.* This number does not include genera distributed in Eastern Asia or Asia but found largely in China, such as *Archineottia* (Plate 2-7), *Hemipilia, Amitostigma, Androcorys, Neofinetia,* and *Bletilla.* Phytogeographycally, China appears to be far richer in endemic genera than North America, where there are only four endemics, namely *Aplectrum, Calopogon, Isotria,* and *Hexalectris* (Correll, 1950). This is probably due to the more complicated climate and topographical features of eastern Asia and to the lack of a complete glacial cover since the Tertiary.

Another remarkable character of the orchid flora of China is that there are more "saprophytic" genera and genera with "saprophytic" species than anywhere else in the world. They amount to over one-tenth of the total number of orchid genera and include: *Sinorchis* (1 species), *Aphyllorchis* (3), *Diplandrorchis* (1), *Archineottia* (2), *Neottia* (7), *Hetaeria* (1), *Epipogium* (2), *Galeola* (4), *Sterosandra* (1), *Didymoplexis* (1), *Gastrodia* (3), *Lecanorchis* (2), *Risleya* (1), *Corallorhiza* (1), *Eulophia* (1), *Yoania* (1), and *Cymbidium* (1). The total is 17 genera and 33 species, found in all tribes and subfamilies of the Monandreae with the exception of Orchidioideae. If the "saprophytic" genera, as well as the tribes and subtribes to which they belong, are taken into account, their number exceeds that in any other region or country in the world. Furthermore, the Orchidaceae is also the largest of all families of angiosperms with "saprophytic" genera in China since it contains seven genera more than the saprophytic family Orobanchaceae, which in China includes only ten genera with a slightly larger number of species (38).

It is well known that in the majority of orchids, notably the terrestrials, seed germination is dependent on specific fungi. In fact some species are partially "saprophytic" or semi-"saprophytic." Mycorrhiza of orchids is in fact one of the most interesting subjects for many botanists, and is of phylogenetic importance. It is noteworthy that in some of the more primitive subtribes in the Monandrae, such as Neottiinae, Limodorinae, Vanil-

[4]We agree with some recent orchidologists, such as Dressler and Dodson (1960), Garay (1960, 1972), van der Pijl and Dodson (1966), and de Vogel (1969), that the Apostasieae cannot logically be excluded from the Orchidaceae without also excluding the Cypripedieae, unless further evidence of sharp distinction is brought forward.

[5]Many so-called saprophytic orchids are in fact parasitic on their fungi.—Ed.

Table 2-1. The genera of Orchidaceae in China

Taxa	Total number of species	Recognized	Distribution[a]	Remarks
Subfam. Apostasioideae				
Tribe Apostasieae				
1. *Neuwiedia*	8	1	Ha, Y	
2. *Apostasia*	6	2	Ha, Ks, Y	
Subfam. Cypripedioideae				
Tribe Cypripedieae				
3. *Cypripedium*	40	23	A, Che, Chi, He, Hop, Ka, Kia, Kr, Kw, L, M, N, Sa, Se, Si, St, Sz, Ta, Ti, Y	
4. *Paphiopedilum*	50	8	Ha, Ks, Kt, Kw, Y	
Subfam. Neottioideae				
Tribe Neottieae				
Subtribe Limodorinae				
5. *Tangtsinia*	1	1	Sz	Endemic
6. *Cephalanthera*	14	6	A, Che, Hon, Hun, Hup, Ka, Kia, Kiu, Kr, Ks, Kt, Se, Sz, Ta, Y	
7. *Sinorchis*	1	1	Kt	Endemic
8. *Aphyllorchis*	20	3	Ha, Ks, Kt, Ta, Ti, Y	
9. *Epipactis*	20	6	Che, Hon, Hop, Hun, Hup, Ka, Kia, Kr, L, Sa, Si, St, Sz, Y	
Subtribe Neottiinae				
10. *Diplandrorchis*	1	1	L	Endemic
11. *Archineottia*	4	2	Sa, Se, Sz	Largely in China
12. *Neottia*	10	7	Chi, He, Hop, Ka, Kia, Kr, L, M, Sa, Se, Si, Sz, Ti, Y	
13. *Listera*	30	21	Chi, He, Hop, Hup, Ka, Kia, Kr, L, Sa, Si, Sz, Ti, Y	
Subtribe Cloraeinae				
14. *Corybas*	50	3	Ks, Ta, Y	
Subtribe Cryptostylidinae				
15. *Cryptostylis*	20	1	Kt, Ta	
Subtribe Prasophyllinae				
16. *Microtis*	10	1	A, Che, F, Hun, Kia, Kt, Sz, Ta	
Subtribe Spiranthinae				
17. *Goodyera*	40	25	In whole country except He, L, M, Si	
18. *Erythrodes*	100	2	Kt, Ta, Y	
19. *Hylophila*	5	1	Ta	Incl. *Dicerostylis*
20. *Herpysma*	2	1	Y	Hu (1974a)
21. *Vexillabium*	5	1	Ta	Incl. Chinese species of *Pristiglottis*
22. *Ludisia*	1	1	Ha, Kt, Y	Incl. *Haemaria*
23. *Cheirostylis*	22	11	F, Ha, Hun, Ks, Kt, Kw, Sz, Ta, T	Incl. *Arisanorchis*
24. *Myrmechis*	7	5	Ha, Hup, Ks, Sz, Ta, Y	
25. *Zeuxine*	46	14	F, Ha, Hun, Hup, Ks, Kt, Kw, Sz, Ta, Y	
26. *Odontochilus*	20	6	Sz, Ti, Y	
27. *Anoectochilus*	25	6	Che, F, Ha, Kia, Ks, Kt, Ta, Y	
28. *Vrydagzynea*	40	2	Ha, Kt, Ta	
29. *Hetaeria*	20	7	Ha, Hup, Kt, Sz, Ta	
30. *Spiranthes*	50	2	In whole country	
31. *Manniella*	2	1	Hongkong	
32. *Tropidia*	20	3	Ha, Ta, Ti, Y	
33. *Corymborkis*	10	1	Y	

(*continued*)

Table 2-1.—Continued

Taxa	Total number of species	Recognized	Distribution[a]	Remarks
Subtribe Epipogiinae				
34. *Epipogium*	5	2	Ha, Kia, Kt, Sz, Ta, Ti, Y	
35. *Sterosandra*	1	1	Ta	Garay and Sweet (1974)
Subfam. Orchidioideae				
Tribe Orchideae				
Subtribe Orchidinae				
36. *Orchis*	100	17	In whole country except A, Che, F, Ha, Hun, Kia, Kiu, Ks, Kt, Kw	Incl. *Chusua, Galeorchis, Ponerorchis, Aceratorchis*
37. *Hemipilia*	13	9	Hup, Ks, Kw, Sa, Se, Sz, Ta, Y	Largely in China
38. *Bracycorythis*	32	2	Hun, Ks, Kt, Kw, Ta, Y	Syn. *Phyllomphax*
39. *Plantanthera*	200	40	In whole country except He, L	
40. *Coeloglossum*	2	1	Chi, He, Hon, Hop, Ka, Kr, L, M, Sa, Se, Si, Sz, Ta, Ti, Y	
41. *Tulotis*	4	3	In whole country except A, F, Ha, He, Ks, Kt, Kw, Si	Syn. *Perularia*
42. *Smithorchis*	1	1	Y	Endemic
43. *Diphylax*	1	1	Y	
44. *Herminium*	40	35	In whole country	
45. *Amitostigma*	21	18	In whole country except F, Ha, He, Ks, Kt, M, Si	Largely in China
46. *Neottianthe*	6	4	A, Hon, Hop, Hup, Kr, M, Sa, Se, Si, Sz, Ti, Y	Incl. *Symphosepalum*
47. *Gymnadenia*	10	3	Chi, He, Hop, Ka, Kia, Kr, L, M, N, Sa, Se, Si, Sz, Y	
48. *Tsaiorchis*	1	1	Y	Endemic
49. *Pecteilis*	4	2	Ha, Kia, Ks, Kt, Kw, Sz, Y	
50. *Diplomeris*	3	1	Kw, Ti, Y	
51. *Peristylis*	60	6	Hun, Hup, Ks, Kt, Kw, Sz, Ti, Y	
52. *Habenaria*	600	70	In whole country except Si	Incl. Hemihabenaria
53. *Porolabium*	1	1	Chi, Sa	Endemic
54. *Androcorys*	5	3	Kw, Se, Ta, Ti	Largely in China
Subtribe Disinae				
55. *Satyrium*	110	2	Kw, Sz, Ti, Y	
Subtribe Coryciinae				
56. *Disperis*	75	2	Ta, Hong Kong	
Subfam. Epidendroideae				
Tribe Gastrodieae				
Subtribe Vanillinae				
57. *Vanilla*	70	2	F, Ks, Kt, Kw, Ta, Y	
58. *Galeola*	20	4	Ha, Ks, Kt, Kw, Se, Sz, Ta, Ti, Y	
Subtribe Gastrodinae				
59. *Didymoplexis*	20	1	Ha, Ta	
60. *Gastrodia*	20	3	Hop, Hun, Hup, Kia, Kr, Kw, L, Se, Sz, Ta, Ti, Y	
Subtribe Pogoniinae				
61. *Pogonia*	7	3	Che, F, He, Hun, Hup, Kia, Ks, Kt, Kw, St, Sz, Ta, Y	
62. *Nervilia*	80	3	Kt, Sz, Ta, Ti, Y	
63. *Lecanorchis*	5	2	Ta	

(*continued*)

Table 2-1.—Continued

Taxa	Total number of species	Recognized	Distribution[a]	Remarks
Tribe Epidendreae				
Subtribe Thuniinae				
64. *Thunia*	5	1	Sz, Ti, Y	
65. *Arundina*	5	2	Che, F, Ha, Kia, Ks, Kt, Kw, Sz, Ta, Ti, Y	
Subtribe Arethusinae				
66. *Bletilla*	6	4	A, Che, Hon, Hun, Hup, Ka, Kia, Kiu, Ks, Kw, Se, Sz, Ta, Ti, Y	Largely in China
Subtribe Bletiinae				
67. *Pachystoma*	11	2	Ha, Ks, Kt, Kw, Ta	
68. *Acanthephippium*	15	5	Kt, Ta, Y	
69. *Tainiopsis*	1	1	Y	
70. *Calanthe*	100	41	A, Che, F, Ha, Hon, Hun, Ka, Kia, Kiu, Ks, Kt, Kw, Se, Sz, Ta, Ti, Y	
71. *Cephalantheropsis*	6	3	Ha, Kt, Ta	Incl. *Paracalanthe*
72. *Phaius*	50	8	F, Ks, Kt, Kw, Sz, Ta, Y	
73. *Anthogonium*	1	1	Ks, Ti, Y	
74. *Ania*	11	3	Ha, Kt, Y	
75. *Spathoglottis*	46	2	Ks, Ta, Ti, Y	
Subtribe Collabiinae				
76. *Collabium*	8	2	Kt, Ta, Y	Incl. *Collabiopsis*
77. *Mischobulbum*	7	2	F, Ks, Kt, Ta, Y	
78. *Nephelaphyllum*	12	1	Ha	
79. *Hancockia*	1	1	Y	Subendemic
80. *Chrysoglossum*	12	4	Ks, Kt, Ta, Y	
81. *Tainia*	25	7	Che, F, Kia, Ks, Kt, Sz, Ta, Y	
82. *Diglyphosa*	4	1	Y	
Subtribe Coelogyninae				
83. *Coelogyne*	200	16	Ha, Kia, Ks, Kt, Kw, Y	
84. *Pleione*	10	6	A, Che, F, Hun, Hup, Kia, Kw, Sz, Ta, Y	
85. *Panisea*	7	4	Ha, Ks, Kw, Y	
86. *Pholidota*	55	12	Che, F, Ha, Hun, Hup, Kia, Ks, Kt, Kw, Sz, Ta, Ti, Y	
87. *Otochilus*	4	4	Ti, Y	
88. *Neogyna*	1	1	Y	
89. *Bulleyia*	1	1	Y	Subendemic
90. *Ischnogyne*	1	1	Kw, Se, Sz	Endemic
91. *Dendrochilum*	75	1	Ta	
Subtribe Eriinae				
92. *Cryptochilus*	2	1	Y	
93. *Eria*	370	36	F, Ha, Ks, Kt, Kw, Sz, Ta, Ti, Y	
94. *Appendicula*	100	4	Ha, Kt, Ta	
95. *Ceratostylis*	60	3	Ha, Kt, Y	
96. *Agrostophyllum*	60	2	Ha, Ta, Y	
97. *Polystachya*	210	1	Y	
98. *Podochilus*	60	1	Ha, Ks, Kt, Y	
Subtribe Thelasiinae				
99. *Phreatia*	190	4	Ta, Y	
100. *Thelasis*	20	2	Ha, Kt, Ta, Y	
Tribe Malaxideae				
Subtribe Malaxidinae				
101. *Liparis*	250	45	Che, F, Ha, He, Hon, Hun, Hup, Ka, Kia, Ks, Kt, Kw, L, Sa, Se, Sz, Ta, Y	

(*continued*)

Table 2-1.—Continued

Taxa	Total number of species	Recognized	Distribution[a]	Remarks
102. *Malaxis*	300	15	In whole country except He, L, Si	Syn. *Microstylis*
103. *Oberonia*	300	20	Ha, Kia, Ks, Kt, Kw, Sz, Ta, Ti, Y	
104. *Hippeophyllum*	5	1	Ta	Masamune (1974)
105. *Risleya*	1	1	Sz, Ti, Y	
Subtribe Dendrobiinae				
106. *Bulbophyllum*	1000	36	A, Che, F, Ha, Hun, Hup, Ks, Kt, Kw, Se, Sz, Ta, Ti, Y	
107. *Cirrhopetalum*	100	30	Ha, Ks, Kt, Kw, Sz, Ta, Y	
108. *Dendrobium*	1400	63	A, Che, F, Ha, Hun, Hup, Ka, Kia, Ks, Kt, Kw, Se, Sz, Ta, Y	
109. *Epigeneium*	35	5	Ha, Ks, Kt, Y	Syn. *Sarcopodium*
110. *Flickingeria*	70	3	Ha, Kt, Ta, Y	Syn. *Ephemerantha*
Subtribe Genyorchidinae				
111. *Sunipia*	22	7	Ta, Ti, Y	Incl. *Ione*
112. *Monomeria*	4	1	Ti, Y	
Tribe Vandeae				
Subtribe Cymbidiinae				
113. *Cymbidium*	50	20	A, Che, F, Ha, Hon, Hun, Hup, Ka, Kia, Kiu, Ks, Kt, Kw, Se, Sz, Ta, Ti, Y	Incl. *Cyperorchis, Arethusantha*
Subtribe Cyrtopodiinae				
114. *Corallorhiza*	14	1	Hop, Se, Si, Sz	
115. *Eulophia*	200	12	Ha, Hup, Ks, Kt, Kw, Sz, Ta, Y	
116. *Geodorum*	6	4	Ha, Ks, Kt, Sz, Ta, Y	
117. *Yoania*	3	1	Ta	
118. *Tipularia*	5	2	Ka, Se, Sz, Ta	
119. *Calypso*	1	1	Kr, Sz, Ti	
120. *Changnienia*	1	1	Che, Hun, Hup, Kiu, Sz	Endemic
121. *Cremastra*	3	1	A, Che, F, Hun, Hup, Ka, Kiu, Ks, Kt, Kw, Sa, Se, Ta, Y	
122. *Oreorchis*	14	7	Che, He, Hup, Ka, Kia, Se, Sz, Ta, Ti, Y	Incl. *Kitigorchis*
Subtribe Vandinae				
123. *Chamaeanthus*	6	1	Ta	Hsieh (1955)
124. *Thrixspermum*	100	10	Ha, Ks, Kt, Sz, Ta, Y	
125. *Pteroceras*	30	4	Ha, Ta, Y	Incl. Chinese species of *Sarcochilus*
126. *Chiloschista*	3	2	Ta, Y	
127. *Doritis*	2	1	Ha	
128. *Phalaenopsis*	35	4	Ha, Sz, Ta, Ti, Y	
129. *Kingidium*	5	2	Ha, Y	Incl. Chinese species of *Burmannia*
130. *Ornithochilus*	1	1	Ks, Kt, Y	
131. *Aërides*	40	1	Ks, Ti, Y	
132. *Rhynchostylis*	15	1	Y	
133. *Luisia*	40	4	Che, Ha, Ks, Kt, Sz, Ta, Y	
134. *Diploprora*	5	1	F, Ha, Ks, Kt, Ta, Y	
135. *Stauropsis*	1	1	Y	
136. *Vanda*	60	8	Ha, Ks, Kt, Kw, Sz, Ta, Y	
137. *Vandopsis*	20	1	Ks, Kt, Y	
138. *Arachnis*	11	1	Ha, Ks, Ta, Y	
139. *Esmeralda*	2	1	Ha, Y	Tan (1976)
140. *Renanthera*	13	1	Ha	
141. *Ascocentrum*	5	2	Ta, Y	Incl. *Ascolabellum*

(*continued*)

Table 2-1.—Continued

Taxa	Total number of species	Recognized	Distribution[a]	Remarks
142. *Trichoglottis*	60	2	Ta	Incl. Chinese species of *Staurochilus*
143. *Holcoglossum*	3	3	Hup, Sz, Ta, Y	Subendemic
144. *Neofinetia*	1	1	Che, Kia	Largely in China
145. *Schoenorchis*	20	3	Ha, Ta, Y	
146. *Pelatanthera*	8	1	Y	
147. *Saccolabium*	6	1	Ta	Incl. *Tuberolabium*
148. *Gastrochilus*	20	16	Kt, Kw, Sz, Ta, Ti, Y	Incl. *Haraella*
149. *Malleola*	20	1	Ti	
150. *Uncifera*	5	1	Kw, Y	
151. *Robiquetia*	20	2	Ha, Kt, Ta	
152. *Pomatocalpa*	60	2	Ha, It, Ta	
153. *Cleisostoma*	100	16	Che, F, Ha, Kiu, Ks, Kt, St, Ta, Ti, Y	Syn. *Sarcanthus*
154. *Sarcophyton*	3	1	Ta	
155. *Acampe*	15	2	Ha, Ks, Kt, Ta, Y	
156. *Anota*	2	2	Ha, Y	
157. *Taeniophyllum*	100	4	Hun, Sz, Ta	
158. *Microtatorchis*	20	1	Ta	

[a] *Abbreviations:* A = Anhwei; Che = Chekiang; Chi = Chinghai; F = Fukien; Ha = Hainan and South China Sea Islands; He = Heilungkiang; Hon = Honan; Hop = Hopeh; Hun = Hunan; Hup = Hupeh; Ka = Kansu; Kia = Kiangsi; Kiu = Kiangsu; Kr = Kirin; Ks = Kwangsi; Kt = Kwangtung; Kw = Kweichow; L = Liaoning; M = Inner Mongolia; N = Ningsia; Sa = Shansi; Se = Shensi; Si = Sinkiang; St = Shantung; Sz = Szechuan; Ta = Taiwan; Ti = Tibet; Y = Yunnan.

linae, and Pogoniinae, "saprophytes" are especially numerous. This is doubtless due to the origin and evolution of the orchids, a topic to which we call special attention in the forthcoming *Orchid Flora of China.*

Recently Discovered Primitive Orchids

In the process of studying the orchid flora of China we have discovered four very primitive genera of systematic and phylogenetic significance. They are *Tangtsinia* (Chen, 1965), *Sinorchis* (Chen, 1978), *Diplandrorchis* (Chen, 1979a), and *Archineottia* (Chen, 1979b), all monotypic and native to China except the last, which contains four species, two in Sikkim and India and two in China.

Tangtsinia nanchuanica S. C. Chen (Plate 2-4) appears closely similar in habit to *Cephalanthera,* especially to *C. falcata* (Thunb.) Bl. It is considered to be one of the most primitive members of the Monandrae because of several very interesting floral features. Its flowers are erect, hardly twisted, and have a regular perianth. The column is trigonous-cylindric, possessing neither clinandrium nor rostellum but an erect anther with a short filament, a terminal stigma, and five staminodes. Three staminodes are larger, ligulate-oblong, about 0.6 mm long, opposite to the petals, representing the stamens of the inner whorl. The remaining two are very small and located at the apex of the lateral ribs of the column, of which they are elongated parts. These staminodes are located opposite the lateral sepals, standing for the three stamens of the outer whorl. Obviously they are relics which degenerated from the two fertile stamens during the course of

Plate 2-4. Tangtsinia nanchuanica S. C. Chen (drawn by C. R. Liu). **1:** Plant with flowers and fruits, × 1. **2:** Flower with sepals and petals spread open, × 4. **3:** Upper part of column, front view, × 10. **4:** Column, side view, × 10. **5:** Pollinia, side view, × 10. **6:** Seeds, × 10. **7:** Diagram.

evolution. This also indicates that the early column of orchids has developed from the fusion of six stamens with a central style and stigma.

Tangtsinia nanchuanica is only found at the edges of sparse forests and shrubbery as well as on the grassy slopes at Mt. Chin Fu and its surrounding regions in Nanchuan County of southeastern Szechuan at an altitude between 700 and 2100 m within an area of 250 square kilometers. It is of interest to note that the so-called living fossil *Cathaya argyrophylla* Chun et Kuang is also found there.

A second genus, *Sinorchis* (Plate 2-5), was formerly included in *Aphyllorchis,* from which it does not differ in habit. The difference is that it has a regular perianth with the lip and two lateral petals quite similar, as well as a nearly terminal stigma; both characters are rather rare in the family. Evidently this is also a very primitive orchid, although there is a difference of opinion as to whether it should be treated as a distinct genus. The only species, *S. simplex* (Tang et Wang) S. C. Chen, grows in Mei Hsian County, Kwangtung Province in southern China.

The third is *Diplandrorchis* (Plate 2-6), a very interesting genus with only one species, *D. sinica* S. C. Chen, found in Huan Ren County of Liaoning Province in northeastern China. Its flowers are erect with the perianth nearly regular. The stigma is disklike at the apex of the column. There is neither rostellum nor other appendages. Of special interest is the occurrence of two fertile stamens, borne on the upper part of the column near the terminal stigma, one of which is opposite the dorsal sepal. The other is opposite the median petal (labellum or lip). Thus they represent two median stamens of both inner and outer whorls. This is, indeed, quite unique in the family. In the Diandrae, as in *Apostasia* and *Cypripedium,* there are also two fertile stamens, but they are opposite the lateral petals and thus represent the lateral stamens of the inner whorl. There is no doubt that *Diplandrorchis* is a very primitive or relic genus and of great phylogenetic significance. It has been placed in the subtribe Neottiinae of the Monandrae simply because of its habit. Of course, further study is needed.

Archineottia (Plate 2-7) is also a "saprophytic" genus and, like *Diplandrorchis,* belongs to the subtribe Neottiinae. It is characterized by an incomplete or a very primitive column, which has a terminal stigma and an erect stamen with a free filament attached to the back of the column; there is no rostellum. On the back of the column there is a thick ridge of the same texture and appearance as the filament to which it is joined at its upper end. Apparently, the ridge is the lower part of the filament adnated to the compound style or column. This is another feature of interest perhaps not occurring in any other living orchids.

This primitive genus consists of four species, of which *A. gaudissartii* (Hand.-Mzt.) S. C. Chen and *A. smithiana* (Schltr.) S. C. Chen are found in China. The former, with its regular perianth of six quite similar segments (except that the three petals are somewhat narrower), may be one of the most primitive types in the Monandrae. This species is native to Yue Shan of southern Shansi, while the other is found in southwestern Szechuan and southern Shensi.

It should be noted here that the occurrence of these primitive genera in China is not surprising since the subtribes Limodorinae and Neottiinae to which they belong are largely found in this region and have long been considered the most primitive. For

Plate 2-5. Sinorchis simplex (Tang et Wang) S. C. Chen (drawn by C. R. Liu). **1:** Plant, × 1. **2:** Flower with sepals and petals spread open, × 2. **3:** Upper part of column, side view, × 2. **4:** Diagram.

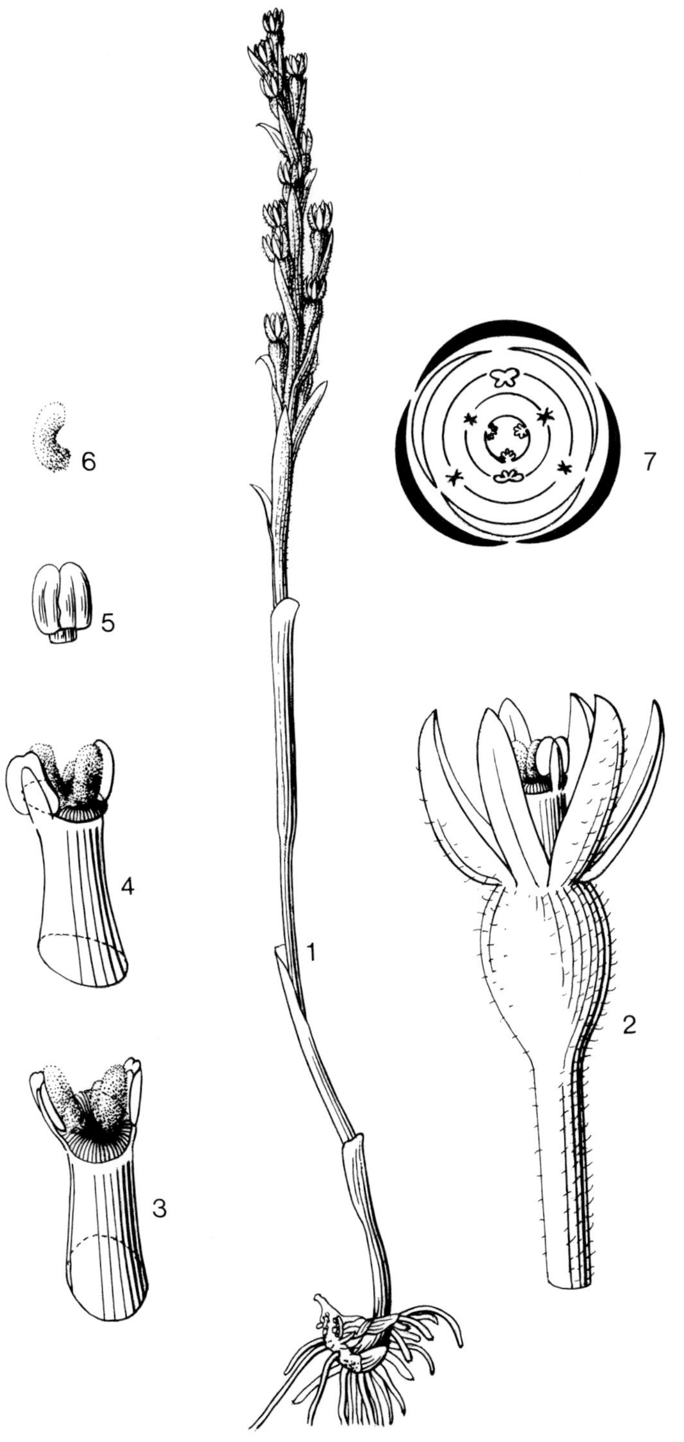

Plate 2-6. Diplandrorchis sinica S. C. Chen (drawn by C. R. Liu). **1:** Plant, × 1. **2:** Flower, × 7. **3:** Column, side view, × 10. **4:** Column, front-side view, × 12. **5:** Anther, × 12. **6:** Pollinium, × 12. **7:** Diagram.

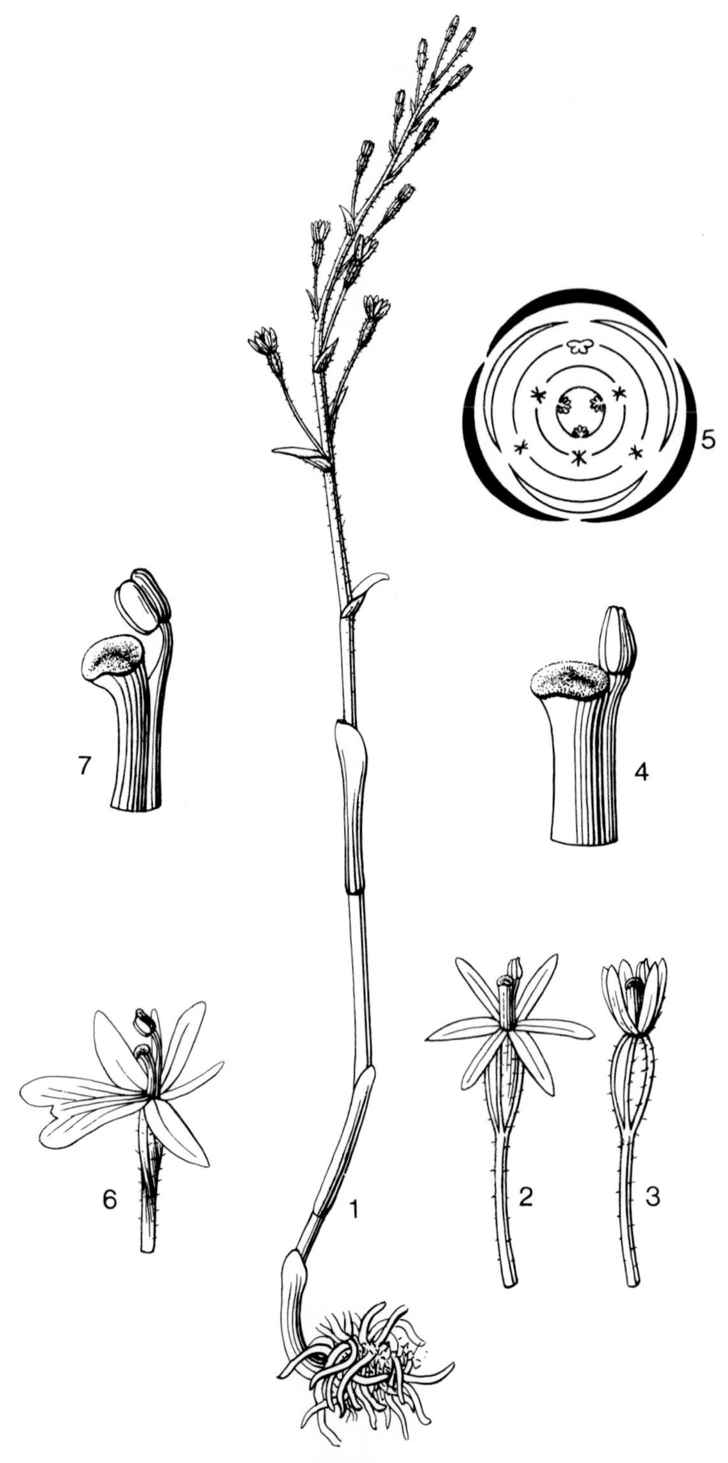

Plate 2-7. **1-5,** *Archineottia gaudissartii* (Hand.-Mzt.) S. C. Chen (drawn by C. R. Liu). **1:** Plant, × 1. **2:** Flower with sepals and petals spread open, × 3. **3:** Flower, × 3. **4:** Column, × 12. **5:** Diagram. **6-7,** *A. smithiana* (Schltr.) S. C. Chen (drawn by C. R. Liu). **6:** Flower, × 3. **7:** Column, × 12.

example, the affinity of these subtribes to *Cypripedium* (Diandrae) was noticed by many botanists. Early in 1937, the German orchidologist Mansfeld noted several resemblances between *Cypripedium* and *Epipactis*. It is difficult, we know, to distinguish the specimens of these genera from each other in herbaria when flowers and fruits are lacking. Embryological studies (Swamy, 1949) indicate that *Cypripedium, Neottia,* and *Listera* all have suspensorless embryos. Duncan (1959) indicated that *Cypripedium, Neottia, Listera, Cephalanthera,* and *Limodorum* have comparatively few and relatively long chromosomes. This suggests a closer affinity of these two subtribes to *Cypripedium* than to any other members of the Monandrae. In fact, many orchidologists have considered the primitive genus *Cephalanthera* of the subtribe Limodorinae as the progenitor of the Monandrae. Consequently it is by no means fortuitous that we have discovered several more primitive types which are closely related to it. This discovery, as well as the coexistence of several very primitive characteristics, may be good indications that they do not seem to be derivative or peloric, but surviving ancient types.

Clearly, China is rich in primitive orchids. In addition to the genera just mentioned, its orchid flora also includes *Neuwiedia, Apostasia, Vanilla,* and *Pogonia,* which are all of great phylogenetic interest. Altogether, only a few primitive orchid types have been found to date. For instance, only seven genera[6] possess a regular or nearly regular perianth: *Neuwiedia, Apostasia, Tangtsinia, Sinorchis, Diplandrorchis, Archineottia* (sect. *Archineottia*), and *Thelymitra.* Of these, except *Thelymitra,* which is only found outside of China (in Australia), all occur in China (three endemic and three extending to Malaysia, India, and Sikkim). This has attracted, and will certainly continue to attract, the attention of an increasing number of orchidologists.

Range and Geographical Distribution

Orchid fruits contain numerous minute seeds, which can be effectively dispersed by wind, often over long distances. Ridley (1930), having analyzed the flora of a large number of oceanic islands, concluded that orchids are capable of surviving in all islands provided ecological circumstances are suitable. This view is no doubt correct in the case of widespread species which do not have strict environmental requirements. However, the distribution of orchids often depends upon many factors. For instance, their seeds are almost entirely dependent on specific fungi for germination and development; seedling growth is rather slow, usually requiring several years to reach blooming size; with a few exceptions orchids are pollinated by specific insects. Hence, although orchid seeds are minute and numerous, their dispersal, germination, and growth are not always assured. In other words, the distribution of orchids, like that of angiosperms, is not simply dependent upon the dispersal of the seeds.

According to R. Takhtajan's *The Floristic Regions of the World* (1969) [which is based mainly on Engler's *Syllabus der Pflanzenfamilien* (Mattick, 1964) and on R. Good's *The Geography of the Flowering Plants* (1964)], China may be divided into three floristic re-

[6]This number does not include *Aceratorchis,* which we treat here as being a congenus of *Orchis.* This is a primitive taxon, possessing a nearly regular perianth, by which it differs from *Orchis* and other members of the subfamily *Orchidioideae.* We still question the propriety of our treatment of this interesting orchid.

gions: the Indo-Chinese Region (southern Tibet, a greater part of Yunnan, Kwangsi, Kwangtung and Taiwan, southeastern Kweichow, southern Fukien, and Hainan and the South China Sea Islands); the Eastern Asian Region (all provinces except tropical arid parts); and the Western and Central Asiatic Region (Sinkiang, Inner Mongolia, Ningsia, the Chinghai-Tibetan Plateau, a greater part of Kansu, western Szechuan, and arid parts of Shensi and Shansi). This division is reasonable on the whole, even though the boundaries are still a subject of disagreement among botanists (Handel-Mazetti, 1931; Li, 1944; Wu, 1979).

The Indo-Chinese Region has the richest orchid flora in China. Of the 158 Chinese genera, with the exception of a few endemics of the Eastern Asian Region and several temperate members, almost all occur in this region, especially in Yunnan and Taiwan. As far as we know at present, there are 117 genera and about 380 species in Yunnan and 96 genera and 289 species in Taiwan.

The orchid flora of the region, lying on the northern fringe of tropical Asia, has a close affinity to that of Malaysia. Many genera with a distribution center in Malaysia can also be found here. These include *Neuwiedia, Apostasia, Paphiopedilum, Cryptostylis, Agrostophyllum, Podochilus, Phreatia, Oberonia,* and *Sunipia,* as well as many members of the subtribe Vandinae. Almost all of them occur in Indo-China and probably entered the southern part of China through there. In Yunnan, where the Hengtuan Mountains extend from north to south, several tropical genera have spread along the Nuchiang and Lantsang rivers. They reach the northwestern part of Yunnan and extend upward to the southeastern part of Tibet. Examples are *Gastrochilus, Phalaenopsis, Vanda, Eria,* and *Oberonia.* It is likely that a number of eastern Asian and even temperate genera, e.g., *Hemipilia, Bletilla, Cephalanthera,* and *Epipactis,* have reached Thailand via the same route.

The orchid distribution in Taiwan is of considerable interest to us. In 1919, when R. Schlechter investigated the orchids of China and Japan, he indicated that the orchid flora of Taiwan differs greatly from that of the continent. He based his argument on the fact that no less than 13 Taiwanese genera are not found on the continent. In addition, only 19 Taiwanese species are common to the continent and most of them are widespread. He considered the orchid flora of Taiwan to be more closely related to that of the Philippines, because the previously mentioned 13 genera all occur there and the *Dendrobium* species in Taiwan show a close relationship to those in the Philippines. This view is in close agreement with Engler's suggestion that phytogeographically Taiwan and the Philippines should be merged into a single region called "Provinz der Philippinen und Formosa" belonging to "Monsungebiet" (Monsoon Region).

Since 1919, Merrill (1923), Engler (1923), Masamune (1937), Li (1957), and others have thoroughly investigated (from different viewpoints) the distribution of seed plants and their affinity to the flora of Taiwan. Their conclusions do not support the view that the flora of Taiwan is more closely related to that of the Philippines than to that of the continent. However, the possibility that the orchids may be an exception due to their minute seeds has not been considered.

The 13 genera that Schlechter listed as not having been recorded from the continent are *Didymoplexis, Microtis, Cryptostylis, Vanilla, Arisanorchis* (= *Cheirostylis*), *Aphyllorchis, Mischobulbum, Collabium, Chrysoglossum, Dendrochilum, Agrostophyllum, Taeniophyllum,* and *Tri-*

choglottis. Since then, however, all of them except *Dendrochilum* and *Trichoglottis* have been found on the mainland. On the other hand, ten additional genera (*Lecanorchis, Sterosandra, Hylophila, Vexillabium, Hippeophyllum, Yoania, Chamaenanthus, Sarcophyton, Saccolabium*, and *Microtatorchis*) must be added to those not found on the continent. With the exception of *Yoania* these genera are also found in the Philippines. Among them, *Trichoglottis luchuensis* (Rolfe) Garay et Sweet, *Microtatorchis compacta* (Ames) Schltr., and *Chamaeanthus wenzelii* Ames occur in both Taiwan (most in Lu Tau and Lan Hsu) and the Philippines, but not on the continent or in Indo-China. This can be taken to suggest that a relationship might exist between the orchid floras of Taiwan and the Philippines, even though several genera also occur in Burma, Thailand, and Indo-China (*Lecanorchis, Sterosandra, Hylophila, Hippeophyllum, Dendrochilum, Sarchophyton*, and *Saccolabium*, for example).

In the genus *Dendrobium*, we find three species (*D. ventricosum* Kränzl., *D. miyakei* Schltr., and *D. chameleon* Ames) which are common only to Taiwan and the Philippines. The first two grow in Lan Hsu, where the climate is similar to that of the Philippines because of its geographical proximity. *Dendrobium* is one of the largest genera in Orchidaceae, consisting of about 1400 species with a distribution center in Malaysia. There are 77 species in the Philippines (Ames, 1924) and 63 in China (including 17 species in Taiwan). A comparison of the species from Taiwan with those in continental China and the Philippines is of phytogeographical interest (Table 2-2).

Among the 17 species, six are found on the continent and three in the Philippines, and one widespread species (*D. crumenatum*) occurs in both places. There are seven Taiwan endemics. These may be important considerations in efforts to explain the affinity between the orchid flora of Taiwan and those of the continent and the Philippines.

It should be pointed out that no orchid genera are endemic to Taiwan (two genera, *Arisanorchis* and *Haraella*, were previously described as new but are now considered to be

Table 2.2. Dendrobium species in Taiwan

Species of Taiwan	Synonyms	Endemics	Philippines	Continental China
1. *D. alboviride* Hay.		+		
2. *D. chameleon* Ames	*D. randaiense* Hay.		+	
	D. longicalcaratum Hay.			
3. *D. crumenatum* Sw.	*D. Kwashotense* Hay.		+	+
4. *D. denneanum* Kerr.	*D. clavatum* Lindl. ex Paxt.			+
	D. aurantiacum Rchb. f.			
	D. flaviflorum Hay.			
5. *D. falconeri* Hook.	*D. erythroglossum* Hay.			+
6. *D. furcatopedicellatum* Hay.		+		
7. *D. goldschmidtianum* Kränzl.		+		
8. *D. leptoclandum* Hay.	*D. tenicaule* Hay., non Hook. f.	+		
9. *D. linawianum* Rchb. f.				+
10. *D. miyakei* Schltr.	*D. pseudo-hainanense* Masam.		+	
	D. irayense Ames et Qusumb.			
11. *D. moniliforme* (L.) Sw.	*D. heishanense* Hay.			+
12. *D. nakaharai* Schltr.		+		
13. *D. nobile* Lindl.	*D. formosanum* (Rchb. f.) L.			+
14. *D. sanseiense* Hay.		+		
15. *D. somai* Hay.		+		
16. *D. tosaense* Makino	*D. pere-fauriei* Hay.			+
17. *D. ventricosum* Kränzl.	*D. equitans* Kränzl.		+	

congenera of *Cheirostylis* and *Gastrochilus*, respectively). This is probably due to the nature and location of Taiwan (a continental island). In contrast, the number of endemic species is large, reaching 130 and constituting up to 45 per cent of the total, according to *Flora of Taiwan* V (Liu and Su, 1978). This percentage appears to be much higher than in Hainan, where there are only 16 endemic species (or less than 10 per cent of the total).

It is notable that many genera endemic or subendemic to the Eastern Asia Region—e.g., *Amitostigma, Hemipilia, Androcorys, Bletilla, Oreorchis*, and *Cremastra*—have representatives in Taiwan. Even genera which are distributed mainly in the temperate region, such as *Cypripedium, Orchis, Coeloglossum, Tulotis, Cephalanthera, Epipactis*, and *Listera*, are found in Taiwan. They are found there probably because Taiwan was connected with the continent during the Quaternary period and also because of the many high mountains (altitude over 3,000 m) on this island. In the Philippines there are also mountains of over 2,500 m in elevation, extending from north to south, yet not a single representative of either eastern Asian or temperate genera can be found there. Even epiphytes like *Holcoglossum* and *Pleione*, which are very common in the Asiatic continent and Taiwan, are not indigenous to the Philippines. Moreover, genera with disjunctive distribution in North America and eastern Asia (*Tipularia* and *Pogonia*) show a similar pattern. These facts indicate that the orchid flora of Taiwan (like the seed plants in general) bears a closer relation to that of the continent than to the orchid flora of the Philippines.

The area of Hainan is almost equal to that of Taiwan, but orchids there are rarer and fewer, especially endemics. As stated above, the number of genera in Hainan is less than two-thirds of that in Taiwan, and the number of species is even lower. All genera and the majority of species found there also occur in Indo-China and the continent. Of the total (63 genera and 150 species) a high proportion of the genera (over 50 per cent) and species (about 60 per cent) are epiphytic. This percentage (probably the highest in the country) is much higher than that in Taiwan, where epiphytic genera and species account for a little less than 40 and 39 per cent, respectively. All orchid genera in Hainan are tropical. Those that are endemic or subendemic to Eastern Asian Region or are widely distributed in the northern temperate zone (of which no less than a dozen can be found in Taiwan) are not present in Hainan. In Vietnam and Thailand, which are at the same latitude as Hainan, a few temperate genera do occur (e.g., *Cephalanthera, Epipactis, Listera, Pogonia*, and *Hemipilia*). The reason for this may be that there are no mountains higher than 2,000 m in Hainan.

Phytogeographically, one of the most complex regions in China is Yunnan. The flora of this region, including the orchids, is far richer than that of any other region in China. Many orchid genera with distribution centers in the north temperate zone occur there, especially in the northwestern part (*Orchis* and *Listera*, for example). Tropical genera are also found in considerable numbers in the southern and western parts. Some of them have not been recorded elsewhere in China, as for example *Corymborkis, Neogyna, Polystachys, Rhynchostylis, Stauropsis*, and *Pelatantheria*. Of the 117 genera found there, 47 are epiphytic. This amounts to 40 per cent of the total number of genera and is almost the same proportion as in Taiwan. Endemism is remarkably high. In the southern and western parts of Yunnan there are four endemic or subendemic genera, namely

Smithorchis, Tsaiorchis, Hancockia, and *Bulleyia.* These appear to be the only endemic orchid genera in the whole Indo-Chinese Region. Clearly, then, Yunnan has a rich orchid flora and is of great phytogeographical interest. The fact that more than two-thirds of the orchid genera in China occur in Yunnan is of particular significance in studies of the Orchidaceae of China.

It is interesting to point out that a species of *Manniella, M. hongkongensis* Hu et Barr., has been recently reported from Hong Kong (Hu and Barretto, 1975). This genus includes only three species, distributed disjunctively in western Africa and northern South America (Garay, 1964), and now in eastern Asia. If *M. hongkongensis* is a true native species in Hong Kong (rather than a chance introduction), it would be of considerable phytogeographical importance.

The second region is the Eastern Asian. It has more endemic genera but contains fewer taxa (74 genera and 195 species) than the Indo-Chinese Region. Terrestrials constitute a large percentage, with 54 genera. Only a few epiphytes can be found in this region, mostly rock-dwellers (lithophytes) such as *Dendrobium, Liparis, Cymbidium, Bulbophyllum, Cirrhopetalum, Eria, Pleione, Pholidota, Coelogyne, Ischnogyne, Oberonia,* as well as a few members of the subtribe Vandinae.

On the whole, our recent knowledge indicates that in the northeast the epiphytes are limited to Tsingtao and Lienyunkang; in middle China to the south slope of the Chinling Mountains; in western China to the east slope of Tsetuo Mountain and Tahsueh Mountain of Szechuan, and in Tibet to Kyerong, Nyalam, Yatung, Tingkye, Pome, Tsona, Zayu, and Markham (Table 2-3). Plate 2-8 shows the northern and western limits of the epiphytic orchids in China.

The available information (Table 2-3) indicates that the northern and western limits of epiphytic orchids in China almost correspond with those of the subtropics. At Laushan Mountain in Tsingtao, where the climate is oceanic, several subtropical plants (e.g., *Lindera, Machilus, Trachelospermum,* and bamboo) grow very well, even though in January the weather is very cold, with an average minimum temperature of −4° to −5°C (on January 26, 1913, a record low −12°C was recorded; Chu *et al.,* 1940). On the southern slope of the Chinling Mountains we can find orange trees and bamboo, but in winter temperatures there can drop to −1° to −2°C in January (the record low was −5.2°C on January 3, 1936; Chu *et al.,* 1940). Further north in Taipai Mountain and Huihsien County of Kansu Province it is still colder. Epiphytic orchids can grow there only in microclimate areas. Few genera and species can be found there. In addition, there is a paucity of individuals; most of them grow on rocks and have small pseudobulbs.

In the west, epiphytic orchids are distributed primarily along the eastern and southern borders of the Chinghai-Tibetan Plateau. However, some can be found further northward along rivers. For instance, in the Yikong-Thagmad region (30°10′N, 95°E) of Pome we find a considerable number of epiphytic orchids (10 genera and 14 species), whereas in the remaining parts of Tibet at the same latitude none have been recorded. This is obviously due to the influence of the Yalu Tsangpo River and its tributaries.

China is the only country in the northern hemisphere with large areas extending from the north edge of the tropical region to the temperate zone. In North America and

Table 2.3. The northern and western limits of epiphytic orchids in China

Epiphytic species	Northern and western limits					Remarks
	Provinces or regions	Localities	Lat.	Long.	Alt. (m)	
Cleisostoma scolopendrifolium	Shantung	Tsingtao	36°4′N	120°19′E		On rocks
(Makino) Garay	Kiangsu	Lienyunkang	34°37′N	119°14′E		On rocks
Luisia teres	Szechuan	Kangting	30°3′N	102°13′E	1600	On rocks
(Thunb.) Bl.						
Phalaenopsis wilsonii	Szechuan	Tienchuan	30°5′N	102°45′E	1500	
Rolfe	Szechuan	Muli	28°15′N	100°55′E	2150	On tree trunks
Gastrochilus distichus	Tibet	Thagmad	30°6′N	95°E	2300	On cliffs
(Lindl.) O. Kuntze	Tibet	Zayu	28°36′N	97°24′E	2400	On cliffs
Gastrochilus calceolaris	Tibet	Nyalam	28°6′N	85°54′E	2000	On tree trunks or rocks
(Smith) D. Don						
Aërides uniflora	Tibet	Nyalam	28°6′N	85°54′E	2100	On trees
(Lindl.) Summerh.						
Kingidium taenialis	Tibet	Kyerong	28°54′N	85°12′E	2000	On tree trunks
(Lindl.) P. F. Hunt						
Ischnogyne mandarinorum	Shensi	Chinling				On rocks
(Kränzl) Schltr.	Szechuan	Tienchuan	30°5′N	102°45′E	1500	On rocks
	Szechuan	Wenchuan	31°21′N	103°24′E		On rocks
	Szechuan	Paohsing	30°25′N	102°50′E		On rocks
Pleione bulbocodioides	Shensi	Taipaishan			1780	On rocks
(Franch.) Rolfe	Szechuan	Wenchuan	31°21′N	103°24′E	1800	On rocks
	Tibet	Markham	29°36′N	98°30′E	2900	On rocks
	Tibet	Chawaron	28°30′N	98°18′E		On rocks
	Tibet	Zayu	28°36′N	97°24′E		On rocks
Pleione hookerianum	Tibet	Nyalam	28°6′N	85°54′E	3100	On tree trunks or rocks
(Lindl.) O. Kuntze						
	Tibet	Yatung	27°24′N	88°54′E	2800	On rocks
Coelogyne punctata	Tibet	Nyalam	28°6′N	85°54′E	2800	On tree trunks
Lindl.	Tibet	Yatung	27°24′N	88°54′E	2680	On rocks
	Tibet	Tingkye	28°18′N	87°42′E	2350	On trees
Coelogyne occultata	Tibet	Thagmad	30°6′N	95°E	2100	On tree trunks
Hook. f.	Tibet	Pome	29°48′N	95°42′E	2400	
	Tibet	Linchih	29°30′N	94°18′E	2300	On rocks
	Tibet	Tingkye	28°18′N	87°42′E	2400	
	Tibet	Zayu	28°36′N	97°24′E	2400	On rocks
Coelogyne cristata	Tibet	Nyalam	28°6′N	85°54′E	1900	On trees or rocks
Lindl.						
Pholidota rupestris	Tibet	Thagmad	30°6′N	95°E	2000	
Hand.-Mzt.	Tibet	Zayu	28°36′N	97°24′E	2100	
Liparis caespitosa	Tibet	Thagmad	30°6′N	95°E	2000	On rocks
(Thouars) Lindl.						
Liparis fargesii	Kansu	Huihsien	33°50′N	105°56′E	1300	On rocks
Finet	Shensi	Shanyang	33°38′N	109°54′E	900	On rocks
	Szechuan	Paohsing	30°25′N	102°50′E		On rocks
	Szechuan	Lushan	30°3′N	103°6′E		On rocks
Oberonia caulescens	Szechuan	Tienchuan	30°5′N	102°45′E	1400	On tree trunks
Lindl.	Tibet	Thagmad	30°6′N	95°E	2500	
	Tibet	Yikong	30°12′N	94°48′E	2200	On rocks
	Tibet	Nyalam	28°6′N	85°54′E	3710	
	Tibet	Zayu	28°36′N	97°24′E	2900	
	Tibet	Tsona	27°54′N	91°6′E	2530	
Dendrobium hancockii	Kansu	Huihsien	33°50′N	105°56′E		On rocks
Rolfe	Shensi	Shanyang	33°38′N	109°54′E		On rocks
	Shensi	Ningshan	33°16′N	108°32′E		On rocks
	Szechuan	Kangting	30°3′N	102°13′E	1600	On rocks
Dendrobium denneanum	Szechuan	Hsiaochin	30°58′N	102°28′E	3400	On rocks
Kerr						
Dendrobium candidum	Tibet	Zayu	28°36′N	97°24′E		
Wall. ex. Lindl.	Tibet	Tingkye	28°18′N	87°42′E	2500	On rocks
Dendrobium hookerianum	Tibet	Thagmad	30°6′N	95°E	2080	
Lindl.	Tibet	Linchih	29°30′N	94°18′E	2300	On rocks
Dendrobium mondicola	Tibet	Nyalam	28°6′N	85°54′E	2300	On rocks
P. H. Hunt et Summerh.	Tibet	Kyerong	28°54′N	85°12′E	2380	On rocks

(*continued*)

Table 2-3.—Continued

Epiphytic species	Northern and western limits: Provinces or regions	Localities	Lat.	Long.	Alt. (m)	Remarks
Eria alba Lindl.	Tibet	Thagmad	30°6′N	95°E	2080	
Eria excavata	Tibet	Thagmad	30°6′N	95°E	2000	
Lindl. ex Hook. f.	Tibet	Nyalam	28°6′N	85°54′E	2400	On rocks
Eria graminifolia	Tibet	Thagmad	30°6′N	95°E	2100	
Lindl.	Tibet	Zayu	28°36′N	97°24′E	2300	
	Tibet	Tingkye	28°18′N	87°42′E	2400	On trees
	Tibet	Nyalam	28°6′N	85°54′E	2350	On trees
Eria spicata (D. Don) Hand.-Mzt.	Tibet	Kyerong	28°54′N	85°12′E	2380	On trees or rocks
Bulbophyllum kwangtungense Kerr?	Shensi	Taipaishan				On rocks
Bulbophyllum cariniflorum Rchb. f.	Tibet	Kyerong	28°54′N	85°12′E	2200	On rocks
Bulbophyllum leopardinum (Wall.) Lindl.	Tibet	Nyalam	28°6′N	85°54′E	1700	On rocks
Bulbophyllum reptans Lindl.	Tibet	Yikong	30°12′N	94°48′E	2500	
Cirrhopetalum elatum Hook. f.	Tibet	Tingkye	28°18′N	87°42′E	2500	
Cirrhopetalum bomiense (Tsi) Tsi	Tibet	Thagmad	30°6′N	95°E	2030	On rocks
Cirrhopetalum wallichii	Tibet	Thagmad	30°6′N	95°E	2080	On rocks
Lindl.	Tibet	Tingkye	28°18′N	87°42′E	2500	
	Tibet	Nyalam	28°6′N	85°54′E	2400	On trees
	Tibet	Zayu	28°36′N	97°24′E	2500	
	Tibet	Tsona	27°54′N	91°6′E	2510	
Sunipia candida	Tibet	Kyerong	28°54′N	85°12′E	2100	
(Lindl.) P. F. Hunt	Tibet	Nyalam	28°6′N	85°54′E	2300	On trees
	Tibet	Tingkye	28°18′N	87°42′E	2400	
Cymbidium iridioides	Tibet	Tingkye	28°18′N	87°42′E	2400	
D. Don	Tibet	Thagmad	30°6′N	95°E	2100	On rocks
	Tibet	Zayu	28°36′N	97°24′E	2100	On rocks
Cymbidium erythraecum	Tibet	Thagmad	30°6′N	95°E	2100	On rocks
Lindl.	Tibet	Zayu	28°36′N	97°24′E	2100	On rocks
Cymbidium lancifolium Hook.	Tibet	Zayu	28°36′N	97°24′E	2400	On trees

Africa there are no parallels because of the oceans or deserts. Consequently, further phytogeographical and ecological studies of the epiphytic orchids in China would be of much interest.

Among the best known and most important mountains in China are the Chinling Mountains, the watershed of northern and southern China. From the north of the Chinling Mountains to the Northeast China Plain there are no epiphytic orchids, and such common subtropical terrestrials as *Bletilla* and *Calanthe* are also not found there. Temperate orchids replace them, but the number of both genera and species is small. For instance, in Shensi-Kansu-Ningsia Basin only three genera have been recorded (*Platanthera, Epipactis,* and *Neottianthe*). In the Northeast China Plain (i.e., the provinces Liaoning, Kirin, and Heilungkiang) only 23 genera and 33 species have been recorded (Liou, 1959). Therefore, it is very remarkable that despite a poor orchid flora and low levels of endemism such primitive types as *Diplandrorchis sinica* and *Archineottia gaudissartii* were preserved in the forests of this area.

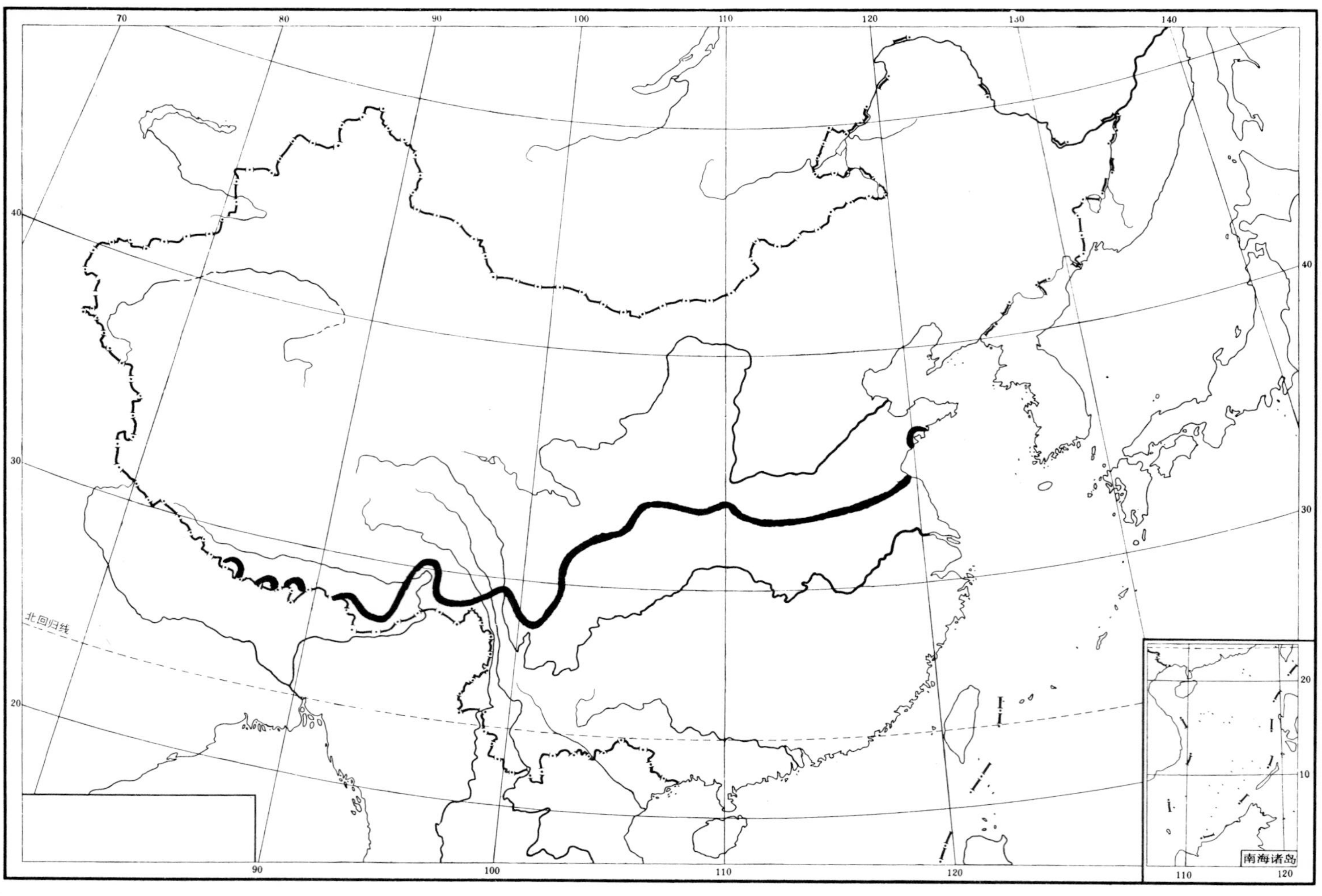

Plate 2-8. The northern and western limits of the epiphytic orchids in China (drawn by S. C. Chen).

Table 2-4. The number of orchid genera in each province or region of China

Province or region	Terrestrials	Epiphytes	"Saprophytes"	Total
Anhwei	14	3		17
Chekiang	22	7		29
Chinghai	13			13
Fukien	19	8		27
Hainan[a]	26	33	4	63
Heilungkiang	12		1	13
Honan	13	1		14
Hopeh	13		3	16
Hunan	21	6	1	28
Hupeh	25	6	1	32
Kansu	21	2	1	24
Kiangsi	22	6	3	31
Kiangsu	13	1		14
Kirin	18		2	20
Kwangsi	33	21	2	56
Kwangtung	36	25	4	65
Kweichow	27	14	2	43
Liaoning	12		1	13
Inner Mongolia	11		1	12
Ningsia	10			10
Shansi	18		2	20
Shensi	21	4	5	30
Sinkiang	11		1	12
Shantung	10	1		11
Szechuan	41	16	7	64
Taiwan	51	37	8	96
Tibet	29	12	6	47
Yunnan	64	47	6	117

[a] Including the South China Sea Islands.

The Western and Central Asiatic Region, including mainly Inner Mongolia, the Chinghai-Tibetan Plateau, and Sinkiang, contains an even smaller number of orchids. In the entire region only the following can be found: *Cypripedium, Coeloglossum, Herminium, Neottianthe, Gymnadenia, Orchis, Tulotis, Platanthera, Goodyera, Spiranthes, Neottia, Listera, Epipactis, Malaxis,* and *Corallorhiza,* as well as the monotypic genus *Porolabium* which is endemic to Chinghai and Shansi. The region occupies the largest area of China but has the smallest number of orchids (16 genera and 24 species).

The number of genera in each province or region of China is shown in Table 2-4.

Utilization and Cultivation

Utilization of orchids as folk drugs in China began early in ancient times. The well-known crude drugs *chih chien, shih hu,* and *pai chi* were recorded 2000 years ago in the *Materia Medica of the Mythical Emperor* of the Han Dynasty and still occupy an important place in traditional Chinese medicine. During the sixth century, Tao Hun-chin's *Ming Yi Pieh Lu* (Extra Accounts of Renowned Drugs) reported that *shih hu* grows on the rocks by the waterside in the valleys of Lu An. This is probably the earliest record regarding the habitat of epiphytic orchids in China. The collection and utilization of this drug during that period appear to have been extensive.

Chih chien ("red arrow," alluding to the color and shape of new shoots in spring), also named *tien ma* ("heavenly hemp," referring to the partially disintegrated scape in autumn, first mentioned in *Kai Pao Pen Tsao* (Kai Pao's Materia Medica, published in 973 A.D.), is the dry tubers of *Gastrodia elata* Bl. This drug is assumed to increase strength and vitality, to improve circulation, and to strengthen the memory. It is prescribed for rheumatism, neuralgia, paralysis, headaches, and other nervous disorders.

Gastrodia elata is widely distributed in Hupeh, Szechuan, Yunnan, Hunan, Kiangsi, Kweichow, Shensi, Hopeh, Liaoning, Kirin, Tibet, and Taiwan, growing in thick humus forest soils. The drug comes mainly from Szechuan, Hupeh, Yunnan, and Kweichow. Generally it is collected in winter when the plants have not yet sprouted or in the spring when the flower buds have just emerged from the soil. The tuber called *tong ma* (winter *Gastrodia*) is of better quality than that collected in spring, which is called *chun ma* (spring *Gastrodia*).

At present cultivation of *Gastrodia elata* has been successful in Yunnan, Shensi, Shantung, Honan, and Hunan. Histological observations indicate that the plant becomes infected by fungus (*Armillaria mellea* Fr.) during seed germination. Yunnan, where tree stumps of *Castanopsis platyacantha* R. et W. and *Schima crenata* Korth. become infected with *Armillaria mellea* following deforestation, is a good locality for production of *G. elata* tubers (Chou, 1974). The time required for plants to reach harvest size is three years or longer.

Shih hu ("rock-living"), a term presently applied to almost all species of *Dendrobium* and sometimes to a few species of *Flickingeria* in China, is descriptive of the habitat of several lithophytic species. Their stems are used as a drug which is reported to have tonic, stomachic, and antiphlogistic properties. It is prescribed for patients who are suffering from high fever, thirst, general weakness, and excessive perspiration.

Commercial production of *shih hu* is chiefly in Kwangsi, Yunnan, Kweichow, and Szechuan. The common types of *shih hu* and their venacular names on the Chinese market are as follows (see Table 2-1 for list of abbreviations):

I. Chin chai shih hu (golden hairpin *Dendrobium*)
   1. *D. nobile* Lindl. (Ha, Hup, Ks, Kt, Kw, Sz, Ta, Y)
   2. *D. linawianum* Rchb. f. (Ks, Ta)

II. Er huan shih hu (earring *Dendrobium*)
   1. *D. candidum* Wall. ex Lindl. (Ks, Kw, Ti, Y)
   2. *D. hercoglossum* Rchb. f. (Ha, Hun, Kia, Ks, Kt, Kw, Y)
   3. *D. moniliforme* (L.) Sw. (Che, Hun, Kia, Ks, Kt, Kw, Ta, Y)

III. Ma pien shih hu (horse whip *Dendrobium*)
   1. *D. fimbriatum* Hook. var. *oculatum* Hook. (Ks, Kt, Kw, Y)
   2. *D. denneanum* Kerr (Ks, Kt, Kw, Sz, Ta, Ti, Y)
   3. *D. hancockii* Rolfe (Hup, Ka, Ks, Se, Sz, Y)
   4. *D. chrysanthum* Wall. ex Lindl. (Ks, Kw, Ti, Y)

IV. Huan tsao shih hu (yellow herb *Dendrobium*)
   1. *D. loddigesii* Rolfe (Ha, Ks, Kt, Kw)
   2. *D. tosaense* Makino (A, Che, Hup, Ki, Ta)
   3. *D. aduncum* Wall. ex Lindl. (Ks, Kt, Kw, Y)

4. *D. wilsonii* Rolfe (Hun, Hup, Ks, Kt, Kw, Sz)
5. *D. crepidatum* Lindl. ex Paxt. (Kw, Y)
6. *D. lohoense* Tang et Wang (Hup, Ks, Kt, Kw, Y)
7. *D. devonianum* Paxt. (Kt, Kw, Y) and also *D. dennearum, D. moniliforme, D. hercoglossum, D. chrysanthum,* and *D. fimbriatum* var. *oculatum*

V. Yu kua shih hu (melon *Flickingeria*)

1. *Flickingeria lonchophylla* (Hook. f.) A. D. Hawkes (Kw, Y)
2. *F. fimbriata* (Bl.) A. D. Hawkes (Ha, Kt, Y)

Pai chi ("white chicken," referring to the delicate fleshy white rhizomes of the plant, which resemble a cute chicken toy according to Hu, 1971) is a name for the rhizomes of *Bletilla striata* (Thunb.) Rchb. f., *B. yunnanensis* Schltr., and *B. ochracea* Schltr. It is credited with the power of strengthening weak lungs and dissolving extravasated blood. These species are widely distributed in the provinces south of the Chinling Mountains, but commercial drug production is centered in Kweichow (where large quantities are produced), Hupeh, Hunan, Honan, Chekiang, and Shensi. It is also said that the drug from Szechuan is of high quality. Digging and collecting are usually done from August to November. In addition to being used as medicine, *pai chi* is also used as a sort of glue.

Tien ma, shih hu, and *pai chi* are elements of traditional Chinese medicine, which we call *tsong yao.*

Many orchids are also used in Chinese folk medicine as herbs, known as *tsao yao* (herbal medicine). They are usually used in the countryside by local doctors. Among them *Galeola faberi* Rolfe and *G. lindleyana* (Hook. f. et Thoms.) Rchb. f. are used for curing prolapse of the uterus; *Gymnadenia conopsea* (L.) R. Br., *Habenaria delavayi* Finet, *Herminium monorchis* (L.) R. Br., and *Spiranthes sinensis* (Pers.) Ames are credited with the power of strengthening and nourishing a weakening body; *Ludisia discolor* (Ker-Gawl.) A. Rich., *Goodyera schlechtendaliana* Rchb. f., *Pholidota chinensis* Lindl., *P. yunnanensis* Schltr., *Epigenium fargesii* (Finet) Gagnep., *Cymbidium ensifolium* (L.) Sw., *C. floribundum* Lindl., *C. goeringii* (Rchb. f.) Rchb. f., *C. pendulum* (Roxb.) Sw., *Dendrobium jenkinsii* Wall. ex Lindl., and *Nervilia fordii* (Hance) Schltr. are prescribed to improve and strengthen weak lungs and for the relief of coughs; *Eria graminifolia* Lindl., *Coelogyne occultata* Hook. f., and *Oberonia iridifolia* Lindl. are good for the stomach and improve digestion; *Eria pannea* Lindl., *Calanthe discolor* Lindl., *Neottianthe cucullata* (L.) Schltr., *Epipactis mairei* Schltr., *Thunia alba* (Lindl.) Rchb. f., *Cypripedium japonicum* Thunb., and *C. macranthum* Sw. are taken to dissolve extravasated blood and to improve circulation; *Goodyera procera* (Ker-Gawl.) Hook. and *Cirrhopetalum andersonii* Hook. f. are said to be good for relieving rheumatic pains; *Cleisostoma scolopendrifolium* (Makino) Garay, *Arundina chinensis* Bl., and *Calanthe fimbriata* Fr. are prescribed for the relief of internal heat or fever; *Habenaria dentata* (Sw.) Schltr., *Cypripedium margaritaceum* Fr., and *Platanthera chlorantha* Cust. ex. Rchb. are believed to nourish lungs and kidneys; *Epipactis helleborine* (L.) Cranz. and *Cypripedium franchetii* are said to have the ability to regulate the flow of vital energy, and remove obstruction to it, and to improve blood circulation; *Liparis nervosa* (Thunb.) Lindl., *L. japonica* (Miq.) Maxim., *Tulotis ussuriensis* (Reg. et Maack) Hara, *Pleione bulbocodioides* (Fr.) Rolfe, and *Oberonia myosurus* (Forst.) Lindl. are reported to reduce inflammation, dissolve extravasated blood, and cause swellings to subside. The active principles, if any, in these folk herbal drugs are not known and require further investigations.

Although *tien ma* is cultivated for medicinal uses, other orchids such as *shih hu* and *pai chi* are grown in China chiefly as ornamentals. Historically the cultivation of orchids in China presumably began with cymbidiums, since very few other species were recorded as being under cultivation in Chinese literature. This may be due to the long tradition of the Chinese people of enjoying the delicate fragrance and simple but elegant shape of *Cymbidium* flowers, and the fact that terrestrial cymbidiums are easier to grow than other orchids.

At present cymbidiums are the favorite cultivated ornamental plants in the south of China. There they are found not only in botanical gardens and parks, but also in many private yards. Orchid exhibitions (primarily cymbidiums) are held frequently and are welcomed by the populace. Currently the most popular orchids under cultivation are *chun lan* [spring orchid = *Cymbidium goeringii* (Rchb. f.) Rchb. f.; Plate 2-9], *chien lan* [Fukien orchid = *C. ensifolium* (L.) Sw.; Plate 2–10], *mo lan* [ink orchid = *C. sinense*

Plate 2-9. Chun lan [spring orchid = *Cymbidium goeringii* (Rchb. f.) Rchb. f.], one of the most popular orchids under cultivation in China (drawn by C. R. Liu). **1:** Plant, × ½. **2:** Column and lip, × 1. **3:** Lip, × 1. **4:** Petal, × 1. **5-6:** Lateral sepals, × 1.

Plate 2-10. Chien lan [Fukien orchid = *Cymbidium ensifolium* (L.) Sw.], one of the most popular orchids under cultivation in China (drawn by C. R. Liu). **1:** Plant, × ½. **2:** Flower, × 1.

(Andr.) Willd.], *hui lan* (Hui orchid = *C. faberi* Rolfe), and *han lan* (cold orchid = *C. kanran* Makino), and a great many of their varieties, forms, and cultivars. In the Peking Botanical Garden, South China Botanical Garden, Kunming Botanical Garden, and Hanchow Botanical Garden cymbidiums have long been cultivated. In addition to terrestrial species some epiphytic ones such as *C. pendulum* (Roxb.) Sw., *C. floribundum* Lindl., *C. dayanum* Rchb. f., and *C. hookerianum* Rchb. f. are also being grown. Few of the other tropical and subtropical orchids, such as *Dendrobium, Phaius, Calanthe, Paphiopedilum, Vanda, Liparis, Renanthera, Acampe, Pholidota, Coelogyne,* and *Bletilla,* are under cultivation, but they are not very popular. Presently in the Peking Botanical Garden and Hanchow Botanical Garden special attention is being paid to the cultivation of cymbidiums, whereas the South China Botanical Garden and Kunming Botanical Garden specialize in tropical orchids. Some botanical gardens in Taiwan specialize in the introduction of exotic species and hybrids.

During the rule of the "Gang of Four" floriculture was severely criticized and attacked cruelly. Cultivated orchids were wantonly destroyed. In the Peking Botanical Garden the situation is the worst, but restoration has now started. It is fortunate that at present our work is being supported, and that both orchid cultivation and research have improved greatly.

Summary and Conclusions

China is a country with a magnificent ancient civilization in which the cultivation and utilization of orchids have a long history. The available data, historical records, medicinal uses, and cultivation as ornamentals indicate that their appreciation started in China much earlier than in any other country.

The orchid flora of China has several outstanding characteristics. First, it is very rich, and all five subfamilies and eight tribes of the family are represented here. In America and Africa only four and three subfamilies, respectively, are found. Second, it is distinguished by its remarkably large number of primitive types. In the Diandrae only *Selenipedium* and *Phragmipedium* are absent, whereas in the Monandrae almost all primitive genera are found in China. This is especially true for the subtribes Limodorinae and Neottiinae. Among them some are very rare, for example *Tangtsinia, Diplandrorchis,* and *Archineottia.* Third, "saprophytes" are numerous. Of the total 158 genera and 966 species in China, 82 genera and 550 species are terrestrials, and 64 genera and 416 species are epiphytes. The remaining 17 genera and 33 species are "saprophytes." Among them the genera *Hetaeria, Eulophia,* and *Cymbidium* include both "saprophytic" and autophytic members. The number of "saprophytic" orchids, genera and species in China exceeds that in any other country or region.

Phytogeographically, China includes three floristic regions. Most of the orchids of China are found in the Indo-Chinese Region, particularly in Yunnan, Taiwan, Kwangtung, and Hainan. Only six provinces or regions in China have more than 50 genera of orchids (Table 2-4). Among them, Yunnan ranks first with 117 genera (70 per cent of the total number of genera in China). Taiwan is in second place with 96 genera. Kwangtung (65) is third, followed by Szechuan (64), Hainan (63), and Kwangsi (56). Except for Szechuan, all of these areas belong to the same floristic region.

Taiwan and Hainan are continental islands separated from the mainland since the Quaternary period. Their orchid floras resemble that of the continent, but Taiwan's orchids also bear a limited relationship to those in the Philippines.

Most endemic genera occur in the Eastern Asian Region, where terrestrials are in the majority and epiphytes are relatively few (only 19 genera). The northern boundary for epiphytes runs along the southern slope of the Chinling Mountains, whereas the western boundary extends along the southern and western margins of the Chinghai-Tibetan Plateau. The Chinling Mountains are like a partition that divides China into northern and southern parts. Neither epiphytes nor subtropical terrestrials have crossed over these mountains. On the northern side only temperate genera are found. This is quite similar to the situation in the Western and Central Asiatic Region. One major difference is that some noteworthy endemics were preserved here. In the latter region, almost all orchids are temperate and have a wide distribution range.

The orchid flora of southern China is closely related to that of Malaysia, because of its geographical position. Many tropical genera also appear here, notably those belonging to the subtribes Vandinae, Dendrobiinae, Eriinae, and Malaxidinae. The orchid flora of the Eastern Asian Region is more closely related to that of North America, especially in the distribution patterns of *Tipularia, Pogonia, Listera, Tulotis,* and *Cypripedium.* In addition, many genera with a transoceanic distribution range are represented in China, such as the transpacific genera *Tropidia* and *Erythrodes;* paleotropical genera like *Disperis, Satyrium,*[7] *Brachycorythis, Didymoplexis, Zeuxine, Cheirostylis, Galeola, Nervilia, Oberonia, Phaius, Agrostophyllum,* and *Acampe;* pantropical genèra such as *Habenaria, Goodyera, Corymborkis, Vanilla, Liparis, Malaxis, Calanthe, Eulophia, Polystachya, Bulbophyllum,* and the somewhat doubtful *Manniella*—showing that historically this region has had a close connection with a different continent.

The collection of orchids in China is still far from complete and extensive field work is needed, especially in the southern part of Tibet, western and southern Yunnan, southern Kwangsi, the Chinling Mountains, western Szechuan, Hunan, and Kweichow. New taxa have been described and new records have been established recently. Nevertheless, with more detailed field studies and comparative examinations of the ever increasing number of specimens, it is likely that the number of species may be reduced. For instance, the number of orchid species in Taiwan is 366 according to "An Enumeration of the Formosan Orchidaceae" by A. T. Hsieh (1955), but it was reduced to 282 in the *Flora of Taiwan* V (Liu and Su, 1978). In our recent studies of the orchid flora of China (especially *Cymbidium, Calanthe,* and *Dendrobium*), we have come to conclusions which support a similar reduction.

In summary, the orchids of China are very numerous, but their collection and cultivation are inadequate. At present volumes 17–19 (Orchidaceae) of *Flora Reipublicae Popularis Sinicae* are in preparation. Several botanists have started to work on tissue culture of orchids, and plan to initiate research on cytology, physiology, embryology, horticultural methods, and other areas. A great deal of work still remains to be done, and the coopera-

[7]Recently a species of this genus, namely *Satyrium ciliatum* Lindl., has been reported to have female, male, and hermaphrodite forms (Chen, 1979c). This species is distributed in Bhutan, Sikkim, and China (extending from Tibet to Szechuan). However, it is of interest that the unisexual forms are only found in northwestern Yunnan, western Szechuan, and southwestern Kweichow.

tion of botanists from different fields is needed in studying this large and complex family.

Acknowledgments

Professor Fu-hsiung Wang, head of the laboratory of phytomorphology of our institute, and Professor Joseph Arditti read the manuscript and made many valuable suggestions. Professor Chun-jo Lui translated the two poems on the paintings. Mr. Kiu-an Liu, a staff member at the Palace Museum in Peking, assisted us in the selection of an ancient painting. Mr. Zhan-huo Tsi helped us to type the manuscript and with other materials. Mrs. G. Barretto and Mrs. Jiao-lan Li provided us with some of the relevant literature. Mrs. Chun-rung Liu prepared the diagnostic drawings. To all of them we express our deepest appreciation.

Literature Cited

Ames, O. 1924. Apostasiaceae and Orchidaceae, p. 252–458. *In* E. D. Merrill (ed.), An enumeration of Philippine flowering plants. I. Bureau of Printing, Manila.

Arditti, J. (ed.). 1977. Orchid biology: Reviews and perspectives. I. Cornell Univ. Press, Ithaca and London.

Blume, C. L. 1858. Florae Javae et insularum adjacentium nova series. I (Orchidaceae). C. G. Sulpke, Amsterdam.

Bretschneider, E. V. 1880. Early European researches into the flora of China. J. North China Branch Roy. Asiat. Soc. n.s. 15:1–194.

Chen, S. C. 1965. A primitive new orchid genus Tangtsinia and its meaning in phylogeny. Acta Phytotax. Sinica 10(3):193–207.

——. 1978. Sinorchis—a primitive new genus of Orchidaceae from China. Acta Phytotax. Sinica 16(4): 82–85.

——. 1979a. On Diplandrorchis, a very primitive and phylogenetically significant new genus of Orchidaceae. Acta Phytotax. Sinica 17(1):1–6.

——. 1979b. The column types of Neottia and Archineottia of the family Orchidaceae and their taxonomic and phylogenetic significance. Acta Phytotax. Sinica 17(2):9–22.

——. 1979c. Notes on bisexual and unisexual forms of Satyrium ciliatum Ldl. Acta Phytotax. Sinica 17(4):54–60.

Chou, H. 1974. In propagation of Gastrodia elata Bl. Acta Botanica Sinica 16(3):288–289.

Chu, C., J. Lee, and P. K. Chang. 1940. The temperature of China. National Research Institute of Meteorology, Nanking.

Correll, D. S. 1950. Native orchids of North America. Chronica Botanica, Waltham, Mass.

Dressler, R. L., and C. H. Dodson. 1960. Classification and phylogeny in the Orchidaceae. Ann. Missouri Bot. Gard. 47:25–68.

Duncan, R. E. 1959. Orchids and cytology, p. 189–314. *In* C. L. Withner (ed.), The orchids: A scientific survey. Ronald Press, New York.

Engler, A. 1923. Zustimmende Bemerkungen zu Herrn Elmer D. Merrill's Abhandlung über die pflanzengeographische Scheidung von Formosa und den Philippinen. Bot. Jahrb. Engler 58:605–606.

Fukuyama, N. 1932. Neue Orchideen von Formosa. Trans. Nat. Hist. Soc. Formosa 22:413–416.

——. 1933. Studia Orchidacearum Japonicarum. II. Ann. Rep. Bot. Gard. Taih. Imp. Univ. 3:81–86.

——. 1934a. Studia Orchidacearum Japonicarum. I. Bot. Mag. Tokyo 48: 297–308.

——. 1934b. Studia Orchidacearum Japonicarum. III. Bot. Mag. Tokyo 48:429–442, 504–509.

——. 1935a. Studia Orchidacearum Japonicarum. IV. Bot. Mag. Tokyo 49:290–297, 340–341.

——. 1935b. Studia Orchidacearum Japonicarum. V. Bot. Mag. Tokyo 49:438–444, 480–481.

——. 1935c. Studia Orchidacearum Japonicarum. VI. Bot. Mag. Tokyo 49:663–670, 733–734.

——. 1935d. Studia Orchidacearum Japonicarum. VII. Bot. Mag. Tokyo 49:757–764, 825–826.
——. 1936a. Studia Orchidacearum Japonicarum. VIII. Bot. Mag. Tokyo 50:16–24, 57–58.
——. 1936b. Orchid flora of Formosa, p. 22–26. *In* Y. Yamamoto *et al.*, Materials for the floras of Formosa and Micronesia. Taihoku Univ., Taihoku.
——. 1937. Miscellaneous notes on Formasan species of Phalaenopsis. J. Taiwan Mus. Assoc. 5:246–260.
——. 1938. Studia Orchidacearum Japonicarum. X. Bot. Mag. Tokyo 52:242–247, 272–273.
——. 1944. On the genus Anota Schltr. Trans. Nat. Hist. Soc. Formosa 34:103–114.
Garay, L. A. 1960. On the origin of the Orchidaceae. Bot. Mus. Leafl. Harvard Univ. 19:57–96.
——. 1964. Evolutionary significance of geographical distribution of orchids. *In* Yeoh Bok choon (ed.), Proc. 4th World Orch. Conf., Singapore, pp. 170–187.
——. 1972. On the origin of the Orchidaceae. II. J. Arnold Arboretum 53:202–215.
Garay, L. A., and H. R. Sweet. 1974. Orchids of Southern Ryukyu Islands. Harvard Bot. Museum, Cambridge, Mass.
Good, R. 1964. The geography of the flowering plants. Longmans, Green, London.
Handel-Mazzetti, H. 1931. Die pflanzengeographische Gliederung und Stellung Chinas. Bot. Jahrb. Engler 64:309–323.
Hayata, B. 1906. Supplements to the Enumeratio Plantarum Formosanarum. Bot. Mag. Tokyo 20:76.
——. 1911. Materials for a flora of Formosa. J. Coll. Sci. Imp. Univ. Tokyo 30:309–355.
——. 1912. Orchidaceae, p. 131–145. *In* Icones Plantarum Formosanarum. II. Bureau of Productive Industries, Taihoku.
——. 1913. Orchidaceae, p. 194. *In* Icones Plantarum Formosanarum. III. Bureau of Productive Industries, Taihoku.
——. 1914. Orchidaceae, p. 23–129. *In* Icones Plantarum Formosanarum. IV. Bureau of Productive Industries, Taihoku.
——. 1915. Orchidaceae, p. 213. *In* Icones Plantarum Formosanarum. V. Bureau of Productive Industries, Taihoku.
——. 1916. Orchidaceae, p. 66–94. *In* Icones Plantarum Formosanarum. VI. Bureau of Productive Industries, Taihoku.
——. 1918. Orchidaceae, p. 40–41. *In* Icones Plantarum Formosanarum. VII. Bureau of Productive Industries, Taihoku.
——. 1919. Orchidaceae, p. 130–132. *In* Icones Plantarum Formosanarum. VIII. Bureau of Productive Industries, Taihoku.
——. 1920. Orchidaceae, p. 108–118. *In* Icones Plantarum Formosanarum. IX. Bureau of Productive Industries, Taihoku.
——. 1921. Orchidaceae, p. 32–35. *In* Icones Plantarum Formosanarum. X. Bureau of Productive Industries, Taihoku.
How, F. C. 1958. A dictionary of the families and genera of seed plants in China. Science Press, Peking.
Hsieh, A-tsai. 1955. An enumeration of the Formosan Orchidaceae. Quart. J. Taiwan Mus. 8:213–282.
Hu, S. Y. 1965. Whence the Chinese generic names of orchids? Amer. Orchid Soc. Bull. 54:518–521.
——. 1970. Dendrobium in Chinese medicine. Econ. Bot. 24:165–174.
——. 1971a. Orchids in the life and culture of the Chinese people. Chung Chi J. 10:1–26.
——. 1971b. The Orchidaceae of China. Quart. J. Taiwan Mus. 24:68–103.
——. 1971c. The Orchidaceae of China. II. Quart. J. Taiwan Mus. 24:182–255.
——. 1972a. The Orchidaceae of China. III. Quart. J. Taiwan Mus. 25:40–67.
——. 1972b. The Orchidaceae of China. IV. Quart. J. Taiwan Mus. 25:199–230.
——. 1973a. The Orchidaceae of China. V. Quart. J. Taiwan Mus. 26:131–165.
——. 1973b. The Orchidaceae of China. VI. Quart. J. Taiwan Mus. 26:373–406.
——. 1974a. The Orchidaceae of China. VII. Quart. J. Taiwan Mus. 27:155–190.
——. 1974b. The Orchidaceae of China. VIII. Quart. J. Taiwan Mus. 27:419–467.
——. 1975. The Orchidaceae of China. IX. Quart. J. Taiwan Mus. 28:125–182.
——. 1976. The genera of Orchidaceae in Hong Kong Illustrated. Chinese Univ., Hong Kong.
——. 1977. The origin and meaning of the generic names of Chinese orchids. Quart. J. Taiwan Mus. 30:123–186.
Hu, S. Y., and G. Barretto. 1975. New species and varieties of Orchidaceae in Hong Kong. Chung Chi J. 14:1–34.
Lang, K. Y., and Z. H. Tsi. 1976. Orchidaceae, p. 602–772, 960–994, t. 8034–8374. *In* Iconographia Cormophytorum Sinicorum. V. Science Press, Peking.

Li, H. L. 1944. The phytogeographical divisions of China, with special reference to the Araliaceae. Proc. Acad. Nat. Sci. Philadelphia 96:249–277.
——. 1957. The genetic affinities of the Formosan flora. In Proc. 8th Pacif. Sci. Congr. Quezon. 4:189–195.
Lin, T. P. 1975. Native orchids of Taiwan. I. Chan Ta Press, Chiaya.
——. 1978. Native orchids of Taiwan. II. Chan Ta Press, Chiaya.
Lindley, J. 1830–40. The genera and species of orchidaceous plants. Ridgway, London.
Linnaeus, C. 1753. Species Plantarum. II (ed. 1), p. 939–954. Impensis Laurertii Salvii, Stockholm.
Liou, T. N. 1959. Claves Plantarum Chinae boreali-orientalis (Orchidaceae), p. 587–598. Science Press, Peking.
Liu, T. S., and H. J. Su. 1978. Orchidaceae, p. 859–1137, pl. 1541–1653. *In* Flora of Taiwan. V. Epoch Publishing Co., Taipai.
Mansfeld, R. 1937. Über das System der Orchidaceae. Blumea. Suppl. 1:25–37.
Masamune, G. 1932a. Notes on the Formosan orchids. Jap. J. Bot. 8:257–263.
——. 1932b. Symbolae florae Australi-Japonicae. I. J. Soc. Trop. Agr. 4:194–195.
——. 1933. A list of orchidaceous plants indigenous to Formosa. Trop. Hort. 3:29, 33–37.
——. 1934a. Beiträge zur Kenntnis der Flora von Südjapan. II. Trans. Nat. Hist. Soc. Formosa 24:208–214.
——. 1934b. Beiträge zur Kenntnis der Flora von Südjapan. III. Trans. Nat. Hist. Soc. Formosa 24:280.
——. 1935. Beiträge zur Kenntnis der Flora von Südjapan. IV. Trans. Nat. Hist. Soc. Formosa 25:14.
——. 1937. Le caractere et les affinites de la flore alpine de Taiwan. Bot Mag. Tokyo 51:232–235.
——. 1965–78. Icones Plantarum Asiaticum. XXVI–LXXI. J. Geobot. 14–26: pl. 107–255.
Mattick, F. 1964. Übersicht über die Florenreiche und Florengebiete der Erde, p. 626–629, fig. *In* A. Engler, Syllabus der Pflanzenfamilien. II. (12th ed., H. Milchior, ed.). Gebruder Borntraeger, Berlin-Nikolassee.
Merrill, E. D. 1923. Die Pflanzengeographische Scheidung von Formosa und Philippinen. Bot. Jahrb. Engler 58:599–604.
Miyoshi, M. 1932. On the manuscripts written by Matsuoka Joan. Honzu 6:43–55.
Pijl, L. van der, and C. H. Dodson. 1966. Orchid flowers: Their pollination and evolution. Univ. Miami Press, Coral Gables, Fla.
Reichenbach, H. G. 1852. Gartenorchideen. Bot. Zeit. 10:761–772, 927–937.
——. 1855. Orchideae Hongkongense a cl. Hance et cl. Seemann lectae. Bonplandia 3:249–250.
——. 1858–1900. Xenia Orchidacea. Beiträge zur Kenntniss der Orchideen. I–III. Brockhaus, Leipzig.
——. 1861. Orchidaceae, p. 167–933. *In* Walpers, Ann. Bot. Syst. VI. Sumptibus Ambrosii Abel, Leipzig.
——. 1884. Über das System der Orchideen. Bull. Cong. Int. Bot. Hort. St. Petersburg 39–58.
Ridley, H. N. 1930. The dispersal of plants throughout the world. L. Reeve, London.
Rolfe, R. A. 1903a. Orchidaceae in Forbes and Hemsley: An enumeration of all plants known from China proper, Formosa, Hainan, Corea, the Luchu Archipelago, and the Island of Hongkong, together with their distribution and synonymy. J. Linn. Soc. London Bot. 36:5–67.
——. 1903b. Orchids of China. Orchid Rev. 11:103.
Schlechter, R. 1919. Orchideologiae Sino-Japonicae prodromus, eine kritische Besprechung der Orchideen Ost-Asiens. Fedde Repert. Beih. 4:1–219.
——. 1921. Additamenta ad Orchideologiam chinensem. Fedde Repert. 17:22–28, 63–72.
Swamy, B. L. G. 1948a. Vascular anatomy of orchid flowers. Bot. Mus. Leafl. Harvard Univ. 13:61–95.
——. 1948b. Embryological studies in the Orchidaceae. I. Gametophytes. Amer. Midland Naturalist 41:184–201.
——. 1949. Embryological studies in the Orchidaceae. II. Embryogeny. Amer. Midland Naturalist 41:202–232.
Swartz, O. 1800. Afhandling om Orchidernes Slägter och deras systematiska indelning. Vet. Akad. Nya Handl. 21:115–138, 202–254.
Takhtajan, A. 1969. Flowering plants, origin and dispersal (translated by C. Jeffrey). Oliver and Boyd, Edinburgh.
Tan, K. W. 1976. Taxonomy of Arachnis, Aromodorum, Esmeralda and Dimorphorchis (Orchidaceae). II. Selbyana 1:365–373.
Tang, T., and S. C. Chen. 1977. Orchidaceae, p. 185–264. *In* Flora Hainanica. IV. Science Press, Peking.

Tang, T., and F. T. Wang. 1936. Notes on Orchidaceae of China. I–II. Bull. Fan Mem. Inst. Biol. Bot. 7:1–10, 127–144.
——. 1940. Contributions to the knowledge of eastern Asiatic Orchidaceae. I. Bull. Fan Mem. Inst. Biol. Bot. 10:21–46.
——. 1947. A survey of the classification of the Orchidaceae. Science (Sci. Soc. China) 29:138–141.
——. 1951a. Contributions to the knowledge of eastern Asiatic Orchidaceae. II. Acta Phytotax. Sinica 1(1):23–102.
——. 1951b. Corybas Salisb., a new addition to the orchid flora of China. Acta Phytotax. Sinica 1(2):185–187.
——. 1951c. On the identity of eight Gagnepain's orchidaceous genera from Indo-China. Acta Phytotax. Sinica 1(3–4):257–267.
——. 1954. Orchidaceae, Key to the subfamilies, tribes, subtribes and genera. Acta Phytotax. Sinica 2(4):456–470.
——. 1974. Plantae novae Orchidacearum Hainanensium. Acta Phytotax. Sinica 12(1):35–49.
Tsi, Z. H., and K. Y. Lang, 1979. Orchidaceae, p. 521–534. *In* Claves Familiarum Generumque Cormophytorum Sinicorum. Science Press, Peking.
Vogel, E. F. de. 1969. Monograph of the tribe Apostasieae (Orchidaceae). Blumea 17:313–350.
Withner, C. L. (ed.). 1959. The orchids: A scientific survey. Ronald Press, New York.
Wu, C. Y. 1979. The regionalization of Chinese flora. Acta Botanica Yunnanica 1(1):1–22.
Ying, S. S. 1977. Coloured illustrations of indigenous orchids of Taiwan. I. Taiwan Univ., Taipei.

Additional Literature

Banerji, M. L., and B. B. Thapa. 1969. Orchids of Nepal. J. Bombay Nat. Hist. Soc. 66:286–296.
——. 1969. Orchids of Nepal. II. J. Bombay Nat. Hist. Soc. 66:575–583.
——. 1970. Orchids of Nepal. III. J. Bombay Nat. Hist. Soc. 67:139–152.
——. 1971. Orchids of Nepal. IV. J. Bombay Nat. Hist. Soc. 68:29–36.
——. 1971. Orchids of Nepal. V. J. Bombay Nat. Hist. Soc. 68:660—665.
——. 1973. Orchids of Nepal. VI. J. Bombay Nat. Hist. Soc. 69:283–289.
——. 1973. Orchids of Nepal. VII. J. Bombay Nat. Hist. Soc. 70:25–35.
——. 1973. Orchids of Nepal. VIII. J. Bombay Nat. Hist. Soc. 70:330–338.
——. 1975. Orchids of Nepal. IX. J. Bombay Nat. Hist. Soc. 72:30–42.
——. 1976. Orchids of Nepal. X. J. Bombay Nat. Hist. Soc. 73:149–156.
Bretschneider, E. V. 1895. Botanicon sinicum. III. J. North China Branch Roy. Asiat. Soc. n.s. 29:1–623.
Brieger, F. G. 1975. On the orchid system: General principles and the distinction of subfamilies. *In* K. Senghas (ed.), Proc. 8th World Orch. Conf., Frankfurt, p. 448–504.
Butzin, F. 1978. In Berlin existing types of Schlechter's orchid species. Willdenowia 8:401–408.
Cadbury, W. W. 1945. Recollections of my orchid garden in Canton, China. Amer. Orchid Soc. Bull. 13:410–412.
——. 1946. Dendrobiums and habenarias in South China. Amer. Orchid Soc. Bull. 15:3–8.
Chiang, Y. L., and L. R. Chen. 1968. Observations on Pleione formosana Hayata. Taiwania 14:271–301.
Chien, S. S. 1930. Three new species of orchids from Chekiang. Contr. Biol. Lab. Sci. Soc. China Bot. Ser. 6:23–32, pl. 1-3.
——. 1931. Studies of the Chinese orchids. I. Contr. Biol. Lab. Sci. Soc. China Bot. Ser. 6:79–110, f. 1–13.
——. 1935. A new genus of orchids from eastern China. Contr. Biol. Lab. Sci. Soc. China Bot. Ser. 10:89–91, pl. 12.
Cooray, D. A. 1940. Orchids in oriental literature. Orchidologia Zeylanica 11:3, 73.
Duthie, J. F. 1906. The orchids of the north-western Himalaya. Ann. Bot. Gard. Calcutta 9:81–211.
Finet, E. A. 1897. Sur le genre Oreorchis Lindley. Bull. Soc. Bot. France 44:69–74, pl. 3.
——. 1897. Orchidées nouvelles de la Chine. Bull. Soc. Bot. France 44:419–422, pl. 13-14.
——. 1898. Orchidées recueillies au Yunnan et au Laos, par le Prince Henri d'Orléan. Bull. Soc. Bot. France 45:411–414.
——. 1898. Orchidées nouvelles ou peu connues. J. de Bot. 12:340–344, pl. 5–6.

——. 1899. Sur quelques espècies nouvelles du genre Calanthe. J. de Bot. 46:434–437, pl. 9-10.
——. 1901. Les Orchidées de l'Asie orientale. Rev. Gen. Bot. 13:497–534, pl. 12-18.
——. 1903. Énumération des espècies du genre Dendrobium (Orchidées) formant la collection du Muséum de Paris. Bull. Mus. Hist. Nat. Paris 9:295–303.
——. 1903. Dendrobium nouveaux de l'herbier du Muséum. Bull. Soc. Bot. France 50:372–383, pl. 9-14.
——. 1908. Orchidées nouvelles ou peu connues II. J. de Bot. 55:333–343, po. 10-11.
——. 1909. Orchidées nouvelles ou peu connues III. J. de Bot. 56:97–104, pl. 1-2.
——. 1910. Orchidees du Su-tchuan. Not. Syst. Lecomte 1:260–261.
Gagnepain, F. 1931. Pholidota nouveaux d'Asie. Bull. Mus. Hist. Nat. Paris, ser. 2,3:145–147.
——. 1931. Pleione nouveaux de Chine. Bull. Soc. Bot. France 78:25–26.
——. 1932. Huit genres nouveaux d'Orchidées Indochinoises. Bull. Mus. Hist. Nat. Paris, ser. 2,4:591–601.
——. 1932. Orchidacees nouvelles ou critiques. I. Tainia ou Nephelaphyllum? Bull. Mus. Hist. Nat. Paris, ser. 2,4:705–712.
Gagnepain, F., and A. Guillaumin. 1932–34. Orchidacées in Lecomte, Fl. Gén. Indo-Chine 6:142–654, Paris.
Garay, L. A. 1972. On the systematics of the monopodial orchids. I. Bot. Mus. Leafl. Harvard Univ. 23:149–212.
——. 1973. Systematics of the genus Angraecum (Orchidaceae). Kew Bull. 28:495–516.
——. 1974. On the systematics of the monopodial orchids. II. Bot. Mus. Leafl. Harvard Univ. 23:369–375.
Grant, B. 1895. The orchids of Burma. Hanthawaddy Press, Rangoon.
Guillaumin, A. 1960. Plantes nouvelles, rares ou critiques des serres du Muséum. Bull. Mus. Hist. Nat. Paris, ser. 2,32:188–189.
Hance, H. F. 1876. Two new Hongkong orchids. J. Bot. Brit. For. 14:44–45.
——. 1877. A second Hongkong Cleisostoma. J. Bot. Brit. For. 15:38.
——. 1883. Orchidaceas quattuor novas sinenses proponit. J. Bot. Brit. For. 21:231–233.
——. 1884. Orchidaceas epiphyticas binas novas describit. J. Bot. Brit. For. 22:364–365.
Handel-Mazzetti, H. 1936. Symbolae Sinicae. VII (Orchidaceae), 1323–1360. Julius Springer, Vienna.
——. 1937. Kleine Beiträge zur Kenntnis der Flora von China. VI. Österr. Bot. Zeitschr. 86:302–303.
Hara, H. 1955. On Asiatic species of Tulotis Rafin. Jap. J. Bot. 30:72.
Hara, H., W. T. Stearn, and H. T. Williams. 1978. An enumeration of the flowering plants of Nepal. I, 30–58. Staples Printers Kettering, Northamptonshire.
Hautzinger, L. 1976. Nomenclatural and systematic contributions to the family of the Orchidaceae. Herh. Zoo-Bot. Ges. Wien 115:40–54.
——. 1977. The genus Orchis L. (Orchidaceae); Section Robustocalcare Hautzinger. Ann. Naturhist. Mus. Wien 81:31–74.
Hawley, W. 1949. Chinese folk design. Privately printed, Hollywood, Calif.
Holttum, R. E. 1953. A revised flora of Malaya I (Orchids of Malaya). 3d ed., 1964. Government Printing Office, Singapore.
Hooker, J. D. 1888–1890. The flora of British India (Orchidaceae) 5:667–864; 6:1–198. L. Reeve, London.
How, F. C. 1956. Flora of Kwangchow (Orchidaceae, p. 723–734). Science Press, Peking.
Hu, H. H. 1925. Nomenclatorial changes in Chinese orchids. Rhodora 27:105–107.
Hu, S. Y. 1971. Anoectochilus yungianus, a new species of Orchidaceae. Quart. J. Taiwan Mus. 24:257–262.
——. 1972. Floristic studies in Hong Kong. Chung Chi J. 11:1–25.
Hunt, P. F. 1970. Notes on Asiatic orchids. V. Kew Bull. 24:75–99.
——. 1972. Notes on Asiatic orchids. VI. Kew Bull. 26:171–185.
Hunt, P. F., and V. S. Summerhayes. 1961. Notes on Asiatic orchids. III. Taxon 10:101–110.
——. 1966. Notes on Asiatic orchids. IV. Kew Bull. 20:51–61.
Hunt, P. F., and G. C. Vosa. 1971. The cytology and taxonomy of the genus Pleione D. Don (Orchidaceae). Kew Bull. 25:423–432.
Ishii, Y. 1935. Colored illustrations of eastern Asiatic orchids. Seibundo, Tokyo.
Kimura, K. 1935. Note on the botanical origin of the Chinese medicine shih hu. Abstr. Ann. Conf. 6 Sci. Inst. (Kwangsi) 118–120.

——. 1936–37. Pharmacognostical study of Dendrobiinae plants as the Chinese drug shih hu. Bull. Shanghai Sci. Inst. 6(1936):1–60, p. 1-16; 7(1937):11–46, pl. 1-87.
King, G., and R. Pantling. 1897. Some new Indo-Malayan orchids. J. As. Soc. Beng. 66:578–605.
——. 1898. The orchids of the Sikkim Himalaya. Ann. Bot. Gard. Calcutta 8:1–342, pl. 1-448.
Kränzlin, F. 1891. Beiträge zu einer Monographie der Gattung Habenaria Willd. (I. Allgemeiner Teil) 1–41, Inaug. Dissert., Berlin.
——. 1892. Beiträge zu einer Monographie der Gattung Habenaria Willd. (II. Systematischer Teil). Bot. Jahrb. Engler 16:52–223.
——. 1897–1901. Orchidacearum genera et species. I. Mayer and Müller, Berlin.
——. 1903–04. Orchidacearum genera et species. II. Mayer and Müller, Berlin.
——. 1907. Die Gattung Coelogyne and ihre Verwandten. Orchis 2:1–6.
——. 1908. Orchidaceae quaedam Tibeticae, quas exposuit Fr. Kränzlin. Fedde Repert. 5:196–200.
——. 1910–11. Orchidaceae-Monandrae-Dendrobiinae. *In* Engler, Pflanzenreich. 45(IV.50.II.B.21): 1–382, f. 1–35; 50(IV.50.II.B.21):1–182, f. 1–35.
——. 1911. Orchidaceae-Monandrae-Thelasinae. *In* Engler, Pflanzenreich. 50(IV.50.II.B.23):1–46, f. 1–5.
——. 1921. Orchidaceae Ténianae Yunnanenses. Fedde Repert. 17:99–112.
Kudo, Yakoru. 1934. A list of exotic orchids in Taiwan. Trop. Hort. 4:248–254, 334–338, 374–381.
Kudo, Yushun. 1930. Haraella, a new genus of orchids from Formosa. J. Soc. Trop. Agr. 2:26–28.
——. 1930. Material for a flora of Formosa. I–II. J. Soc. Trop. Agr. 2:147, 237.
Lang, K. Y., and Z. H. Tsi. 1978. Some new taxa of Orchidaceae from Tibet, China. Acta Phytotax. Sinica 16(4):126–129.
Lin, T. P. 1975. New additions to the orchid flora of Taiwan. Taiwania 20:162–164.
Lin, T. P., and C. C. Hsu. 1976. Orchid genera, Anoectochilus and Odontochilus of Taiwan. Taiwania 21:229–236.
——. 1977. The genus Thrixspermum Lour. of Taiwan (Orchidaceae). Taiwania 22:59–72.
Lindley, J. 1852–59. Folia orchidacea, an enumeration of the known species of orchids. J. Matthews, London.
Liu, T. S., and H. J. Su. 1973. New additions to the orchidaceous flora of Taiwan, China. II. Quart. J. Taiwan Mus. 26:443–447.
——. 1975. New additions to the orchidaceous flora of Taiwan, China. III. Quart. J. Taiwan Mus. 28:269–276.
Liu, Y. 1936. Cypripedium in China. J. Bot. Soc. China 3:1051–1061.
Loureiro, J. 1790. Flora Cochinchinensis. II, 518–525. Academy of Science, Lisbon.
Luer, C. A. 1972. The native orchids of Florida. W. S. Cowell, Ipswich.
——. 1975. The native orchids of the United States and Canada. N. Y. Bot. Garden.
Maekawa, F. 1965. On the differentiation of the genus Cymbidium sensu lato. Jap. J. Bot. 40:121–326.
——. 1971. The wild orchids of Japan in colour. Shibundo, Tokyo.
Matsumura, J., and B. Hayata. 1906. Enumeratio Plantarum Formosanarum (Orchidaceae), p. 406–421. Univ. Tokyo, Tokyo.
Maximowicz, C. J. 1886. Diagnoses plantarum novarum Asiaticarum. VI. Bull. Acad. Sci. St.-Petersburg 31:102–108.
Merrill, E. D. 1935. A commentary on Loureiro's "Flora Cochinchinensis." Trans. Amer. Phil. Soc. 24:1–445.
Nackejima, F. 1975. Preliminary notes on the noteworthy Orchidaceae from Formosa, Ryukyu, Bonin Island and Southern Japan (4). Biol. Mag. Okinawa 13:24–37.
Nagano, Y. 1955. Miniature cymbidiums in Japan. Amer. Orchid Soc. Bull. 24:735–743.
Nevski, S. A. 1935. Orchidaceae in Flora of USSR. IV, 589–730. Academiae Scientiarum USSR, Leningrad.
Ohwi, J. 1978. Flora of Japan. Rev. ed. Shibundo, Tokyo.
Panigrahi, G., and P. Taylor. 1975. A new species of Listera (Orchidaceae) from the Himalayas. Kew Bull. 30:559–561.
Panigrahi, G., and J. J. Wood. 1974. A new species of Listera (Orchidaceae) from Asia. Kew Bull. 29:731–733.
Pfitzer, E. 1889. Orchidaceae. *In* Engler and Prantl, Die naturlichen Pflanzenf. II, 6:52–224, f. 41–237. Wilhelm Engelmann, Leipzig.

——. 1903. Orchidaceae-Pleionandrae. *In* A. Engler, Pflanzenreich. 12(IV.50):1–132, f. 1–41. Wilhelm Engelmann, Leipzig.
Pfitzer, E., and F. Kränzlin. 1907. Orchidaceae-Monandrae-Coelogyninae. *In* A. Engler, Pflanzenreich, 32(IV.50.II.B.7):1–169, f. 1–54.
Reinikka, M. A. 1972. A History of the orchids. Univ. Miami Press, Coral Gables, Fla.
Rolfe, R. A. 1896. The Cypripedium group. Orchid. Rev. 4:327–334, 363–367.
——. 1903. The genus Pleione. Orchid Rev. 11:289–292.
——. 1913. Plantae chinenses Forrestianae: Enumeration and description of species of Orchideae. Notes Bot. Gard. Edin. 8:19–29, pl. 9-12.
——. 1913. Chinese cypripediums. Orchid Rev. 21:80–83, f. 21.
Sasayama, M. 1932. Treatise on orchids. I–II. Seimido, Tokyo.
Sasayama, M., and Y. Nagano. 1957. Oriental miniature orchids. Kashimashoden, Tokyo.
Schill, R., and W. Pfeiffer. 1977. Untersuchungen an Orchideen-pollinien unter besonderer Beruecksichtigung ihrer Feinskulturen. Pollen et Spores 19:5–118.
Schlechter, R. 1906. Orchidaceae novae et criticae. III. Fedde Repert. 2:166–171.
——. 1906. Orchidaceae novae et criticae. IV. Fedde Repert. 3:15–20.
——. 1911. Die Polychondreae (Neottiinae Pfitz.) und ihre systematische Einteilung. Bot. Jahrb. Engler 45:375–410.
——. 1911. Die Gattung Thrixspermum Lour. Orchis 5:46–48.
——. 1911. Die Gattung Bletilla. Fedde Repert. 10:254–256.
——. 1912. Enumeration and description of species of Orchidaceae. Notes Bot. Gard. Edin. 5:93–113.
——. 1914. Die Gattung Pleione und ihre Arten. Orchis 8:72–80, pl. 2.
——. 1914–15. Die Orchideen, ihre Beschreibung, Kulture and Züchtung. Handbuch fur Orchideenliebhaber, Zuchter und Botaniker. (2d ed., 1927). Paul Parey, Berlin.
——. 1920. Orchidaceae novae et criticae. LXV. Fedde Repert. 16:353–358.
——. 1920. Eine zweite Art der Gattung Androcorys. Notizbl. Bot. Gart. Berlin 7(68):52–53.
——. 1922. Orchidaceae in Pax, Aufzählung der von Dr. Limpricht in Ostasien gesammelten Pflanzen, in W. Limpricht Botanische Reise. Fedde Repert. Beih. 12:326–352.
——. 1924. Orchidaceae in H. Smith, Plantae Sinenses. Acta Hort. Gothob. 1:125–155.
——. 1924. Die Gattung Cymbidium Sw. und Cyperorchis Bl. Fedde Repert. 20:96–110.
——. 1924. Orchidaceae novae et criticae. LXXV–LXXVI. Fedde Repert. 19:372–381.
——. 1924. Orchidaceae novae et criticae. LXXVII. Fedde Repert. 20:378–384.
——. 1926. Das System der Orchidaceen. Notizbl. Bot. Gart. Berlin 9:563–591.
Schultes, R. E., and A. S. Pease. 1963. Generic names of orchids, their origin and meaning. Academic Press, New York and London.
Schweinfurth, C. 1929. Orchids collected by J. F. Rock on the Arnold Arboretum expedition to northwestern China and northeastern Tibet. J. Arnold Arboretum 10:167–174.
Seidenfaden, G. 1968. The genus Oberonia in mainland Asia. Dansk Bot. Arkiv 25:1–125.
——. 1969. Notes on the genus Ione. Bot. Tidsskr. 64:205–238.
——. 1969. Contributions to the orchid flora of Thailand. I. Bot. Tidsskr. 65:100–162.
——. 1970. Contributions to the orchid flora of Thailand. II. Bot. Tidsskr. 65:313–370.
——. 1971. Contributions to the orchid flora of Thailand. III. Bot. Tidsskr. 66:303–356.
——. 1971. Notes on the genus Luisia. Dansk Bot. Arkiv 27:11–101.
——. 1972. Contributions to the orchid flora of Thailand. IV. Bot. Tidsskr. 67:76–127.
——. 1973. Cirrhopetalum Lindl. Dansk Bot. Arkiv 29:1–260.
——. 1973. Contributions to the orchid flora of Thailand. V. Bot. Tidsskr. 68:41–95.
——. 1973. An enumeration of Laotian orchids. Bull. Mus. Hist. Nat. Paris, ser. 3,71:101–152.
——. 1975. Contributions to the orchid flora of Thailand. VI. Bot. Tidsskr. 70:64–97.
——. 1975. Contributions to a revision of the orchid flora of Cambodia, Laos and Vietnam I. Manuscript, Fredenborg.
——. 1975. Orchid genera in Thailand. I. Dansk Bot. Arkiv 29(2):1–50.
——. 1975. Orchid genera in Thailand. II. Dansk Bot. Arkiv 29(3):1–80.
——. 1975. Orchid genera in Thailand. III. Dansk Bot. Arkiv 29(4):1–94.
——. 1976. Contributions to the orchid flora of Thailand. VII. Bot. Tidsskr. 71:1–30.
——. 1976. Orchid genera in Thailand. IV. Dansk Bot. Arkiv 31(1):1–105.
——. 1977. Contributions to the orchid flora of Thailand. VIII. Bot. Tidsskr. 72:1–14.
——. 1977. Orchid genera in Thailand. V. Dansk Bot. Arkiv 31(3):1–149.

——. 1978. Orchid genera in Thailand. VI. Dansk Bot. Arkiv 32(2):1–195.
——. 1978. Orchid genera in Thailand. VII. Dansk Bot. Arkiv 33(1):1–94.
Seidenfaden, G., and T. Smitinand. 1959–65. The orchids of Thailand (A preliminary list). Siam Society, Bangkok.
Smith, J. J. 1905–12. Die Orchideen von Java. (I. Text, 1905; II–IV. Figuren-Atlas, 1908–12). E. J. Brill, Leiden.
Smith, W. W. 1921. New orchids from Yunnan and northern Burma. Notes Bot. Gard. Edin. 13:189–222.
Soo, R. 1966. Die sog. Orchis Arten der ostasiatisch-nordamerkanischen Flora. Act. Bot. Acad. Sci. Hung. 12:351–352.
——. 1969. A short survey of the orchids of the Soviet Union. Ann. Univ. Sci. Budapest Rolando Edtvo Nominatae Sect. Biol. 11:53–74.
——. 1974. The currently valid names and recent systematic position of species previously relegated to the genus Orchis in the East and South-east Asia and in North America. Act. Bot. Acad. Sci. Hung. 20:349–353.
Stuart, G. A. 1911. Chinese Materia Medica. American Presbyterian Mission Press, Shanghai.
Su, H. J. 1971. Native Vanilla of Taiwan. Taiwan Orchid Soc. Bull. 10:159–163.
——. 1971. Three new species of Cirrhopetalum from Taiwan, China. Quart. J. Taiwan Mus. 24:173–180.
——. 1972. New additions to the orchidaceous flora of Taiwan, China. I. Quart. J. Taiwan Mus. 25:149–155.
——. 1973. New additions to the orchidaceous flora of Taiwan, China. II. Quart. J. Taiwan Mus. 26:433–437.
——. 1974. The native orchids of Taiwan. Harvest Farm Press, Taipei.
Summerhayes, V. S. 1955. A revision of the genus Brachycorythis. Kew Bull. 1955:221–264.
——. 1955. Notes on Asiatic orchids. I. Kew Bull. 1955:587–589.
——. 1957. Notes on Asiatic orchids. II. Kew Bull. 1957:259–268.
Sweet, H. R. 1969. A revision of the genus Phalaenopsis Bl. IV and VII. Amer. Orchid Soc. Bull. 38:225–239, 681–694.
Tso, C. L. 1933. Notes on the orchid flora of Kwangtung. Sunyatsenia 1:131–156.
Tung, H. T. 1966. Orchid culture. 2d. ed. Yi-fen Lu, Hongkong.
Tuyama, T. 1941. Notes on the genus Gastrodia of southeastern Asia. Jap. J. Bot. 17:579–586.
——. 1967. On Epipogium roseum (D. Don) Lindl. in Japan and its adjacent regions, with remarks on other species of the genus. Jap. J. Bot. 42:295–311.
Veitch, H. J. 1887–94. A manual of orchidaceous plants. J. Veitch and Sons, London.
Vermeulen, P. 1959. The different structure of the rostellum in Ophrydeae and Neottieae. Acta Bot. Neerl. 8:338–355.
——. 1965. The place of Epipogium in the system of Orchidales. Acta Bot. Neerl. 14:230–241.
——. 1966. The system of Orchidales. Acta Bot. Neerl. 15:224–253.
Wu, Y. S., and S. C. Chen. 1966. Tres species novae generis Cymbidii e Provincia Szechuan. Acta Phytotax. Sinica 11(1):31–34.
Yamamoto, Y. 1924. Genus novum Orchidacearum ex Formosa. Bot. Mag. Tokyo 38:209.
Yen, T. K. 1964. Icones Cymbidiorum Amoyensium. Committee of Science and Technology of Amoy, Fukien.
Ying, S. S. 1974. Some new species and new combinations of Taiwan orchids, Bull. Exp. Forest Nat. Taiwan Univ. 114:154–157.

SYMBIOSIS

3

Orchid Mycorrhiza*

GEOFFREY HADLEY†

*The literature survey pertaining to this chapter was concluded in March 1980; the chapter was submitted in April 1980, and the revised version was received in May 1980.

†This review is dedicated to the memory of Dorothy G. Downie, who unknowingly led me into her footsteps, and to the many research students who have contributed to my understanding of the topic.

Introduction and History

It is well established that orchids have fungi growing in their roots. Cells of the root cortex are occupied, to a greater or lesser extent, by clusters of fungal hyphae, and these occur under both natural and horticultural conditions. The orchid-fungus relationship is one of many known forms of mycorrhiza (the term means fungus-root), a common form of symbiosis between fungi and plants.

Although the presence of fungi in orchid roots was noticed in the early years of the nineteenth century it was the extensive studies of many European and exotic orchids by Reissek (1847) that established their occurrence as a normal phenomenon. Reissek in fact attempted but failed—understandably, in view of the lack of knowledge of microorganisms at that time—to isolate the fungus from the roots. Wahrlich (1886) also examined numerous exotic orchids in conditions of cultivation and was the first investigator to describe the changes in the intracellular fungus as it became digested by the host cell. Subsequently, in a detailed investigation of *Neottia nidus-avis,* Magnus (1900) recognized in addition to the "digestion cells" zones of "host cells" in which the fungus remained alive and probably hibernated.

But the recognition of the importance of orchid fungi in seed germination came with the brilliant pioneering of Noel Bernard. Having established (Bernard, 1903) that seeds germinated successfully only when infected by a suitable fungus, he carried out extensive investigations into all aspects of the symbiosis and eventually published his results in a paper (Bernard, 1909) subsequently recognized as a classic study of mycorrhiza. He hypothesized that the fungus enhanced the nutrient status of the seedling and was kept in check by it, i.e., as a form of controlled parasitism.

Contemporary with Bernard was the early work of Hans Burgeff (1909). He isolated and named many fungi, studied them in culture, and developed the concept of digestion of the fungus as the principal means of nutrition for the orchid. Although the evidence for this came from experiments with germinating seeds and the infection of protocorms, an obvious extension of the hypothesis was to explain the occurrence of fungi in the roots of adult plants similarly. Burgeff subsequently investigated patterns of root infection and described many series of orchids in which increasing dependence on the fungus for a source of carbon was paralleled by decreasing leaf area, sometimes coupled with a decrease in pigmentation (Burgeff, 1936, 1943, 1959).

Both Bernard and Burgeff, in studying seed germination, conducted many experiments using various combinations of nutrients in the absence of a fungus. Although germination did occur in some cases, they recognized that in general, and particularly under natural conditions, successful germination was virtually impossible in the absence of a fungus. This implies that germination is not successful unless it is succeeded by growth. In the absence of infection, as will be described later, the growth rate is often negligible regardless of the availability of nutrients. Consequently, a symbiotic method for seed germination was developed which involved culture tubes inoculated with both seeds and fungi. Not surprisingly, this procedure was elaborate and difficult for the commercial grower. It was also characterized by a low success rate (Bernard, 1909), although the English nurseryman Joseph Charlesworth used the method fairly extensively until ill health forced him to retire (Ramsbottom, 1923).

In contrast, the nonsymbiotic method of seed germination offered obvious attractions in orchid horticulture. This method was developed by Lewis Knudson (1922, 1924), who concentrated on testing various combinations of nutrients and successfully germinated seeds of many orchids on a medium containing only sugars and mineral salts. He eventually (Knudson, 1930) grew a *Laeliocattleya* hybrid asymbiotically from seed to flower. Many others followed in his footsteps, using a plethora of nutrient media. In fact, the majority of investigations into germination have been concerned with manipulations of nutrient media (Arditti, 1967, 1979), and the nonsymbiotic method is successful with many growers.

Orchidologists, naturally, are more concerned with orchids than with their mycorrhizas. It is not surprising, therefore, that there was a divergence of horticultural and scientific interests following the earlier difficulties with seed germination and the subsequent development of clonal propagation techniques (Vajrabhaya, 1977). To orchid growers mycorrhiza remains a curious natural phenomenon the understanding and exploitation of which is outside their interest and profession. Orchid mycorrhizal studies became the field of the mycologist and the physiologist. Many questions remain unsolved, but in recent years there has developed a much greater understanding of the nature of orchid fungi and the role they play in enhancing germination of the seed. Comparison with other mycorrhizal systems (Smith, 1974) leaves no doubt that orchid mycorrhizas are fairly uniform, but are very different from others and represent a unique, highly evolved symbiotic system.

What follows is a review of past and present research coupled with a synthesis of our current understanding of this system. As a mycologist I obviously see the orchid problem from a special viewpoint, different from that of the orchidologist and the commercial grower. Some personal opinions apparent in this review may reflect an inadequate appreciation of orchids as a whole.

General Organization of Mycorrhizal Tissues

Adult Plants

The Orchidaceae exist in such a wide diversity of forms that it is impossible here to describe the mycorrhizal structure of individual species, or indeed the diversity that occurs in the family, in any detail. Earlier reviews by Ramsbottom (1923), Burgeff (1936), and Harley (1969) provide an excellent distillation of the results and progress made by many investigators over the years. It is unequivocally accepted that the underground absorbing organs of orchids are associated with, and infected by, a particular group of fungi which are adapted to occupy cells of the cortex and there form coiled, branched structures of intracellular mycelium.

Reissek (1847, and quoted by Ramsbottom, 1923) was the first person to investigate root structure of a variety of orchids. He concluded that there was a fundamental difference between native European terrestrial species in which the cortex was extensively infected and exotic tropical orchids in which "the fungal masses were arranged singly at the periphery." Reissek's conclusion probably remains valid as a generalization. To it can be added a further general feature, that there is a remarkable consistency in most species of orchids as far as the organization of the intracellular hyphal mass is concerned. Infection occurs from an adjacent cell by means of a single hypha, which may then coil round the inside of the cell wall several times before it infects the next cell (Plate 3-1-**A**). Subsequently the profuse development of branches and their fusion by anastomosis occurs, forming the typical three-dimensional network (the peloton). This is common to all orchids, and the peloton appears to be an adaptation to the host cell, being different from any structure formed by the fungus outside its host.

Roots of north temperate orchids may show almost complete infection of the cortex, as in *Dactylorhiza purpurella* (Plate 3-1-**B**), which is typical of other dactylorchids. In contrast, the infection among tropical species is usually less dense and may be very sparse (Plate 3-1-**C**). It also varies in relation to the habit. For example, a survey of Malayan orchids (Hadley and Williamson, 1972) revealed that some terrestrial species were heavily infected, although in the majority, infection was characteristically spasmodic and localized, and occasionally absent. Epiphytic orchids, as might be expected, did not usually have their aerial roots infected until they came into contact with a suitable substrate. Even then, infection did not spread into aerial parts of the root. Some epiphytes, particularly the liane-forming Vanilleae (in which there is a difference between long, spreading roots and shorter, branched, absorbing ones in contact with the substrate), usually show a pattern of infection originating from the substrate and in turn leading to variations in root structure such as localized hypertrophy and hyperplasia of tissues. Enlargement of infected cells, hypertrophy of host-cell nuclei (see below), and effects such as a localized proliferation of epidermal hairs (Hadley and Williamson, 1972) are probably quite common.

The roots of terrestrial orchids are normally infected soon after their formation. For instance, many north temperate species, which have an annual cycle of development, form new roots in the spring at the start of the growing season, and the infecting fungi soon occupy most of the cortical cells.

Burges (1939) observed digestion about a month after infection in *Dactylorhiza incarnata,* while other workers have described biannual periods of digestion (Vermeulen, 1946) using other species. Digestion may occur randomly or in a defined zone of cells, the "digestion layer," which in some species can be recognized in the inner cortex and is described in detail by Burgeff (1959).

Cells in which the intracellular hyphae are digested are often reinfected, sometimes even for a third time, and various workers have correlated such activity of the fungus with periods of active growth of the orchid. There is no direct evidence for this, but it seems likely that when root activity declines the endophytic fungus also becomes quiescent and most of the intracellular material is in the form of digested pelotons. North temperate orchids show no evidence from isolation work of any vigorous winter activity of the fungus, although *Goodyera repens,* which is evergreen, has yielded active endophytes in isolation tests carried out at all seasons (Hadley, unpublished).

Orchids having rhizomes which function as absorbing organs usually show localized

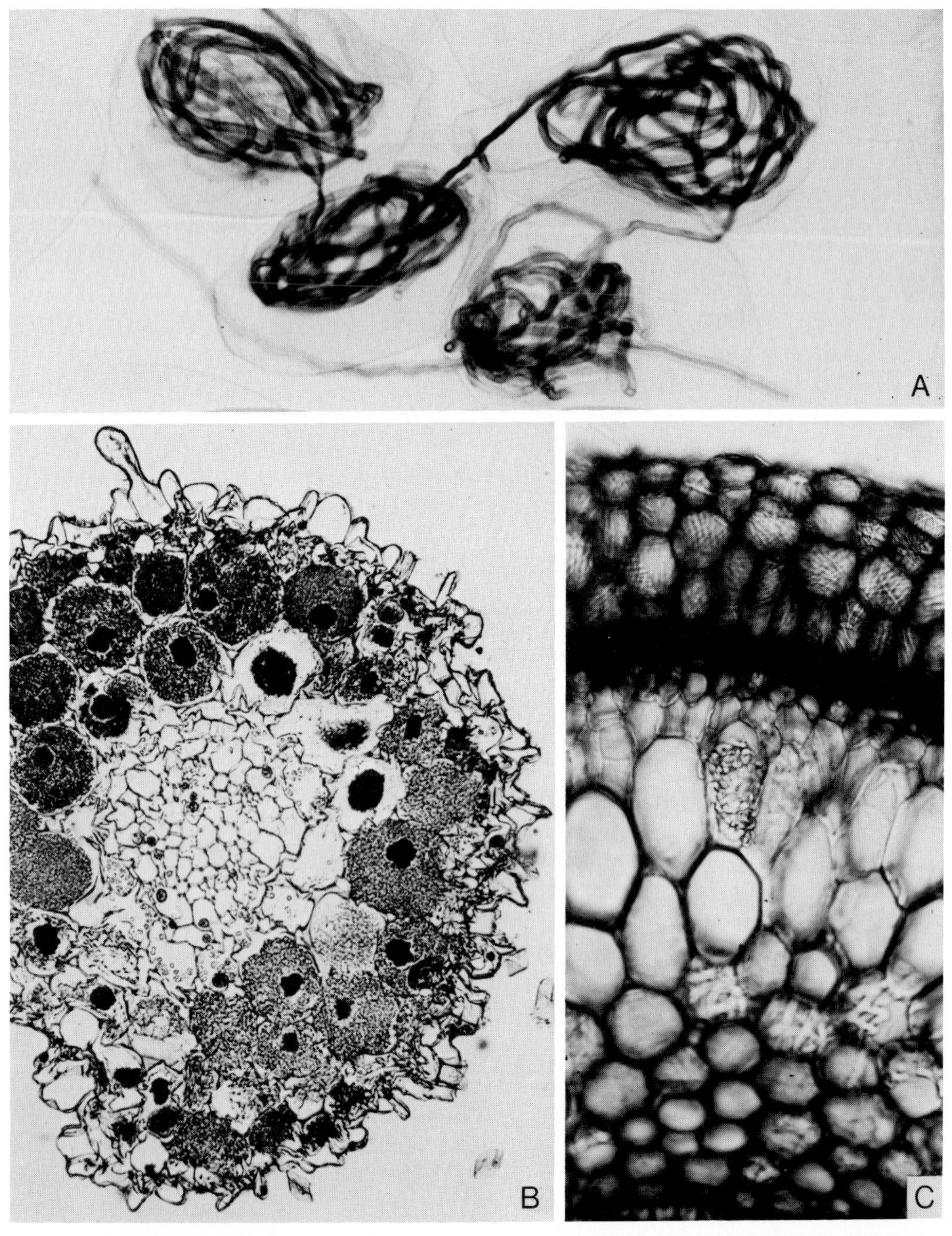

Plate 3-1. Organization of mycorrhizal roots. **A:** Squash preparation of rhizome of *Goodyera repens,* showing infection of cells and development of pelotons; × 300 (photo by Clare Alexander). **B:** Prepared section of *Dactylorhiza purpurella* root, showing dense infection of the cortex with clumps of digested material and secondary pelotons in some cells; × 100 (photo by G. Harvais). **C:** Hand section of *Dendrochilum carnosum* root, showing occasional infection of cells of outer cortex; × 200 (photo by G. Hadley).

zones of infection. Roots which become modified to form storage tubers are not infected, nor are those roots and rhizomes which become green and photosynthetic (Harley, 1969). Stems, leaves, and other aerial parts are free of infection.

In view of these general observations it has been assumed, not surprisingly, that photosynthetic tissue is resistant and never infected by mycorrhizal fungi. However, Hadley and Williamson (1972) found that the infection of chlorophyllous cells was not uncommon in the roots of several epiphytic and terrestrial species of Malayan orchids. Localized zones of photosynthetic tissue, usually on the upper side of the root, were invaded by hyphae which developed pelotons in the normal fashion. Chloroplasts became interspersed with the pelotons and sometimes broke down, leaving fragmented or necrotic particles. Loss of chloroplasts sometimes occurred in adjacent noninfected cells also, and this could explain why other workers have not described the process. It also occurred in protocorms, both in culture and in natural conditions.

Many orchids are robust plants with well-developed green leaves and an extensive root system which may be virtually free of any infection. There appears to be a series ranging from these vigorous autotrophic species through varying degrees of so-called saprophytism[1] (i.e., dependence on increasingly active levels of mycorrhizal association which is often associated with a decrease in pigmentation and/or proportionate area of leaf surface) to the other extreme of total lack of chlorophyll and complete dependence on the mycorrhizal fungus.

Achlorophyllous orchids such as the temperate terrestrial genera *Neottia* and *Corallorhiza,* and the robust tropical epiphytic species of *Gastrodia,* always show very dense infection. In *G. elata* there is a differentiation of layers of infected cells the innermost of which act as the site of digestion. Since it can be assumed that the host plant is unable to obtain carbon by photosynthesis, it follows that the fungus mediates in the supply of all organic material, and it may be expected that the series of layers of healthy and digested intracellular coils of hyphae represent an organized system for transferring material to the host. The distinction between "digesting cells" in which the fungus always degenerates and the "host cells" where the fungus remains alive was originally described for *Neottia* by Magnus (Ramsbottom, 1923). But it was the extensive work of Hans Burgeff that led him to describe several forms of nutritional relationship among the many different known forms of mycorrhizal systems. Two of these, tolypophagy and ptyophagy (Burgeff, 1959), were recognized in orchids. Tolypophagy is the situation found in *Neottia* (above), but also described for many other species, in which definite layers of host cells and digestion cells occur (Fig. 3-1-**A**). It is the common, normal form of relationship that is characteristic of the majority of orchids. Ptyophagy occurs less frequently and is more characteristic of saprophytic orchids such as species of *Gastrodia.* It involves the breakdown of individual hyphae, rather than clumps, as they penetrate into a special "phagocyte layer" of host cells (Fig. 3-1-**B**). The hyphae burst at their tips and the liberated cytoplasm rounds off (a ptyosome) and is digested, finally leaving small clumps of material analogous to the larger digested masses in tolypophagy.

[1]So-called but incorrect, since the fungus is the saprophyte and the orchid obtains its nutrients from the fungus.

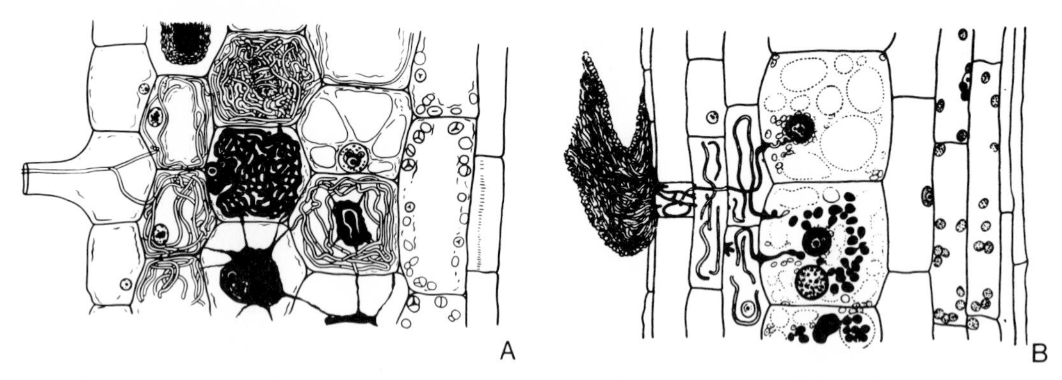

Fig. 3-1. Mycorrhizal infection in orchid roots (diagrammatic, modified from Burgeff, 1959). **A:** Tolypophagy in *Platanthera chlorantha,* showing one layer of host cells adjacent to the epidermis and two layers of digestion cells. **B:** Ptyophagy in *Gastrodia callosa,* showing two layers of passage cells and the phagocyte layer with hyphae liberating the cytoplasm, which is absorbed into spherical ptyosomes.

Burgeff (1959) also describes hyphal coils which in some cells function as storage organs and accumulate glycogen, protein, and fat before their tolypophagous digestion commences. Certain hyphae (*Eiweisshyphen*) take longer to digest and may be highly refractive, until eventually they are absorbed into the clump of digested material.

Despite the great detail of such observations, there is still no direct evidence from modern work of the means by which orchid cells actually carry out the digestion, or of the extent to which the process is essential for adult green orchid plants. Indeed, Bernard (1909) regarded the digestion process as a host defense mechanism in which the function of the phagocytic host cells is to check the spread of the fungus. Another possibility is that the disorganization of intracellular hyphae is not due to digestion by the host but to a process of autolysis, in which the fungus itself provides the enzymes of self-destruction. The products would then be available to the host. These possibilities are discussed later in relation to function at the interface.

Protocorms and Seedlings

The structure and organization of the orchid seedling and the development of its mycorrhizal infection are very relevant to an understanding of the orchid-fungus relationship. However, it is not appropriate at this point to describe the initial infection process. That is best dealt with under seed germination, while the relation between infection and the stimulation of growth is considered in the section on orchid-fungus relationships.

Germinating seeds develop into protocorms, rather structureless masses of cells in which there is little differentiation of tissues, apart from recognizable basal and apical regions. They ultimately give rise to shoot initials and one or more roots, leading to the development of a juvenile plant or plantlet.

In north temperate species germination and early development occur in darkness underground so that the protocorms (like those of some Lycopodiaceae) are completely heterotrophic. The mycorrhizal fungus occupies the basal cells of young protocorms

(Fig. 3-2-**A**) but later, as the vascular system and shoot begin to develop, the pattern of infection relates more closely to that of adult roots, being restricted to parenchyma cells in the cortical tissues. This can be seen in protocorms from the field (Fig. 3-2-**B**), as well as in those grown in culture. As the shoot develops, it produces chlorophyll and is able to photosynthesize, leading to a measure of autotrophy.

Infection follows a similar pattern in species with protocorms which are potentially autotrophic and become green at an early stage. The basal portion of the protocorm usually carries typical hyphal coils in most of the parenchyma cells, and the chlorophyllous cells occupy its upper part. Bernard (1909) and Burgeff (1959) both described infection in protocorms in relation to further development, as leaves and roots form.

In north temperate species, as protocorms develop into plantlets, they produce one or more root initials which become infected from the environment or culture medium. The infected zone of the protocorm does not extend into the root (Fig. 3-2-**C,D**). Infection of roots in culture appears to be similar to that in adult plants from the field. In *Dactylorhiza purpurella* (Plate 3-1-**B**) nearly all of the cortex is infected. On the other hand, infection was spasmodic in *Spathoglottis plicata* roots in culture (Williamson, 1970) which reflects the situation in field material of this orchid (Hadley and Williamson, 1972). Likewise in those orchids which form rhizomes, such as *Goodyera repens,* it seems from the limited evidence so far available that the distribution of infection is similar in field and culture material.

In general, therefore, orchid plantlets rapidly assume the features of adult plants in terms of their mycorrhizal organization.

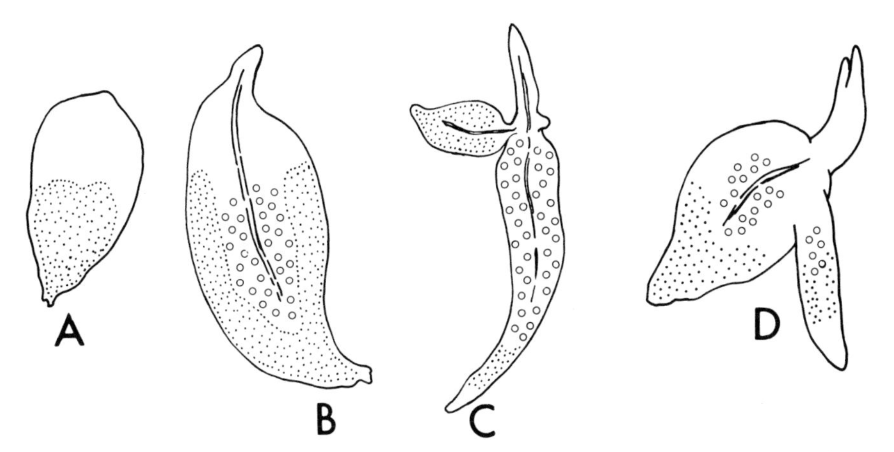

Fig. 3-2. Distribution of infection in *Dactylorhiza purpurella.* (Diagrammatic median sections; infected zone shaded, circles indicate region of starch storage tissue in inner cortex. All diagrams drawn by the author.) **A:** Culture-grown protocorm, 21 days after infection; × 40. **B:** Protocorm from field, probably several months old; × 15. **C:** Plantlet from field with remains of protocorm (left), well-developed shoot with a root initial (right), and first-year tuber packed with starch, terminating in a young infected root; ×5. **D:** Culture-grown protocorm, one year old, with young shoot and the first root. Infection in the protocorm is not continuous into the root; × 5.

Fine Structure and Intracellular Effects of the Fungus

Intracellular infections of plants by fungi have provided exciting material for electron-microscope studies, but the investigation of fine structure of orchid mycorrhiza has lagged behind that of obligate parasites and other plant-fungus systems. Dörr and Kollmann (1969) briefly described the structure of infected roots of *Neottia nidus-avis,* and material of *Habenaria dentata* was looked at by Borriss, Jeschke, and Bartsch (1971). Nieuwdorp (1972) examined roots of eight European orchids but most appeared to be showing only the later stages of digestion of the endophytic fungi. Later, Strullu and Gourret (1974), also using field material, investigated the sequence of infection stages in young roots of *Dactylorhiza maculata* with a series of excellent micrographs.

The first comprehensive study of the fine structure of laboratory-grown material infected with an endophytic fungus in controlled conditions was carried out by Hadley, Johnson, and John (1971), using protocorms of *Dactylorhiza purpurella* infected with a fungus which had been isolated from roots of the same host. The fungus stimulated growth, and details of the relationship in the early stages of infection were obtained.

It is not the purpose of this review to go into extensive detail of the fine structure of the orchid-fungus relationship, and little can be added to the paper by Hadley (1975) and the summary review by Arditti (1979). But two aspects deserve emphasis: firstly, the essential details of the system as regards the development and digestion of the fungus, and, secondly, the relationship at the interface between orchid cell cytoplasm and fungal hyphae, which offers interesting comparisons with other host-fungus symbioses. The term "host" is used with reference to the orchid in a general sense and does not imply parasitism by the fungus.

Infected Cells

Thin sections show that the hyphae forming the pelotons of the fungus are often closely packed and host cytoplasm is distributed among them, which creates a large interface area between the two partners of the symbiosis. The young hyphae stain densely but those of older pelotons are vacuolate, stain less densely, and assume bizarre shapes (Plate 3-2-**A**). Hyphae are always enclosed in a thin layer of host-cell cytoplasm with the plasmalemma adjacent to them, so that they do not actually penetrate the cytoplasm. Larger amounts of cytoplasm occur in association with the host-cell organelles, plastids, and nucleus (Plate 3-2-**B**).

The walls of hyphae are electron-dense and always seem to be associated with a granular outer layer of less dense material beyond which is the host-cell plasmalemma (Plate 3-3-**A**). Hadley, Johnson, and John (1971) interpreted this granular material as a fungal wall layer, but later studies (Strullu and Gourret, 1974; Hadley, 1975) suggested that it is probably an encasement layer (Bracker and Littlefield, 1973; but see also Hadley, 1975) originating from the host. Furthermore, there is increasing evidence from work with vesicular-arbuscular mycorrhizas and obligate parasites that the deposition of host-originated material as an encasement or encapsulating layer is a common feature of biotrophic host-fungus systems. The early stages of infection in orchids may or may not be biotrophic.

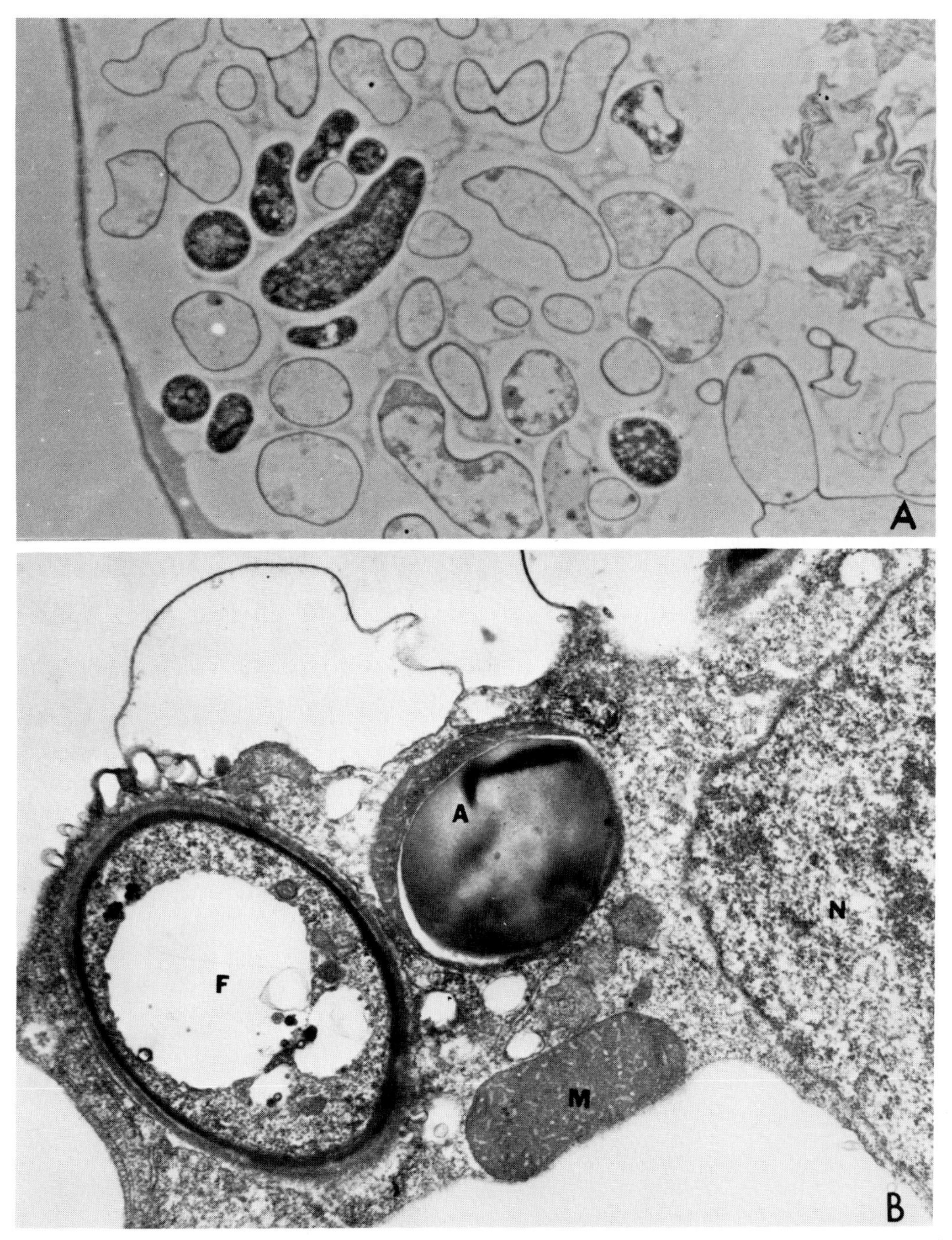

Plate 3-2. Intracellular structure of orchid mycorrhiza. **A:** Thin section of hyphae causing secondary infection in cell of *Dactylorhiza purpurella* protocorm, with remains of digested peloton (right) from primary infection. Some hyphae are rich in cytoplasm and stain densely; × 1500 (phase-contrast photograph by B. Williamson). **B:** Section of hypha of fungus (F) surrounded by host-cell cytoplasm containing a mitochondrion (M), an amyloplast (A), and part of the nucleus (N); × 17,000 (electron micrograph by G. Hadley and R. P. C. Johnson).

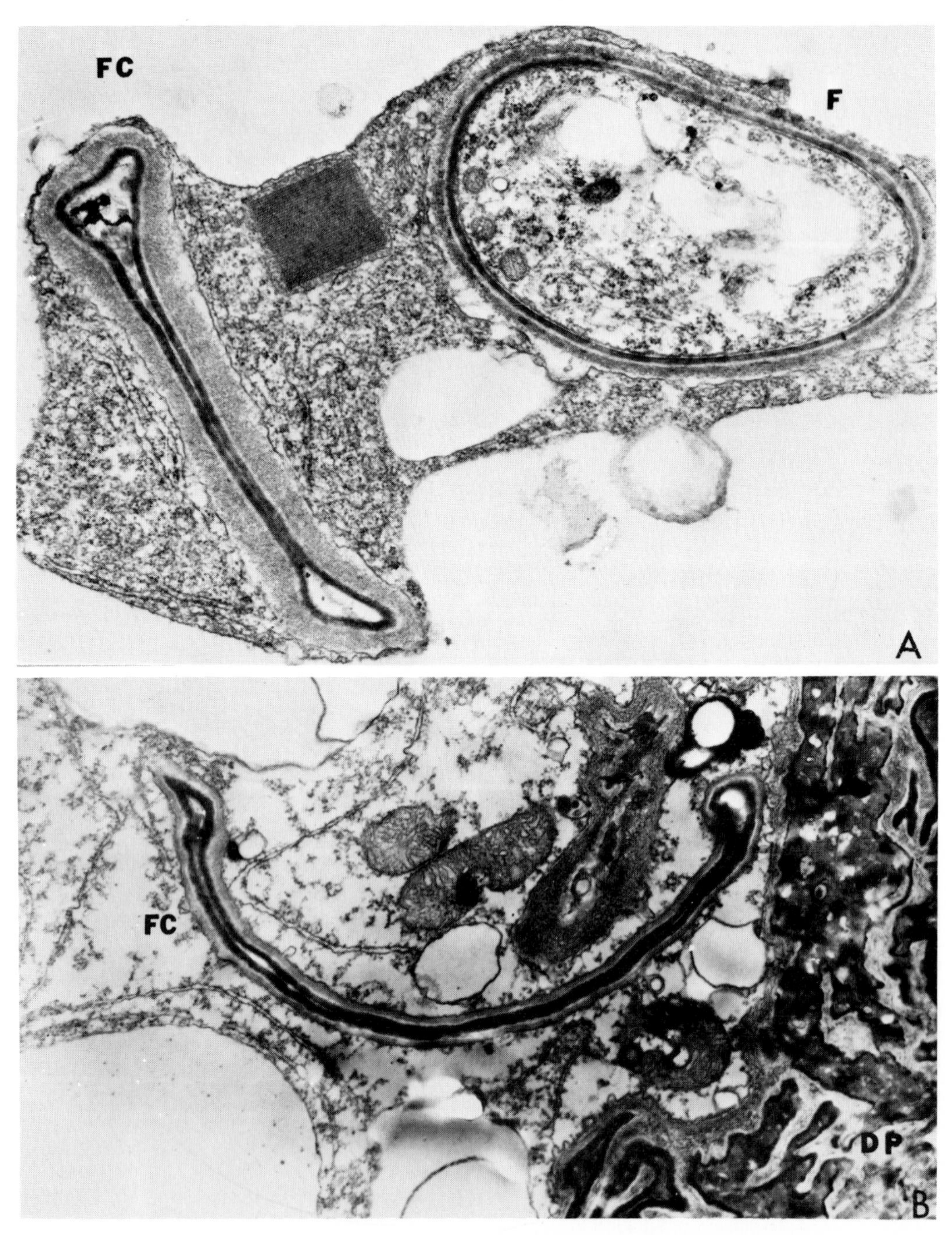

Plate 3-3. Intracellular structure of orchid mycorrhiza (electron micrographs by G. Hadley and R. P. C. Johnson). **A:** Hyphae in host cytoplasm. The granular encasement layer is thicker around the collapsed, flattened hypha (FC) than that around the normal hypha (F); × 19,000. **B:** Collapsed hypha (FC) with host cytoplasm and organelles at the edge of a digested peloton (DP) in which remains of hyphal walls can be seen; × 17,000.

As development proceeds the hyphae may lose their normal structure and collapse so that they appear completely flattened (Plate 3-3-**A**), often with a thicker encasement layer. The remains of the walls become grouped together and often fold in complex patterns as they congeal into a matrix of material of uncertain origin (Plate 3-3-**B**) recognizable as the digested peloton under the light microscope.

Nuclear Hypertrophy

One of the characteristics of infected cells, recognized and illustrated by many earlier investigators, is the enlargement of the host-cell nucleus, which readily takes up nuclear stains even after the intracellular hyphae are digested, and is a prominent feature of all mycorrhizal tissues. Where infection is spasmodic, as in *Spathoglottis plicata* roots (Hadley and Williamson, 1972), the uninfected cells near to infected ones also contain enlarged nuclei while more distant ones do not.

Nuclear hypertrophy is well known in infected tissues and is associated with increase in DNA content. By measuring DNA levels Williamson (1970) showed that endoreduplication of DNA occurred naturally, but nuclear sizes in infected tissue (using roots of *S. plicata* and *Dactylorhiza purpurella*) were always larger than in control uninfected material. The same was found in cultures of young symbiotic protocorms of *D. purpurella*, compared with asymbiotic ones (Williamson and Hadley, 1969).

The effect of nuclear hypertrophy on metabolism, if any, is unclear. Williamson (1970) suggested that it might be related to enhanced protein synthesis since infected cells may also be rich in ribosomes and rough endoplasmic reticulum (Dörr and Kollmann, 1969).

Function at the Interface

There is little evidence from fine structure as to whether metabolites are transferred biotrophically, i.e., from fungus to host across a living interface. Hadley, Johnson, and John (1971) observed protuberances on hyphae and complexes of vesicles in some hyphae but these may not be connected with nutrient transfer. Nieuwdorp (1972) hypothesized that the fungus prevents deposition of pectin and cellulose by the host, and becomes enveloped in these materials as the hyphae degenerate. Although Nieuwdorp found no evidence of "fungus-digesting enzymes," it is important to remember that the material was roots, collected from the field at a late stage of development, and may have been infected by various fungi not all of which would necessarily be true mycorrhiza-formers (Warcup and Talbot, 1967). Williamson (1973), using cytochemical techniques, investigated the possibility that acid phosphatase enzymes may be specifically involved in the digestion process, but was unable to identify particular sites and could not resolve the question of whether breakdown of the hyphae occurs autolytically or as a result of digestion by host enzymes.

In general, therefore, although investigation of fine structure has provided much new information about the organization of the orchid-fungus system, it has not yet revealed much about functional characteristics.

Orchid Fungi

Taxonomic Status and General Features

The majority of the fungi associated with orchids are considered to be members of the form-genus *Rhizoctonia,* and numerous species were described by earlier workers. Bernard (1903, 1904) made the first pure culture isolation and subsequently (Bernard, 1909) described three species, *R. mucoroides,* (Fig. 3-3-**A**), *R. repens* (Fig. 3-3-**B**), and *R. lanuginosa.* Several others were described, particularly by Burgeff (1936), who advanced the general hypothesis of specificity: "one orchid–one fungus." Curtis (1939) added to the list of known species, summarized the state of knowledge of orchid fungi at that time, and showed that some orchids harbored more than one fungus, i.e., that the relationship is not specific (see below).

The connection between *Rhizoctonia* species and some basidiomycete genera such as *Corticium* made possible the description of *C. catonii* (Burgeff, 1936). Several other orchid fungi originating from both temperate and tropical species were recognizable as Basidiomycetes by the presence of clamp connections or fruiting bodies. These were usually associated with achlorophyllous ("saprophytic") orchids and are known to include *Armillaria mellea,* which is symbiotic with *Gastrodia* species (Kusano, 1911; Campbell, 1962) and with *Galeola septentrionalis* (Hamada, 1940); *Hymenochaete crocicreas,* which is symbiotic with *Galeola altissima* (Hamada and Nakamura, 1963); and one or more species of *Fomes,* which are symbiotic with *Galeola hydra* (Burgeff, 1936) and with *Gastrodia sesamoides* (Campbell, 1964). These fungi may occur in triple symbioses involving other plants, and this is discussed below.

Orchid rhizoctonias are recognized by the general features of the mycelium in culture, the presence in most isolates of short inflated segments which resemble spores (Fig. 3-3), and the formation of loose aggregates of hyphae regarded as poorly developed sclerotia or resting bodies. However, many isolates have been induced to form sexual stages in culture which has made possible their classification in the basidiomycete genera *Cer-*

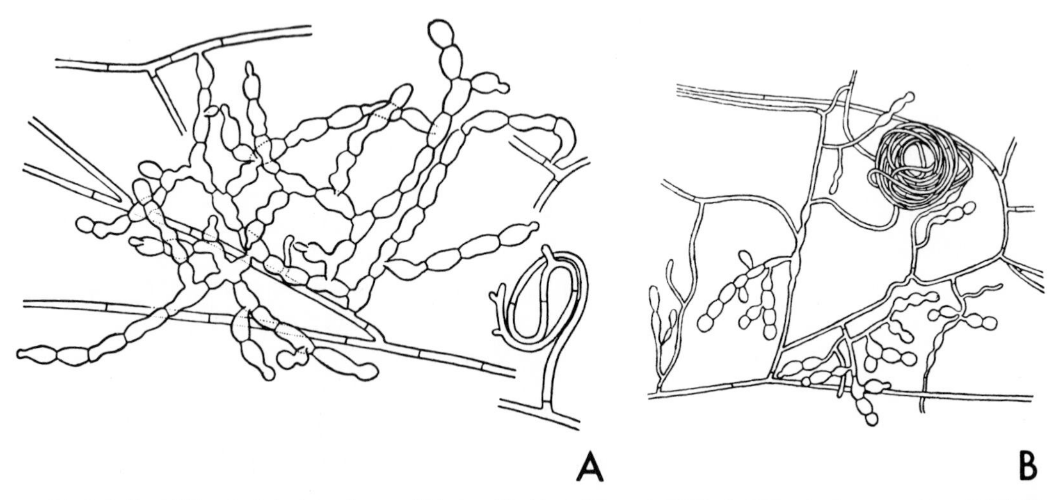

Fig. 3-3. Two *Rhizoctonia* isolates from orchids (modified from Bernard, 1909). **A:** *Rhizoctonia mucoroides;* × 160. **B:** *R. repens;* × 160.

atobasidium, Sebacina, Thanatephorus, and *Tulasnella* (Warcup and Talbot, 1967, 1971). Indeed, the endophyte originally described by Bernard as *Rhizoctonia repens* is of common occurrence and is now known to be synonymous with *Tulasnella calospora.* It is one of several orchid fungi including *Ceratobasidium cornigerum* (synonymous with *Rhizoctonia goodyerae-repentis*) which have also been found in soil free of orchids, suggesting that they may be widespread as soil saprophytes.

The form-genus *Rhizoctonia* also includes several species which are widespread and often aggressive parasites on a variety of hosts. Of these, *R. solani* was isolated from orchid roots by Downie (1957) and shown by her (1959a,b) to be symbiotic with protocorms of *Dactylorhiza purpurella.* Later Harvais and Hadley (1967a) extended the host range of *R. solani.* Hadley (1970b) found that isolates of *R. solani* varied in their symbiotic ability and one strain of *Thanatephorus cucumeris* (synonymous with *R. solani*) known to be a pathogen of Cruciferae was symbiotic with several of the orchids he tested.

The unifying feature of most isolates of fungi from orchid roots remains their vegetative form, i.e., the features that enable them to be recognized as *Rhizoctonia* species. It is certain that only a small proportion of the many fungi isolated have been given published descriptions, and the understanding of what constitutes a species remains vague. Furthermore, from culture and symbiosis tests it seems likely that there may be numerous races or strains, varying in physiological features and (less so) in morphology, within the spectrum of one species or a group of similar isolates.

In an extensive series of isolations from *Dactylorhiza purpurella* and other north British species, Harvais and Hadley (1967a) obtained 244 rhizoctonias (out of 420 root isolates). These were separated into 15 main groups possibly corresponding to species, yet only three of the 15 agreed with previously published descriptions. The latter were *R. repens, R. goodyerae-repentis,* and *R. solani.* Harvais and Hadley found that isolates in any one group were rarely obtained from more than one habitat, apart from *R. repens* which was widespread. Curtis (1939) also found a wide distribution of this fungus.

Rhizoctonia species are known to be physiologically variable but there has been no comprehensive study of variation in orchid fungi, perhaps because of difficulties of identification at the species level. However, recent experiments with fresh isolates (Alexander and Hadley, in press) indicate that considerable differences in growth-stimulating ability exist among strains of *R. goodyerae-repentis* in tests with germinating seed of *Goodyera repens.*

Specificity

Early workers assumed a specific relationship between host and fungus (i.e., one host–one fungus). Specificity was disproved, however, by Curtis (1939), who argued that ecological distribution of the fungi was related to habitat rather than host. The results of Harvais and Hadley (1967b) supported this idea and, further, showed that *Dactylorhiza purpurella* was symbiotic with nearly all isolates tested. Subsequently Hadley (1970b), in an extensive series of cross-inoculation experiments using orchids and fungi of worldwide distribution, found that some orchids, particularly *D. purpurella,* became infected by many strains of fungi originating from a variety of habitats, while others were more exacting (Table 3-1).

Table 3-1. Interactions between germinating orchid seeds and fungi from various sources (condensed from Hadley, 1970b)

Fungus	Isolate	Source (host)	Orchids: *Dactylorhiza purpurella*	*Goodyera repens*	*Cymbidium canaliculatum*	*Epidendrum ibaguense*	*Epidendrum radicans*	*Spathoglottis plicata*
Ceratobasidium cornigerum	Rgr	*Goodyera repens*	s	S	s	O	s	O
	Thr1	*Thrixspermum amplexicaule*	SP	sP	—	P	s	O
	O393	*Pterostylis pedunculata*	S	O	s	S	S	—
	O479	*Pterostylis curta*	S	s	s	s	—	—
C. obscurum	O8	*Acianthus reniformis*	P	—	P	—	s	P
Ceratobasidium sp.	F1	*Dactylorhiza purpurella*	S	S	S	S	S	—
	Cs4	Strawberry; root pathogen	SP	s	S	S	S	—
Thanatephorus orchidicola	De1	*Dactylorhiza elata*	sP	P	—	P	—	P
T. sterigmaticus	O60	*Thelymitra antennifera*	S	s	—	O	s	O
T. cucumeris	O269	*Pterostylis vereenae*	S	S	s	O	s	O
	Rs1	Tomato; root pathogen	SP	s	P	O	—	—
	W48	Soil; crucifer pathogen	S	S	s	—	O	O
	W82	Soil; nonpathogenic	S	S	s	—	s	O
Tulasnella calospora	Amo4	*Arachnis* cv Maggie Oei	S	s	S	S	S	S
	Pb47	*Platanthera bifolia*	S	s	S	S	S	S
	RrA	*Dactylorhiza purpurella*	S	s	S	S	S	S
	O388	*Diuris longifolia*	S	s	S	S	S	S
T. cruciata	O296	*Thelymitra* sp.	sP	O	s	O	s	s
Rhizoctonia solani	RS16	Soil; pathogen (?)	SP	SP	s	—	—	—
	RS94	Potato tuber	SP	O	O	—	—	O
Rhizoctonia sp.	T	*Dactylorhiza purpurella*	S	S	s	S	S	—
	Rs10	Rice (pathogen)	S	SP	s	S	S	s

Abbreviations: O = no infection; s = compatible infection but no growth stimulus; S = growth stimulus, i.e., symbiosis; P = parasitism of orchid by fungus. Parasitism after a compatible phase is indicated by sP, SP, etc.

Goodyera repens, for example, is a species which in root-isolation studies is usually found to be infected with one fungus, *Rhizoctonia goodyerae-repentis* (*Ceratobasidium cornigerum*), yet in seed-germination experiments forms mycorrhizal relationships with several different isolates (Hadley, 1970b). It may be an orchid which, because it occurs in a restricted habitat, is not naturally in contact with a variety of fungi in field conditions.

One difficulty in investigating specificity is that the isolation techniques normally used to obtain fungi from roots may yield isolates that are growing on or in the root but are not forming typical intracellular structures and therefore may not be true mycorrhizal endophytes. Warcup and Talbot (1967) suggest that only isolates growing from recognizable pelotons (obtained from macerated roots) should be regarded as mycorrhizal, and can then be tested on germinating seeds.

Using these techniques, evidence of possible specificity has been obtained for some orchids. Warcup (1971) found that field plants of five species of *Diuris* yielded only *Tulasnella calospora* (*Rhizoctonia repens*), while many species of *Caladenia* and related genera yielded almost entirely (94%) *Sebacina vermifera* isolates. In seed-germination tests two *Diuris* species were stimulated only by *Tulasnella calospora,* suggesting a strong degree of specificity. Further studies (Warcup, 1973) found that two species of *Pterostylis* were stimulated to germinate only by *Ceratobasidium cornigerum.* Results with germinating seed of *Caladenia* species were inconclusive.

Work with seven species of *Thelymitra* (Warcup, 1973) showed that several species of *Tulasnella* may be present as root endophytes and these varied in the efficiency with which they stimulated germination of the host orchids. These results probably reflect the previously mentioned problem of physiological variation between isolates. Specificity, therefore, as far as it occurs, may be reflected in both root infection and stimulation of germination.

Nutrition

Many orchid fungi have been isolated and grown in culture, although there are few detailed studies of their nutrition. Most grow readily on undefined media such as potato-dextrose agar. Growth occurs on a wide range of carbohydrates, including cellulose (Smith, 1966; Hadley, 1969), and other naturally occurring materials such as pectin (Pérombelon and Hadley, 1965), suggesting that there is no exacting requirement for glucose or other simple carbon sources. Nor do many isolates require organic nitrogen compounds or vitamins. In other words, some isolates grow freely on defined media consisting of a carbon source with mineral salts including nitrate ions as a source of nitrogen. Others benefit from the inclusion of yeast extract in the medium so that Dox-yeast agar is advantageous or necessary for those isolates which are more exacting. A slower growth rate is also characteristic of such fungi (Fig. 3-4), described by Harvais and Hadley (1967a) as "typical orchid endophytes" among the wide spectrum of isolates they obtained.

More recently, nutritional studies have shown conclusively that some orchid endophytes require certain organic nitrogen compounds. Hadley and Ong (1978) found that four isolates of *Tulasnella calospora* of different (worldwide) origin have exacting requirements for nitrogen, being unable to use nitrate and ammonium ions. Asparagine, glycine, and urea were the best sources; glutamine, arginine, and alanine were less adequate. All isolates were also heterotrophic, but to different degrees, for the vitamins thiamine and *p*-amino benzoic acid.

Earlier, Stephen and Fung (1971a,b), working with two fungi which were different (but both unidentified) *Rhizoctonia* isolates from *Arundina chinensis*, found similar vitamin

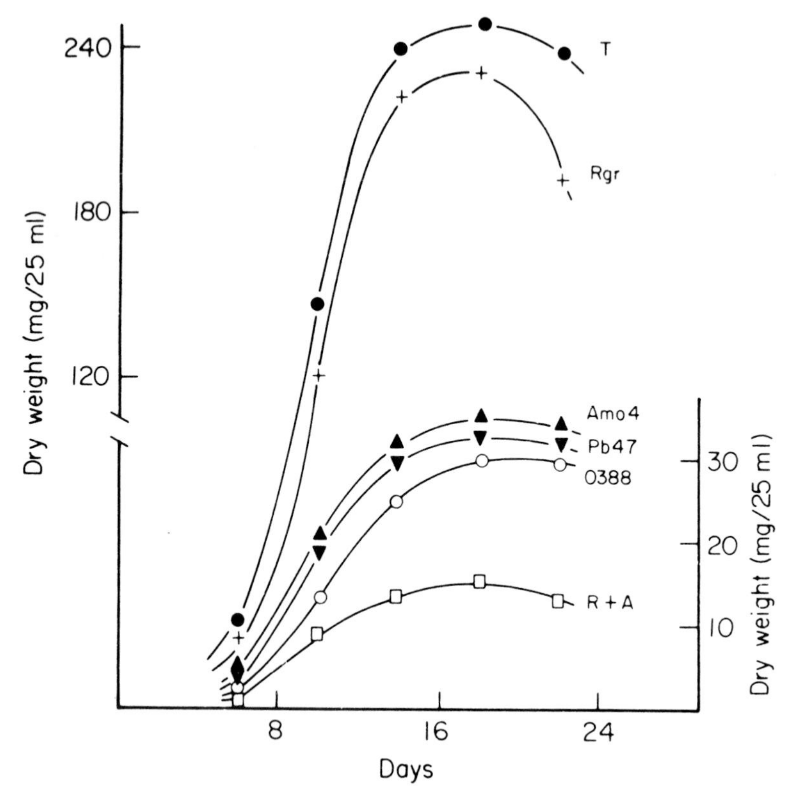

Fig. 3-4. Growth of *Rhizoctonia* sp. (isolate T, from *Dactylorhiza purpurella*) and *R. goodyerae-repentis* (Rgr) on a glucose-nitrate medium, and four strains of *Tulasnella calospora* on the same medium supplemented with 0.1% yeast extract (reproduced from Hadley and Ong, 1978).

requirements. Their isolates were also unable to use nitrate but grew reasonably on ammonium compounds and even better on glutamic acid, urea, arginine, and asparagine. The same vitamin requirements were also reported for a *Rhizoctonia* (species unknown) studied by Hijner and Arditti (1973). However, an isolate of *Rhizoctonia repens* originating from *Orchis militaris*, although heterotrophic for nitrogen, did not appear to require vitamins (Powell and Arditti, 1975). It grew well on a medium containing Knudson C salts (with glucose) and any one of several amino acids.

Natural Occurrence and Distribution

Little is known about the ecology of orchid fungi apart from their occurrence in roots. Circumstantial evidence from isolation work (Curtis, 1939; Harvais and Hadley, 1967a) suggests that they may be widespread in the soil, and in fact *Ceratobasidium cornigerum* (*Rhizoctonia goodyerae-repentis*) and *Tulasnella calospora* (*R. repens*) have both been isolated from soil (Warcup and Talbot, 1967) where orchids are not growing. Bearing in mind that *Rhizoctonia* species are generally known to be able to grow saprophytically, Smith (1974) suggests that orchid endophytes have the enzymatic abilities and the competitive saprophytic advantage to exist independently in the soil.

Among the few orchid fungi of known taxonomic designation, *Tulasnella calospora* and *Ceratobasidium cornigerum* appear to have a worldwide distribution, and this may be true of others.

Seed Germination and Seedling Growth

The historical development of an understanding of orchid seed germination is linked almost inextricably to the investigation of the part played in the process by fungi, yet much more work has been done with nonsymbiotic than with symbiotic systems. Fungi, of course, are only one of the many "factors affecting seed germination," and mycorrhizal symbiosis, for example, occupies only a small portion of a review (Arditti, 1967) of the topic. Mycorrhizal work is described in a survey of the factors affecting symbiotic germination (Warcup, 1975) and a further review by Arditti (1979).

It is my intention in this section merely to summarize our present knowledge of the process of seed germination and the involvement of fungi in it. But I must also draw attention here, for comparative reasons, to the numerous studies of the effect of nutrients in stimulating (i.e., initiating the irreversible stages of) germination and the subsequent development of the protocorm in the absence of any fungus.

The part played by fungi in translocating nutrients into the germinated seed, the subsequent development of infected seedlings, and other aspects of the mycorrhizal relationship will be dealt with under the heading Orchid-Fungus Relationships.

Nonsymbiotic Germination

The foundation of the nonsymbiotic method of germinating orchid seeds was laid by Knudson (1922, 1924), although his bold statement (Knudson, 1925) that the fungus was "not required" for germination brought him severe criticism at the time. Since then, nonsymbiotic germination has become a standard procedure for orchid horticulturists. Nevertheless, at the present time it is probably true to state that many trials and experiments on germinating orchid seed have resulted in failure. The challenge still attracts numerous hopeful investigators despite the variable record of their predecessors and the disjointed evidence from positive results. Even the most comprehensive review of germination, covering the effects of fungi, mineral nutrition, carbohydrate and nitrogen sources, other nutrients, and all physical factors of the environment (Arditti, 1967), cannot pinpoint any one factor as being critical. Although suggesting that orchids are unique in their requirements for certain factors, Arditti acknowledges that "there is no strictly specialized orchid factor." Withner (1974) stated that "the germination of orchid seed has changed little over the years." More recently Arditti (1979) suggests that exacting germination requirements are characteristic of terrestrial species while tropical epiphytes germinate easily if simple sugars are provided. It is not necessary here to review all the recent work on nonsymbiotic germination, but with north temperate species there is little evidence of dramatic improvements in our understanding.

Two factors deserve comment. One is that undefined additives such as coconut milk, peptone, tomato juice, etc., frequently enhance germination and/or development. Even Malaysian beer (Yeoh, 1962) is acclaimed. But none of these materials, despite their convenience and potential value to the orchid breeder, is likely to help in the evaluation of what is happening in nature. It is reasonable to assume that the active components of such undefined mixtures are vitamins and growth substances, and both groups of compounds have been used with positive results, although the effects vary with the species and the stage of development of the seedling (Withner, 1974). Nevertheless, the careful use of combinations of indoleacetic acid, gibberellic acid, adenine, and kinetin did not have as beneficial an effect on *Dactylorhiza purpurella* as did potato extract or coconut milk (Hadley and Harvais, 1968).

The second important point, also evident from the same work and emphasized by Hadley (1970a), is that symbiotic germination and growth are always much more rapid than that with any combination of defined or undefined nutrients. For instance, symbiotic development of *D. purpurella* and other north temperate species was rapid on defined media consisting only of sugar and mineral nutrients; robust protocorms only two or three months old were developing green shoots and were more advanced than noninfected protocorms up to two years old supplied with the most beneficial combination of growth factors.

In a more comprehensive study of *D. purpurella,* Harvais (1972) found that on media containing yeast extract and the amino acids and vitamins present in casein hydrolysate, coupled with various growth-factor treatments involving aminopurines and indoleacetic acid, development of protocorms was "in every respect as good as symbiotic cultures." (However, comparable symbiotic controls were not included in the experiments, nor were there comparisons with field material.) Further treatment with growth factors and a supply of iron also gave good results; plantlets were up to 5 cm tall when about 19 months old. Good development of *Cypripedium reginae* was obtained in similar studies using suitable combinations of nutrients, although vitamins and growth substances were not beneficial (Harvais, 1973).

It must be remembered that *Dactylorhiza purpurella* germinates easily on simple defined media, and manipulating the nutrients primarily affects development rather than stimulating germination. Therefore, although it is possible by careful experimentation to simulate the effects of symbiosis in this species, it is doubtful whether such development is likely in many other orchids where germination does not occur so easily.

Nonsymbiotic germination, then, is subject to many variables and its investigation is related largely to the needs of the grower. It is a process of manipulation of an artificial environment and despite all the investigations undertaken in the past is unlikely to provide a complete scientific understanding of the process in nature. The involvement of a fungus, which is the result of long processes of evolution, is never totally replaced by manipulation of nutrients.

Symbiotic Germination

Under this heading several aspects need to be considered. Firstly, can germination and subsequent growth be separated as different processes? Again, is infection actually necessary before germination occurs, and is germination *per se* therefore enhanced by symbiotic infection?

The initial stage in germination is the uptake of water leading to a swelling of the embryo. If the seed is viable, swelling is followed by the onset of some metabolic activity, including the disappearance of lipid reserves and the formation of starch. In many species these stages readily occur in the absence of infection, and they often precede infection in symbiotic germination. Burgeff (1936, 1959) describes infection of a *Laeliocattleya* hybrid in detail. Fig. 3-5 (taken from his work) shows the early stages consisting of penetration of the cells at the suspensor (base) end of the embryo and the formation of the first pelotons. But it was Bernard (1904) who first described infection, using a *Cypripedium* hybrid and *Bletia* (*Bletilla*) *hyacinthina*. Later Bernard (1909) de-

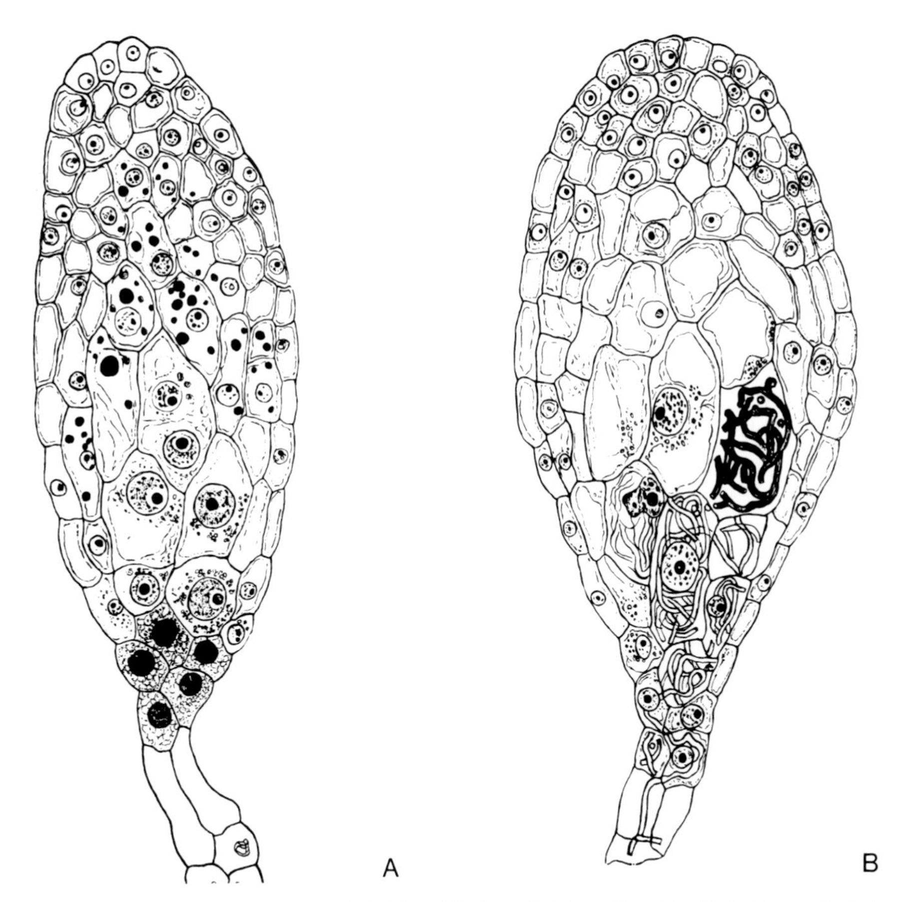

Fig. 3-5. Infection of *Laeliocattleya* hybrid, × 250 (modified from Burgeff, 1936). **A:** germinated, swollen embryo with starch grains (black) before infection. **B:** embryo infected through suspensor cells, with first pelotons visible.

scribed and figured infection of *Odontoglossum* and *Phalaenopsis amabilis*. Penetration of the basal cells of the embryo occurs in most orchids, including north temperate species such as *Goodyera repens* (Mollison, 1943, and Fig. 3-6). In some species infection normally occurs via epidermal hairs, and, clearly, is preceded by an appreciable amount of development, as in *Dactylorhiza purpurella* (Williamson and Hadley, 1970). All these observations suggest that germination, in the sense of at least some metabolic activity, precedes infection.

It is certain that many investigators describing germination and the effects of fungi (and other factors) on the process have been looking at the post-germination growth phase and have not separated the two stages. However, Downie (1940) gave a very lucid description of *Goodyera repens* and showed clearly that germination occurred in water and was enhanced by the endophytic fungus, while growth did not occur in the absence of various nutrients and was considerably improved when the protocorms were infected.

It is possible that most mycorrhizal fungi enhance germination as well as having positive effects on growth, but it is difficult to separate the two processes. For example, percentage germination of *Dactylorhiza purpurella* (Harvais and Hadley, 1967b; Hadley, 1969) and *Goodyera repens* (Purves and Hadley, 1976) was higher in the presence of certain fungi, especially where conditions (e.g., low nutrient levels and low temperatures) were adverse to germination. *G. repens* in fact was stimulated by infection (even on water agar) or by the presence of potato extract, or by high sugar levels, although a striking difference between these treatments was that the stimulation due to the fungus occurred only in the first three or four weeks, regardless of the medium, whereas percentage germination in the (asymbiotic) nutrient media increased over a longer period throughout the experiment (Table 3-2). This is probably due to changes in the status of the medium caused by the fungus and the inhibition of germination by its metabolic products.

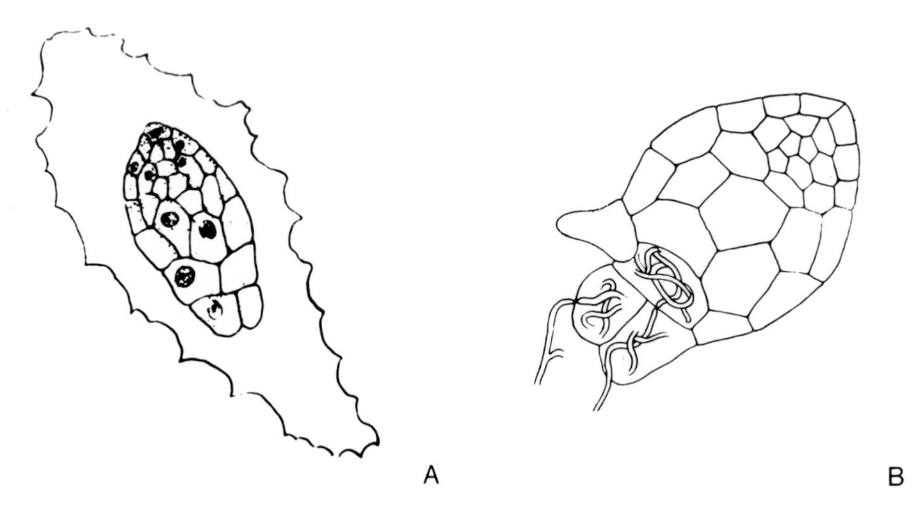

Fig. 3-6. Germination and infection of *Goodyera repens,* × 150 (modified from Mollison, 1943). **A:** Ungerminated seed with outline of seed coat. **B:** Young protocorm showing infection of suspensor cells.

Table 3-2. Percent germination of *Goodyera repens* on various media under asymbiotic conditions and in symbiosis with *Rhizoctonia goodyerae-repentis* (from Purves and Hadley, 1976)

Asymbiotic treatments:

	Medium			
Weeks	Water agar	Pfeffer + 0.1% glucose	Pfeffer + potato extract	Pfeffer + 2% glucose
3	0	6	23	44
6	0	26	60	73
9	<1	31	82	87

Symbiotic treatments:

	Medium		
Weeks	Water agar	Pfeffer + 0.1% glucose	Pfeffer + 1% cellulose
4	11	26	20
8	7	25	29
16	10	a	a

a. No further germination evident, but growth of fungus too dense for accurate assessment.

Results with *D. purpurella* (Harvais and Hadley, 1967b) indicate that certain fungi have very marked beneficial effects on germination while others are not effective, even though they may stimulate growth. Work with Malaysian orchids and their endophytes (Hadley, unpublished) gave similar results.

In general germination is stimulated or accelerated in the presence of mycorrhizal fungi, but the effect is not very pronounced, and it is less apparent in tropical species, which germinate rapidly, than in north temperate ones where germination occurs over a period of many weeks.

Post-germination Growth

Uninfected seedlings of many orchids increase in mass slowly, the rate of growth depending on the type and level of nutrients supplied. Development of seedlings of terrestrial species has been extensively reviewed by Stoutamire (1974), and there are numerous papers in orchid journals reporting germination and growth of horticultural varieties and natural species. Many grow fairly easily even on simple defined media, others respond to the application of vitamins (see Withner, 1974, for a summary), and some species, such as *Dactylorhiza purpurella,* germinate easily but appear to virtually stop growing unless programmed with complex growth factors (Harvais, 1972).

Infection of *D. purpurella* does not affect growth on deionized water or Pfeffer solution (Hadley and Williamson, 1971), but it causes a dramatic enhancement of growth (Fig. 3-7) on water agar or Pfeffer medium with 0.1% glucose. Growth continued for several weeks when cellulose was used to supplement the limited carbon supply. Similar results were obtained with *Spathoglottis plicata* (Hadley, 1969 and Fig. 3-8). Many other orchids used in symbiosis tests (Hadley, 1970b and unpublished results) grow fairly rapidly on simple defined media when infected with a compatible fungus.

As has been stated earlier, development of noninfected protocorms adjacent to infected ones does not occur. Clearly the symbiotic fungus is not merely providing specific nutrients. Infection may be leading to a more complex process of initiating metabolic

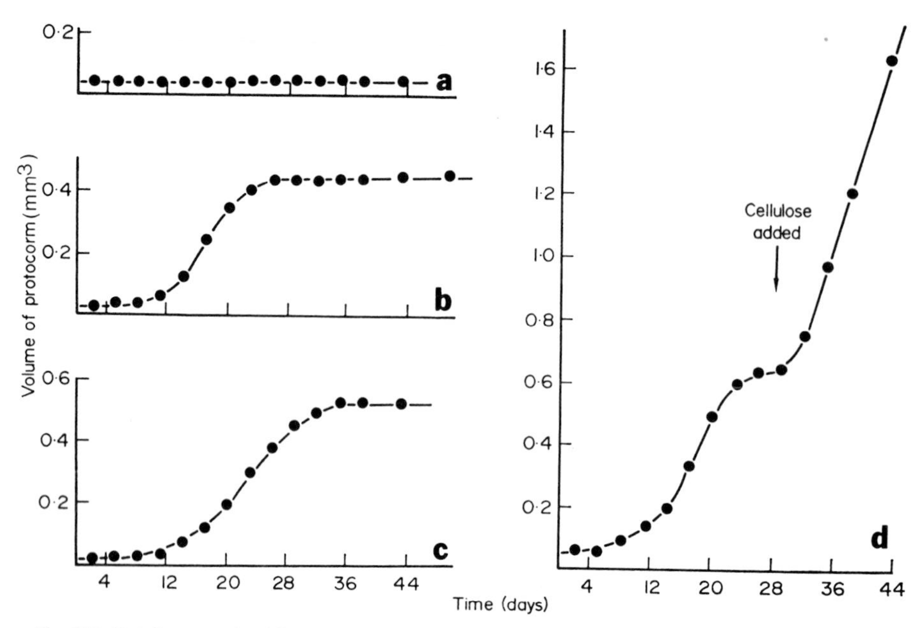

Fig. 3-7. Relative growth of *Dactylorhiza purpurella* protocorms when (a) noninfected, on water agar; and infected with a symbiotic *Rhizoctonia* sp. on (b) water agar, (c) Pfeffer plus 0.1% glucose agar, and (d) the same medium with added cellulose (reproduced from Hadley and Williamson, 1971).

pathways, or providing particular metabolic intermediates, which the orchid then utilizes to attain, eventually, a greater degree of independence.

In general, the protocorms of temperate orchids are completely heterotrophic and develop underground until the shoot initial forms, extends, and produces chlorophyll. The process may take many months or years in asymbiotic culture (Stoutamire, 1974)

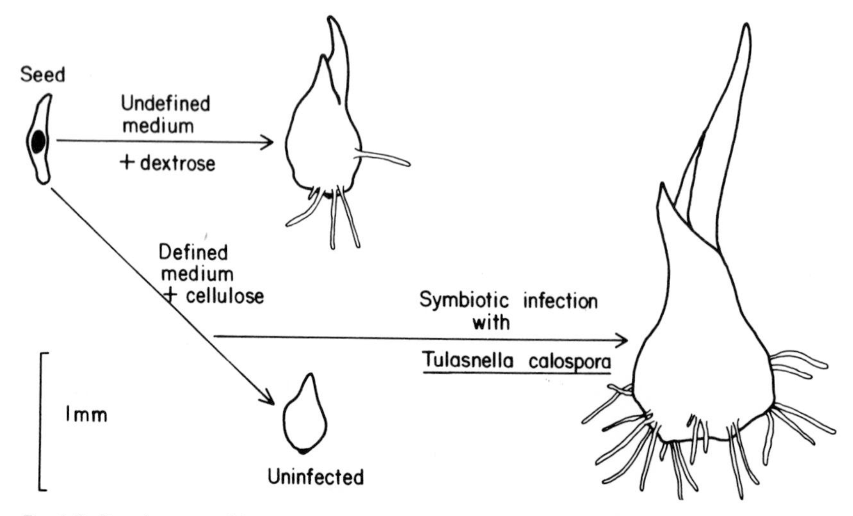

Fig. 3-8. Development of *Spathoglottis plicata* after five weeks' growth on two carbon sources in relation to infection by *Tulasnella calospora* (modified from Hadley, 1969).

compared with a few weeks in symbiotically infected individuals. However, heterotrophic development also occurs in tropical species, many of which readily germinate in light on the surface of a substrate and immediately produce chlorophyll as the embryo commences growth. The presence of chlorophyll does not necessarily lead to independence of organic nutrients (Withner, 1959). Infection leads to a dramatic stimulation of growth, associated with increase in chlorophyll, whereas noninfected seedlings are unable to function completely. Although it has been hypothesized that this inability to photosynthesize is related to a vitamin deficiency (Arditti, 1967), the important point is that in the presence of a symbiotic infection metabolism proceeds in a more normal pattern.

As they enlarge, therefore, orchid seedlings pass into a mature phase. The roots develop and in turn become mycorrhizal, the infected zone being discontinuous with that of the protocorm (Fig. 3-2-**C,D**). Little is known about the importance of infection at this stage due to a lack of investigation of adult plants, but there are indications that in *Goodyera repens* infected plantlets differ from noninfected ones mainly in their rate of nutrient uptake (Purves and Hadley, 1975 and unpublished results) rather than in fundamental features of metabolism. This is considered in more detail in the following section.

Orchid-Fungus Relationships

The understanding of the orchid-fungus relationship, because of its complexity, demands considerable scientific investigation, as has occurred with some other mycorrhizas and obligate parasites. Present knowledge of seed germination, summarized in the preceding section, leads naturally to questions of how the fungus integrates with its host, whether metabolites move in both directions between the partners, and how the system is controlled. Some basic work has been carried out on nutrient exchange in dual culture and, more extensively, in relation to the movement of material from the environment via the fungus to the orchid. Little research has been done to investigate the possibility of photosynthate moving in the reverse direction. The demonstration of phytoalexinlike substances in orchid tissues has suggested a mechanism by which the host can control the infection; the possibility of triple associations involving fungi which are pathogens on other plants is also recognized. These aspects are considered here.

Extracellular Movement of Nutrients from Fungus to Orchid

Most of the horticultural and scientific investigations of germination and growth have assumed that the basic problem is one of providing the nutrients, at the correct level and the appropriate stage of development, which would otherwise originate from the mycorrhizal fungus. As stated earlier, however, there is no specific factor involved, and numerous defined and undefined substances have been used as additives to culture media, with varying success.

Since the fungal partner in the relationship occupies both the root and the immediate external environment, it is not possible with dual cultures to distinguish the direct nutritional effect (i.e., of materials produced by a fungus in culture) from the effect, if any, of metabolites obtained intracellularly from the fungus across a live interface.

The protagonists of nonsymbiotic germination, led by Lewis Knudson and influenced by his findings, argue that the provision of nutrients in the immediate environment of orchid seeds and seedlings, whether by the fungus or in the growth medium, is sufficient to enable development to proceed, and that infection is not essential.

There have been some investigations of whether fungi produce in culture specific compounds which may be required for growth of orchid seedlings. Harvais and Raitsakas (1975), in one of the few studies using a proven symbiotic orchid-fungus pairing, found that the fungus, an isolate of *Rhizoctonia* tentatively identified as *R. solani*, produced in its mycelium (but did not release into the medium) glutamine, glutamic acid, aspartic acid, and small quantities of other amino acids. By growing protocorms of *Dactylorhiza purpurella* on limiting concentrations of amino acids Harvais and Raitsakas (1975) concluded that this orchid was unable to synthesize an adequate amount of glutamic acid and that in nature it might "depend on symbiotic fungi for such supplies." In further work with the same fungus Harvais and Pekkala (1975) found that thiamine and nicotinic acid were released into the culture medium and may be "responsible for the frequently reported stimulation of germination before infection."

Another (unidentified) *Rhizoctonia* species was shown to produce niacin (Hijner and Arditti, 1973), which is the only vitamin repeatedly shown to enhance germination and seedling growth (Arditti, 1967).

However, it is significant that in numerous tests with proven symbiotic fungi growing with orchid seeds in culture, a stimulation of growth (i.e., size increase) has never been observed in noninfected seeds lying adjacent to infected, growing seedlings. No direct (*in vitro*) nutritional stimulation occurs due to the fungus. Growth of noninfected seeds or protocorms where the fungus was present was no better than in fungus-free controls on the same media (Hadley, 1969 and unpublished work). Even the stimulation caused by mycelial extracts of a symbiotic fungus (Burgeff, 1936; Downie, 1949) was no more significant than that found in many experiments with vitamins and other nutrients.

Experiments using trehalose and mannitol, which are both characteristic fungal carbohydrates, show that germination of *Dactylorhiza purpurella* and *Bletilla hyacinthina* (Smith, 1973) and *Goodyera repens* (Purves and Hadley, 1976) occurs only on the former. Trehalose can also be metabolized by orchid leaves (Smith and Smith, 1973). Germination and uptake were no better than on glucose, but this work is interesting since trehalose may be the transfer carbohydrate in infected tissues.

It would seem, therefore, that orchids are unlike many other mycorrhizal plants, in that their requirements cannot be provided completely and efficiently by an external supply of nutrients. In nature they are ecologically obligate symbionts, dependent on infection. The situation is analogous to that of obligate parasitic fungi, although the role of the two partners is reversed and in that situation the fungus is dependent on the higher plant. Obligate parasitic fungi, like orchids, respond only to a very limited degree to exogenous nutrients in axenic culture conditions.

At present, then, the evidence is that movement of nutrients from fungus to orchid occurs only in infected cells.

Translocation of Substances via the Fungus

The symbiotic theory of nutrition, developed originally through the observations of Noel Bernard and Hans Burgeff, argues that mycorrhizal fungi break down complex carbohydrates, absorb the hydrolysates together with other nutrients, and transport them into the orchid tissue. The decomposition of cellulose by orchid fungi was first demonstrated by Smith (1966), and she also showed, using ^{14}C and ^{32}P labeled compounds, that mycelium of *Rhizoctonia repens* translocates the label across a diffusion barrier into protocorms of *Dactylorhiza purpurella.* Hadley (1969) confirmed that cellulose was utilized, leading to rapid and considerable growth, in *D. purpurella, Goodyera repens,* and *Spathoglottis plicata* protocorms infected with suitable endophytic fungi. Growth was much better than on media containing glucose.

In a later paper Smith (1967) obtained evidence that the fungal sugar trehalose was translocated into seedlings, after which the ^{14}C label appeared in other carbohydrates, notably sucrose, which does not occur in fungal hyphae and is characteristic of green plants.

It is not known whether translocation involves transfer across the living fungus-host interface. Hadley and Williamson (1971) postulated that such movement could occur at least in the early stages of infection. However, since the intracellular hyphae are ultimately digested it has been generally assumed that such breakdown is the main means of transfer of nutrients. Some workers (Smith *et al.,* 1969; Lewis, 1973) regard the nutritional pattern as one of necrotrophic parasitism of the fungus by the orchid.

Movement of Nutrients to the Fungus

Since orchid endophytic fungi grow at least to a limited extent in soil, their nutrient requirements (see above) are probably not limiting factors in the natural environment. In experimental conditions there is at present no direct evidence from dual cultures that the orchid provides nutrients for the fungus. Suggestions that "orchids and fungi may have coevolved with respect to vitamin requirements" (Hijner and Arditti, 1973) must be regarded with caution. Also, the size of orchid seeds is such that it is unlikely that one or even a few germinating seeds would release nutrients to an extent that might stimulate the growth of fungi in natural conditions.

As has been stated earlier, many orchid endophytes grow easily on simple media, but those which are heterotrophic for certain nitrogen compounds and/or vitamins, such as *Tulasnella calospora,* may reflect a degree of evolutionary adaptation to the root-infecting habit (Hadley and Ong, 1978), their requirements being satisfied by the host when growing in tissue. Specific amino acid requirements may also be satisfied in this way.

Whether the endophyte obtains major nutrients such as a supply of carbon and nitrogen compounds from the orchid is uncertain. Smith (1967) suggested that hyphae may obtain carbohydrates from orchid tissue. But Hadley and Purves (1974) found little evidence for movement of $^{14}CO_2$ photosynthate from *Goodyera repens* into the endophytic fungus when the latter was allowed to grow out over a nutrient-free agar medium. However, the fungus they used was *Ceratobasidium cornigerum,* which is not nutritionally exacting; similar experiments have not been carried out with *Tulasnella calospora.* Indeed, it would be surprising if an endophytic fungus such as *T. calospora* was not receiving at least some nutrients from its host.

More vigorous, aggressive fungi may upset the delicate balance and become parasitic on their host protocorms (Hadley, 1970b). At this stage they are clearly obtaining nutrients by necrotrophic means.

Infection: The Interrelationship and the Growth Stimulus

It is evident that although germination may be enhanced in the presence of suitable fungi, this may be unrelated to and independent of infection. Infection is not a prerequisite for germination, and some development of the embryo cells usually occurs before the fungus enters. As Burgeff (1959) stated, infection "can take place in the swollen or the germinated seed."

Infection is probably controlled by the host since the fungus is able to penetrate only at a particular site. This is usually the suspensor cells of the embryo (see Figs. 3-5 and 3-6). Epidermal cells of some species may be susceptible to infection (Fig. 3-9-**A**), especially where they form epidermal hairs which appear to be the specific site of infection in some tropical species (Fig. 3-9-**B**) and many north temperate species (Burgeff, 1936; Hadley, unpublished).

Infection follows a consistent pattern on the same host, regardless of the strain or species of fungus involved. Williamson and Hadley (1970) clearly showed that a range of *Rhizoctonia* species including known pathogens of crop plants all infected epidermal hairs of *Dactylorhiza purpurella* in exactly the same fashion; single hyphae constricted to form a narrow penetration peg, passed through the wall, and extended in the lumen of the cell. The cell remained alive and the cytoplasm continued streaming. The sequence of events was different when some of the same fungi were allowed to infect susceptible nonorchid hosts; they formed infection cushions and appressoria before penetrating. These results are clear evidence that the host cell controls the process of infection and the fungi are highly adapted to it.

At what point is the interrelationship apparent in terms of an effect on the germinating seed? Bernard (1909) found that a stimulation of growth occurred "immediately"

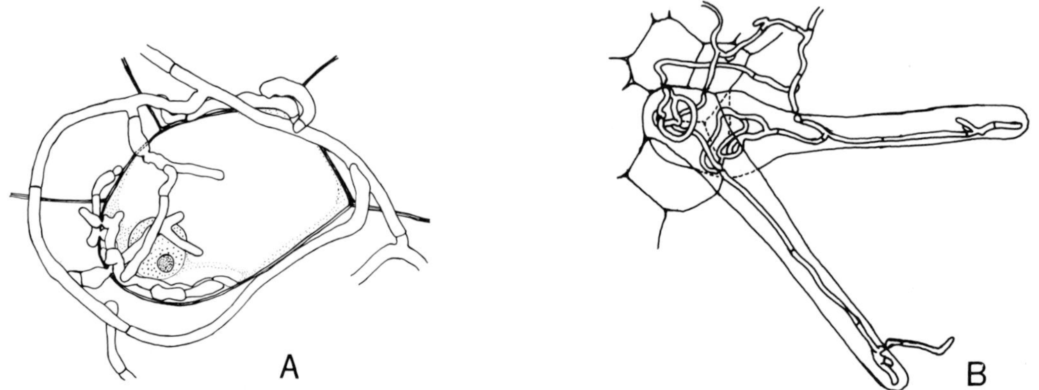

Fig. 3-9. Infection of (**A**) epidermal cell of *Dactylorhiza purpurella;* × 400 and (**B**) epidermal hairs of *Spathoglottis plicata;* × 200. (Drawings by G. Hadley.)

after infection of *Phalaenopsis* and some other orchids, and many other workers have reported an increase in size within a few days. Hadley and Williamson (1971) studied the time scale of the infection process in a selected uniform population of pregerminated *Dactylorhiza purpurella* protocorms. By careful observation and measurement they found that epidermal hairs were penetrated within about 15 hours of contact by hyphae. Intracellular pelotons were visible in the subepidermal cells a few hours later, and the digestion of these first-formed pelotons was observed between 30 and 40 hours, after which the number increased linearly (Fig. 3-10). A measurable increase in the size of protocorms occurred within three days (Fig. 3-11) when only a few pelotons were present (Fig. 3-10).

The growth stimulus therefore occurs very rapidly and it was occurring in some protocorms which were not seen to contain digested pelotons. The possibility that growth is enhanced before any digestion takes place merits further investigation.

Hadley and Williamson (1971) also observed that infection stimulated the hydrolysis of stored starch in *D. purpurella* protocorms, as in other orchids (Burgeff, 1959). It may be, therefore, that the effect of the fungus on the orchid is one in which some metabolic deficiency is corrected immediately after infection occurs, triggering a series of metabolic processes and leading to mobilization of the reserves of the protocorm. In experiments with *Goodyera repens*, which grows fairly well in nonsymbiotic cultures, infected protocorms accumulated less starch than uninfected ones on the same medium (Purves and Hadley, 1976) and the latter seemed totally unable to carry out the metabolic

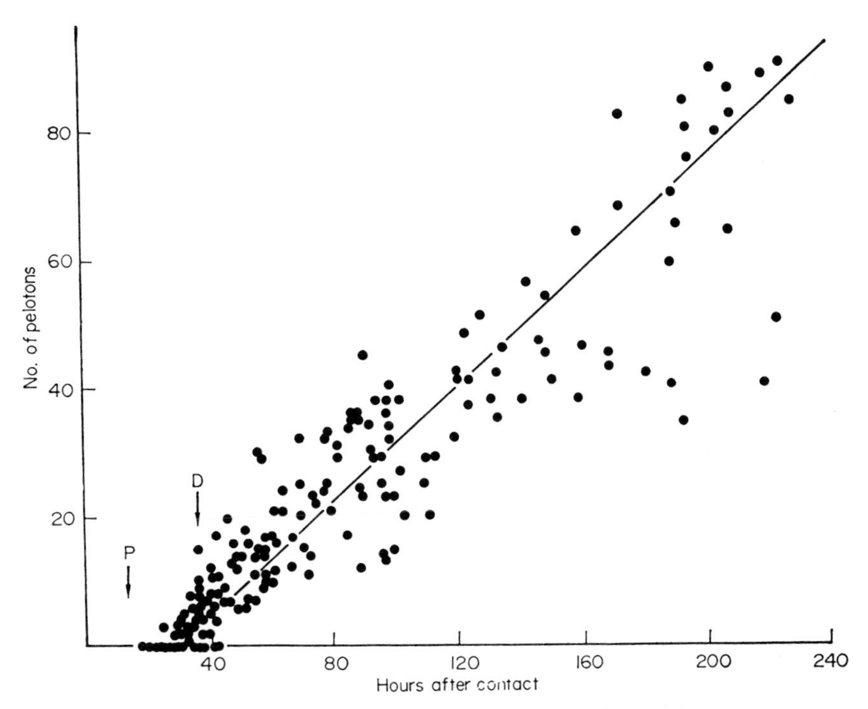

Fig. 3-10. Infection of *Dactylorhiza purpurella* expressed as number of pelotons per protocorm in relation to the initial penetration time (P, 14.5 hours after contact) and the time (D, 30–40 hours) when the first digested pelotons appeared (reproduced from Hadley and Williamson, 1971).

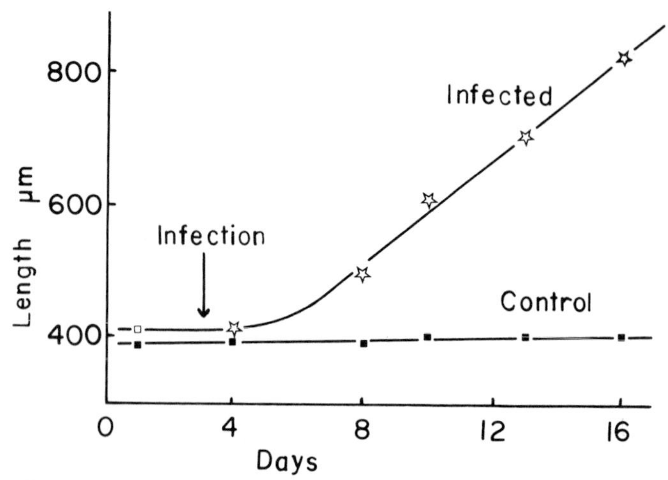

Fig. 3-11. Growth of *Dactylorhiza purpurella* protocorms in response to infection. (Data from G. Hadley.)

processes that would utilize the starch reserves. When protocorms were infected after a period of starch accumulation the resulting growth stimulus correlated with a decline in the starch reserves. The starch was not necessarily utilized by the fungus, which in fact never penetrates starch-containing cells of the inner cortex. However, nonsymbiotic protocorms growing on media lacking nutrients also showed starch depletion and the results were inconclusive as to whether the endophytic fungus is causing metabolic changes whereby the protocorm utilizes its reserves more rapidly. Nevertheless, since some depletion of starch occurred it seems unlikely that the host is totally dependent on nutrients obtained necrotrophically by digestion of the fungus.

It has been generally assumed that orchid cells are phagocytic and digestion is carried out by enzymes produced by the host cell (Burgeff, 1959). Hadley and Williamson (1971) emphasized that digestion may be an autolytic process initiated by fungal enzymes. Williamson (1973) subsequently investigated the location of acid phosphatase enzymes during infection and lysis but was unable to resolve whether they originated from fungus or host.

Infection and Symbiosis in Mature Plants

Most of the available information about the pattern of symbiosis in roots of mature plants comes from microscopic anatomical observations and isolation work with field-grown plants. Plants cannot easily be grown to the adult root-bearing stage in asymbiotic culture, and consequently there is little knowledge of the development of root infections and the physiological (especially the nutritional) significance of the endophytic fungus and its function in the mature plant.

Protocorms, as they develop, eventually produce one or more roots which are always infected afresh from the environment or substrate; the infected region of the root is not continuous with that of the protocorm. The roots of *Dactylorhiza purpurella* and *Goodyera repens* rapidly become heavily infected in symbiotic cultures (Hadley, unpublished and

Fig. 3-2-**D**). In general, starch-filled storage cells of the inner cortex of roots or protocorms, like tubers and bulbs of field material, remain uninfected.

The general pattern of root infection has been described earlier, but the significance of symbiosis in mature plants is not understood. Are nutrients provided via the infecting fungus, as in protocorms? There is no evidence as yet, but it is conceivable that even in those orchids that are largely autotrophic the fungus mediates in the uptake of inorganic nutrients and/or carbon compounds.

Alternatively, is infection of the roots a feature of a highly evolved system in which the fungus and host are maintaining a compatible relationship, which does not involve any movement of nutrients but is an inevitable consequence if germinating seeds are able to accept the potential symbionts?

Finally, the factors controlling infection of roots are not well understood. All evidence relating to this problem comes from work with tubers (discussed in the following section), and it is an assumption that root infections may be controlled in a similar manner.

The Control of Infection: Orchid Phytoalexins

The absence of infection in tubers suggests that they contain a substance toxic to potential invaders; Bernard (1911) first showed this to be so using tubers of a *Loroglossum* species. Subsequently Burges (1939) confirmed these findings using a microdissection technique, but it was the extensive research carried out by Gäumann and his colleagues (for a review see Nüesch, 1963) that established the presence of a specific substance, probably the first recorded phytoalexin, in roots and tubers of *Orchis militaris*. This compound was called orchinol (Gäumann and Kern, 1959). It was subsequently shown to be a widely active fungistatic compound, produced in tubers as a defense reaction in response to contact with mycorrhizal and other fungi. Many European orchids were shown to produce it, and a related compound, hircinol, was later isolated from *Loroglossum hircinum* (Urech *et al.*, 1963; see Arditti, 1966, for a short review). Fisch, Flick, and Arditti (1973) subsequently established the structure of hircinol. Arditti (1979) gives an extensive review of orchid phytoalexin studies.

It is possible that orchid roots produce small quantities of these fungistatic substances but the concentration may be much lower than in the tubers so that infection by some fungi—with high levels of resistance to orchinol—still occurs. Tubers, in contrast, are protected from virtually all fungi, although some aggressive pathogens such as *Rhizoctonia solani* and *Fusarium solani* may be pathogenic to them.

Knowledge of orchid phytoalexins does not explain why some orchids are much more resistant to infection than others or why there is a variation in the spectrum of fungi which can be symbiotic with them.

Triple Symbioses

Many fungi which are mycorrhizal with orchids have the potential for relationships with other plants. *Rhizoctonia solani,* for example, is a well-known pathogen of many hosts as well as an orchid endosymbiont. In fact, it may be both on the same host, as was reported by Alconero (1969) in greenhouse experiments with *Vanilla planifolia* and *V.*

phaeantha. There was a constant fluctuation in the kind of association, and cells with pathogenic infections could occur adjacent to cells showing a mycorrhizal interaction. There are no recorded examples of *Rhizoctonia* species occurring on an orchid and another host simultaneously but the situation has been described with other fungi in many instances.

Ruinen (1953), working with epiphytic orchids, reported that "the mycorrhizal fungus of the epiphyte is potentially parasitic on the supporting trees." The same could be true of *Gastrodia elata,* the underground tuber of which obtains its nutrients from the rhizomorphs of *Armillaria mellea,* a well-known parasite of many trees and shrubs (Kusano, 1911). *Gastrodia cunninghamii* is similarly mycorrhizal with *Armillaria* (Campbell, 1962), and two other *Gastrodia* species have been described by the same author (1963, 1964) as being linked through their mycorrhizal fungi (identity unknown) to tree species on which they were therefore indirect epiparasites. This aspect of orchid mycorrhizal relationships deserves more extensive exploration.

Horticultural Relevance of Mycorrhiza

It is difficult to make a realistic assessment of the importance and usefulness of mycorrhiza to the horticulturist without having experience of the problems at first hand. An examination of the literature suggests that there was a period before and during the early years of this century when symbiotic germination was commonly used, but the practice declined after the development of the asymbiotic method (Knudson, 1922, 1924).

Following Reissek's work (1847), orchid growers were aware of the fungi present in roots and of the need to germinate seeds on "orchid soil," and this led to the method of sowing seeds on soil at the base of the parent plant. Most of the new hybrids obtained during the latter part of the nineteenth century were raised in this way. Subsequently, when Bernard and Burgeff succeeded in isolating orchid fungi and germinating seeds symbiotically in pure culture, many horticulturists tried to apply their techniques. Ramsbottom (1923) developed an interest in symbiotic methods and also described in detail the work of Joseph Charlesworth, who had remarkable success in "raising *Odontoglossum* and its allies by sowing seeds on nutrient media in which the appropriate fungus was growing. His culture flasks were sufficient testimony to the success of the laboratory method when placed upon a commercial scale."

It is interesting to find that Charlesworth, in conjunction with Ramsbottom who subsequently recorded the observations, had the same experience as Bernard in finding that only certain strains of a fungus were active in stimulating germination. Ramsbottom (1929) stated, "The occurrence of different strains of these orchid fungi raises many problems." With present knowledge of the loss of activity of orchid fungi in culture and the general technical difficulties associated with maintenance of cultures in sterile conditions it is perhaps not surprising that horticulturists found Knudson's asymbiotic methods a more attractive prospect.

Symbiotic germination methods, therefore, have been used only occasionally. Westerman (1959) suggested that *Rhizoctonia repens* and *R. mucoroides* would be helpful to the amateur grower but the only recent report is by Clements and Ellyard (1979), who used fungi with some success in the propagation of naturally occurring terrestrial species in Australia.

The difficulty with many orchids is that not only are they difficult to germinate, but they are often resistant to infection, for reasons unknown. Unpublished results in the author's laboratory show that even with fungi such as *Tulasnella calospora,* which is a very common symbiont, many orchids (especially hybrids) give totally negative results and do not become infected.

Therefore, although mycorrhizal fungi could in theory be applied to horticultural practice it would probably be essential to go back to the seminatural situation, germinating seed in association with the parent plant or at least on natural substrates, and develop the method in that way. Symbiotic methods may not offer any advantage with orchids which readily germinate on simple media such as Knudson C. But the accelerated growth of symbiotically infected protocorms would be an attractive prospect with those species or hybrids which do not germinate freely or which have complex nutrient requirements coupled with slow rates of growth in asymbiotic culture.

Acknowledgments

I am grateful to Dr. Ian Alexander and Dr. Brian Williamson for commenting on the draft of this paper, to Eddie Middleton for photographic work, to Gerard van Warmelo for the arduous task of checking the references, and to Kathy Stephen for an excellent performance with her typewriter.

Literature Cited

Alconero, R. 1969. Mycorrhizal synthesis and pathology of Rhizoctonia solani in vanilla orchid roots. Phytopath. 59:426–430.
Alexander, C., and G. Hadley. 1982. Variable symbiotic activity in Rhizoctonia isolates from Goodyera repens. Trans. Brit. Mycol. Soc. (in press).
Arditti, J. 1966. The production of fungal growth regulating compounds by orchids. Orchid Digest 30:88–90.
——. 1967. Factors affecting the germination of orchid seeds. Bot. Rev. 33:1–97.
——. 1979. Aspects of the physiology of orchids. Adv. Bot. Res. 7:421–655.
Bernard, N. 1903. La germination des orchidées. Comp. Rend. Acad. Sci. Paris 137:483–485.
——. 1904. Recherches expérimentales sur les orchidées. Rev. Gen. Bot. 16:405–451, 458–476.
——. 1909. L'évolution dans la symbiose. Ann. Sci. Nat. Bot. 9:1–196.
——. 1911. Sur la fonction fungicide des bulbes d'Ophrydées. Ann. Sci. Nat. Bot. 14:223–234.
Borriss, H., E. M. Jeschke, and G. Bartsch. 1971. Elektronenmikroskopische Untersuchungen zur Ultrastruktur der Orchideen-Mykorrhiza. Biol. Rundsch. 9:177–180.
Bracker, C. E., and L. J. Littlefield. 1973. Structural concepts of host-pathogen interfaces, p. 159–318. *In* R. J. W. Byrde and C. V. Cutting, (eds.), Fungal pathogenicity and the plant's response. Academic Press, London.
Burgeff, H. 1909. Die Wurzelpilze der Orchideen. Fischer, Jena.
——. 1936. Samenkeimung der Orchideen. Fischer, Jena.
——. 1943. Problematik der Mycorrhiza. Naturwiss. 31:558–567.
——. 1959. Mycorrhiza of orchids, p. 361–395. *In* C. L. Withner (ed.), The orchids: A scientific survey. Ronald Press, New York.
Burges, A. 1939. The defensive mechanism in orchid mycorrhiza. New Phytol. 38:273–283.

Campbell, E. O. 1962. The mycorrhiza of Gastrodia cumminghamii Hook. Trans. Roy. Soc. N.Z. 1:289–296.
——. 1963. Gastrodia minor Petrie an epiparasite of manuka. Trans. Roy. Soc. N.Z. 2:73–81.
——. 1964. The fungal association in a colony of Gastrodia sesamoides R. Br. Trans. Roy. Soc. N.Z. 2:237–246.
Clements, M. A., and E. K. Ellyard. 1979. The symbiotic germination of Australian terrestrial orchids. Amer. Orchid Soc. Bull. 48:810–816.
Curtis, J. T. 1939. The relation of specificity of orchid mycorrhizal fungi to the problem of symbiosis. Amer. J. Bot. 26:390–399.
Dörr, I., and R. Kollmann. 1969. Fine structure of mycorrhiza in Neottia nidus-avis (L.) L. C. Rich (Orchidaceae). Planta 89:372–375.
Downie, D. G. 1940. On the germination and growth of Goodyera repens. Trans. Bot. Soc. Edin. 33:36–51.
——. 1949. The germination of Goodyera repens (L.) R. Br. in fungal extract. Trans. Bot. Soc. Edin. 35:120–125.
——. 1957. Corticium solani—an orchid endophyte. Nature 179:160.
——. 1959a. Rhizoctonia solani and orchid seed. Trans. Bot. Soc. Edin. 37:279–285.
——. 1959b. The mycorrhiza of Orchis purpurella. Trans. Bot. Soc. Edin. 38:16–29.
Fisch, M. H., B. H. Flick, and J. Arditti. 1973. Structure and antifungal activity of hircinol, loroglossol and orchinol. Phytochem. 12:437–441.
Gäumann, E., and H. Kern. 1959. Über chemische Abwehrreaktionen bei Orchideen. Phytopath. Z. 36:1–26.
Hadley, G. 1969. Cellulose as a carbon source for orchid mycorrhiza. New Phytol. 68:933–939.
——. 1970a. The interaction of kinetin auxin and other factors in the development of north temperate orchids. New Phytol. 69:549–555.
——. 1970b. Non-specificity of symbiotic infection in orchid mycorrhiza. New Phytol. 69:1015–1023.
——. 1975. Organization and fine structure of orchid mycorrhiza, p. 335–351. *In* F. E. Sanders, B. Mosse, and P. B. Tinker (eds.), Endomycorrhizas. Academic Press, London.
Hadley, G., and G. Harvais. 1968. The effect of certain growth substances on asymbiotic germination and development of Orchis purpurella. New Phytol. 67:441–445.
Hadley, G., R. P. C. Johnson, and D. A. John. 1971. Fine structure of the host-fungus interface in orchid mycorrhiza. Planta 100:191–199.
Hadley, G., and S. H. Ong. 1978. Nutritional requirements of orchid endophytes. New Phytol. 81:561–569.
Hadley, G., and S. Purves. 1974. Movement of ^{14}carbon from host to fungus in orchid mycorrhiza. New Phytol. 73:475–482.
Hadley, G., and B. Williamson. 1971. Analysis of the post-infection growth stimulus in orchid mycorrhiza. New Phytol. 70:445–455.
——. 1972. Features of mycorrhizal infection in some Malayan orchids. New Phytol. 71:1111–1118.
Hamada, M. 1940. Studien über die Mykorrhiza von Galeola septentrionalis Reichb. f.: Ein neuer Fall der Mykorrhizabildung durch intraradikale Rhizomorphen. Jap. J. Bot. 10:151–212.
Hamada, M., and S. I. Nakamura. 1963. Wurzelsymbiose von Galeola altissima Reichb. f., einer chlorophyll-freien Orchidee, mit dem holzstörenden Pilze Hymenochaete crocicreas Berk. et Br. Sci. Rep. Tohoku University 29:227–238.
Harley, J. L. 1969. The biology of mycorrhiza. Leonard Hill, London.
Harvais, G. 1972. The development and growth requirements of Dactylorhiza purpurella in asymbiotic cultures. Can. J. Bot. 50:1223–1229.
——. 1973. Growth requirements and development of Cypripedium reginae in axenic culture. Can. J. Bot. 51:327–332.
Harvais, G., and G. Hadley. 1967a. The relation between host and endophyte in orchid mycorrhiza. New Phytol. 66:205–215.
——. 1967b. The development of Orchis purpurella in asymbiotic and inoculated cultures. New Phytol. 66:217–230.
Harvais, G., and D. Pekkala. 1975. Vitamin production by a fungus symbiotic with orchids. Can. J. Bot. 53:156–163.
Harvais, G., and A. Raitsakas. 1975. On the physiology of a fungus symbiotic with orchids. Can. J. Bot. 53:144–155.

Hijner, J. A., and J. Arditti. 1973. Orchid mycorrhiza: Vitamin production and requirements by the symbionts. Amer. J. Bot. 60:829–835.
Knudson, L. 1922. Nonsymbiotic germination of orchid seeds. Bot. Gaz. 73:1–25.
——. 1924. Further observations on nonsymbiotic germination of orchid seeds. Bot. Gaz. 77:212–219.
——. 1925. Physiological study of the symbiotic germination of orchid seeds. Bot. Gaz. 79:345–379.
——. 1930. Flower production by orchid grown nonsymbiotically. Bot. Gaz. 89:192–199.
Kusano, S. 1911. Gastrodia elata and its symbiotic association with Armillaria mellea. J. Coll. Agric. Imp. Univ. Tokyo 4:1–66.
Lewis, D. H. 1973. Concepts in fungal nutrition and the origin of biotrophy. Biol. Rev. 48:261–278.
Magnus, W. 1900. Studienander endotrophen Mycorrhiza von Neottia nidus-avis L. Jahr. f. wissench. Bot. 205–272.
Mollison, J. E. 1943. Goodyera repens and its endophyte. Trans. Bot. Soc. Edin. 33:391–403.
Nieuwdorp, P. J. 1972. Some observations with light and electron microscope on the endotrophic mycorrhiza of orchids. Acta Bot. Neerl. 21:128–144.
Nüesch, J. 1963. Defence reactions in orchid bulbs. Symp. Soc. Gen. Microbiol. 13:335–343.
Pérombelon, M., and G. Hadley. 1965. Production of pectic enzymes by pathogenic and symbiotic Rhizoctonia strains. New Phytol. 64:144–151.
Powell, K. B., and J. Arditti. 1975. Growth requirements of Rhizoctonia repens M32. Mycopath. 55:163–167.
Purves, S., and G. Hadley. 1975. Movement of carbon compounds between the partners in orchid mycorrhiza, p. 175–194. *In* F. E. Sanders, B. Mosse, and P. B. Tinker (eds.), Endomycorrhizas. Academic Press, London.
——. 1976. The physiology of symbiosis in Goodyera repens. New Phytol. 77:689–696.
Ramsbottom, J. 1923. Orchid mycorrhiza. Trans. Brit. Mycol. Soc. 8:28–61.
——. 1929. Orchid mycorrhiza, p. 1676–1687. *In* B. M. Duggar (ed.), Proc. Int. Congr. Plant Sci. 1926, Ithaca, N.Y. Collegiate Press, Menasha, Wis.
Reissek, S. 1847. Über Endophyten der Pflanzenzelle. Naturwiss. 1:31–46.
Ruinen, J. 1953. Epiphytosis: A second view on epiphytism. Ann. Bogor. 1:101–157.
Smith, D., L. Muscatine, and D. Lewis. 1969. Carbohydrate movement from autotrophs to heterotrophs in parasitic and mutualistic symbiosis. Biol. Rev. 44:17–90.
Smith, S. E. 1966. Physiology and ecology of orchid mycorrhiza fungi with reference to seedling nutrition. New Phytol. 65:488–499.
——. 1967. Carbohydrate translocation in orchid mycorrhizas. New Phytol. 66:371–378.
——. 1973. Asymbiotic germination of orchid seeds on carbohydrates of fungal origin. New Phytol. 72:497–499.
——. 1974. Mycorrhizal fungi. CRC Crit. Rev. Microbiol. 3:275–313.
Smith, S. E., and F. A. Smith. 1973. Uptake of glucose, trehalose and mannitol by leaf slices of the orchid Bletilla hyacinthina. New Phytol. 72:957–964.
Stephen, R. C., and K. K. Fung. 1971a. Nitrogen requirements of the fungal endophytes of Arundina chinensis. Can. J. Bot. 49:407–410.
——. 1971b. Vitamin requirements of the fungal endophytes of Arundina chinensis. Can. J. Bot. 49:411–415.
Stoutamire, W. 1974. Terrestrial orchid seedlings, p. 101–128. *In* C. L. Withner (ed.), The orchids: Scientific studies. Wiley-Interscience, New York.
Strullu, D-G., and J-P. Gourret. 1974. Ultrastructure et évolution du champignon symbiotique des racines de Dactylorchis maculata (L.) Verm. J. de Microscopie 20:285–294.
Urech, J., B. Fechtig, J. Nuesch, and E. Visher. 1963. Hircinol eine antifungisch wirksame Substanz aus Knollen von Loroglossum hircinum (L.). Rich. Helv. Chim. Acta. 46:2758–2766.
Vajrabhaya, T. 1977. Variations in clonal propagation, p. 187–201. *In* J. Arditti (ed.). Orchid biology: Reviews and perspectives. I. Cornell Univ. Press, Ithaca, N.Y.
Vermeulen, P. 1946. Studies on Dactylorchis. Dissertation, Univ. of Amsterdam, Utrecht.
Wahrlich, W. 1886. Beitrag zur Kenntniss der Orchideenwurzelpilze Bot. Zeit. 44:481–488, 497–505.
Warcup, J. H. 1971. Specificity of mycorrhizal association in some Australian terrestrial orchids. New Phytol. 70:41–46.
——. 1973. Symbiotic germination of some Australian terrestrial orchids. New Phytol. 72:387–392.
——. 1975. Factors affecting symbiotic germination of orchid seed, p. 87–104. *In* F. E. Sanders, B. Mosse, and P. B. Tinker (eds.), Endomycorrhizas. Academic Press, London.

Warcup, J. H., and P. H. B. Talbot. 1967. Perfect states of Rhizoctonias associated with orchids. New Phytol. 66:631–641.
——. 1971. Perfect states of Rhizoctonias associated with orchids. II. New Phytol. 70:35–40.
Westerman, E. 1959. Fungus and its association with orchid seed germination. Orchid J. 3:288–289.
Williamson, B. 1970. Induced DNA synthesis in orchid mycorrhiza. Planta 92:347–354.
——. 1973. Acid phosphatase and esterase activity in orchid mycorrhiza. Planta 112:149–158.
Williamson, B., and G. Hadley. 1969. DNA content of nuclei in orchid protocorms symbiotically infected with Rhizoctonia. Nature 22:582–583.
——. 1970. Penetration and infection of orchid protocorms by Thanatephorus cucumeris and other Rhizoctonia isolates. Phytopath. 60:1092–1096.
Withner, C. L. 1959. Orchid physiology, p. 315–360. *In* C. L. Withner (ed.), The orchids: A scientific survey. Ronald Press, New York.
——. 1974. Developments in orchid physiology, p. 129–168. *In* C. L. Withner (ed.), The orchids: Scientific studies. Wiley-Interscience, New York.
Yeoh, B. C. 1962. Beer is best! Malay. Orchid Rev. 6:104–106.

Additional Literature

Arditti, J., and R. Ernst. 1974. Reciprocal movement of substances between orchids and mycorrhizae, p. 299–307. *In* Proc. 7th World Orchid Conf., Medellin, Columbia.
Blakeman, J. P., M. A. Mokahel, and G. Hadley. 1976. Effect of mycorrhizal infection on respiration and activity of some oxidase enzymes of orchid protocorms. New Phytol. 77:697–704.
Blowers, J. W., and J. Arditti. 1970. The importance of orchid mycorrhiza. Orchid Rev. 78:248–249.
Burges, A. 1936. On the significance of mycorrhiza. New Phytol. 35:117–131.
Campbell, E. O. 1970. The fungal association of Yoania australis. Trans. Roy. Soc. N.Z., Biol. Sci. 12:5–12.
Fisch, M. H., Y. Schechter, and J. Arditti. 1972. Orchids and the discovery of phytoalexins. Amer. Orchid Soc. Bull. 41:605–607.
Hadley, G. 1968. Orchids and their symbiotic fungi. Malayan Scientist 4:23–27.
Harvais, G. 1974. Notes on the biology of some native orchids of Thunder Bay, their endophytes and symbionts. Can. J. Bot. 52:451–460.
McLennan, E. I. 1959. Gastrodia sesamoides R. Br. and its endophyte. Aust. J. Bot. 7:225–229.
Nishikawa, T., and T. Ui. 1976. Rhizoctonias isolated from wild orchids in Hokkaido. Trans. Mycol. Soc. Japan. 17:77–84.
Warcup, J. H. 1981. The mycorrhizal relationships of Australian orchids. New Phytol. 87:371–381.
Weber, H. C. 1979. Die Korallenwurz (Corallorhiza trifida Chat., Orchidaceae), ein Saprophyt auf dem Weg zum Parasitismus. Die Orchidee 30:180–183.

POLLINATION ECOLOGY

4

The Biology of Orchids and Euglossine Bees*

NORRIS H. WILLIAMS

*The literature survey pertaining to this chapter was concluded in February 1980; the chapter was submitted in March 1980, and the revised version was received in July 1980.

Introduction

The Orchidaceae, the largest family of flowering plants, exhibits a tremendous amount of floral diversity, and is often thought of as being taxonomically "difficult" as a result of the bizarre modifications of its flowers. The vast majority of these modifications are the result of millions of years of evolution during which orchid flowers have adapted through natural selection to their pollinators. Pollination mechanisms of the orchids vary from the mundane to the most complex to be found among flowering plants. Some of the most interesting systems of pollination involve the euglossine bees and certain orchid species of the American tropics. It has been known for over a century that male euglossine bees visit and often pollinate a wide variety of orchids. Cruger's (1865) original observations (and erroneous interpretations) were widely publicized by Charles Darwin (1877) in the second edition of his classic book on orchid pollination, but almost a century passed before the euglossine bee–orchid relationship began to be investigated in detail. In this chapter I will review the relationships between euglossine bees and the orchids they pollinate.

The Euglossini (Hymenoptera: Apidae) are a tribe of exclusively neotropical bees, usually bright colored. There are three free-living genera and two parasitic ones. The free-living genera are either solitary, communal, or quasi-social, depending on the species. Very little is known about the social life of these bees. However, before discussing the euglossine bee–orchid flower relationship, it is necessary to review some features which concern pollination biology, and each of the organisms.

Morphology and Classification of the Orchids

Both Cronquist (1968) and Takhtajan (1969) place the Orchidaceae in the order Orchidales, an order they consider close to and derived from the Liliales. Thorne (1976) places the family in the inclusive order Liliales, as does Dressler (1974). Garay (1972) appears to place the Orchidaceae in the order Orchidales. Vermeulen (1965) places the family in the order Orchidales and then divides the Orchidaceae into three families (Apostasiaceae, Cypripediaceae, Orchidaceae). Garay (1972) and Dressler (1974) agree that the splitting by Vermeulen is unnecessary and maintain all three groups as members of the family Orchidaceae. Numerous lines of evidence (Rao, 1969, 1974; N. H. Williams, 1979; Newton and Williams, 1978) suggest that this is the best course, although Schill (1978) presents some palynological evidence to suggest that Apostasioideae might be better treated as a separate family. Newton and Williams (1978) came to the conclusion that pollen alone cannot be used to decide the placement of Apostasioideae. On the basis of vegetative (N. H. Williams, 1979) and floral anatomy (Rao, 1969, 1974), comparative chemistry (C. A. Williams, 1979), and other factors (Dressler, 1974; Garay, 1972) it seems best to retain Apostasioideae and Cypripedioideae in the Orchidaceae. The remainder of the family is classified into four more subfamilies according to Dressler (1979a).

Most modern treatments suggest that the Orchidaceae may have arisen from an ancestor of general liliaceous stock (Cronquist, 1968; Takhtajan, 1969; Dressler, 1974). C. A. Williams (1979), however, studied the leaf flavonoids of 142 species of orchids and concluded that the compounds found in Orchidaceae show the closest relationships to those in Commelinaceae, Iridaceae, and Bromeliaceae. Obviously, the relationships of the Orchidaceae need additional investigation.

The general structure of orchid flowers has been discussed by Dressler (1961, 1974), Rao (1974), and van der Pijl and Dodson (1966) among others, and it is not necessary here to engage in the debate concerning the relationships of the various subfamilies of Orchidaceae (Garay, 1972; Dressler, 1979a; Vermeulen, 1965; Rao, 1969, 1974). Orchid flowers consist of three sepals and three petals, any or all of which may be highly modified. Usually one of the petals, variously termed a lip or labellum, differs in appearance from the other two, and is opposite the combined reproductive structure, the column or gynostemium. The column consists of one to three anthers, with two or three in Apostasioideae (de Vogel, 1969), two in Cypripedioideae (Dressler and Dodson, 1960; van der Pijl and Dodson, 1966; Garay, 1960; and citations in these works), and only one in the remainder of the family. The anthers have four microsporangia, some of which may fuse late in development (unpublished work from my laboratory). The pollen may be shed as single grains [as in the Apostasioideae, Cypripedioideae, some Orchidioideae, and some Epidendroideae (Schill and Pfeiffer, 1977; Schill, 1978; Williams and Broome, 1976; Newton and Williams, 1978; Ackerman and Williams, 1980)], tetrads, massulae, or pollinia. Even in Cypripedioideae the single grains are often held together in clumps. In the more advanced orchids the pollen is held in compact units, the pollinia. The number of pollinia is usually constant for a given species and varies from 2 to 12. Development of caudicles (from the tapetum), the rostellum (from a portion of the median stigma lobe), the viscidium (from the rostellum), and the stipe (from the epidermis of the stigmatic area) are important in producing a unit, the pollinarium. The evolution of these structures is discussed by Dressler (1979a). It is the development of the pollinarium that has made possible the very precise and elaborate pollination mechanisms that exist in the orchid family (Dressler, 1968a).

Attraction of Pollinators to Orchids

The most general and primary means of attraction of pollinators to flowers are visual and olfactory cues of one form or another. These often lead to one of the major rewards offered (either nectar or pollen). Nectar guides are often striking in visible light (Faegri and van der Pijl, 1966; Meeuse, 1961; Percival, 1965) and may consist of lines of color or colored hairs, or various ridges or channels which position or lead the insect into the flower and toward the nectary. Jones and Buchmann (1974) discussed the role of ultraviolet patterns in flowers as general orientation cues. Certain orchid flowers have distinctive UV patterns, and those of *Arethusa, Calopogon,* and *Pogonia* may (1) be important in orienting insects on the flowers, and (2) also indicate floral convergence to similar types of pollinators (Thien, 1971; Thien and Marcks, 1972). Nectar is the reward in some insect-pollinated orchids, but apparently not in the three species studied by Thien and

Marcks, who found that *Calopogon* produced none and the other two species manufactured very little. Jeffrey, Arditti, and Koopowitz (1970) analyzed the nectar of both floral and extrafloral nectaries of 72 species of orchids and several hybrids and found that although there are differences in the types of sugars produced by different taxa (but not between floral and extrafloral exudates), no relationship exists between pollinators and saccharide content. Since the pollinia of orchids are hidden in the anther until removed by the pollinator, pollen is almost never a reward for the visiting insects.

Odor has generally been thought to be of some importance in attracting pollinators to plants (Faegri and van der Pijl, 1966; Percival, 1965). The special case of odor as an attractant of euglossine bees will be discussed in detail below. Kullenberg (1956, 1973, 1975) and Kullenberg and Bergstrom (1975, 1976a,b, and references therein) have discussed the odors of *Ophrys* and the roles they play in the pollination of the various species of this genus. Nilsson (1978) showed that the floral fragrances of *Platanthera chlorantha* were active in invoking landing and feeding reactions in several moth species that pollinate this orchid. He also suggested (Nilsson, 1979) that the floral fragrances of *Cypripedium calceolus* are important in manipulating the most frequent visitor, females of *Andrena haemorrhoa,* into the labellum of this slipper orchid. Smith and Snow (1976) presented evidence that night-flying moths find *Platanthera ciliaris* by following an odor trail. Stoutamire (1974a) noted that the purple-fringed *Platanthera psycodes* is pollinated during the day by skippers and hawkmoths, and suggested that since the flowers continue to produce fragrance at night they may also be pollinated by nocturnal moths. Stoutamire (1968), Thien (1969a,b), and Thien and Utech (1970) have shown that *Habenaria obtusata* is pollinated by female mosquitoes of the genus *Aedes* as well as by geometrid moths. Carbon dioxide is postulated as the attractant for the mosquitoes, but the attractant for the moth is unknown. Adams and Goss (1976) and Goss and Adams (1976) presented evidence that male ctenuchid moths (*Lymire edwardsii*) were attracted to the fragrance of *Epidendrum anceps.*

In a number of orchid species the visual and olfactory cues are deceitful, as there is no food reward. In some instances food appears to be present in the form of pollen, as in *Calopogon* (Thien and Marcks, 1972), and it is probable that *Arethusa, Calopogon, Pogonia,* and *Calypso* all have pollination systems based on food deception (Mosquin, 1970; Thien and Marcks, 1972). Van der Pijl and Dodson (1966) and Nierenberg (1972) demonstrated that female bees of the genus *Centris* are deceived by flowers of some species of *Oncidium* that mimic flowers of members of the Malpighiaceae. Vogel (1974) reported on the secretion of oil drops by some members of the Malpighiaceae, and apparently some members of *Oncidium* are visual mimics (especially ultraviolet mimics) of these malpighiaceous flowers. The female *Centris* lands on the *Oncidium* and attempts to collect the oil droplets just as she would on the malpigh. Other members of the genus *Oncidium* appear to be involved in an aggressive mimicry situation in which males of several species of *Centris* attack the flowers, apparently as a type of territorial defense (Dodson, 1962a; van der Pijl and Dodson, 1966; Nierenberg, 1972). Other forms of deceit involve sex and lead to pseudocopulation as in *Ophrys* (Kullenberg and Bergstrom, 1976a,b) and *Trichoceros* (van der Pijl and Dodson, 1966). Stoutamire (1974b, 1975) discussed pseudocopulatory relationships in a number of Australian terrestrial orchids. Additional cases of deceit in the attraction of pollinators are discussed by van der Pijl and Dodson (1966).

Pollination Classes and Syndromes

Although information about the pollination of the vast majority of orchid species is lacking, most can be assigned to one or another of the general classes by virtue of the particular syndromes they possess. The most general class is certainly that involving the attraction of hymenoptera by visual and olfactory cues and a reward of nectar. Flowers in this class usually have the following features: they are zygomorphic, open during the day, generally have a prominent horizontal landing platform, are variously colored but seldom red, nectar guides are often present, odors are pleasant, and the nectar is found in tubular or hidden nectaries (this and the following syndromes are adapted from Vogel, 1954, and van der Pijl and Dodson, 1966). Some flies are also attracted to this class of flower, but others prefer flowers with odors that resemble dung, carrion, or other decaying substances.

Ackerman and Mesler (1979) recently reported on the pollination of *Listera cordata* by fungus gnats. Even though the flowers of *L. cordata* exhibit sapromyophily, Ackerman and Mesler showed that the fungus gnats visited the flowers solely for the nectar.

Various groups of lepidoptera visit and pollinate orchid flowers, some of which are visited by butterflies and others by moths. For the most part skippers have habits that allow them to be grouped with butterflies for pollination studies. The general syndrome of butterfly pollination includes flowers that are open during the day but may not necessarily be zygomorphic, a horizontal landing platform, red or yellow colors (usually, but exceptions occur), fresh and agreeable odors, long and narrow nectar tubes, and simple nectar guides, often consisting of ridges of tissue rather than colored lines. Moth-pollinated flowers, however, are weakly to strongly zygomorphic, usually open and fragrant at night, have landing platforms which are lacking or curved backwards, are whitish or greenish in color, possess sweet and strong odors, contain copious nectar in long tubes, the nectar guides are absent as lines but often present as morphological structures, and the flowers are often pendulous.

Bird pollination occurs to a certain extent, and the syndrome includes flowers which are red, yellow, or purple, open during the day, weakly to strongly zygomorphic, lack landing platforms (lip reduced or recurved), usually have no odor, produce abundant nectar in deep tubes, possess no nectar guide lines, and bear pendent flowers.

With the exception of autogamy, other forms of pollination are essentially lacking in Orchidaceae. Van der Pijl and Dodson (1966) list the pollination spectrum of the family as follows: hymenoptera 50% (not including the euglossine bees), moths 8%, butterflies 3%, birds 3%, flies 15%, mixed agents 8%, autogamous 3%, and euglossine bees 10%. Even though this is a preliminary estimate based on relatively little data, it is still apparent that the euglossine bees constitute an important proportion of the pollination vectors of the Orchidaceae.

Our knowledge of the distribution of pollination types among the subfamilies and tribes of the Orchidaceae is certainly incomplete at this time. Pollination of the Apostasioideae is completely unknown. The Cypripedioideae are pollinated mainly by bees and flies (Dodson, 1966a; Stoutamire, 1967; Nilsson, 1979), but the means of attraction and pollinators of the majority of the species are still unknown. Among the Neottioideae fungus gnat, syrphid, wasp, and bee pollination are known (Ackerman, 1975; Ackerman and Mesler, 1979; Ivri and Dafni, 1977; Nilsson, 1978), as well as various types of deceit based on pseudocopulation (Stoutamire, 1974b, 1975). Van der Pijl and Dodson (1966) list what is generally known of pollination in this group. In the Orchidioideae moth, butterfly, fly, bee, and wasp pollination all occur, as well as pseudocopulation (van der Pijl and Dodson, 1966; Vogel, 1954).

The Epidendroideae is the largest subfamily of the Orchidaceae, and has the widest variety of pollination types, as every pollination mechanism found in the family exists in this subfamily. Euglossine pollination is known in the subtribes Sobraliinae (i.e., *Sobralia,* but it is unclear if the bees visit solely for nectar or if some of them are also attracted by floral fragrances) and Laeliinae (e.g., *Cattleya,* in some cases for nectar, in others for floral fragrances; van der Pijl and Dodson, 1966). In the Vandoideae euglossine syndrome pollination is known among the Cyrtopodiinae (*Cyrtopodium,* for example), all members of the Catasetinae and Gongoreae (Stanhopeinae), and most members of the Lycastinae, Zygopetalinae, and a few genera of the Oncidiinae (*Aspasia, Notylia, Trichocentrum, Trichopilia, Macradenia*). For additional discussion of pollination among the various groups of the Orchidaceae, see van der Pijl and Dodson (1966), which is the source of most of the information above.

Morphology and Classification of the Euglossine Bees

The Euglossini are an exclusively neotropical group of bees. Moure (1967b) reviewed the taxonomy of the group as a whole and some of the subgroups (Moure, 1946, 1947, 1950, 1960a,b,c, 1963, 1964, 1965, 1967a,b, 1969a,b, 1970, 1976). Recent work by Dressler (1978a,b) and Kimsey (1977, 1979a,b) has clarified the nomenclature, taxonomy, and relationships of several genera and subgenera of Euglossini. They are treated as a tribe of the subfamily Bombinae, family Apidae, by Michener (1974). Wilson (1971) follows earlier classifications and treats the group as a subfamily (Euglossinae) of the Apidae. Both authors are quite similar in their treatments (as would be expected, since Wilson is following earlier versions of Michener's classification) and consider the family Apidae to include the euglossine bees, the bumblebees, the stingless bees (*Melipona, Trigona,* and their associated genera), as well as the true or stinging honeybees of the genus *Apis.* The Bombini, Meliponini, and Apini are all eusocial insects, whereas the Euglossini are solitary, communal, or quasisocial depending on the species (Michener, 1974; Dodson, 1966b; Roberts and Dodson, 1967; Zucchi *et al.,* 1969b).

There are three free-living genera (*Eulaema, Eufriesia,* and *Euglossa*) and two parasitic genera (*Exaerete* and *Aglae*). The largest genus is *Euglossa* with approximately 75 species, followed by *Eufriesia* with 60 or so species. *Eufriesia* is the widest ranging of the genera and extends from north-central Mexico to Argentina. Kimsey (1979a) recently reduced *Euplusia* to a synonym of *Eufriesia,* but because so many records exist under the name *Euplusia,* I will refer to the majority of *Eufriesia* as *Euplusia* in this chapter. *Eulaema* is probably the best known of the genera taxonomically with approximately 13 species (Dressler, 1979c). *Exaerete* has five species, and *Aglae* is monotypic (and also the most limited geographically, being found essentially only in the Amazonian area). Kimsey (1979b) has recently revised *Exaerete* and discussed its biology.

The largest of these bees are in the genus *Eulaema*. It consists of large, brown or black, variously striped, hairy bees. *Euplusia* (in the old sense) contains bees that are of more metallic appearance. Some species superficially resemble *Eulaema;* others appear more like very large euglossas. *Eufriesia* (in the restricted sense) is very similar to *Euplusia* and differs only in having a head that is much more metallic in appearance. *Euglossa* consists of small to medium-sized bees colored bright metallic blue, green, bronze, or mixed colored. *Exaerete* is brightly metallic green or blue-green, and *Aglae* is bright metallic blue. All species are very fast flyers and tend to be very wary.

Male euglossine bees are characterized by feathery brushes on the front tarsi and greatly inflated hind tibiae (Plate 4-1-**4**). The most useful taxonomic features have been associated with external morphology and coloration. These features as they pertain to several groups have been reviewed recently by Moure (1970) and Dressler (1978a,b, 1979c). The female bees often present more difficult taxonomic problems than the males of a given species. Therefore, most of what is known about the taxonomy and relationships of the group is based on male bees. The internal morphology of the euglossine bees is still poorly known. Cruz-Landim *et al.* (1965) studied the hind tibiae of several species. Sakagami (1965a) examined the internal morphology of the hind tibiae of *Eulaema nigrita*. Cruz-Landim (1967) studied the salivary gland system (including the mandibular glands) of 78 species of bees, including both males and females of the Euglossini. The female bees lack the front tarsal brushes, and hind tibiae are not inflated, as in the males, but adapted for transporting pollen. Additional morphological studies have been conducted by Cruz-Landim (1963) and Vogel (1963b).

Nests and nesting biology of the euglossine bees have been described by Bennett (1965, 1966, 1972a), Dodson (1966b), Myers and Loveless (1976), Roberts and Dodson (1967), Sakagami (1965a), Sakagami *et al.* (1967), Sakagami and Michener (1965), Sakagami and Sturm (1965), and Zucchi *et al.* (1969b). Michener (1974) and Zucchi *et al.* (1969a,b) have reviewed the nesting biology and social behavior of the euglossine bees. In general *Euplusia* and *Eufriesia* are nonsocial, but one sometimes finds nests aggregated in a particularly favorable habitat. Some species of *Euglossa* are solitary, others are somewhat colonial. *Eulaema* species tend to be more colonial, and Michener suggests that perhaps all species of *Eulaema* may have this tendency. Mating behavior has been seen very rarely in euglossine bees. Dodson (1966b) reported mating behavior in two species, and Kimsey (personal communication; Dept. of Entomology, University of California, Davis) has observed mating in two species. At the time of this writing, estimates of population size, longevity, dispersal, and flight ranges are being investigated, but nothing firm has been published (Ackerman, personal communication, Dept. of Biological Science, Florida State University; Kimsey, personal communication). Additional investigations have been reported on temperature while in flight (Inouye, 1975), species diversity (Ricklefs *et al.*, 1969), and ecology (Braga, 1976).

Plate 4-1. **1:** *Euplusia mexicana* visiting *Clowesia russelliana* in Chiapas, Mexico. **2:** *Euplusia* sp. visiting *Stanhopea oculata* in Chiapas, Mexico. **3:** Two species of *Euplusia* attracted to pure cineole in Chiapas, Mexico. Note the pollinarium on the back of one individual. **4:** A species of *Euglossa.* Note the large hind tibia and the "slit" through which the floral fragrances are absorbed. (All photos by N. H. Williams.)

Food-Plant Relationships of Euglossine Bees

Male and female euglossine bees forage widely for nectar. Females apparently have a large foraging area (Janzen, 1971). It has also been postulated that the nomadic males forage over large areas (Williams and Dodson, 1972). Males and females visit orchids such as *Cischweinfia dasyandra* in search of nectar and pollinate them (Williams, unpublished). Several species of *Sobralia, Cattleya, Maxillaria,* and possibly *Rodriguezia* are visited and pollinated by both male and female euglossine bees, but orchid flowers are not major food sources of these bees. Major food sources of the euglossine bees among the monocotyledons include members of the Costaceae and Marantaceae; several families of dicotyledons are also major sources of food.

Kennedy (1978) recently presented the most detailed account of field observation of euglossine bees visiting *Calathea* (Marantaceae). Additional observations and information are given by Kennedy (1973, 1977, 1978) for Marantaceae, Maas (1972) for Zingiberaceae (Costoideae), and Ayensu (1973) for Velloziaceae. Zucchi *et al.* (1969b) listed the known euglossine visitors in search of pollen and nectar for various species of Bromeliaceae, Liliaceae, Amaryllidaceae, Musaceae, Zingiberaceae, Cannaceae, Marantaceae, and Orchidaceae. Dodson (1966b) listed the food plants of a number of species of euglossine bees (Dodson's observations were summarized by Zucchi *et al.*, 1969b). Zucchi *et al.* (1969b) also summarized previous records of visits of male and female euglossine bees to dicotyledons. They listed 24 families to which males and females are attracted for food, either as nectar or pollen (for females), or which attract only females for pollen gathering.

Male and female euglossine bees visit a number of tubular-flowered species for nectar. The females often visit various species of dicotyledons with porous anthers to collect pollen. Michener (1962) and Wille (1963) have discussed the method used by females to collect pollen from anthers with porous openings ("buzz" flowers). Janzen (1968) suggested that several species of Passifloraceae are pollinated by nectar-feeding euglossine bees. Prance (1976) reported that euglossine bees visit several species of the Lecythidaceae, with males and females apparently collecting nectar and females collecting pollen. Mori and Kallunki (1976) cited a personal communication of Dressler (Dressler in Mori, 1970) that *Eulaema* visits *Eschweilera* and gathers an aromatic substance, but closer observations by Dressler revealed that no floral fragrances were collected (Dressler, personal communication, Smithsonian Tropical Research Institute). The family Gesneriaceae contains two distinct types of flowers visited and pollinated by euglossine bees. One type attracts only males and will be discussed later (the andro-euglossophilous species, as Wiehler, 1976, 1978, called them). The other group attracts both sexes, with males and females feeding on nectar; the latter also collect pollen (the gynandro-euglossophilous species in Wiehler's terminology). The gynandro-euglossophilous species appear to be typical bee-pollinated flowers; however, for some reason they do not seem to attract bees in general, but rather attract primarily members of the Euglossini. The euglossine bees that are associated with this phenomenon are rather specific in their choice of flowers, and their specificity seems to be important in maintaining the integrity of numerous species of the neotropical Gesneriaceae (see Wiehler, 1978, for a more complete discussion).

Janzen (1971) suggested that female euglossine bees "trapline," possibly over distances of 20 km or more, as they search for nectar and pollen, and that this type of foraging behavior is an important factor in the long-distance movement of pollen in the tropics. We have suggested that the action of males may also be important in long-distance pollen flow (see below and Williams and Dodson, 1972). The fact that some of the males seem to be somewhat nomadic indicates that they might also be important as long distance pollinators of their food plants (Dodson, 1966b; Williams and Dodson, 1972).

Euglossine Pollination of Orchids

The relationship between male euglossine bees and orchid flowers was first reported by Cruger (1865), whose work was described in great detail by Darwin (1877). Unfortunately, although Cruger had the opportunity to observe euglossine behavior on orchids in their natural state, he apparently was not a keen observer and his interpretation of the activity of the male bees is at variance with all later observations. Cruger reported that the bees gnawed the flowers. Even though Darwin spent some time in South America, he apparently never saw euglossine bees visiting orchid flowers. Therefore, he only repeated what Cruger had published. This is unfortunate, because Darwin was a very careful observer, and if the actions of the male bees had been carefully noted, the biological relationships between orchids and euglossine bees might have been understood earlier.

Allen (1950, 1952) reported on the visits of male euglossine bees to *Coryanthes* and *Cycnoches*, but thought the bees were gnawing on the lips, as Cruger had reported. Later, Allen (1954) described the actions of male euglossine bees on flowers of *Gongora* and did not suggest that they were gnawing on the tissues, but unfortunately he did not note that they were actually doing something far different. The first accurate reports of what the male euglossine bees were doing on the flowers were presented by Dodson and Frymire (1961a), who described the pollination of several species of *Stanhopea*. Later they also described the pollination of several additional genera (Dodson and Frymire, 1961b). This was followed by the work of Vogel (1963b, 1966a,b, 1967), numerous papers by Dodson (1962a,b, 1963, 1965a,b, 1966a,b,c, 1967a,b, and others mentioned below), and Dressler (1966, 1967, 1968a,b, 1978a,b, 1979c).

The relationship of male euglossine bees to orchid flowers has been called "euglossine pollination" (Dressler, 1968a) and the "euglossine syndrome" (Williams and Dodson, 1972). These terms are synonymous, and I will use them to describe the special relationship between male euglossine bees and the flowers of certain species of orchids (and other plants, see below). The orchid flowers that are pollinated exclusively by male euglossine bees are very fragrant, lack nectar, and attract no other groups of bees or insects. It must be emphasized that no food is present in the flowers for the male bees, and that the female bees are not attracted to those flowers that have the "euglossine syndrome." Dodson and Frymire (1961a,b) suggested that the male bees are attracted to the flowers solely by their fragrance, and this has been verified (Dodson *et al.*, 1969; Hills *et al.*, 1972; Williams and Dodson, 1972; Dodson, 1970, 1975a; also others cited below).

Pollination of the flowers is dependent on the actions of the bees as they collect the floral fragrance compounds. It has been clearly demonstrated that the floral fragrance components of orchid flowers are the attractants for the male euglossine bees and evoke a definite routine (Dodson, 1970, 1975a; Dodson *et al.*, 1969; Hills *et al.*, 1968, 1972; Williams and Dodson, 1972). The bees (1) are attracted to the fragrance, (2) land on the flower, (3) move to the area of maximum odor production (usually the base of the labellum), (4) brush the surface of the labellum with the tarsal brushes on their front legs, (5) launch into the air, (6) transfer the fragrances they have collected to the inflated hind tibiae, and (7) return to the flower (Dodson and Frymire, 1961a,b; Dodson *et al.*, 1969; Evoy and Jones, 1971; Michener *et al.*, 1978). Some of the behavioral movements of the bees as they transfer the collected materials to their hind legs are still not completely known, but work is currently in progress on their motor activities in my laboratory. It is during the brushing (or scratching, as it has sometimes been called), or during the preparation for the launch, or during the actual launching into the air to transfer the floral fragrance components that the pollinarium is removed from the flower and deposited on the bee. These mechanisms vary from the simple system of backing out of the flower in some species of *Trichocentrum* to the more elaborate ones in the Catasetinae and Gongoreae.

Fragrances in some species of *Cattleya* and *Meiracyllium* indicate that they might be pollinated by male euglossine bees, although we do know that some species of *Cattleya* are pollinated by both sexes searching for nectar (van der Pijl and Dodson, 1966). With these two possible exceptions (and see *Sobralia,* above), all other species of euglossine-pollinated orchids are found in the vandoid tribes and subtribes (the classification of Dressler, 1974). The Dichaeinae, Catasetinae, and Gongoreae are exclusively euglossine-pollinated, and pollination by these bees is almost the rule in Zygopetalinae. Euglossine pollination is found in at least seven genera of Oncidieae and two genera of Cyrtopodiinae. The euglossine-pollinated orchids will be discussed in taxonomic order below (following the taxonomic treatment of Dressler, 1974), but are listed in alphabetical order in Table 4.1 at the end of this chapter.

Zygopetalinae

The Zygopetalinae can be divided into at least three generic alliances: the *Lycaste* alliance, the *Zygopetalum* alliance, and the *Chondrorhyncha* alliance. Within the *Lycaste* alliance, *Anguloa* is completely euglossine-pollinated, and the same is true for most species of *Lycaste. Neomoorea* is probably euglossine-pollinated. Dodson (1966c) has described the pollination of *Anguloa clowesii* by *Eulaema boliviensis.* The flowers of *Anguloa* are somewhat tulip-shaped, with the sepals and lateral petals forming the cup. The column and lip are articulated at the bottom of the cup, and the major area of odor production is the callus at the base of the lip. An insect entering the flower headfirst cannot reach the odor-producing area; it must back into the flower. The bee holds to the edges of the sepals and petals with the midlegs and holds the lip with the hind legs, and brushes with the front legs on the surface of the callus. Its hold on the sepals and petals prevents the lip from rocking. As the bee prepares to leave the flower to hover and transfer, he releases his hold on the flower with his midlegs, grabs the lip with all six legs, and puts all of his weight on the articulated lip, which then rocks backward toward the column. The viscidium is then slipped between the abdomen and thorax and becomes attached as the bee tries to escape. If a bee with a pollinarium attached were to visit a flower, after being thrown against the column, the pollinia would be in the correct position to be deposited in the stigma. Dodson (1966c) suggests that all species of *Anguloa* are pollinated in the same way, and *Eulaema cingulata* has been reported as the pollinator of *A. ruckeri.* We have examined the floral fragrances of three species of *Anguloa,* and have found that they contain alpha- and beta-pinene, cineole, and carvone in their fragrances (Williams *et al.,* 1981). All four compounds have been found previously in orchid floral fragrances (Hills *et al.,* 1968, 1972; Dodson *et al.,* 1969; Williams and Dodson, 1972).

Van der Pijl and Dodson (1966) reported euglossine pollination in three species of *Lycaste,* and the floral fragrances suggest that more species are pollinated by these bees. Methyl cinnamate, a known euglossine attractant (Dodson *et al.,* 1969), is present in the fragrance of *L. cruenta* (Williams *et al.*, 1981). In *Lycaste* the simplest type of pollination mechanism is found. The male bee enters the flower on the lip, whose base it brushes, and as he backs out the visicidium is placed between his thorax and abdomen. If a bee has a pollinarium on its back, the pollinia would be inserted into the stigma as the bee backs out of the flower. A single bee can therefore pollinate one flower and remove the pollinia for a subsequent pollination of another flower all in one visit.

The *Chondrorhyncha* alliance exhibits a simple system of attracting the male bee to the source of fragrance at the base of the lip. *Eulaema polychroma* lands on the lip of *Pescatoria wallisii* and works his way to the base of the lip. Because of the close position of lip and column, the pollinarium is placed on the end of the scutellum as the bee backs out to hover (Dodson and Frymire, 1961b). In *Cochleanthes aromatica* the system is also simple, but the column arches over the head of the bee as he is in the brushing position. As he leaves the lip, the pollinarium is deposited on the top of his head (van der Pijl and Dodson, 1966). A similar pollination mechanism exists in *Huntleya meleagris* (van der Pijl and Dodson, 1966). In *Kefersteinia* the bee lands on the lip and makes his way to its base. The viscidium and the base of the stipe are wrapped around the base of the antenna. It is not clear just how the pollinarium becomes attached to the base of the antenna, but that it does is obvious from the large number of male bees found carrying the pollinaria of *Kefersteinia* (Dressler, 1968b). This differential placement of the pollinaria is probably an efficient method of mechanical isolation among sympatric species of this group.

One species each of *Zygopetalum* and *Zygosepalum* (*Menadenium*) was reported to be euglossine-pollinated (Dodson, 1965a; van der Pijl and Dodson, 1966). A simple mechanism is probable in these two genera, in which the pollinarium is simply placed on the head of the visiting bee. Dodson and Frymire (1961a) found a pollinarium of a species allied to *Zygopetalum* on the abdomen of a specimen of *Euglossa hemichlora*, and they suggested the species in question might be either a *Chondrorhyncha* or a *Pescatoria*.

Dichaeinae

The Dichaeinae are, as far as is known, pollinated exclusively by male euglossine bees. Dressler (1968b) reported finding pollinaria of *Dichaea panamensis* on the face of *Euglossa cordata*. No species of *Dichaea* is known to produce nectar, nor do any have the syndromes for any other pollination system. This group is thought by Dressler (1974) to be related to the *Kefersteinia* complex of the Zygopetalinae, a group known to be euglossine-pollinated (see discussion above).

Cyrtopodiinae

Dodson and Frymire (1961b) found pollinaria of a *Cyrtopodium* species (possibly *C. punctatum*) on the metathorax of *Euglossa hemichlora* in Ecuador, but the same species is visited by a *Centris* in Florida. It is possible that *Cyrtopodium* may not really be a euglossine-pollinated orchid in the strict sense, but no nectar has been found in the flowers. The scents of *C. andersonii* and *C. punctatum* flowers have not been studied by gas chromatography yet, but they do not seem to have a strong euglossine-syndrome fragrance. Dressler (personal communication) suggests that *Galeandra* is euglossine-pollinated.

Catasetinae

All species of Catasetinae are euglossine-pollinated (Dodson and Frymire, 1961b; Dodson, 1962a,b, 1975c; Hills *et al.,* 1972; Dressler, 1968a,b). Dodson (1975c) divided the genus *Catasetum* into more natural groups by segregating from it the small genera *Clowesia* and *Dressleria.* Allen (1952) misinterpreted what was actually happening in the pollination of *Cycnoches,* but a correct interpretation was given by Dodson (1962b). The current taxonomic status of the genera is now stable, but the species of some of the groups (*Mormodes, Cycnoches*) are still poorly defined and their taxonomy is in a state of change. The evidence suggests that this group for the most part is still in a state of active evolution and the species may or may not be completely reproductively isolated from one another (Dodson, 1978b). An attempt to group these genera from primitive to advanced would arrange them in the sequence *Dressleria, Clowesia* (both genera are bisexual), *Mormodes, Catasetum,* and *Cycnoches* (or vice versa, since it would be difficult to decide if either *Catasetum* or *Cycnoches* is more advanced than the other).

Dodson (1975c) showed that not only are the species of *Dressleria* structurally distinct, but each species is pollinated by a different group of euglossine bees; they are clearly reproductively isolated on the basis of the pollinators they attract. There are 10–20 flowers on the raceme of *Dressleria eburnea* (*C. suave*); the males of *Eulaema cingulata* land on any open surface of the flower and attempt to make their way to the source of the fragrance, which is produced in a cavity formed by the lip. As the male bee tries to make his way into the cavity, he brushes against the anther cap, and the pollinarium (which is under tension) is released and flung so that the viscidium strikes the bee either on the trochanters of his legs or on the front of the forelegs (Dodson, 1962b). Dodson mentioned that the stipe is curled as it is first released, and it is not possible for the bee to immediately pollinate the same flower. Furthermore, the viscidium covers the stigma, and pollination cannot occur until the pollinarium has been removed. If the bee enters a flower from which the pollinarium has been removed, the pollinia are forced into the stigma, and pulled from the bee as he exits. Dodson (1975c) reported that although some of the species of *Dressleria* may be sympatric, each species has a rather specific pollinator, or group of pollinators, so that interspecific pollination is prevented, or reduced. The small amount of information that exists on the floral fragrances of this genus indicates that each species sampled has a very distinctive complement of fragrance compounds (Hills *et al.,* 1972).

Clowesia consists of five species with small to medium-sized bisexual flowers (Dodson, 1975c). Pollinators are known for two species (*C. warczewitzii* and *C. russelliana;* Dodson, 1975c), and the floral fragrances of four of the five species have been analyzed (Hills *et al.,* 1972). Dodson (1962b) described the pollination of *C. russelliana* by captive *E. cingulata* bees (not known to be a natural pollinator). Dodson and Williams (Dodson, 1975c) observed natural pollination of *C. russelliana* in Chiapas, Mexico, by *Euplusia mexicana* (Plate 4-1-**1**). The bee landed on the lip of the flower, made his way into the saccate portion of the base of the lip where the fragrance is produced, brushed on the lip, and backed out of the flower. As the bee entered the flower, he touched the end of the viscidium, which was released and the pollinarium was attached to his metathorax as he left the flower. It takes some time for the anther cap to dry and fall off the pollinarium, and if the bee then subsequently enters another flower, the pollinia (one or both) are left behind in the stigmatic cavity as the bee exits.

Dressler (1968a,b) noted the first incidence of "bees-knees pollination" in *C. warczewitzii* by *Eulaema nigrita* and *E. bombiformis.* After landing on the flower and orienting himself on the lip (usually at an angle perpendicular to the long axis of the lip), the bee thrusts his front legs into the cavities of the lip. As he brushes at the source of odor production, the bee touches the viscidium with his "knee" (the femur) and dislodges the pollinarium. The viscidium becomes attached to the leg just above the joint between the femur and the tibia. Dodson (1975c) suggested that this curious type of pollination is not known in other orchids, but is possible that some kind of modified "elbow pollination" may also occur in *Sievekingia* (in the Gongoreae) (Dressler, 1968a; van der Pijl and Dodson, 1966). The most important floral fragrances identified to date in *Clowesia* include alpha-pinene, cineole, 2-phenyl ethyl acetate, and methyl cinnamate. Methyl cinnamate is obviously an important component of the attractant in *C. roseum,* although the pollinator has not yet been found. Cineole comprises approximately 78% of the fragrance of *C. russelliana,* and its pollinator (*Euplusia mexicana*) is attracted to the pure chemical (Plate 4-1-**3**; Hills *et al.*, 1972). Although we do not yet know the pollinator of *C. thylaciochila,* the presence of 2-phenyl ethyl acetate in its fragrance (over 90% of the total) and its geographic range suggest that *Euplusia concava* may be its pollinator, since it is strongly attracted to this compound (Hills *et al.,* 1972).

In contrast to *Dressleria* and *Clowesia,* which have perfect flowers, *Mormodes* has flowers that are functionally male until the pollinarium has been removed (Plate 4-2-**E**). Dressler (1968b) indicated that there are two distinctive types of flowers, at least in the Panamanian species he studied. One type has functional, short-lived male flowers with a narrow column, which are more numerous than the functionally female flowers. The second type has female flowers that (1) are usually produced by larger plants than the male flowers (see Dodson, 1962b, and Gregg, 1975, 1978, for similar observations in *Catasetum* and *Cycnoches*), (2) have a much wider column, which results in a much larger stigma, and which (3) straightens out into the female position even if the flower is not visited and the pollinia are not disturbed by an insect visitor (with the result that the pollinia are discharged and lost).

The column in *Mormodes* is usually curved over the lip and terminates in a long slender appendage which touches or almost touches the lip. As he begins brushing the lip, the bee often slightly touches the apex of the column, thereby causing the pollinarium to be discharged and attached to his back. After the pollinarium has been removed (either by the action of the bee in the male flowers or spontaneously in the functionally female flowers), the column straightens and the stigmatic surface is positioned more or less parallel to the lip surface. The pollinarium is held erect on the thorax of the bee, and if he should happen to walk on the lip of a flower in the female condition, the pollinia (one or both) will be deposited in the stigma. It takes about 30 minutes for the stipe of the pollinarium to straighten from its coiled condition to the erect position necessary for pollination (Dodson, 1962b).

Dodson (1962b) described pollination of one species of *Mormodes* in Ecuador, and Dressler (1968b) discussed the bees which are attracted to and pollinate seven species of *Mormodes* in Panama and Costa Rica. For the most part, each species of *Mormodes* for

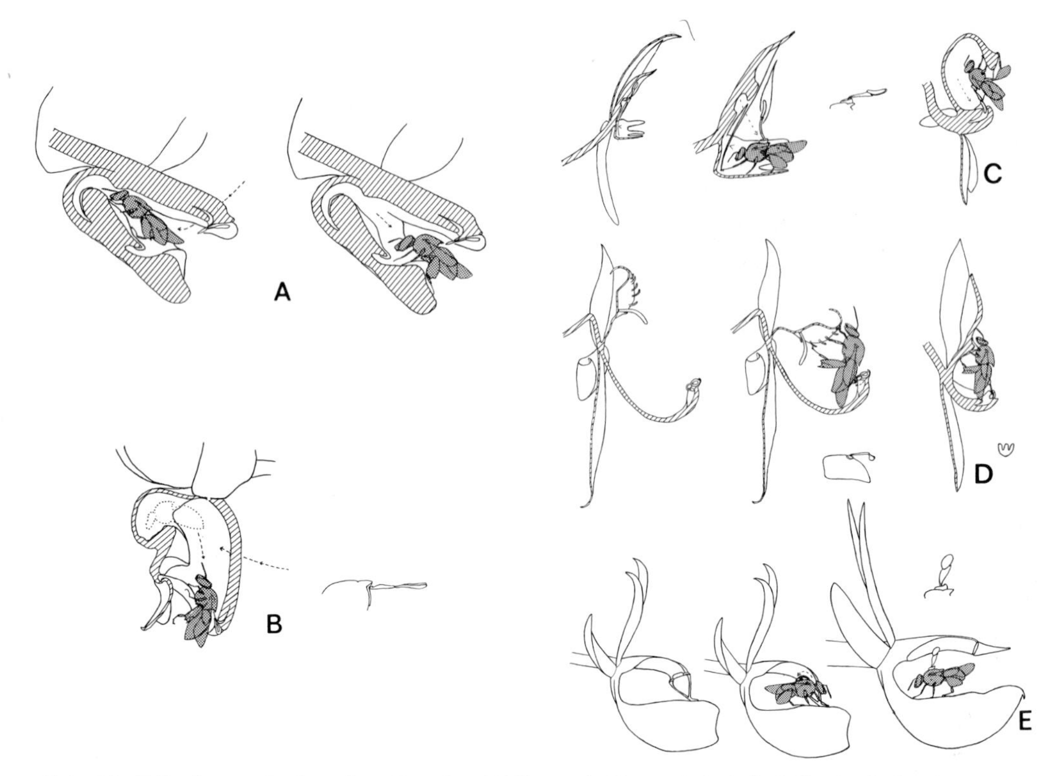

Plate 4-2. Pollination mechanisms (from Dressler, 1968b; used with permission). **A:** Pollination in *Stanhopea ecornuta.* Left, longitudinal section of flower with the bee brushing near base of lip (having entered from side); right, the bee backing out beneath the apex of the column. The placement of the pollinarium is probably as in **B,** though the stipe is much shorter. **B:** Pollination in *Stanhopea oculata* and allies. Left, the bee enters from the side (arrows), brushes in the hypochile (dotted outline), and, on withdrawal, may slip and fall through the flower, the pollinarium then being placed beneath the scutellum; right, placement of *S. costaricensis* pollinarium on *Euplusia schmidtiana,* the viscidium (thickness exaggerated) is largely placed on the propodeum. **C:** Pollination in *Catasetum bicolor.* Left, longitudinal section of male flower with pollinarium in place. Center, same showing *Euglossa cordata* touching the antenna and receiving the pollinarium on the thorax. Right, longitudinal section of female flower with bee backing out of flower and inserting a pollinium in the stigma. **D:** Pollination in *Cycnoches aureum.* Left, longitudinal section of male flower with pollinia in place. Center, the weight of *Eulaema nigrita* pulls the lip down and the abdomen of the bee strikes the apex of the column, causing the pollinarium to be expelled and placed near the apex of the abdomen. Right, longitudinal section of female flower with the bee inserting a pollinium in the stigma. Lower right, apex of the female column, showing the slits which catch the pollinia. **E:** Pollination in *Mormodes igneum.* Left, staminoid flower of the truncate extreme, with the pollinarium in place. Center, with *Euglossa igniventris* touching the apex of the column and causing the pollinarium to be expelled. The viscidium usually strikes the thorax of the bee. Right, pistilloid flower of the elliptic extreme, with a bee passing beneath the stigma and effecting pollination.

which pollination information is available attracts a specific pollinator, although a slight sharing of pollinators appears to exist (Dressler, 1968b). This sharing may be responsible for the apparent hybridization among several species of *Mormodes* and the resulting taxonomic confusion (Dodson, 1962b). One species of *Mormodes* analyzed by gas chromatography had alpha- and beta-pinene and cineole in its fragrance (Hills *et al.,* 1968), but comparative data on other species of *Mormodes* are lacking.

The pollination biology of *Cycnoches* species which produce large male flowers is fairly well known, but the species with small male flowers present a problem in that the taxonomy of the group is poorly known and the female flowers are particularly difficult to identify. It seems apparent that a number of sibling species exist in this group. Allen (1952) reported the pollination of *C. warszewiczii* (= *ventricosum*) by *Eulaema cingulata* (the bee was reported as a synonym, *Centris fasciata*). Dodson and Frymire (1961b) reported that *E. cingulata* pollinated *C. lehmannii,* and Dressler (1968b)discussed the pollination of six species of *Cycnoches* from Panama and Costa Rica. Dressler correctly pointed out that Allen (1952) was overenthusiastic in lumping species of *Cycnoches,* and additional research will probably show that more species exist in the *C. egertonianum* complex than previously assumed (Dressler, 1968b). Allen (1952) divided the genus into two sections, *Eucycnoches* (nomenclaturally this should be section *Cycnoches*) with male and female flowers of almost equal size and appearance, and section *Heteranthae* with strikingly dissimilar male and female flowers. Allen (1952) and Dressler (1968b) mentioned that the female flowers of the dimorphic species are nearly indistinguishable, and suggested that this has hampered studies of both taxonomy and pollination biology. Although floral fragrance analyses are available for only a few species of *Cycnoches* (Hills *et al.,* 1968), it appears that each species has a species-specific quantitative and qualitative blend of floral scent compounds. Both male and female flowers of a species produce the same floral fragrances (see discussion below). Alpha- and beta-pinene and benzyl acetate have been identified from the floral fragrances of four species of *Cycnoches* (Hills *et al.,* 1968).

Dodson and Frymire (1961b) and van der Pijl and Dodson (1966) have described the pollination of species in both sections of the genus *Cycnoches. C. lehmannii* (section *Cycnoches*) is visited and pollinated by *Eulaema cingulata* in western Ecuador. The male bee approaches the flower and lands grasping the large callus of the lip. In the male flowers the callus is slightly larger and more extended than in the female flowers. The male bee approaches the flower, lands on the lip, and gradually maneuvers himself into the proper position to collect the fragrances which are produced by the callus tissue. On the male flower, as the bee attempts to get into position, he must release the grasp of his hind legs in an attempt to get closer to the source of the odor. As he does so his abdomen swings down, and its tip brushes against the tip of the column and the trigger mechanism of the anther, which is under tension, discharges the viscidium violently and it sticks to the end of the abdomen as the stipe is released from the anther. The stipe dries and straightens in approximately 40 minutes. After a few hours the anther cap dries and falls from the pollinia. The female flower is generally similar to the male, but the anther is vestigial, the column is shorter and stouter, and the callus is slightly smaller and not so extended. The male bee approaches the flower, lands and orients himself on the lip, and begins brushing at the source of the odor. In order to transfer the collected floral fragrance, he must launch into the air, and as he releases his grip on the callus and base of the lip, he falls a short distance and his abdomen swings upward. If a pollinarium is attached to the abdomen, the pollinium (usually only one) is caught in the stigmatic cleft, and pollination is effected as the pollinium is pulled from the stipe. Van der Pijl and Dodson (1966) reported the pollination of three species in this section of the genus.

The section *Heteranthae* contains species of *Cycnoches* with markedly dissimilar male and female flowers (Plate 4-2-**D**). The female flowers are fairly large and quite similar to those of the section *Cycnoches.* The male flowers are quite small, with a lip that is thin, flexible, usually with numerous fleshy projections, and borne in large numbers on pendent inflorescences. Unlike members of the section *Cycnoches,* which are pollinated by bees of the genus *Eulaema,* species in the section *Heteranthae* are pollinated by smaller bees of the genus *Euglossa* as well as by *Eulaema* (in one species). Male bees land on the blade of the lip and crawl to the underside toward the source of the odor. The lip is quite flexible, and an insect crawling to the underside causes the tip of his abdomen to come into contact with the apex of the column. This brings about the discharge of the pollinarium; the viscidium is attached to the tip of the abdomen. As the stipe dries and straightens, the anther cap falls from the pollinia in approximately one hour. Transfer of the pollinia to the stigma occurs as described for the other section of the genus. Although Allen (1952) reduced a number of taxa to synonymy with *C. egertonianum,* Dressler (1968b) and Dressler and Dodson (cited in van der Pijl and Dodson, 1966) point out that a number of distinct morphological species have been lumped into *C. egertonianum,* together with a number of sibling species which are isolated from one another by their floral fragrances. They found four forms of *C. egertonianum* that were species-specific in the pollinators they attracted, and that failed to attract the pollinators of the other forms even when placed together in the same area. Gregg (1979) presented evidence that *C. dianae* has an odor profile distinct from both *C. densiflorum* and *C. stenodactylon* and suggested that it is reproductively isolated from these species on the basis of its distinct floral fragrance. *C. densiflorum* and *C. stenodactylon* are easily separated on the basis of morphological structures. She also provided evidence that *C. dianae* is polymorphic in its production of floral fragrances, and that this polymorphism may be leading to sympatric speciation in the populations she studied. It should be emphasized that male and female flowers of a particular species (morphological or sibling) produce the same floral fragrances. This has been demonstrated by Hills (1968), Hills *et al.* (1972), and Dodson, Gregg, Hills, and Williams (unpublished). Dodson (1962b), Gregg (1975, 1978), and van der Pijl and Dodson (1966) have discussed the ecology and physiology of sex expression in *Cycnoches* and *Catasetum,* at least in part in relation to pollination biology.

With the removal of *Dressleria* and *Clowesia,* the genus *Catasetum* becomes a much more homogeneous group (Dodson, 1975c, 1978a,b). Hills *et al.* (1972, following Mansfeld, 1932a,b) divided the true catasetums into two sections (*Pseudocatasetum* and *Catasetum*) and the section *Catasetum* into two alliances (based on the taxonomic work of others cited in Hills *et al.,* 1972, and Hills, 1968). The section *Catasetum* contains the most diverse types of flowers found in the Catasetinae, at least in the male flowers. The female flowers are rather uniform, although Hills *et al.* (1972) indicated that it is possible (but difficult) to distinguish the female flowers of the various species of the true catasetums. The pollination of various species of *Catasetum* has been mentioned or described by Cruger (1865), Hoehne (1933), Allen (1952), Dodson (1965a, 1978b), Dodson and Frymire (1961b), Dressler (1968b), Porsch (1955), and van der Pijl and Dodson (1966). The floral fragrances of several species have been studied (Hills, 1968; Hills *et al.,* 1968, 1972). Alpha- and beta-pinene, cineole, linalool, methyl benzoate, benzyl acetate, carvone, methyl salicylate, methyl cinnamate, myrcene, ocimene, nerol, 2-phenyl ethyl acetate, and 2-phenyl ethyl alcohol have been identified in the floral fragrances of various species of *Catasetum* (Hills *et al.*, 1968, 1972). Dodson (1978b) reports piperitone in the fragrance of one species of *Catasetum* from Guyana.

Male flowers produced by different species are quite distinct morphologically. Male flowers are usually found on small plants, often growing in shade or in nutritionally poor habitats (Dodson, 1962b, 1978b; Gregg, 1975, 1978). Two or three days after anthesis the flowers begin producing odor (Hills, 1968; van der Pijl and Dodson, 1966), and attract male bees which land on them, usually on the lip. The column of the male flowers of the section *Catasetum* (which contains all but two species of the genus) contains one or two long, equal or unequal antennae which are very sensitive to movement. After landing on the flower the male bee begins brushing or scratching at the source of odor production, which is at the base of the lip near the ends of the antennae. As the bee moves slightly while scratching, he comes into contact with the antennae, against which he need only brush slightly with his thorax or head. The anther is held in place under great tension, which is released if the antennae are touched even slightly. This release discharges the anther violently. The viscidium is forcefully attached to the thorax of the bee by an efficient glue which sets rapidly and cements the pollinarium to the back of the bee. As the stipe dries the pollinia are elevated somewhat above the back of the bee. In approximately 20 minutes the anther cap dries and falls from the pollina, which are now in the correct position for pollination of the female flower. Fragrance production by the male flowers from which the pollinaria have been removed usually ceases in less than an hour, and is not resumed on following days (Dodson, in van der Pijl and Dodson, 1966; Hills, 1968).

Female flowers of species in the section *Catasetum* are remarkably uniform, and careful observations are necessary to distinguish between those of various species (Hills *et al.*, 1972). These flowers start to produce fragrance later than the male flowers, usually three to four days after anthesis. Female flowers are remarkably long-lived, and often live up to a month if not pollinated. Dodson (van der Pijl and Dodson, 1966) reports that once pollination has occurred, however, production of floral fragrance ceases in a matter of hours. Bees are attracted to the female flowers and land on the large hooded lip. If a bee with a pollinarium on his thorax lands on the female flower, and enters the hooded lip upside down, he carries the pollinarium into the lip as he goes to brush at the source of odor production (which is well inside the hooded lip—the bee will often disappear into the lip). As the bee exits the flower to transfer the collected floral fragrance compounds to his hind tibiae, the pollinia are slipped into the stigmatic cleft, the pollinium is pulled from the stipe as the caudicles stretch and eventually break, and pollination is accomplished. Usually only one pollinium of a pollinarium is left in the stigma (Plate 4-2-**C**).

Male and female flowers of the same species, although remarkably dimorphic (Darwin, 1877), produce the same floral fragrances, and the male bees find both by the scent. Hills *et al.* (1972) discussed the importance of floral fragrances in this genus as isolating mechanisms, along with other important factors such as geographical and mechanical isolating mechanisms (Fig. 4-1). It seems quite probable that subtle changes in floral fragrances might be important in sympatric speciation of this genus, as has been suggested in *Cycnoches* (Gregg, 1979) and *Gongora* (Dodson *et al.*, 1969).

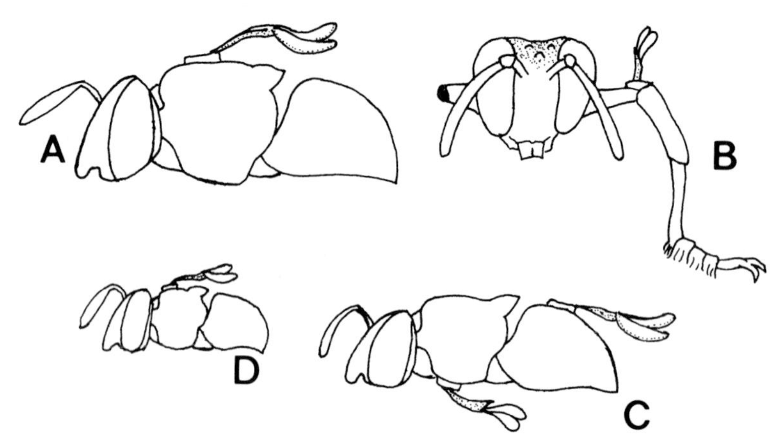

Fig. 4-1. Various places on a bee to which *Catasetum* and *Clowesia* pollinaria are attached (redrawn from Dressler, 1968a, with permission). **A:** Pollinarium of *Catasetum macrocarpum* on *Eulaema meriana.* **B:** Pollinarium of *Clowesia warczewitzii* on *E. meriana.* This is the "elbow" system of pollination. **C:** Pollinaria of *Catasetum saccatum* (dorsal) and *Catasetum discolor* (ventral) on *Eulaema cingulata.* **D:** Pollinarium of *Catasetum barbatum* on *Euglossa cordata.* These sympatric species of the Catasetinae are mainly isolated by the mechanical placement of the pollinaria at different locations on the pollinators. Ethological isolation is also an important aspect of the population biology of these species.

The two species of section *Pseudocatasetum* lack well-developed antennae on the column of the male flowers; instead they have small flaps of tissue (Dodson, 1978b). Floral fragrances of these species (*C. discolor* and *Ç. longifolium*) are very similar (Hills *et al.,* 1972), and Hills (1968) had predicted that they would have the same pollinators. Williams and Dodson (1972) confirmed this prediction. Dodson (1978b) discussed the pollination of the two species in Guyana and the hybridization which occurred in some of the disturbed habitats near Dawa. The two species appear to be ecologically isolated from each other in their natural habitats, but may hybridize when the habitat is disturbed.

Gongoreae

As with Catasetinae, all genera and species of Gongoreae (sometimes referred to as the subtribe Stanhopeinae) are pollinated exclusively by male euglossine bees (Plate 4-1-**2**, Fig. 4-2, Plate 4-2-**A,B**, Plate 4-3). Flowers of the approximately 18 genera of this group are usually very fragrant, and their pollination mechanisms vary from rather simple to the bizarre. The pollination biology of this group has been described by Dodson (1962b, 1963, 1965a,b, 1967b, 1970, 1975a,b), Dodson and Frymire (1961a,b), Dressler (1966, 1968a,b), Allen (1954), van der Pijl and Dodson (1966 and references therein), Vogel (1963b), and Schmid (1969). At this time the taxonomic status of the genera is fairly stable, but the species and their relationships are for the most part not well understood. Recent efforts at improving the taxonomy of various groups in this tribe have been made by Dressler (1977a), Dodson (1975a,b), Dodson and Frymire (1961a), and Garay (1970), among others, but the tribe is in need of a great deal of taxonomic work. The flowers are often large, fleshy, and intricate. Such characteristics do not lead to easy taxonomic

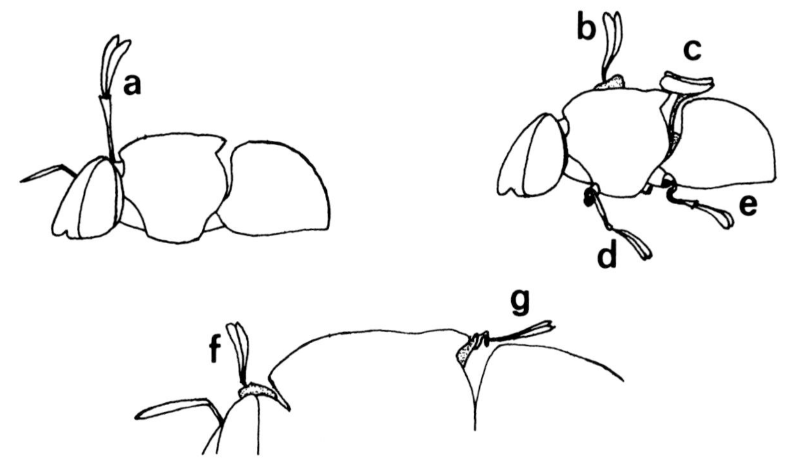

Fig. 4-2. Pollinaria placement by several members of the Gongoreae (redrawn from Dressler, 1968a, with permission). *Lacaena* (a), *Peristeria* (b), *Coryanthes* (c), *Cirrhaea* (d), *Sievekingia* (e), *Peristeria elata* (f), *Acineta* (g).

work, since many critical features are altered or destroyed on pressing. The phylogenetic relationships of the various genera of the tribe are not clear. Dodson and Frymire (1961a) have offered a preliminary, tentative suggestion of the generic relationships within the tribe, but they point out that much more work needs to be done.

Because published phylogenetic schemes for this tribe are limited, I will treat the pollination biology alphabetically. Most of the following discussion is abridged from Dressler (1966, 1968a,b), Dodson (1962a, 1963, 1965a,b, 1967b, 1975a,b), van der Pijl and Dodson (1966), Dodson and Frymire (1961a,b), Dodson and Hills (1966), and Hills *et al.* (1968).

So far as is known, the genus *Acineta* is pollinated exclusively by bees of the genus *Euplusia* (van der Pijl and Dodson, 1966; Dressler, 1968b). All available evidence suggests that the pollinators are *E. concava* or *E. venezolana* (R. L. Dressler, personal communication). As Dressler (1968b) has stated, the pollination of this genus is fairly simple. The bee enters the campanulate flower to scratch at the base of the lip. The only way to exit is to back out (Plate 4-3-**A**), and in so doing the bee touches the rostellum, which places the pollinarium on his back (on the scutellum). If the bee enters and leaves another flower, the pollinia are placed on the stigma as he leaves. *Euplusia concava* is attracted to cineole (Williams, Dressler, and Dodson, unpublished), and *Acineta chrysantha* produces this compound as part of its fragrance (Williams, Whitten, and Dodson, in preparation). Dodson (1965a) has observed the pollination of *A. chrysantha* by *E. concava,* as has Dressler (1968b). Dressler (1968b) suggests that the isolating mechanism between *A. superba* and *A. chrysantha* is solely geographic.

Hoehne (1933) and van der Pijl and Dodson (1966) have reported on the pollination of *Cirrhaea.* The flowers are erect on a pendent inflorescence, and the male bee lands on the lip of the flower. As he touches the rostellum, the viscidium is attached to the leg. Hoehne (1933) reported *Euplusia violacea* as a pollinator of the genus, and van der Pijl and Dodson (1966) found six specimens of *Euglossa mandibularis* with a total of 17

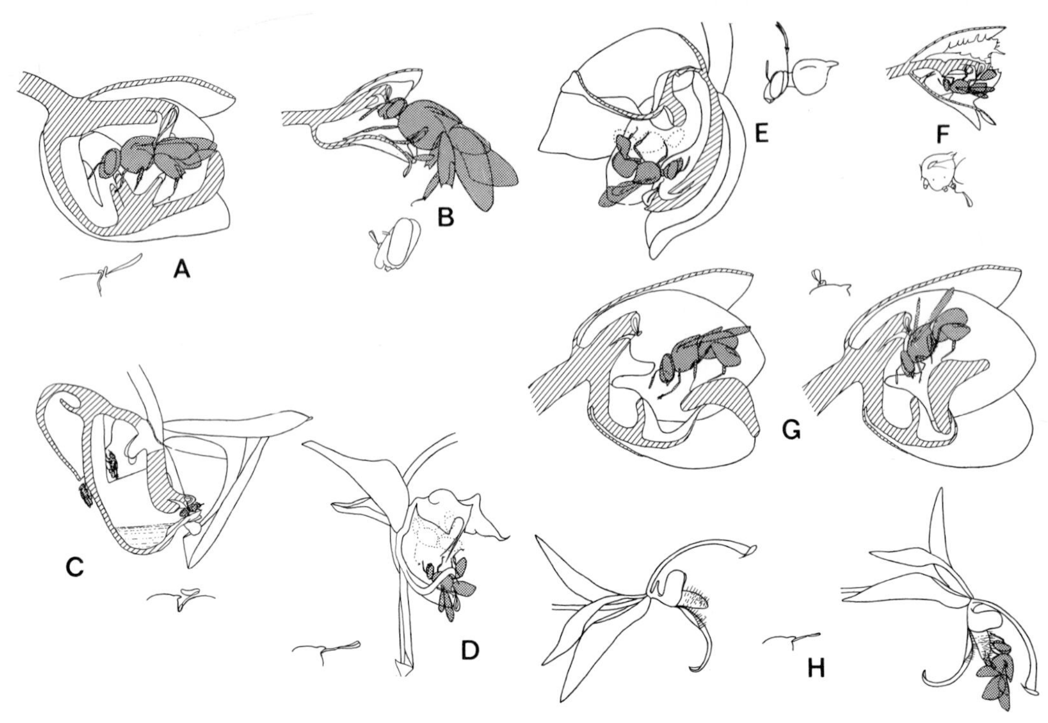

Plate 4-3. Pollination systems of Gongoreae (from Dressler, 1968b; used with permission). **A:** Pollination in *Acineta superba,* showing *Euplusia concava* as it backs out and receives the viscidium beneath the scutellum. Inset shows the placement of the pollinarium on and beneath the scutellum; the thickness of the viscidium is exaggerated. **B:** Pollination in *Coeliopsis hyacinthosma,* showing *Eulaema meriana* receiving a pollinarium on the clypeus. Inset shows the placement of the pollinarium on the head of the bee. **C:** Pollination in *Coryanthes maculata,* showing two euglossas brushing on the lip near the hood and a third emerging beneath the apex of the column and receiving a pollinarium on the base of the abdomen. Inset shows the placement of the pollinarium between the thorax and the abdomen. **D:** Pollination in *Gongora tricolor;* the position of the bee while brushing is shown by dotted lines. When the bee falls its course is guided by the column and petals, and the pollinarium is placed beneath the scutellum. Inset shows placement of pollinarium; the thickness of the viscidium and stipe is exaggerated. **E:** Pollination in *Lacaena spectabilis,* the position of the bee while brushing is shown by dotted lines; when the bee falls, the pollinarium is placed between the head and prothorax. Inset shows placement of the pollinarium. **F:** Pollination in *Sievekingia fimbriata,* with *Euglossa sapphirina* entering the flower upside-down. The hooked viscidium catches on the trochanter of the mid or hind leg. Inset shows the placement of the pollinarium on the trochanter of the hind leg. **G:** Pollination in *Peristeria* species; left, longitudinal section of the flower with *Euglossa dodsoni* on the hinged lip; right, the lip overbalanced and the bee thrown against the column. Inset shows placement of the pollinarium on the thorax. **H:** Pollination in *Polycycnis gratiosa;* left flower in normal position; right flower with *Euglossa villosa* on lip; the weight of the bee causes the flower to bend downward, the pollinia then being hooked beneath the bee's scutellum. Inset shows placement of pollinarium.

pollinaria attached to their legs. Nothing is known about the floral fragrances of *Cirrhaea.*

Possibly the simplest pollination mechanism in the tribe is found in the genus *Coeliopsis.* Dressler, Dodson, and Hills (Dressler, 1968b) observed the pollination of *Coeliopsis hyacinthosma* by *Eulaema cingulata, Eulaema meriana,* and *Euplusia schmidtiana.* The bees try to scratch at the base of the lip, and in doing so press their faces against the column, causing the pollinarium to be deposited on the clypeus (Plate 4-3-**B**). If the bee were to visit another flower, the pollinia would slide into the stigmatic cleft. Dressler (1968b) and van der Pijl and Dodson (1966) report that the species sets numerous seed capsules.

The pollination mechanisms of *Coryanthes* are among the most bizarre in the flowering plants (Plate 4-3-**C**). The action of the male euglossine bees on *Coryanthes* was first reported by Cruger (1865), who thought they were chewing the flower, but his reports have never been verified by other investigators. Darwin (1877) reported Cruger's observations but never observed euglossine pollination in the wild himself. Allen (1950) reported on pollination of *Coryanthes,* but he also suggested that the bees were feeding on the flowers. Dodson (1962a) was the first to notice that the bees were not feeding on the flowers, but were actually collecting the fragrances with their front tarsal brushes. Vogel (1963b) and Dodson (1965b) elaborated on the pollination of *Coryanthes,* with the latter reporting on the pollination of six species. All sepals and two petals are reflexed upward and out of the way of the lip and the column. The lip is rather intricate in structure with a large hood-shaped hypochile, an elongate tubular mesochile, and a large bucket-shaped epichile which partly surrounds the column. The column is bent sharply at a right angle near its apex where it is surrounded by the epichile, with the anther at the very tip of the column and the stigma located next to the anther. Two glands are located at the basal junction of the lip and the column. They secrete water (or a very watery fluid, but not nectar) into the bucket. The male bees which are attracted to the floral fragrances land on the hypochile and begin scratching near the mesochile. In the process of collecting the fragrances a bee will often slip and fall into the bucket. Bees are unable to climb out of the bucket because of the steep, waxy sides, and apparently are unable to fly with wet wings. The only way out of the bucket is through the small opening at the apex of the lip where it is close to the tip of the column. If the anther has not been removed, it takes the bee 15–30 minutes to work his way out of the bucket. In leaving the bucket, the bee first passes the stigma, then the anther. The viscidium of the pollinarium is crescent-shaped and attaches at the base of the abdomen between it and the thorax. This positions the pollinia on the top of the abdomen. Once the anther has been removed, the bees are able to move out of the flower with very little trouble in approximately a minute. If a bee has a pollinarium on his back, the pollinia will be deposited in the stigma as he leaves the bucket. Each species for which pollination information is available has a distinct pollinator (or group of pollinators), and it is obvious from the work of Dodson (1965b) that the floral fragrances are the source of attraction for the male bees, which are species-specific in their visits to *Coryanthes.* Species of *Coryanthes* are sympatric in several areas, and the species-specific pollinators are the main isolating mechanism among them.

Pollination of *Gongora* has been described correctly by Dressler (1966, 1968b), Allen (1954), Dodson and Frymire (1961b), and very incorrectly by Cruger (1865, and repeated by Darwin, 1877). Flowers are pendent on a long raceme. The lip is above the column. The two small petals are adnate to the column and have the appearance of two small wings. The sepals are placed away from the column and lip. Floral fragrances are produced on the underside of the base of the lip. Therefore, the bee must land on the top of the lip and crawl to the underside to collect the fragrance. As the bee releases his hold on the lip and begins his efforts to right himself and transfer the collected material to his hind legs, he slips and falls down the column (Plate 4-3-**D**). The petals and column form a chute down which he slides, first passing the stigma and then the anther. As he slips past the anther, the viscidium is hooked under his scutellum. After the pollinarium has been removed the stigma opens and is receptive for pollination. Dressler (1966, 1968b) described the pollination of several *Gongora* species in detail, and Dodson and Hills (1966) and Hills *et al.* (1968) have discussed the floral fragrances of a number of species. Dodson *et al.* (1969) discussed the role of the floral fragrances as isolating mechanisms and their role in sympatric speciation. It appears that a number of species (however recognized) have specific pollinators (or groups of pollinators) and that ethological isolation is the most important isolating mechanism in the genus.

On the basis of the structure of the pollinarium found on a specimen of *Eulaema meriana,* Dressler (1976a) showed that *Houlletia tigrina* was probably pollinated by this bee. We have found that *H. tigrina* produces mainly methyl salicylate in its fragrance (Williams, Whitten, and Dodson, in preparation), and that *E. meriana* is attracted to this compound. Although we do not have observations of pollination of *Houlletia,* chemical analysis of floral fragrances, the use of chemical baits in attracting male euglossine bees, and studies of pollinaria structure have made it possible to study its pollination indirectly, at least to the point of determining the most likely pollinator(s). This approach has also been used with other species (Dressler, 1976a, 1977b).

Pollination has not been observed in *Kegeliella,* but Dressler (1968b) observed that *Euplusia concava* is attracted to flowers of *K. atropilosa* and found pollinaria of this species on several specimens of this bee. The pollinaria are placed on the back of the head between it and the thorax.

Pollinators have been recorded for at least two species of *Lacaena* (Dressler, 1968b; van der Pijl and Dodson, 1966). The pendent flowers require that the bee enter the flower upside down to brush on the midlobe of the lip. As the bee releases his grip and starts to fall, his head passes the viscidium and the pollinarium becomes attached to the back of the head between it and the thorax (Plate 4-3-**E**). Nothing is known regarding the floral fragrances of species in this genus.

Nothing is known about the pollination of either *Lueddemannia* or *Lycomormium. Lueddemannia pescatorei* and *Lycomormium squalidum* both contain cineole in their floral fragrances, as well as alpha- and beta-pinene and additional unidentified compounds, but neither scent is distinctive enough to predict the probable pollinator (Williams, Whitten, and Dodson, in preparation). Sweet (1974) discussed the taxonomy of the four species of *Lycomormium.*

Dressler (1968b) observed four species of *Euglossa* visiting *Paphinia clausula,* and *E. gorgonensis* removed pollinaria. The bees enter the non-resupinate flowers right side up and the pollinaria become attached to their legs as they leave the flower. This is probably an atypical system in this genus, but nothing is known of pollination in other species. Nothing is known about the floral fragrances of *Paphinia.*

Pollinators are known for three species of *Peristeria* (Dressler, 1968b; van der Pijl and Dodson, 1966; Dodson, 1965a). The lip is hinged, and as the bee tries to advance on it to the source of the fragrance, the hinge tips and throws the bee against the column. The side lobes of the lip and wings on the column hold the bee, and the pollinarium is deposited on the thorax of the bee as it tries to leave the flower (Plate 4-3-**G**). Each species of *Peristeria* has a specific set of pollinators, and it appears likely that both floral fragrances and size serve to isolate the species from each other.

Two species of *Polycycnis* have been studied: *Polycycnis barbata* is pollinated by *Eulaema speciosa* (Schmid, 1969) and *P. gratiosa* is pollinated by *Euglossa villosa* (Dressler, 1968b). Dressler (1977a,b) has described several new species of *Polycycnis,* and has added to the known pollinators of *P. gratiosa.* In addition to *E. villosa,* he has found *Euplusia anisochlora, E. chrysopyga,* and *E. rufocauda* carrying pollinaria of this species. He has also found two undescribed species of *Euglossa* carrying pollinaria of *P. muscifera* in Panama (Dressler, 1977b), one of which was later described as *E. oleolucens* (Dressler, 1978a). Pollination mechanisms are apparently the same in all species. The bee lands on the lip and rapidly moves to its base to begin scratching. His weight pulls the flower down, and as he starts to hover to transfer the collected floral fragrance, the viscidium becomes attached to the scutellum and the pollinarium is removed from the flower (Plate 4-3-**H**). If the bee already has a pollinarium on his back, the pollinia may be inserted into the stigma as he leaves the flower. Floral fragrance data are not available for *Polycycnis.*

Dodson (van der Pijl and Dodson, 1966) observed several bees of *Euplusia* cf. *purpurata* visiting *Schlimia trifida.* Dodson suggested that the pollinarium is attached under the scutellum. No more is known about pollination in *Schlimia,* nor is any information available about the floral fragrances of the genus. Sweet (1973) discussed the taxonomy of the genus but had no observations on pollination.

Pollination information is available for five of the twelve species of *Sievekingia.* Dressler (1968b) observed bees visiting both *S. fimbriata* and *S. suavis,* and Dodson (1965a; van der Pijl and Dodson, 1966) observed pollination in *S. jenmanii.* Bees land on the petals and enter the flower upside down and as they attempt to brush and then leave it, the viscidium is attached to the trochanter of either the midleg or the hind leg. The same type of pollination mechanism occurs in all species which have been observed. Dressler (1976b) has noted that *S. fimbriata* attracts both small euglossas and the larger *Euplusia duckei* (Plate 4-3-**F**). The large bees cannot get as far into the flower as the smaller ones. Pollinaria are deposited on the foreleg of *Euplusia* whereas on the smaller bees the pollinaria attached to the midleg or the base of the hind leg. Dressler also noted that bees carrying pollinaria of both *S. fimbriata* and *S. reichenbachiana* have been collected quite far from known localities of the orchids. This indicates either that the orchids have a larger range than previously thought, or that the bees are very long-distance flyers. Dressler (1979b) reported on the pollination of *S. butcheri* by *Euglossa cyanura.*

More information is available on the pollination biology of *Stanhopea* than on any other genus of Gongoreae. Natural pollination has been described by Dodson and Frymire (1961a,b), Dodson (1962a, 1965a, 1967b,c, 1970, 1975a,b), and Dressler (1968a,b). Floral fragrances have been analyzed for a number of species (Dodson and Hills, 1966; Hills *et al.,* 1968; Williams, Whitten, and Dodson, in preparation). Dodson (van der Pijl and Dodson, 1966) has pointed out that there are two pollination mechanisms in *Stanhopea,* although they do show some intergradation. Primitive species of *Stanhopea* (such as *S. ecornuta, S. annulata,* and *S. pulla*) have small, rather closed flowers. The bees enter the flower, brush at the base of the lip, then back out and pass by the apex of the column, at which time the viscidium is attached to the underside of the scutellum (Plate 4-2-**A**). Dressler (1968b) noted that the bees enter the flower of *S. ecornuta* from the side and may leave by the same route. However, in many cases the bees back out, with the projections on each side of the apex of the lip keeping them on the center of the lip. This positions the bee correctly for exiting against the column and having the pollinarium placed on his back.

The advanced species of *Stanhopea* have what has been called a "drop flower" or a "fall-through" flower (Dressler, 1968b). Dodson and Frymire (1961a) described the pollination of *S. jenishiana* (reported originally as *S. bucephalus*) by *Eulaema bomboides*. The bees do not land on the flower and crawl to the lip, but rather fly directly into the flower and alight either on the front or on the back of the hypochile. They move into the base of the lip to get to the source of the fragrance, but often lose their footing and fall through the flower. The horns of the mesochile keep an individual from slipping out of the side of the flower and direct him down through the chute past the apex of the column. In the process the projecting viscidium slips under the scutellum as the bee passes out of the flower. Apparently in all species of *Stanhopea* the pollinarium must be removed for the stigma to become receptive, and if a bee visits a receptive flower carrying pollinia, they are stripped from the stipe of the pollinarium as the bee falls through the chute while passing the stigma (Plates 4-1-**2**, 4-2-**B**).

Different species of the advanced stanhopeas attract large bees of the genus *Eulaema* as well as the smaller euglossas, and in some cases it appears that the size of the visitor as well as ethological isolating mechanisms are important. A particular species of *Stanhopea* may attract both large and small bees, but the small bees might simply fall through the flower without coming into contact with the viscidium. Bees that enter the flower and do not immediately fall through will brush in the inflated basal portion of the lip where the fragrance production is strongest. As the bee attempts to launch into the air to hover and transfer the collected floral fragrance, he will often fall through the flower. The waxy surface of the lip makes it difficult for him to climb out of the side of the flower, and the arching column interferes with his ability to fly. The only way out is to slide down the chute past the rostellum with the projecting viscidium, which slips under and attaches to the scutellum. A large number of floral fragrance compounds have been identified in *Stanhopea* (Dodson and Hills, 1966; Hills *et al.*, 1968; Williams, Whitten, and Dodson, in preparation): alpha- and beta-pinene, cineole, citronellal, linalool, geraniol, benzyl acetate, methyl salicylate, 2-phenylethyl acetate, methyl cinnamate, alpha-phellandrene, alpha-terpineol, and methyl benzoate. Pollinator specificity and attraction to these compounds will be discussed below.

The genus *Trevoria* comprises five species, none of which is common (Garay, 1970). Nothing is known about the pollination biology, pollinators, or floral fragrances of the genus. One plant found in flower in Panama had a definite "euglossine" fragrance, but no pollinators were observed (Dressler and Williams, unpublished). Dressler (1976b) suggests that its pollination mechanism may be similar to that of *Sievekingia* on the basis of the shape of the viscidium.

Oncidieae

The tribe Oncidieae (subtribe Oncidiinae of some classifications) contains six genera that are euglossine-pollinated. The tribe as a whole has the most diverse types of pollination systems found in the Orchidaceae. To my knowledge, Oncidieae are not adapted to pollination by moths, flies, or bats. However, the orchids in this tribe have adapted to all other pollinating vectors of orchids (bat pollination is unknown in the Orchidaceae).

A few observations are available on pollination of *Aspasia.* Dodson and Frymire (1961b) listed *A. psittacina* as being visited by *Eulaema cingulata* and *E. polychroma* (the orchid was mistakenly identified as *A. epidendroides* and one bee reported as *E. tropica,* now correctly known as *E. polychroma;* both names were corrected in van der Pijl and Dodson, 1966). Dressler (personal communication) has seen *Exaerete smaragdina* visiting *A. principissa* in central Panama. He has also caught males of *E. frontinalis, E. smaragdina, Eulaema cingulata,* and *E. meriana* carrying pollinaria of an *Aspasia* in central Panama. The only *Aspasia* in the area is *A. principissa,* so one can infer that these species are pollinators of *A. principissa* (Williams, 1974). The pollination system of *Aspasia* is still unclear.

Aspasia psittacina has a strong odor of clove oil (eugenol), a compound known to attract males of *Eulaema cingulata* (Dodson and Frymire, 1961b; Williams and Dodson, 1972), *Exaerete frontalis,* and *E. smaragdina* (Dodson, Dressler, and Williams, unpublished) as well as other species. Flowers of *A. epidendroides, A. principissa,* and *A. psittacina* all have a definite aroma of eugenol, but we have not yet verified this (Williams and Dodson, unpublished). Examinations of hundreds of flowers during the last two decades has shown that nectar is not produced, although there is a definite nectar tube or chamber in the flower formed by the union of the bases of the lip and column. I have suggested that *Aspasia* flowers produced nectar at one time and attracted both male and female euglossines as food plants, but that they have stopped producing it and are now well along the path of evolving the euglossine syndrome (Williams, 1974).

Macradenia is pollinated by male euglossines and has a characteristic "euglossine" aroma. *Euglossa hemichlora* and *E. villosiventris* have been observed pollinating and collected carrying pollinaria of *M. brassavolae* (Dodson, Dressler, and Williams, unpublished), but no other information on pollination is available. The flowers lack any sign of a nectary, but do produce abundant, unidentified fragrance.

Dodson (1967a) and Dressler (1968b) studied euglossine pollination in *Notylia,* a genus which is exclusively pollinated by these bees. The source of the fragrance is at the base of the lip, and as the bee scratches at this area the pollinarium is deposited on the front of its head (frons). After the pollinarium has been withdrawn from the anther, the stipe dries and wilts slightly, with the result that the pollinia are in the correct position to be inserted into the stigma. The very small size of the stigma results in the deposition of only one of the two pollinia in a stigma. Bees of the same species may pollinate more than one species of *Notylia.* However, when this occurs the species of *Notylia* are usually geographically isolated from one another (Dodson, 1967a). Dressler (1968b) and Dodson (1967a) reported on the visits of nonpollinating insects of the wrong size, and presented evidence that sympatric species of *Notylia* have distinct pollinators. Dressler (1968b) observed a male *Eulaema cingulata* visiting a plant of *N. pentachne,* and noted that the bee had several pollinaria on its head.

I have observed *E. cingulata* visiting *N. pentachne* in Panama, and the insects removed a number of pollinaria as they crawled and brushed on the flowers. All pollinaria were placed on the front of the head. Dodson and Frymire (1961b) reported finding pollinaria of an orchid thought to be allied to *Notylia* with the viscidium curved around the trochanters and femurs of the second pair of legs of *Euplusia surinamensis* (originally reported as *Euplusia smaragdina*). This was tentatively identified as the genus *Pterostemma* in van der Pijl and Dodson (1966). Further, more detailed observations show the pollinaria to be of a *Sievekingia* (Dressler, personal communication).

At least one species of *Rodriguezia, R. leeana,* is euglossine-pollinated, but some other species in the genus depend on hummingbirds (van der Pijl and Dodson, 1966; Dodson, 1965a, 1967b). It is not clear if the flowers are true euglossine-syndrome flowers, or if they produce a small amount of nectar.

Trichocentrum is pollinated exclusively by euglossine bees. There is a nectarylike structure formed by the extended base of the lip, but no nectar has ever been found, which indicates this is a nonfunctional nectary (Williams, unpublished). Flowers do produce abundant fragrance at times, and numerous males of *Euglossa* have been seen visiting *Trichocentrum* (Dressler, personal communication; van der Pijl and Dodson, 1966; Dodson, 1965a, 1967b; Williams, unpublished). The bees enter the flowers, brush at the base of the lip, launch into the air to transfer the collected fragrances, and manage to get the pollinaria attached to the frons, or the tops of their heads. Bees collected in central Panama through the use of chemical baits often carry one to several pollinaria of *Trichocentrum* (Williams, Dressler, and Dodson, unpublished). Flowers of *T. panamensis* and *T. tigrinum* have strong, distinctive "euglossine" fragrances, but I have not yet been able to confirm the identity of the floral fragrance compounds.

Dodson (1962a) reported that *Trichopilia rostrata* is pollinated by *Euglossa viridissima* (corrected to *E. hemichlora* in van der Pijl and Dodson, 1966). We have collected *Euglossa dressleri* carrying pollinaria of *Trichopilia subulata,* and *Euglossa heterosticta* with pollinaria of *T.* cf. *leucoxantha* (Dressler and Williams, unpublished). A number of males of several unidentified species of *Euglossa* were observed visiting and removing pollinaria from *T. subulata* in Panama (Dressler and Williams, unpublished).

The great diversity of pollination systems in the Oncidieae (as opposed to exclusive euglossine pollination in Gongoreae and Catasetinae) make it tempting to propose an evolutionary sequence for pollination systems in the tribe. I postulate that the most primitive type of pollination in the Oncidieae is based on the attraction of nectar-feeding bees and wasps to flowers such as *Leochilus, Mesospinidium,* etc. Various types of spurred nectaries have evolved from this basic type, these being found in such genera as *Comparettia, Ionopsis,* and *Rodriguezia. Symphyglossum, Rodriguezia,* and *Cochlioda* are all thought to have evolved hummingbird pollination from the basic bee-nectar flowers (van der Pijl and Dodson, 1966). Euglossine bees do visit various genera for nectar (i.e., *Cischweinfia*), but they also visit a number of genera to collect the floral fragrances (*Notylia, Macradenia,* etc.). I postulate that male euglossine pollination is derived from bee-nectar flowers, with such genera as *Aspasia, Trichocentrum,* and *Trichopilia* currently intermediate between nectar-producing flowers and "pure" euglossine flowers. In the intermediate-stage flowers, a nectarylike structure is formed, but no nectar is produced. In the pure euglossine-syndrome flowers, such as *Notylia* and *Macradenia,* not even a trace of a nectarylike structure remains (Williams, 1974; Williams and Dressler, unpublished).

Relationships of Male Euglossine Bees to Other Plant Families

Male euglossines visit several species in a few families other than orchids to collect floral fragrances. Dodson (1966b) noted that males visited *Anthurium, Spathiphyllum,* and *Xanthosoma* of the Araceae. Dressler (1967) reported that males visit and brush on the spadix of *Anthurium* and *Spathiphyllum,* and Williams and Dressler (1976) discussed euglossine visitation of *Spathiphyllum* more extensively. Additional reports of euglossine visits to members of the Araceae are given by Dressler (1968a) and Zucchi *et al.* (1969b). It appears that some species of *Spathiphyllum* are also visited and pollinated by members of the genus *Trigona,* as well as by male euglossine bees (Williams and Dressler, 1976; Ackerman and Montalvo, personal communication). The trigonas are collecting pollen from the inflorescences, whereas the male euglossine bees collect the floral fragrances and behave on the spadix as they do on orchid flowers they pollinate. If male euglossine bees are the primary pollinators of *Spathiphyllum,* then each species of *Spathiphyllum* seems to have its own pollinator or group of pollinators. Those species which share pollinators are either allopatric or in different sections of the genus (Williams and Dressler, 1976). The fragrances of *S. cannifolium* have been analyzed, and they contain alpha- and beta-pinene, cineole, benzyl acetate, and several undetermined compounds. Cineole and benzyl acetate are known male euglossine bee attractants (Williams and Dodson, 1972, and unpublished).

Buchmann (1980) observed both male and female *Euglossa imperialis* visiting flowers of *Xiphidium caeruleum* Aubl. in Panama. The flowers are "buzz" flowers from which the females collect pollen. The males collect floral fragrances from the tepals and become lightly dusted with pollen. Buchmann did not know for sure if the males actually pollinated the flowers they visited. Kimsey (cited in Buchmann, 1980) also observed numerous male *E. imperialis* visiting flowers of *X. caeruleum.* The floral fragrances were not identified.

No other members of the monocotyledons are known to attract male euglossine bees. However, a species of *Dicranopygium* (Cyclanthaceae) in Panama produces abundant fragrance that appears to be composed mainly of benzyl acetate, but no bees have ever been observed visiting the flowers. A species of *Carludovica* (also Cyclanthaceae) in western Panama has an aroma of methyl cinnamate, but no bees were observed visiting it either (Williams, unpublished).

In the dicotyledons male euglossine bees have been observed visiting and brushing on flowers of several families. In the Gesneriaceae, Vogel (1966b) described the visits of male euglossine bees to *Gloxinia perennis.* Dressler (1967, 1968a) discussed the attraction of male euglossine bees to flowers of *Drymonia* and suggested that some species of this genus appear to be borderline cases of euglossine pollination. Dressler found only males visiting *D. turrialvae,* but Dodson (cited in Dressler, 1968a) observed both sexes on *D. mollis.* I have seen males and females visiting *D. macrantha.* The females land with their tongues extended and collect pollen. Males visit, brush, and exhibit the floral-fragrance-transferring action (Williams, unpublished). Some species of *Drymonia* produce flowers which have a distinctive "euglossine" aroma (*D. killippii,* for example), but others have no noticeable fragrance. Dressler is probably correct in suggesting that members of this genus (as well as other genera) contain species that have evolved the euglossine syndrome within groups that already attracted males and females foraging for either nectar or pollen (or both). Analyses of floral fragrances are not available for any member of the Gesneriaceae, but *Gloxinia perennis*, several species of *Drymonia*, and some species of *Paradrymonia* have flowers which produce a distinct "euglossine" aroma.

Cyphomandra hartwegii (Solanaceae) attracts males of *Eulaema bombiformis* in central Panama. Its flowers are pendent (when they produce the most fragrance), and the bees must land and turn upside down to collect the fragrances. The anthers have porous openings, and as the bee prepares to hover and transfer the collected material, the pollen is shaken through the pores onto the underside of the thorax and abdomen. These flowers have a definite aroma of benzyl acetate, a compound known to attract *E. bombiformis*. No females have been observed visiting them (Dressler, personal communication; Dressler and Williams, unpublished). The mechanism of pollen deposition is similar to that described for other flowers with porous anthers (Michener, 1962; Wille, 1963).

Armbruster and Webster (1979) reported that male euglossine bees pollinate *Dalechampia spathulata* (Euphorbiaceae), whereas females pollinate a related species, *D. magnistipulata* (Webster and Armbruster, 1979). *Eulaema polychroma, E. luteola,* and *E. cingulata* males seem to be the major pollinators of *D. spathulata,* from which they collect fragrance compounds that are produced by a large gland in the inflorescence. The bees brush the gland, and the fragrance is reported to be reminiscent of wintergreen. This would imply either methyl salicylate or methyl benzoate in the fragrance, although none of these species is known to collect either of these compounds. Armbruster and Dunn (in Armbruster and Webster, 1979) report that the floral glands produce at least four as yet unidentified monoterpenes.

Dressler (1967) observed males of *Euglossa cordata* (since identified as *E. variabilis*) and *Eufriesia pulchra* brushing on the limb of flowers of a cultivated bignoniaceous vine in Panama (reported as *Bignonia magnifica,* now known as *Saritaea magnifica*). The flowers were also visited and entered by females of a number of species.

Dodson (1965a, cited in Dressler, 1967) reported male euplusias brushing on an unidentified member of the Myrtaceae and on an unidentified legume of the Mimosaceae. The bees were not identified to species.

In addition to the brushing behavior exhibited by bees on flowers, some species collect organic compounds from other sources (Dressler, 1967; Dressler and Williams, unpublished; Dodson, Dressler, and Williams, unpublished). Dressler reports several instances of *Euplusia* and *Euglossa* species brushing on rotten logs. Dodson and Dressler saw numbers of *Euglossa purpurea* and *Euplusia schmidtiana* brushing on a tree trunk. They also observed *Euplusia purpurata* brushing on the sides of boards in Peru (Dressler, 1967). The compounds the bees collect may be natural products of the logs and roots, and Dressler has also suggested that the bees are possibly attracted to fungal products from the rotting logs. We have also seen males of *Euplusia concava* brushing on roots of an unidentified tree in Panama, both on Cerro Campana and on Cerro Tute. In both cases the roots had an aroma similar to safrole or anethole, but we have not yet been able to determine the compounds present in them (Dressler and Williams, unpublished).

The Attraction of Male Euglossine Bees to Orchid Floral Fragrances

From the time of Cruger (1865) through the time of Allen (1950, 1952), the attraction of male euglossine bees to orchid flowers was erroneously assumed to be due to food the male bees were assumed to be obtaining from the flowers. The first workers to suggest that the bees were visiting the flowers exclusively because of the floral fragrances were Dodson and Frymire (1961a,b). This suggestion was later elaborated by Dodson in a series of publications over almost two decades (see Literature Cited). Vogel made significant contributions to our knowledge of the biology of euglossine bees (Vogel, 1963a,b, 1966a,b), as did Dressler (1967, 1968a,b). Dodson and Hills (1966) first demonstrated that different orchid species have distinctive odors. However, the major advances in our understanding of the biology of the male euglossine bees were (1) the identification of a number of the floral fragrance components in several species of orchids pollinated by these bees and (2) the resulting demonstration that at least some of the floral fragrance compounds are strong attractants for them (Adams, 1968; Hills *et al.*, 1968, 1972; Dodson *et al.*, 1969; Dodson, 1970; Williams and Dodson, 1972).

Floral fragrance components have been tentatively identified (Hills *et al.*, 1968, 1972; Williams *et al.*, 1981). Tentatively identified compounds were used in bioassays in the field in a variety of areas in Central and South America (Plate 4-1-**3**). Details of the gas-chromatographic techniques may be found in Hills *et al.* (1968, 1972). The identities of several compounds identified by our techniques have been confirmed by combined gas chromatography–mass spectrometry by Holman and Heimermann (1973) and Holman (personal communication). Holman and his coworkers have not refuted any of our identifications (Holman, personal communication).

Substances we have identified in orchid floral fragrances are monoterpenes or simple aromatic compounds. They are: alpha-pinene, beta-pinene, myrcene, alpha-phellandrene, cineole, ocimene, p-cymene, citronellal, linalool, geraniol, methyl benzoate, alpha-terpineol, benzyl acetate, piperitone, d-carvone, citronellol, methyl salicylate, nerol, 2-phenylethyl acetate, 2-phenylethanol, methyl cinnamate, eugenol, vanillin, and skatole.

The bioassay for biological activity consists of presenting the compounds to bees in the field in the following manner. Clean blotter pads (new herbarium blotters) are cut into 5 × 5 cm squares, tacked onto trees or posts, and saturated with the compound in question. The compounds used were purchased from a number of organic chemical supply houses and tested for purity by GLC analyses. See Williams and Dodson (1972) for more discussion. We have also used small vials with wicks with similar results. Whenever a field bioassay was conducted with a previously untested compound, the bees it attracted were allowed to land on the compound, brush and collect it, hover and transfer it to the hind legs, and return to the pad before being captured (Plate 4-1-**3**,**4**). During the first few days of sampling in any given area, all bees which come to the fragrance pads are collected as vouchers. Extensive screening was carried out in Mexico, Guatemala, El Salvador, Nicaragua, Costa Rica, Panama, Colombia, Ecuador, and Guyana. Additional tests were performed in Honduras, Peru, Venezuela, and Brazil. We have found that some of the compounds, such as cineole, benzyl acetate, methyl salicylate, skatole, and vanillin, are good general euglossine attractants. Others, such as methyl benzoate, ocimene, methyl cinnamate, d-carvone, 2-phenylethyl acetate, 2-phenylethanol, and linalool, are more limited attractants. Some compounds do not attract male euglossine bees, but seem to serve as modifiers or repellents. Examples are alpha- and beta-pinene, alpha-terpineol, and to a lesser extent ocimene (Bennett, 1972b; Dodson *et al.*, 1969; Dodson, 1970; Williams and Dodson, 1972).

We have tested structurally similar compounds in a number of areas (Williams and Dodson, 1972; Dodson *et al.*, 1969). In one field test in Guyana we compared eugenol with methyl isoeugenol and found that although over 300 male euglossine bees (representing seven species) were attracted to eugenol, only one came to methyl isoeugenol. Compounds such as bornyl acetate, anethole, linalyl acetate, limonene, menthone, menthol, and numerous other monoterpenes were tested and found to attract a few individuals of a very few species (Dodson *et al.*, 1969; Williams and Dodson, 1972). In general, it is possible to conclude that, with a few exceptions, substances which are structurally similar to compounds identified in orchid fragrances are not very good attractants for male euglossine bees. Anisyl acetate, which is not known to occur naturally, is a very good attractant for a new species of *Euglossa* from Ecuador (Williams, unpublished). Alpha- and beta-ionone are not known from orchid fragrances but are often good attractants (Lopez, 1963; Dodson *et al.*, 1969). Methyl ionone, however, is a very poor attractant (Williams and Dodson, 1972). We have found that naturally occurring compounds with little superficial structural differences, such as methyl salicylate and methyl benzoate (which may have different biosynthetic pathways), differ greatly in their attraction potential and effects. Eleven species of euglossines were attracted to methyl salicylate in central Panama, and only two came to methyl benzoate (Dodson, 1970). Alpha-ionone attracted only one species, whereas beta-ionone attracted five (Dodson, 1970).

We have observed concentration effects with some compounds (Dodson *et al.*, 1969). Both alpha- and beta-ionone attract more bees if the pad is unsaturated. When pads are saturated, several days to a few weeks must pass before these compounds attract large numbers of bees. Similar results have been obtained with anethole. In general, however, a freshly saturated pad attracts more bees than a dry one.

A few preliminary experiments have been carried out on the effects of combinations of compounds. Dodson (1970) showed that pure cineole and benzyl acetate attracted 433 bees (27 species) and 36 bees (4 species), respectively. A mixture of cineole : benzyl acetate (1 : 39) attracted only 49 bees (8 species). This indicates that some of the species attracted to cineole were either repelled by benzyl acetate or were unable to sense cineole in the presence of benzyl acetate. When alpha- and beta-pinene were added to the cineole–benzyl acetate mixture, the number of bees was reduced to 7 (4 species). Williams and Dodson (1972) tested several mixtures, and found that they attracted fewer individuals and (almost always) a smaller number of species than pure compounds. Pure benzyl acetate attracted 40 bees (4 species), while benzyl acetate : beta-pinene (10 : 1, 4 : 1, and 1 : 1) attracted 14, 3, and 6 individuals (4, 2, and 2 species). We have found that both methyl salicylate and eugenol are good attractants separately, but methyl salicylate renders pure eugenol less attractive (or it is possible that bees which are attracted to eugenol are repelled by methyl salicylate). Eugenol has the same effect on all bees attracted to methyl salicylate (only one specimen was attracted to a 1 : 1 mixture). These preliminary experiments indicate (1) that the addition of repellents to attractants reduces the effects of the latter, and (2) that increased amounts of repellent reduce the attraction even more. Furthermore, if two attractants are mixed, the attraction effect of each is reduced.

No information is available on biosynthetic pathways of the compounds found in orchid fragrances. However, information does exist on the genetics of monoterpenes and on the biosynthesis of terpenes and simple aromatics in mints, pines, and other plants. This will be reviewed here briefly and only as it pertains to orchid floral fragrances, pollination biology, and speciation (a detailed treatment is beyond the scope of this review).

It is generally assumed that monoterpenes are derived in a pathway which leads from mevalonic acid through geranyl pyrophosphate to monoterpene (Runeckles and Mabry, 1973). The monoterpenes can be grouped into three classes: acyclic, monocyclic, and bicyclic compounds. Irving and Adams (1973) have presented data which show that the biosynthetic pathways for each of these classes is inherited as a unit. Hybrids made from parents containing different pathways contain both parental groups of compounds, with some of them being inherited transgressively. Irving and Adams were able to estimate the minimum number of genes involved with the production of various terpenes. In general the smallest number of genes seems to range from one to seven, with most compounds having an average of two genes. Hefendehl and Murray (1976) have found that only a few genes are necessary for the conversion of some monoterpenes to other monoterpenes. In *Pinus* monoterpene synthesis is under the strong control of one to several genes (Hanover, 1971). Extrapolation from work with mints and pines suggests that floral fragrance differences among closely related species of orchids (as in the euglossine-pollinated Gongoreae or Catasetinae) are due to only a few genes. Such minor gene differences might explain the apparent sympatric speciation of *Gongora* in Costa Rica and other areas (Dodson *et al.*, 1969).

The aromatic compounds found in orchid floral fragrances seem to be produced from shikimic acid via several different pathways, a discussion of which is beyond the scope of this review. It seems probable, however, that small genetic changes may lead to very important changes in the floral fragrances, and thus in the attraction of pollinators. Most of the species sampled to date seem to be capable of producing monoterpenoids and a variety of other aromatic compounds (Hills *et al.*, 1968, 1972; Williams, Atwood, and Dodson, 1981; Williams, Whitten, and Dodson, in preparation).

The floral fragrance compounds are produced almost exclusively by the labellum in glandular structures called osmophores (Dodson and Williams, unpublished; Vogel, 1963a,b). Vogel (1963a, 1966a,b) discussed the structure and function of the osmophores ("glandular, multicellular and clearly differentiated tissue within the floral region, which is well exposed to the atmosphere") of a variety of orchids. The tissues consume reserve carbohydrates very rapidly and produce strong floral odors, which are often highly specific attractants of insects (Vogel, 1963a).

In some orchids (*Stanhopea*) the osmophore region is very rugose, whereas in others (*Cycnoches*) the surface is very smooth (Vogel, 1963b; Williams and Whitten, unpublished). Vogel has suggested that the osmophores produce the scent as a microliquid layer that accumulates in the capillary spaces between the rugose projections of their tissues. The greatly divided, villilike appearance of the osmophorous region on the labellum in some species certainly provides a large surface area from which the fragrances can evaporate. Transmission and scanning electron miscroscope studies of osmophores are not available, but are necessary for a better understanding of these important structures. Vogel's work suggests that precursors are translocated to just below the epidermis of the osmophores, where the fragrances are synthesized and rapidly translocated to the surface where they volatilize (or they may be released as a microliquid layer). The compounds do not seem to be manufactured and stored for later release, since we have not been able to detect any of them in subtending tissues (Williams, Hills, and Dodson, unpublished).

Our observations on field and greenhouse-grown plants suggest that air temperature plays an important role in the production or release of scents. On cool or very overcast days, the flowers produce very little or no fragrances, whereas during sunny, warm (or hot) periods scent production is abundant. Vogel (1963a) reports that in some non-orchidaceous plants an increase in floral temperature volatilizes the floral fragrances. This is also the case in some of the Araceae (see Meeuse, 1978, for a review). Evidence for elevated floral temperature is lacking in the euglossine-pollinated orchids, and the lack of fragrance production on cool days suggests that these plants do not have the ability to increase the labellar temperature sufficiently above ambient to volatilize the odoriferous compounds. We have noted that male euglossine bees usually do not fly during inclement weather. Therefore, it is not surprising that the flowers that depend on them for pollination lack the ability to produce floral fragrances during such periods. There would be no selective advantage to the production of floral fragrances when the pollinators are inactive. I might mention that (1) floral fragrance production decreases during the afternoon and stops at night, and (2) male euglossine bees are usually not attracted in large numbers to isolated compounds during these periods (Adams, 1966; Braga, 1976; Adams, Dodson, Dressler, and Williams, unpublished).

Production of fragrance appears to be under some type of hormonal control. Dodson (1962b) and Vogel (1963b), among others, have noted that scent evolution is reduced or ceases entirely in male flowers of *Catasetum* following release of the pollinarium from the flower. In most flowers of the Gongoreae fragrance production stops once the pollinarium has been removed, or after the flower has been pollinated.

All evidence to date indicates that the bees sense the floral fragrances with their antennae. We have tried field experiments with several species in which various body parts were excised from the insect (Adams, Dodson, Dressler, and Williams, unpublished). Removal of the hind legs did not stop the male bees from collecting floral fragrances (as isolated compounds), and the insects actually tried to complete the hovering and transferral portion of the sequence. Removal of the front legs prevented them from being able to collect the compounds, although they were obviously able to perceive them. Evoy and Jones (1971) tried to invoke the behavioral sequence by applying eugenol directly to the front tarsal brushes, but instead of completing the sequence, the bees vigorously wiped the front tarsi with the midlegs in an apparent attempt to clean them. Evoy and Jones also applied eugenol to the antennae, and the bees tried to clean it off. In no case were Evoy and Jones able to get the bees to transfer the eugenol to the hind legs by direct application of the compound to the insects. They suggested that it is necessary for the bees to intiate the sequence with the brushing reaction, and that the mere presence of the compound on their forelegs is not sufficient to invoke the transfer. Evoy and Jones noted that transfer has never been seen without prior brushing, and brushing has not been observed without subsequent transfer. They concluded that the entire operation (brushing, launch into flight, and transfer while hovering) represents closely linked activities of the neural pathways. We have seen male bees approach saturated pads in flight, and as they get closer move their antennae and keep them pointed in such a fashion that at all times they point toward the source of the fragrance (even when the bees are hovering to one side of the pad). We have also noticed that bees often rub their front tarsi over the antennae after they have landed on a pad and brushed a few times. Quite often the bee will land, brush for a minute or two, rub its front tarsi over the antennae, launch into the air, transfer to the hind legs, land again, and repeat the sequence. Not all individuals seem to rub the tarsi over the antennae, nor is the reaction necessarily confined to a particular compound.

Williams and Dodson (1972) and Hills *et al.* (1972) have mentioned the role of the floral fragrances in long-distance attraction of pollinators. In a population of tropical epiphytes which is widely dispersed over a very large area, where individuals are not dispersed in close proximity, such attraction is important. Visual cues, even at close range, are relatively unimportant in attracting male euglossine bees, especially when compared to floral fragrances (Dressler, 1968b; Williams, unpublished). Olfactory cues could therefore serve to increase the area of recognition and could be an extremely efficient means of attracting a pollinator over long distances (Williams and Dodson, 1972). I believe that this increase in the area from which pollinators are attracted is an important factor in the population biology and evolution of the orchids of tropical America.

Attraction of pollinators over long distances is also related to long-distance dispersal or flow of pollen (gametes) (Janzen, 1971; Williams and Dodson, 1972). Janzen (1971) suggested that female euglossine bees follow a regular "trap line" during their foraging and that this is important in long-distance pollen flow in tropical plants. We have suggested that male euglossine bees, which are not tied to a particular nesting spot but sleep in a variety of locations, might be as important (or even more important) as long-distance pollinators and vectors in long-distance pollen flow (Williams and Dodson, 1972). For the most part, we have had very limited success with mark-recapture experiments with male euglossine bees in the tropics. Kroodsma (1975) suggested that individual males reside in a given locality throughout one morning, and may remain at one site over several days. This is probably true, but evidence based on more detailed, longer-period sampling suggests that males of at least some species may move over fairly long distances

(Ackerman, personal communication). Certainly a male euglossine bee has the potential to travel over fairly great distances during its lifetime, and thus to be an important agent in long-distance pollen flow.

The role of orchid floral fragrances as isolating mechanisms between closely related species has been studied in greatest detail in the genus *Catasetum* (Hills, 1968; Hills *et al.*, 1972; Dodson, 1962b, 1978b). In *Catasetum*, forms which produce different fragrances and attract different pollinators have been considered distinct species (Hills *et al.*, 1972). Differential fragrance production is one of the primary barriers to gene exchange among otherwise interfertile entities in this genus. In some cases geographical, mechanical, and seasonal isolation seem to be as important as the ethological or ecological isolation which result from species-specific (1) floral fragrance production and (2) attraction of a pollinator or group of pollinators (Hills *et al.*, 1972). Dodson (1978b) suggested that local ecological (and ethological) conditions are important in the isolation of the two closely related species of *Catasetum* in Guyana; however, it seems that, in general, floral fragrances and the more or less specific pollinators attracted are one (or the main) type of isolation between closely related sympatric species of *Catasetum*. The data are less complete on *Gongora* and *Stanhopea*, but again it seems that species-specific floral fragrances and the resulting species-specific pollinator (or group of pollinators) specificity are of major importance as isolating mechanisms between closely related species (Dodson, 1962a,b; Dodson and Frymire, 1961a; Dodson *et al.*, 1969; Williams, Whitten, and Dodson, in preparation). It seems that the combination of specific attractants and repellents and general attractants and repellents are actually the means for species-specific attraction of pollinators. There are certainly numerous instances of visitors that are attracted but fail to serve as effective pollinators because they are simply of inappropriate size. In these cases the floral fragrances have served not to repel a nonpollinating visitor, but rather to limit the number of species which are attracted and then screened further by size, placement of pollinaria, and the like.

Why Do Male Euglossine Bees Visit Orchid Flowers? Evidence and Speculation

The behavior of the pollinators on orchid flowers is interesting. So are the adaptations for pollination among the Orchidaceae in general and in the euglossine-pollinated species in particular. With all that, the overriding question is, Why do male euglossine bees visit orchid flowers and collect the floral fragrances? A number of suggestions have been made to explain this peculiar behavior of the male bees, but clear answers are still not available (Dodson, 1975d; Dodson *et al.*, 1969; Vogel, 1963b, 1966a,b). The early suggestions of Vogel (1963b) that the orchid flowers were essentially mimicking the nests or odors of the females have been effectively put to rest (Dodson *et al.*, 1969). Vogel (1966b) suggested that the collected floral fragrances were borrowed pheromones used by the males to mark their territories or to attract females. Since females are not attracted to the flowers, nor to the isolated floral fragrance compounds, the males must modify them in some manner if they are to serve as pheromones. Dodson *et al.* (1969) suggested three possible reasons for the collection of floral fragrances.

1. The bees require the floral fragrance compounds because of their inability to synthesize a necessary metabolite, and therefore live longer than they would if they did not have access to these substances. Dodson (1966b) presented some preliminary evidence that male bees lived only 13–14 days if not provided with orchid flowers, whereas those supplied with appropriate orchid flowers lasted approximately 31 days. However, Ackerman (personal communication) has kept captive bees hatched from nests alive in captivity for over two months, even in the absence of orchid flowers or floral fragrance compounds. The possibility still exists that one of the major reasons for collection of floral fragrances by male bees is their need for compounds directly or as precursors for some necessary metabolite. However, this hypothesis requires further investigation.

2. Male bees may use the floral fragrance compounds to attract other males of the same species to a mating site. We have suggested this as a possibility, on the assumption that females may be attracted by the loud buzzing produced by the males as they collect the floral fragrances and later as they display and buzz at the mating site (Dodson *et al.*, 1969; Dodson, 1975d). The problem with this hypothesis is that the females do not seem to be attracted to buzzing. Dodson (1975d) reported that no females were attracted to recordings of buzzing of males of several species.

Dodson (1975d) proposed lek formation as a possible reason for the collection of floral fragrances. His hypothesis, loosely stated, is that a particular male bee visits a number of different orchids to obtain a wide range of floral fragrance compounds. Certain individual bees eventually obtain the proper blend of compounds to become what he calls attractor males. These attractor males then attract other males of the same species to their display sites. Dodson suggests that the leks are established only by attractor males which have collected the proper blend of compounds. The active male swarm is assumed to attract females which happen to be in the general vicinity of the lek. Kimsey (personal communication) has observed mating in several species of euglossines, but reports no evidence to support the lek hypothesis. Dodson has indicated that gas-chromatographic analyses of the hind legs of the attractor males contain more compounds than those of nonattractors. We have analyzed several hundred male hind legs, and also find variations in the number and amount of compounds they contain. The lek hypothesis may be the correct one, and the reason why male euglossine bees visit orchid flowers to collect floral fragrances. However, at this writing other investigators have not substantiated it. Furthermore, mating behavior may differ among the species and genera, and it should not be surprising that investigators studying different species report different behavioral observations.

3. We (Dodson *et al.*, 1969) suggested that male euglossine bees may convert the floral fragrance compounds into female sex attractants. Females are not attracted to the pure fragrances, but it is possible that they would be attracted by modified compounds. This hypothesis has not been investigated during the past decade, but last year I accumulated data which suggest that this may be the best of the three existing hypotheses. An investigation of this hypothesis is at its inception, and data from both field and laboratory are meager at this point; however, the preliminary results are promising. To clarify this hypothesis, it is first necessary to summarize a certain amount of information about

related bees, particularly bumblebees and those that pollinate *Ophrys*. Information about other insects is also important to the development of the theory.

Techniques for the study of natural odors have been described by Bergstrom (1973, 1975), Stallberg-Stenhagen *et al.* (1973), Dodson and Hills (1966), and Hills *et al.* (1968, 1972). Several investigators have found that male bumblebees scent-mark various objects in their flight territories (Bringer, 1973; Bergstrom and Svensson, 1973; Kullenberg *et al.*, 1973; Kullenberg, 1973, 1975; Kullenberg and Bergstrom, 1975). Although earlier reports indicated that the bees were probably marking with their mandibular glands, recent observations are that they are actually using the cephalic portion of the labial glands. The bees mark territories, and it has been demonstrated that different species of *Bombus* produce species-specific blends of pheromones consisting of aliphatics, isoprenoids, or both. It is thought that the females are attracted to the flight territories, specifically to the scented places. The males return to these scented sites and copulate with the females there. Kullenberg and his co-workers have shown that males are attracted to extracts of the heads of various species, and to synthetic attractants.

Tengo and Bergstrom (1976, 1977) found that the cephalic secretions of each species of *Andrena* contain a specific blend of compounds, and Tengo (1979) studied the mandibular gland secretions of *Andrena* male bees (which are attracted to and pollinate *Ophrys* orchids). Males establish a territory and perform a mating flight, and mark parts of the habitat with a scent that attracts both males and females to the site where premating behavior and copulation occur. Tengo tested both natural extracts and synthetic substances. He concluded that (1) the mandibular gland compounds are excitants and attractants in the male mating-flight behavior, (2) straight-chain alkanols are important as releaser compounds in the male mating flight, (3) the mixture of compounds found in the mandibular glands is more important than any individual substance. Bergstrom (1978) studied the compounds found in the labella of *Ophrys* species and found a variety of terpenoids and fatty-acid derivatives. He also listed the compounds which have been found in extracts of their pollinators, and some of them are the same. Kullenberg and Bergstrom (1973, 1976a,b) have discussed the role of male bees as pollinators of *Ophrys*. It is now firmly established that (1) *Ophrys* flowers are pollinated only by male bees that receive no food reward and (2) the labellum stimulates copulatory behavior in the male. Kullenberg and Bergstrom have demonstrated that it is the floral fragrance that stimulates the copulatory behavior, which is enhanced further by visual and tactile factors. Kullenberg has found that cephalic extracts of female bees (*Eucera*) attract males of the species, and the scents of *Ophrys* flowers also attract and excite male bees of the same genus (*Eucera*). He also found that the floral scents are species-specific for the males they attract.

The bumblebee model suggests that male bees mark certain spots with cephalic secretions, and that females are attracted to these areas. Males return to these spots as they travel around this flight territory, then copulation may occur if a female is encountered at the marked spot. The compounds used for marking appear to be long-chain hydrocarbons and several monoterpenes and sesquiterpenes. The *Ophrys* model suggests that the flowers produce compounds which are similar to the cephalic extracts of female bees in some cases, and that these substances attract males. Some of these compounds seem to act as behavioral releasers, and appear to be long-chain hydrocarbons, monoterpenes, and sesquiterpenes. A model for the use of floral fragrances by male euglossine bees may be similar to the two models described above.

My hypothesis is that male euglossine bees (1) collect the floral fragrances of orchids, (2) store them in their hind legs, (3) modify them, (4) translocate some of the modified compounds to the mandibular glands, and (5) use mandibular gland secretions to mark a mating site to which females are attracted. The evidence (mainly preliminary) on which I base this suggestion is presented below.

During the spring of 1978 at the Fortuna dam site in Chiriqui Province, Panama, I caught a number of *Eulaema nigrita* males who were attracted to isolated orchid floral fragrance compounds. In this case, the attractant was cineole. The heads of the males were excised and crushed on the side of a tree. Five bees were attracted to this crude extract, and although none of them was caught, they appeared to be females. None of the five showed any interest in the hind legs of the bees from which the heads were removed. Several males of *E. nigrita* were attracted to the hind legs and were caught. This preliminary field experiment prompted me to analyze the cephalic extracts of a number of species of euglossine bees. We have now analyzed approximately 30 species of euglossine bees, and the results are as follows.

1. The males of each species of euglossine bee have a species-specific blend of compounds in their heads (presumably in the mandibular glands). Some compounds are found in several species, but the qualitative and quantitative blends appear to be species-specific.
2. Variations among individuals of a species are minor in regard to the manibular glands.
3. Analyses of the hind legs indicate (as Dodson, 1975d, suggested) that differences in the amount of the floral fragrance exist between individuals.
4. In addition to the floral fragrance compounds found in the hind legs, several substances found in the cephalic extracts also appear in the hind leg extracts. However, some compounds found in the hind legs do not appear in the cephalic extracts.
5. Analyses of female heads show that they do not contain the same compounds found in the males. The few compounds found in the cephalic extracts of females exist in very low concentrations.

It is certainly tempting to hypothesize that the male bees are using these modified compounds to attract females. The analogy with *Bombus* and *Ophrys* is satisfying, and the hypothesis seems to be good. There are, however, problems with it. First, there is no information on how the male bees translocate the presumably modified floral fragrance compounds to the mandibular glands. They may complex the molecule with some other compounds, such as lipoprotein, then de-complex it when the compound has arrived in the mandibular gland (see Duffey, 1980, for a discussion of sequestration of plant natural products by insects, and Cruz-Landim *et al.*, 1965, for a discussion of modifications of the hind tibia for absorption and release of compounds). Second, a suggestion can be made that the males use these mandibular gland compounds as defensive compounds. The problems with this suggestion are that (1) females do not have similar compounds, (2) each species seems to have a specific blend (qualitatively and quantitatively), and (3) it does not seem logical for each species to have its own extremely distinctive defensive compound blend. I suggest, as a working hypothesis, that the male bees manufacture a blend of compounds, either in the hind legs or in the mandibular glands, which are used as a part of the mating behavior. These blends consist of both generalized attractants, and modifiers or repellents. Some of the compounds may be long-distance attractants (compounds of low molecular weight), whereas others are short-range releasers of mating behavior in the females. They may be used either as species-recognition compounds to ensure that only the appropriate females respond, or as signals which release a particular type of behavior in them. Some of the widespread compounds may be generalized female attractants; others may act to exclude all but conspecific females. Experiments to test these suggestions are in progress.

Speculation on the Origin and Evolution of the Euglossine Syndrome

A commonly held view seems to be that orchid species have species-specific pollinators (Garay, 1972). In instances where pollination has been observed only once or a few times, it does appear that some species of orchids have a species-specific pollinator. However, in cases in which pollination has been observed many times, it appears that a very large number of orchid species do not have one specific pollinator. Rather, a given species of orchid may have several pollinators, one or more of which may be more effective or more common. In many cases involving the euglossine bees, we often find that one bee may visit a number of orchid species, some of which are so distantly related that an interspecific pollination does not lead to fertilization, and others that are closely related but are not cross pollinated because of mechanical barriers (morphology and size of flower or pollinaria, or placement of pollinaria on the insect). There are certainly some euglossine-pollinated orchids that attract only one species of bee. However, there are very few, if any, euglossine bees that visit only one species of orchid. Very few one-to-one relationships of plant to pollinator seem to exist among the euglossine-pollinated orchids. Several species of orchids also attract nonpollinating euglossine visitors, usually bees of an inappropriate size for pollination of the orchid. These species of bees are attracted by the floral fragrance compounds.

Male euglossine bees may be able to obtain some or all of the compounds they need from sources other than orchid flowers. As mentioned earlier, some tree roots produce compounds that attract certain species of euglossine bees. It is known that some fungi produce monoterpenes, and it is possible that the male bees first began collecting organic compounds from fungi or flowers of other plant groups in the evolutionary past. Certainly a larger number of gymnosperms produce abundant terpenoids, and a number of the angiosperms synthesize both terpenes and aromatics in their flowers, leaves, or stems. Although it is known that male bees can obtain at least some of the compounds from other sources, it would appear on the basis of present knowledge that they collect most substances from euglossine-syndrome orchids, a few aroids, and several gesneriads.

The peculiar adaptations for collection and storage of floral fragrance compounds are not found in any other bees, and the high degree of development of these structures in the Euglossini suggests that this is an extremely important aspect of their life. Fossil euglossines are not known (although fossil *Bombus* are reported to be possibly from the

Oligocene; B. S. Simpson, Department of Botany, University of Texas, Austin, personal communication), so it is difficult to estimate when the euglossine-orchid relationship might have begun. Dressler (1968a) believes that it is a relatively ancient relationship. It is obvious that the male bees have specialized adaptations for collecting organic compounds, but it is not clear that there is any co-evolutionary adaptation of the bee to particular orchid flowers. The orchids are obviously adapted to pollination by male euglossine bees, and are dependent on them for their reproductive success. There are no indications of reciprocal modifying influences in the evolution of these two groups, but rather an indication that the bees are adapted to collect fragrances, and orchids have evolved pollination systems which take advantage of the bees' requirements for these compounds.

It is possible to suggest that orchids which are pollinated by male euglossine bees evolved from nectar-producing flowers that produced strong floral fragrances. Some orchid species attract both male and female euglossine bees, which feed on nectar (some species of *Cattleya, Cischweinfia,* etc.). There are also examples of orchid species which appear to have nectaries but fail to produce nectar (*Aspasia, Trichocentrum*). Such species may attract only male euglossine bees. Ackerman (personal communication) reports that in Panama he has collected female eulaemas carrying pollinaria of *Cochleanthes lipscombiae.* This species does not produce a strong euglossine fragrance; it has distinct nectar guides on the lip, but produces no nectar. This is probably a case of a species on the evolutionary pathway to becoming male-euglossine-pollinated (it also attracts male eulaemas). Such species should be expected in a family that still appears to be in a state of rapid evolution and speciation (Dressler and Dodson, 1960).

It is obvious that a satisfactory answer has not yet been provided to the question "Why do male euglossine bees visit orchid flowers?" An answer may become available when we have more information on the chemical content of the hind legs and mandibular glands. Radioactively labeled compounds will have to be used in attempts to trace the metabolism (or lack thereof) of floral fragrance compounds (Dodson, 1970) in the bees. More information is also needed on (1) natural pollination of euglossine-syndrome orchids, (2) chemical constitution of floral fragrances, and (3) the relationships of the various groups of bees to the different chemical and structural properties of the compounds they collect.

Acknowledgments

Portions of this work were supported by grants from the American Orchid Society Fund for Education and Research and from the National Science Foundation, Washington, D.C. (DEB-7911556). I thank Dr. James D. Ackerman, Dr. John T. Atwood, Dr. Robert L. Dressler, Dr. John L. Neff, Dr. Alec M. Pridgeon, Dr. Rudolf Schmid, Dr. Beryl Simpson, Dr. William L. Stern, and Mr. W. Mark Whitten for reading the manuscript and offering numerous constructive criticisms. Dr. Calaway H. Dodson and Dr. Robert L. Dressler have allowed the use of much unpublished information and have spent many days discussing orchids and euglossine bees with me over the past fifteen years. Additional support and encouragement has come from Sr. Ron Matusalem throughout this project.

Table 4-1. Euglossine-pollinated orchids and their pollinators

Orchid	Visitor	Notes	Reference
Acineta			
A. barkeri	*Euplusia concava*	a	Grant, in P & D
A. chrysantha	*Euplusia concava*	a	Dodson, 1965a
A. superba	*Euplusia concava*	a	Dressler, 1968b
Anguloa			
A. clowesii	*Eulaema boliviensis*	a	Dodson, 1966c
A. ruckeri	*Eulaema cingulata*	a	R. Wilson, in P & D
Aspasia			
A. principissa	*Eulaema cingulata*	c	Dressler, private communication
	Eulaema meriana	c	Dressler, private communication
	Exaerete frontalis	c	Dressler, private communication
	Exaerete smaragdina	a,c	Dressler, private communication
A. psittacina	*Eulaema cingulata*	a	Dodson & Frymire, 1961b
(as *A. epidendroides*)	*Eulaema polychroma*	a	Dodson & Frymire, 1961b
Catasetum			
C. barbatum	*Euglossa cordata*	a	Dodson & Dressler, in P & D
C. bicolor	*Euglossa cordata*		Dressler, 1968b
	Euglossa tridentata	a	Dressler, 1968b
	Euglossa cyanaspis	a	Dressler, 1968b
C. cernuum	*Euplusia violacea*	a	Hoehne, 1933
C. costatum	*Euglossa* sp.		Ostlund, in P & D
C. discolor	*Eulaema meriana*		Dressler, in Dodson, 1978b
	Eulaema cingulata	a	Dressler, in P & D
	Euglossa chlorosoma	b	Dodson, 1978b
C. fimbriatum	*Euplusia auriceps*		Dressler, in P & D
C. hookeri	*Euglossa cordata*	a	Dressler & Dodson, in P & D
C. integerrimum	*Eulaema cingulata*	a	Pollard, D. O. Allen, in P & D
	Eulaema polychroma	a	Pollard, D. O. Allen, in P & D
C. longifolium	*Eulaema bombiformis*		Dodson, 1978b
	Eulaema cingulata		Dodson, 1978b
	Eulaema meriana		Dodson, 1978b
C. luridum	*Euglossa cordata*	a	Dressler, in P & D
C. macrocarpum	*Euglossa imperialis*	b	Ducke, 1902, in P & D (original not seen); Dressler, in P & D
	Eulaema cingulata	a	Cruger, 1865; Ducke, 1902 in P & D (original not seen); Dressler, in P & D; Dodson, 1978b
	Eulaema bombiformis		Dodson, 1978b
	Eulaema aff. *luteola*		Dodson, 1978b
	Eulaema meriana	a	Ducke, 1902, in P & D (original not seen)
	Eulaema bennettii	a	Bennett, in P & D
	Eulaema nigrita	a	Dressler, in P & D
C. macroglossum	*Eulaema bomboides*	a	Dodson & Frymire, 1961b
	Eulaema cingulata	a	Dodson & Frymire, 1961b
	Eulaema polychroma	a	Dodson & Frymire, 1961b
	Eulaema speciosa		Dodson & Frymire, 1961b
C. maculatum	*Eulaema cingulata*	a	Allen, 1952; Dodson, 1965a
	Eulaema polychroma	a	Allen, 1952; Dodson, 1965a
C. platyglossum	*Eulaema bomboides*		Dodson & Frymire, 1961b
	Eulaema cingulata	a	Dodson & Frymire, 1961b
	Eulaema polychroma		Dodson & Frymire, 1961b
C. reichenbachianum	*Euglossa* sp.		Dressler, in P & D
C. saccatum	*Eulaema cingulata*	a	Dodson, 1965a
	Euglossa ignita	b	Dodson, 1965a
	Euglossa augaspis	b	Dodson, 1965a
C. tabulare	*Eulaema cingulata*	a	Dodson, in P & D
C. thompsonii	*Euglossa augaspis*		Dodson, 1978a
	Euglossa cognata		Dodson, 1978a
	Euglossa cordata		Dodson, 1978a
	Euglossa liopoda		Dodson, 1978a
	Euglossa mixta		Dodson, 1978a
	Euglossa sp. UM-10		Dodson, 1978a
	Euglossa sp. RD-1215		Dodson, 1978a
	Eulaema cingulata	b	Dodson, 1978a

(*continued*)

Table 4-1.—Continued

Orchid	Visitor	Notes	Reference
C. viridiflavum	*Eulaema cingulata*	a	Dressler, in P & D; Dodson, 1965a; Dressler, 1968b
Chondrorhyncha			
C. sp. (Panama)	*Eulaema speciosa*	c	Dodson & Dressler, unpublished
C. marginata (Panama)	*Eulaema meriana*	c	Dodson, Dressler, and Williams, unpublished
Cirrhaea			
C. sp.	*Euplusia violacea*		Hoehne, 1933
	Euglossa mandibularis	c	P & D
Clowesia			
C. russelliana	*Euplusia mexicana*		Williams & Dodson, in Dodson, 1975c
C. warczewitzii	*Eulaema bombiformis*	a	Dressler & Dodson, in P & D
	Eulaema nigrita	b	Dressler & Dodson, in P & D
Cochleanthes			
C. aromatica	*Eulaema seabrae*		Dodson, 1965a
C. sp. (Ecuador)	*Eulaema meriana*	a	Dodson, 1965a
Coeliopsis			
C. hyacinthosma	*Eulaema cingulata*	a	Dressler, Dodson, & Hills, in Dressler, 1968b
	Eulaema meriana	a	Dressler, Dodson, & Hills, in Dressler, 1968b
	Euplusia schmidtiana	a	Dressler, Dodson, & Hills, in Dressler, 1968b
	Euplusia or *Eulaema* sp.	a	Dressler, Dodson, & Hills, in Dressler, 1968b
	Euglossa dodsoni	b	Dressler, Dodson, & Hills, in Dressler, 1968b
	Euglossa tridentata	b	Dressler, Dodson, & Hills, in Dressler, 1968b
Coryanthes			
C. elegantium	*Euglossa gibbosa*		Dodson & Gentry, 1978
(= *C. wolfii*)	*Euglossa tridentata*		Dodson & Gentry, 1978
	Euglossa hemichlora	a	Dodson, 1965b
C. leucocorys	*Eulaema meriana*		Dodson, 1965b
	Euglossa ignita	b	Dodson, 1965b
C. macrantha	*Eulaema cingulata*	a	Bennett, in P & D; Dodson, 1965a
	Eulaema basalis	a	Bennett, in P & D
C. maculata	*Euglossa azureoviridis*	a	Dressler, 1968b
C. rodriguezii	*Euplusia superba*	a	Dodson, 1965a
	Eulaema meriana	b	Dodson, in P & D
C. speciosa	*Euglossa cordata*		Allen, 1950
C. aff. *speciosa*	*Euglossa alleni*	c	Dodson & Dressler, in P & D
C. trifoliata	*Euglossa ignita*	a	Dodson, 1965a
	Euglossa mixta	b	Dodson, in P & D
Cycnoches			
C. aureum	*Eulaema nigrita*	a	Dressler, in Dodson, 1965a; Dressler, 1968b
	Euglossa tridentata	b	Dressler, in P & D; Dressler, 1968b
	Euglossa cyanaspis	b	Dressler, in P & D; Dressler, 1968b
	Euglossa crassipunctata	b	Dressler, in P & D; Dressler, 1968b
C. egertonianum	*Euglossa ignita*	a	Dressler, 1968b
(typeform)	*Euglossa flammea*	a	Dressler, 1968b
C. cf. *egertonianum* (#1)	*Euglossa cyanura*	a	Dressler, 1968b
C. cf. *egertonianum* (#2)	*Euglossa tridentata*	a	Dressler, 1968b
C. cf. *egertonianum* (#3)	*Euglossa hansoni*	a	Dressler, 1968b
C. lehmannii	*Eulaema cingulata*	a	Dodson & Frymire, 1961a
C. pentadactylon	*Euplusia superba*	c	Dodson, 1965a
C. peruviana (as *C. egertonianum*)	*Euglossa hemichlora*	a	Dodson & Frymire, 1961a
C. ventricosum	*Eulaema cingulata*	a	Dodson, 1965a
C. ventricosum var. *warscewiczii*	*Eulaema cingulata*	a	Allen, 1952; Dressler, 1968b

(*continued*)

Table 4-1.—Continued

Orchid	Visitor	Notes	Reference
Dichaea			
D. panamensis	*Euglossa cordata*	c	Dressler, 1968b
D. riopalenquensis	*Eulaema meriana*		Dodson & Gentry, 1978
Dressleria			
D. dilecta	*Euglossa hansoni*		Dodson, 1975c
D. eburnea	*Eulaema cingulata*		Dodson & Frymire, 1961b
D. helleri	*Euglossa asarophora*		Dressler, in Dodson, 1975c
	Euglossa championi		Dressler, in Dodson, 1975c
	Eulaema nigrita		Dressler, in Dodson, 1975c
	Euplusia schmidtiana		Dressler, in Dodson, 1975c
	Euplusia rufocauda		Dressler, in Dodson, 1975c
	Euplusia anisochlora		Dressler, in Dodson, 1975c
D. suavis	*Euglossa tridentata*		Dressler, in Dodson, 1975c
	Euplusia ornata		Dressler, in Dodson, 1975c
Gongora			
G. armeniaca	*Euglossa* cf. *viridissima*	a	Dodson, 1965a
G. armeniaca var. *bicornuta*	*Euglossa dodsoni*		Dodson, 1965a
G. bufonia	*Euplusia violacea*		Hoehne, 1933
G. grossa	*Euglossa hemichlora*	a	Dodson, 1962a
	Euglossa nigropilosa		Dodson, 1965a
	Euglossa sp. nov.		Dodson & Gentry, 1978
G. quinquenervis (Palmar, Costa Rica)	*Euglossa cordata*		Allen, 1954
G. quinquenervis (Tilaran, Costa Rica)	*Euglossa cordata*		Dressler, 1968b
G. quinquenervis (Panama)	*Euglossa cordata*		Dressler, 1968b
	Euglossa hemichlora	a	Dressler, 1968b
	Euglossa townsendii	a	Dressler, 1968b
	Euglossa tridentata	a	Dressler, 1968b
	Euglossa cyanaspis		Dressler, 1968b
G. quinquenervis (Quevedo, Ecuador)	*Euglossa* cf. *variabilis*		Dodson, 1962a
G. quinquenervis (Iquitos, Peru)	*Euglossa ignita*	a	Dodson, 1962a
	Euglossa augaspis	a	Dodson, 1962a
	Euglossa decorata	a	Dodson, 1962a
	Euglossa cordata		Dodson, 1962a
G. quinquenervis (Rio Palenque, Ecuador)	*Euglossa tridentata*		Dodson & Gentry, 1978
G. sp. (Golfito, Costa Rica)	*Euglossa flammea*	a	Dressler, 1968b
	Euglossa dodsoni		Dressler & Dodson, in P & D
G. sp. ("Guanacaste red")	*Euglossa viridissima*	a	Dressler, 1968b
G. sp. (#1 of Dressler, 1968b)	*Euglossa gorgonensis*	a	Dressler, 1968b
	Euglossa asarophora	b	Dressler, 1968b
	Euglossa nigrosignata		Dressler, 1968b
	Euglossa villosa	b	Dressler, 1968b
G. sp. ("yellowlip" of Dressler, 1968b)	*Euglossa gorgonensis*	a	Dressler, 1968b
	Euglossa hansoni	a	Dressler, 1968b
	Euglossa cybelia	a	Dressler, 1968b
	Eulaema cingulata	b	Dressler, 1968b
	Eulaema nigrifacies	b	Dressler, 1968b
	Eulaema polychroma	b	Dressler, 1968b
	Eulaema speciosa	b	Dressler, 1968b
G. tricolor	*Euglossa cyanura*	a	Dressler, 1968b
	Exaerete smaragdina		Dressler, 1968b
G. unicolor	*Euglossa purpurea*	a	Dressler, 1968b
Houlletia			
H. brocklehurstiana	*Euglossa* sp. RLD-BR7		Dressler, 1968b
Huntleya			
H. meleagris	*Eulaema meriana*	a	Dodson, 1965a
Kefersteinia			
K. graminea	*Eulaema polyzona*	a	Dodson, 1965a
K. sp.	*Euglossa* sp. RLD-206	c	Dressler, 1968b
Kegeliella			
K. atropilosa	*Euplusia concava*	a	Dressler, 1968b

(*continued*)

Table 4-1.—Continued

Orchid	Visitor	Notes	Reference
Lacaena			
L. bicolor	*Euplusia* cf. *caerulescens*		D. O. Allen, in P & D
L. spectabilis	*Euglossa maculilabris*	a	Dressler & Dodson, in P & D; Dressler, 1968b
Lycaste			
L. aromatica	*Euglossa viridissima*	a	Ostlund, Pollard, in P & D
L. consobrina	*Euglossa viridissima*	a	Pollard, in P & D
L. xytriophora	*Euglossa* cf. *variabilis*		Dodson, 1962a
Mormodes			
M. atropurpureum	*Euglossa championi*	a	Dressler, 1968b
	Euglossa mixta	a	Dressler, 1968b
	Euglossa cybelia		Dressler, 1968b
M. cf. *buccinator*	*Euglossa hemichlora*	a	Dodson, 1962a
M. cartonii	*Euglossa cordata*		Dressler, 1968b; Allen, 1954 (as *M. igneum*)
	Euglossa mixta	b	Dressler, in P & D
M. colossus	*Euglossa mixta*		Dressler, 1968b
	Euglossa asarophora	a	Dressler, 1968b
	Euglossa maculilabris		Dodson & Dressler, in P & D
	Eulaema cingulata	b	Dressler, 1968b
	Eulaema meriana	b	Dressler, 1968b
	Eulaema nigrita	b	Dodson & Dressler, in P & D
M. flavidum	*Euglossa viridissima*	a	Dressler, 1968b
M. igneum	*Euglossa igniventris*	a	Dressler, 1968b
	Euglossa mixta	b	Dressler, 1968b
M. lineatum	*Euglossa viridissima*	a	Pollard, in P & D
M. maculatum	*Euglossa viridissima*	a	Pollard, in P & D
M. powellii	*Euglossa tridentata*	a	Dressler, 1968b
M. uncia	*Euglossa* sp.		Ostlund, in P & D
Notylia			
N. cf. *barkeri* (Costa Rica)	*Euglossa erythrochlora*	a	Dressler, 1968b
N. buchtenii	*Euglossa augaspis*	a	Dodson, 1965a
	Euglossa ignita	b	Dodson, 1965a
N. cf. *buchtenii*	*Euglossa ignita*	a	Dodson, 1965a
N. panamensis	*Euglossa hemichlora*	a	Dressler, 1968b
N. pentachne	*Eulaema cingulata*	a	Dressler, in P & D; Williams
N. sp. (aff. *barkeri*)	*Euglossa ignita*		Dressler, 1968b
	Euglossa hansoni		Dressler, 1968b
	Euglossa tridentata	c	Dressler, 1968b
	Euglossa sapphirina	b	Dressler, 1968b
N. xyphorius	*Euplusia surinamensis*	a	Dodson & Frymire, 1961b
N. wulschlegeliana	*Euplusia surinamensis*		Dressler and Dodson, in P & D
Paphinia			
P. clausula (= *P. cristata* var. *modiglianiana*)	*Euglossa gorgonensis*	a	Dressler, 1968b
	Euglossa hansoni	a	Dressler, 1968b
	Euglossa asarophora	a	Dressler, 1968b
	Euglossa hemichlora	a	Dressler, 1968b
Peristeria			
P. elata	*Euplusia concava*	a,c	Dressler, 1968b
	Euglossa crassipunctata	b	Dressler, 1968b
P. pendula	*Euglossa ignita*	a	Dodson, 1965a
	Euglossa mixta	b	Dodson, 1965a
	Eulaema meriana	b	Dodson, 1965a
P. sp. (Panama)	*Euglossa cordata*		Dressler, 1968b
	Euglossa deceptrix	a	Dressler, 1968b
	Euglossa dodsoni	a	Dressler, 1968b
	Euglossa dressleri	b	Dressler, 1968b
	Euglossa igniventris	b	Dressler, 1968b
	Euglossa imperialis	b	Dressler, 1968b
	Euglossa tridentata	b	Dressler, 1968b
	Euglossa heterosticta	b	Dressler, 1968b
	Euglossa cybelia		Dressler, 1968b
	Euglossa sp. RLD-110		Dressler, 1968b
	Euglossa maculilabris	c	Dressler, 1968b
	Euglossa sp. RLD-206		Dressler, 1968b

(*continued*)

Table 4-1.—Continued

Orchid	Visitor	Notes	Reference
	Euglossa bursigera	b	Dressler, 1968b
	Eulaema nigrifacies	b	Dressler, 1968b
	Eulaema nigrita	b	Dressler, 1968b
	Eulaema meriana or *luteola*	b	Dressler, 1968b
	Euplusia schmidtiana	b	Dressler, 1968b
Pescatoria			
P. wallisii	*Eulaema polychroma*	a	Dodson & Frymire, 1961a
Polycycnis			
P. barbata	*Eulaema speciosa*	a,c	Schmid, 1969
P. gratiosa	*Euglossa villosa*	a	Dressler, 1968b
	Euplusia anisochlora	c	Dressler, 1977b
	Euplusia chrysopyga		Dressler, 1977b
	Euplusia rufocauda		Dressler, 1977b
P. muscifera	*Euglossa oleolucens*	c	Dressler, 1977b
	Euglossa sp.	c	Dressler, 1977b (species not named in original report)
Rodriguezia			
R. leeana	*Euglossa nigropilosa*	a	Dodson, 1965a
Schlimia			
S. trifida	*Euplusia* cf. *purpurata*		Dodson, in P & D
	Euglossa townsendii	b	Dodson, in P & D
Sievekingia			
S. butcheri	*Euglossa cyanura*		Dressler, 1979b
S. fimbriata	*Euglossa mixta*	b	Dressler, 1968b, 1976b
	Euglossa sapphirina	a	Dressler, 1968b, 1976b
	Euglossa cybelia		Dressler, 1976b
	Euglossa crassipunctata	b,c	Dressler, 1976b
	Euplusia duckei		Dressler, 1976b
	Euplusia mussitans		Dressler, 1976b
	Euglossa dressleri	b	Dressler, 1968b
S. jenmanii	*Euglossa nigropilosa*	c	Dodson, 1965a
S. reichenbachiana	*Euplusia surinamensis*		Dressler, 1976b
S. rhonhofiae	*Euglossa hansoni*		Dressler, 1976b
	Euglossa tridentata		Dodson & Gentry, 1978
	Euglossa sp. nov.		Dressler, 1976b
	Euglossa heterosticta		Dressler, 1976b
S. suavis	*Euglossa dodsoni*	a	Dressler, 1968b
	Euglossa townsendii	b	Dressler, 1968b
	Euglossa bursigera	c	Dressler, 1976b
Sobralia			
S. decora	*Euglossa viridissima*		Dressler, in P & D
S. leucoxantha	*Eulaema speciosa*	a	Dodson, 1965a
S. rosea	*Eulaema polyzona*	a	Dodson, 1965a
	Euplusia ornata		Dodson, 1965a
S. sessilis	*Euglossa cordata*	a	Ducke, 1902, in P & D (original not seen)
S. violacea	*Euplusia surinamensis*		Dodson, 1962a
	Eulaema cingulata	a	Dodson, 1962a
	Eulaema polychroma	a	Dodson, 1962a
	Eulaema speciosa	a	Dodson, 1965a
S. aff. *weberbaueriana*	*Eulaema polychroma*	a	Dodson, in P & D
Stanhopea			
S. annulata	*Euglossa granti*		Dodson, 1975b
S. candida	*Euglossa chlorosoma*		Dodson, 1975b
(= *S. randii*)	*Euglossa ignita*	a	Dodson, 1965a
	Eulaema meriana	b	Dodson, 1965a
S. cirrhata	*Euglossa* cf. *flammea*	a	Dressler, 1968b
S. connata	*Eulaema speciosa*	a	Dodson, 1965a, 1975b
	Euglossa nigropilosa	b	Dodson, 1965a
S. costaricensis	*Euplusia schmidtiana*	a	Dressler, 1968b
	Eulaema seabrae		Dodson, 1965a
S. ecornuta	*Euplusia schmidtiana*	a	Dressler, 1968b
	Eulaema seabrae		Dodson, in P & D
	Eulaema nigrita	b	Dressler, 1968b
	Euglossa allosticta	b	Dressler, 1968b

(*continued*)

Table 4-1.—Continued

Orchid	Visitor	Notes	Reference
	Euglossa imperialis	b	Dressler, 1968b
	Euglossa tridentata	b	Dressler, 1968b
S. embreei	*Eulaema bomboides*		Dodson, 1975b
S. florida	*Euglossa nigropilosa*	a	Dodson, 1965a, 1975b
	Eulaema meriana	b	Dodson, 1965a
S. frymirei	*Eulaema bomboides*		Dodson, 1975b
S. gibbosa	*Eulaema meriana*	a	Dodson, 1965a
S. grandiflora	*Eulaema meriana*	a	Dressler, in P & D
	Euglossa ignita	b	Ducke, 1902, in P & D (original not seen)
S. impressa	*Euglossa granti*		Dodson, 1975b
S. aff. *jenishiana*	*Eulaema bomboides*		Dodson & Frymire, 1961a
S. cf. *oculata*	*Euglossa crassipunctata*	b	Dressler, 1968b
	Euglossa cyanaspis	b	Dressler, 1968b
	Euglossa flammea	b	Dressler, 1968b
	Euglossa hemichlora	b	Dressler, 1968b
	Euglossa tridentata	b	Dressler, 1968b
S. reichenbachiana	*Eulaema leucopyga*		Dodson, in P & D
S. saccata	*Euglossa viridissima*	a	Dressler, Schwartz, Pollard, in P & D
S. tigrina	*Euglossa viridissima*	a	Friese, 1899, in P & D (original not seen)
S. tricornis	*Eulaema meriana*	a	Dodson & Frymire, 1961a
S. wardii	*Eulaema polychroma*		Dodson, 1965a
S. warscewicziana	*Euplusia macroglossa*	a	Dodson, 1965a
Trichocentrum			
T. panamensis	*Euglossa cordata*		Dressler, in P & D
T. tigrinum	*Eulaema cingulata*	a	Dodson, 1962a
Trichopilia			
T. rostrata	*Euglossa hemichlora*	a	Dodson, 1962a
Zygopetalum			
Z. rhombilabium	*Eulaema cingulata*	a	Dodson, 1965a
Zygosepalum			
Z. labiosum	*Eulaema meriana*	a	Dressler, in P & D

P & D = van der Pijl and Dodson, 1966.
[a] Observed pollinating.
[b] Nonpollinating visitor.
[c] Captured carrying pollinia.

Literature Cited

Ackerman, J. D. 1975. Reproductive biology of *Goodyera oblongifolia* (Orchidaceae). Madroño 23:191–198.

Ackerman, J. D., and M. R. Mesler. 1979. Pollination biology of *Listera cordata* (Orchidaceae). Amer. J. Bot. 66:820–824.

Ackerman, J. D., and N. H. Williams. 1980. Pollen morphology of the Neottieae and its impact on the classification of the Orchidaceae. Grana 19:7–18.

Adams, R. M. 1966. Attraction of bees to orchids. Fairchild Trop. Gard. Bull. 21:6–7, 13.

——. 1968. The attraction of Euglossini (Hymenoptera: Apidae) to fragrance compounds of orchid flowers. Ph.D. diss., Univ. Miami, Coral Gables, Fla.

Adams, R. M., and G. J. Goss. 1976. The reproductive biology of the epiphytic orchids of Florida. III. *Epidendrum anceps* Jacquin. Amer. Orchid Soc. Bull. 45:488–492.

Allen, P. H. 1950. Pollination in *Coryanthes speciosa*. Amer. Orchid Soc. Bull. 19:528–536.

——. 1952. The swan orchids, a revision of the genus *Cycnoches*. Orchid J. 1:173–184, 225–230, 273–276, 349–403.

——. 1954. Pollination in *Gongora maculata*. Ceiba 4:121–125.

Armbruster, W. S., and G. L. Webster. 1979. Pollination of two species of *Dalechampia* (Euphorbiaceae) in Mexico by euglossine bees. Biotropica 11:278–283.

Ayensu, E. S. 1973. Biological and morphological aspects of the Velloziaceae. Biotropica 5:135–149.
Bennett, F. D. 1965. Notes on a nest of *Eulaema terminata* Smith (Hymenoptera). Insectes Sociaux 12:81–92.
——. 1966. Notes on the biology of *Stelis* (*Odontostelis*) *bilineolata* (Spinola), a parasite of *Euglossa cordata* (Linnaeus) (Hymenoptera: Apoidea: Megachilidae). J. New York Entomol. Soc. 74:72–79.
——. 1972a. Observations on *Exaerete* spp. and their hosts *Eulaema terminata* and *Euplusia surinamensis* (Hymen., Apidae, Euglossinae) in Trinidad. J. New York Entomol. Soc. 80:118–124.
——. 1972b. Baited McPhail fruitfly traps to collect euglossine bees. J. New York Entomol. Soc. 80:137–145.
Bergstrom, G. 1973. Studies on natural odoriferous compounds. VI. Use of a pre-column tube for the quantitative isolation of natural, volatile compounds for gas chromatography/mass spectrometry. Chemica Scripta 4:135–138.
——. 1975. Development of an integrated system for the analyses of volatile communication substances in social Hymenoptera, p. 173–187. *In* C. Noirot, P. E. Howse, and C. Le Masne (eds.), Pheromones and defensive secretions in social insects. International Union for the Study of Social Insects, Imprimerie de l'Université de Dijon, Dijon, France.
——. 1978. Role of volatile chemicals in *Ophrys*-pollinator interactions, p. 207–230. *In* G. Harborne (ed.), Biochemical aspects of plant and animal coevolution. Academic Press, New York.
Bergstrom, G., and B. G. Svensson. 1973. 2,3-dihydro-6, trans-farnesol: Main component in the cephalic marker secretion of *Bombus jonellus* Kl. (Hym., Apidae) males. Zoon Suppl. 1:61–65.
Braga, P. I. S. 1976. Atração de abelhas polinizadoras de Orchidaceae com auxílio de iscas-odores na campinarana e floresta tropical úmida de região de Manaus. Ciência e Cultura 288:767–773.
Bringer, B. 1973. Territorial flight of bumble-bee males in coniferous forest on the northern most part of the island of Öland. Zoon Suppl. 1:15–22.
Buchmann, S. L. 1980. Preliminary anthecological observations on *Xiphidium caeruleum* Aubl. (Monocotyledonae: Haemodoraceae) in Panama. J. Kansas Entomol. Soc. 53:685–699.
Cronquist, A. 1968. Evolution and classification of flowering plants. Houghton Mifflin, Boston.
Cruger, H. 1865. A few notes on the fecundation of orchids and their morphology. J. Linn. Soc. London, Bot. 8:127–135.
Cruz-Landim, C. da. 1963. Evolution of the wax and scent glands in the Apinae (Hymenoptera, Apidae). J. New York Entomol. Soc. 71:2–31.
——. 1967. Estudo comparativo de algumas glândulas das abelhas (Hymenoptera, Apoidea) e respectivas impliçóes evolutivas. Arq. Zool. S. Paulo 15:177–290.
Cruz-Landim, C. da, A. C. Stort, M. A. da Costa Cruz, and E. W. Kitajima. 1965. Orgão tibial dos machos de Euglossini. Estudo ao microscópico óptico e electrônico. Rev. Brasil. Biol. 25:323–341.
Darwin, C. 1877. The various contrivances by which orchids are fertilised by insects. 2d ed. D. Appleton, New York (seen as 1884 revised edition).
Dodson, C. H. 1962a. The importance of pollination in the evolution of the orchids of tropical America. Amer. Orchid Soc. Bull. 31:525–534, 641–649, 731–735.
——. 1962b. Pollination and variation in the subtribe Catasetinae (Orchidaceae). Ann. Missouri Bot. Gard. 49:35–56.
——. 1963. The Mexican stanhopeas. Amer. Orchid Soc. Bull. 32:115–129.
——. 1965a. Agentes de polenización y su influencia sobre la evolución en la familia Orquidacea. Univ. Nac. Amaxonia Peruana, Inst. General de Investigaciones. Iquitos, Peru. 128 pp.
——. 1965b. Studies in orchid pollination: The genus *Coryanthes*. Amer. Orchid Soc. Bull. 34:680–687.
——. 1966a. Studies in orchid pollination: *Cypripedium, Phragmopedium* and allied genera. Amer. Orchid Soc. Bull. 35:125–128.
——. 1966b. Ethology of some bees of the tribe Euglossini (Hymenoptera: Apidae). J. Kansas Entomol. Soc. 39:607–629.
——. 1966c. Studies in orchid pollination: The genus *Anguloa*. Amer. Orchid Soc. Bull. 35:624–627.
——. 1967a. Studies in orchid pollination. The genus *Notylia*. Amer. Orchid Soc. Bull. 36:209–214.
——. 1967b. Relationships between pollinators and orchid flowers. Atas do Simpósio sôbre a Biota Amazônica 5:1–72.
——. 1967c. El género *Stanhopea* en Colombia. Orquideología 2:7–27.
——. 1970. The role of chemical attractants in orchid pollination, p. 83–107. *In* K. L. Chambers (ed.), Biochemical Coevolution. Oregon State Univ. Press, Corvallis.

——. 1975a. Clarification of some nomenclature in the genus *Stanhopea* (Orchidaceae). Selbyana 1:46–55.
——. 1975b. Orchids of Ecuador: *Stanhopea*. Selbyana 1:114–129.
——. 1975c. *Dressleria* and *Clowesia:* A new genus and an old one revived in the Catasetinae (Orchidaceae). Selbyana 1:130–137.
——. 1975d. Coevolution of orchids and bees, p. 91–99. *In* L. E. Gilbert and P. H. Raven (eds.), Coevolution of animals and plants. Univ. Texas Press, Austin.
——. 1978a. Three new South American species of *Catasetum* (Orchidaceae). Selbyana 2:156–158.
——. 1978b. The catasetums (Orchidaceae) of Tapakuma, Guyana. Selbyana 2:159–168.
Dodson, C. H., R. L. Dressler, H. G. Hills, R. M. Adams, and N. H. Williams. 1969. Biologically active compounds in orchid fragrances. Science 164:1243–1249.
Dodson, C. H., and G. P. Frymire. 1961a. Preliminary studies in the genus *Stanhopea* (Orchidaceae). Ann. Missouri Bot. Gard. 48:137–172.
——. 1961b. Natural pollination of orchids. Missouri Bot. Gard. Bull. 49:133–152.
Dodson, C. H., and A. H. Gentry. 1978. Flora of the Rio Palenque Science Center. Selbyana 4:1–628.
Dodson, C. H., and H. G. Hills. 1966. Gas chromatography of orchid fragrances. Amer. Orchid Soc. Bull. 35:720–725.
Dressler, R. L. 1961. The structure of the orchid flower. Missouri Bot. Gard. Bull. 49:60–69.
——. 1966. Some observations on *Gongora*. Orchid Digest 30:220–223.
——. 1967. Why do euglossine bees visit orchid flowers? Atas do Simpósio sôbre a Biota Amazônica 5:171–180.
——. 1968a. Pollination by euglossine bees. Evolution 22:202–210.
——. 1968b. Observations on orchids and euglossine bees in Panama and Costa Rica. Rev. Biol. Trop. 15:143–183.
——. 1974. Classification of the Orchid family, p. 259–279. *In* M. Ospina (ed.), Proc. 7th World Orchid Conf., Medellin, Colombia.
——. 1976a. How to study orchid pollination without any orchids, p. 534–537. *In* K. Senghas (ed.), Proc. 8th World Orchid Conf., Frankfurt, Germany.
——. 1976b. Una *Sievekingia* nueva de Colombia. Orquideología 11:215–221.
——. 1977a. Dos *Polycycnis* nuevas de Sur America. Orquideología 12:313.
——. 1977b. El género *Polycycnis* en Panama y Costa Rica. Orquideología 12:117–133.
——. 1978a. New species of *Euglossa* from Mexico and Central America. Rev. Biol. Trop. 26:167–185.
——. 1978b. An infrageneric classification of *Euglossa*, with notes on some features of special taxonomic importance (Hymenoptera; Apidae). Rev. Biol. Trop. 26:187–198.
——. 1979a. The subfamilies of the Orchidaceae. Selbyana 5:197–206.
——. 1979b. Una *Sievekingia* llamativa de Panama. Orquideología 13:221–227.
——. 1979c. *Eulaema bombiformis, E. meriana,* and Mullerian mimicry in related species (Hymenoptera: Apidae). Biotropica 11:144–151.
Dressler, R. L., and C. H. Dodson, 1960. Classification and phylogeny in the Orchidaceae. Ann. Missouri Bot. Gard. 47:25–68.
Duffey, S. S. 1980. Sequestration of plant natural products by insects. Ann. Rev. Entomol. 25:447–477.
Evoy, W. H., and B. P. Jones. 1971. Motor patterns of male euglossine bees evoked by floral fragrances. Animal Behaviour 19:583–588.
Faegri, K., and L. van der Pijl. 1966. The principles of pollination ecology. Pergamon Press, London.
Garay, L. A. 1960. On the origin of the Orchidaceae. Bot. Mus. Leafl. Harvard Univ. 19:57–96.
——. 1970. El género *Trevoria* F. C. Lehmann. Orquideología 5:3–13.
——. 1972. On the origin of the Orchidaceae. II. J. Arnold Arboretum 53:202–215.
Goss, G. J., and R. M. Adams. 1976. The reproductive biology of the epiphytic orchids of Florida. IV. Sexually selective attraction of moths to the floral fragrance of *Epidendrum anceps* Jacquin. Amer. Orchid Soc. Bull. 45:997–1001.
Gregg, K. B. 1975. The effect of light intensity on sex expression in species of *Cycnoches* and *Catasetum* (Orchidaceae). Selbyana 1:101–113.
——. 1978. The interaction of light intensity, plant size, and nutrition in sex expression in *Cycnoches* (Orchidaceae). Selbyana 2:212–223.
——. 1979. Population variation and incipient speciation in *Cycnoches* from Panama. Bot. Soc. Amer. Misc. Series Publ. 157:56.

Hanover, J. W. 1971. Genetics of terpenes. II. Genetic variances and interrelationships of monoterpene concentrations in *Pinus monticola*. Heredity 27:237–245.

Hefendehl, F. W., and M. J. Murray. 1976. Genetic aspects of the biosynthesis of natural odors. Lloydia 39:39–52.

Hills, H. G. 1968. Fragrance analysis in chemotaxonomy of the genus *Catasetum* (Orchidaceae). Ph.D. diss., Univ. Miami, Coral Gables, Fla. 52 pp.

Hills, H. G., N. H. Williams, and C. H. Dodson. 1968. Identification of some orchid fragrance components. Amer. Orchid Soc. Bull. 37:967–971.

——. 1972. Floral fragrances and isolating mechanisms in the genus *Catasetum* (Orchidaceae). Biotropica 4:61–76.

Hoehne, F. C. 1933. Contribução para o conhecimento do gênero *Catasetum* Rich. Bol. Agric., Ser. 33a, Número único: 133–196.

Holman, R. T., and W. H. Heimermann. 1973. Identification of components of orchid fragrances by gas chromatography–mass spectrometry. Amer. Orchid Soc. Bull. 42:678–682.

Inouye, D. W. 1975. Flight temperatures of male euglossine bees (Hymenoptera: Apidae: Euglossini). J. Kansas Entomol. Soc. 48:366–370.

Irving, R. S., and R. P. Adams. 1973. Genetic and biosynthetic relationships of monoterpenes, p. 187–214. *In* V. C. Runeckles and T. J. Mabry (eds.), Terpenoids: Structure, biogenesis, and distribution. Recent Advances in Phytochemistry, vol. 6. Academic Press, New York.

Ivri, Y., and A. Dafni. 1977. The pollination ecology of *Epipactis consimilis* Don (Orchidaceae) in Israel. New Phytol. 79:173–177.

Janzen, D. H. 1968. Reproductive behavior in the Passifloraceae and some of its pollinators in Central America. Behaviour 32:35–48.

——. 1971. Euglossine bees as long-distance pollinators of tropical plants. Science 171:203–205.

Jeffrey, D. C., J. Arditti, and H. Koopowitz. 1970. Sugar content in floral and extrafloral exudates of orchids: Pollination, myrmecology and chemotaxonomy implication. New Phytol. 69:187–195.

Jones, C. E., and S. L. Buchmann. 1974. Ultraviolet floral patterns as functional orientation cues in hymenopterous pollination systems. Animal Behaviour 22:481–485.

Kennedy, H. 1973. Notes on Central American Marantaceae I. New species and records from Panama and Costa Rica. Ann. Missouri Bot. Gard. 60:413–426.

——. 1977. An unusual flowering strategy and new species in *Calathea*. Bot. Notiser 130:333–339.

——. 1978. Systematics and pollination of the "closed-flowered" species of *Calathea* (Marantaceae). Univ. Calif. Publ. Botany 71:1–90.

Kimsey, L. S. 1977. New species of bees in the genera *Euplusia* and *Eufriesia*. Pan-Pacific Entomol. 53:8–18.

——. 1979a. Synonymy of the genus *Euplusia* Moure under *Eufriesia* Cockerell (Hymenoptera, Apidae, Euglossini). Pan-Pacific Entomol. 55:126.

——. 1979b. An illustrated key to the genus *Exaerete* with descriptions of male genitalia and biology (Hymenoptera: Euglossini, Apidae). J. Kansas Entomol. Soc. 52:735–746.

Kroodsma, D. E. 1975. Flight distances of male euglossine bees in orchid pollination. Biotropica 7:71–72.

Kullenberg, B. 1956. Field experiments with chemical sexual attractants on aculeate Hymenopteran males. I. Zool. Bidrag Uppsala 31:253–352, 5 plates.

——. 1973. Field experiments with chemical sexual attractants on aculeate Hymenopteran males. II. Zoon Suppl. 1:31–43.

——. 1975. Chemical signals in the biocoenosis. Biological Signals 73–85. Kungl. Fysiografiska Sallsakapet. Lund, Sweden.

Kullenberg, B., and G. Bergstrom. 1973. The pollination of *Ophrys* orchids, p. 253–258. *In* G. Bendz and J. Santesson (eds.), Chemistry in botanical classification. Nobel Symposium 25. Nobel Foundation, Stockholm.

——. 1975. Chemical communication between living organisms. Endeavour 34:59–66.

——. 1976a. The pollination of *Ophrys* orchids. Bot. Notiser 129:11–20.

——. 1976b. Hymenoptera aculeata males as pollinators of *Ophrys* orchids. Zoological Scripta 5:13–23.

Kullenberg, B., G. Bergstrom, B. Bringer, B. Carlberg, and B. Cederberg. 1973. Observations on the scent marking by *Bombus* Latr. and *Psithyrus* Lep. males and localization of site of production of the secretion. Zoon Suppl. 1:23–30, 2 plates.

Lopez, F. 1963. Two attractants for *Eulaema tropica* L. J. Econ. Entomol. 56:540.

Maas, P. J. M. 1972. Costoideae (Zingiberaceae). Flora Neotropica. Monograph No. 8. Hafner, New York. 140 pp.
Mansfeld, R. 1932a. Über die Heteranthie und das System der Gattung *Catasetum* L. C. Rich. Festschrift Deutsch. Bot. Ges. 50A:92–108.
——. 1932b. Die Gattung *Catasetum* L. C. Rich. Repert. Sp. Nov. 30:257–275, 31:99–125.
Meeuse, B. J. D. 1961. The story of pollination. Ronald Press, New York.
——. 1978. The physiology of some sapromyophilous flowers, p. 97–104. *In* A. J. Richards (ed.), The pollination of flowers by insects. Academic Press, London.
Michener, C. D. 1962. An interesting method of pollen collecting by bees from flowers with tubular anthers. Rev. Biol. Trop. 10:167–75.
——. 1974. The social behavior of the bees. Harvard Univ. Press, Cambridge, Mass.
Michener, C. D., M. L. Winston, and R. Jander. 1978. Pollen manipulation and related activities and structures of bees of the family Apidae. Univ. Kansas Sci. Bull. 51:575–601.
Mori, S. 1970. The ecology and uses of the species of *Lecythis* in Central America. Turrialba 20:344–350.
Mori, S., and J. A. Kallunki. 1976. Phenology and floral biology of *Gustavia superba* (Lecythidaceae) in central Panama. Biotropica 8:184–192.
Mosquin, T. 1970. The reproductive biology of *Calypso bulbosa* (Orchidaceae). Can. Field Nat. 84:291–296.
Moure, J. S. 1946. Notas sôbre as mamangabas. Bol. Agric. Curitiba. 4:21–50.
——. 1947. Novos agrupamentos genéricos e algumas especies novas de abelhas sulamericanas. Museu Paranaense Publ. Avulsas No. 3:1–37.
——. 1950. Contribução para o conhecimento do gênero *Eulaema* Lepeletier (Hymen.-Apidoidea). Dusenia 1:1–18.
——. 1960a. Abelhas da região neotropical descritas por G. Gribodo (Hymenoptera-Apoidea). Bol. Univ. Paraná, Zool., 1:1–18.
——. 1960b. Notes on the types of the neotropical bees described by Fabricius (Hymenotera: Apoidea). Studia Entomol. 3:97–160.
——. 1960c. Notas sôbre os tipos de abelhas do Brasil descritas por Perty em 1833 (Hymenoptera-Apoidea). Bol. Univ. Paraná, Zool., 6:1–23.
——. 1963. Una nueva especie de *Eulaema* de Costa Rica (Hymenoptera, Apoidea). Rev. Biol. Trop. 11:211–216.
——. 1964. A key to the parasitic euglossine bees and a new species of *Exaerete* from Mexico (Hymenoptera, Apidae). Rev. Biol. Trop. 12:15–18.
——. 1965. Some new species of euglossine bees (Hymenoptera: Apidae). J. Kansas Entomol. Soc. 38:266–277.
——. 1967a. Descrição de algumas espécies de Euglossinae (Hym., Apoidea). Atas do Simpósio sôbre a Biota Amazônica 5:373–394.
——. 1967b. A checklist of the known euglossine bees (Hymenoptera, Apidae). Atas do Simpósio sôbre a Biota Amazônica 5:395–415.
——. 1969a. The central American species of *Euglossa* subgenus *Glossura* Cockerell, 1917 (Hymenoptera, Apidae). Rev. Biol. Trop. 15:227–247.
——. 1969b. Abelhas euglossinas e orquídeas. Ciência e Cultura 21:467–468.
——. 1970. The species of euglossine bees of Central America belonging to the subgenus *Euglossella* (Hymenoptera, Apidae). An. Acad. Brasil. Cienc. 42:147–157.
——. 1976. Notas sôbre os exemplares tipos de *Euplusia* descritos por Mocsary (Hymenoptera: Apidae). Studia Entomol. 19:262–314.
Myers, J., and M. D. Loveless. 1976. Nesting aggregations of the euglossine bee *Euplusia surinamensis* (Hymenoptera: Apidae): Individual interactions and the advantage of living together. Can. Entomologist 108:1–6.
Newton, G. D., and N. H. Williams. 1978. Pollen morphology of the Cypripedioideae and the Apostasioideae (Orchidaceae). Selbyana 2:169–182.
Nierenberg, L. 1972. The mechanism for the maintenance of species integrity in sympatrically occurring equitant oncidiums in the Caribbean. Amer. Orchid Soc. Bull. 41:873–882.
Nilsson, L. A. 1978. Pollination ecology and adaptation in *Platanthera chlorantha* (Orchidaceae). Bot. Notiser 131:35–51.
——. 1979. Anthecological studies on the lady's slipper, *Cypripedium calceolus* (Orchidaceae). Bot. Notiser 132:329–347.

Percival, M. S. 1965. Floral biology. Pergamon Press, London.
Pijl, L. van der, and C. H. Dodson. 1966. Orchid flowers: Their pollination and evolution. Univ. Miami Press, Coral Gables, Fla.
Porsch, O. 1955. Zur Biologie der Catasetum-Blüte. Oesterr. Bot. Z. 102:117–157.
Prance, G. T. 1976. The pollination and androphore structure of some Amazonian Lecythidaceae. Biotropica 8:235–241.
Rao, V. S. 1969. The floral anatomy and relationships of the rare apostasias. J. Indian Bot. Soc. 68:374–385.
——. 1974. The relationships of the Apostasiaceae on the basis of floral anatomy. Bot. J. Linn. Soc. 68:318–327.
Ricklefs, R. E., R. M. Adams, and R. L. Dressler. 1969. Species diversity of *Euglossa* in Panama. Ecology 50:713–716.
Roberts, R. B., and C. H. Dodson. 1967. Nesting biology of two communal bees, *Euglossa imperialis* and *Euglossa ignita* (Hymenoptera: Apidae), including description of larvae. Ann. Entomol. Soc. Amer. 60:1007–1014.
Runeckles, V. C., and T. J. Mabry (eds.). 1973. Terpenoids: Structure, biogenesis, and distribution. Recent Advances in Phytochemistry, vol. 6. Academic Press, New York.
Sakagami, S. F. 1965a. Über den Bau der männlichen Hinterschiene von *Eulaema nigrita* Lepeletier (Hymenoptera, Apidae). Zool. Anz. 175:347–354.
——. 1965b. Über den Nestbau von zwei *Euplusia*-Bienen (Hymenoptera, Apidae). Kontyu 33:11–16.
Sakagami, S. F., S. Laroca, and J. S. Moure. 1967. Two Brazilian apid nests worth recording in reference to comparative bee sociology, with description of *Euglossa melanotricha* Moure sp. n. (Hymenoptera, Apidae). Annot. Zool. Japon. 40:45–54.
Sakagami, S. F., and C. D. Michener. 1965. Notes on the nests of two euglossine bees, *Euplusia violacea* and *Eulaema cingulata* (Hymenoptera, Apidae). Annot. Zool. Japon. 38:216–222.
Sakagami, S. F., and H. Sturm. 1965. *Euplusia longipennis* (Friese) und ihre merkwürdigen Brutzellen aus Kolumbien (Hymenoptera: Apoidea). Insecta Matsumurana 28:83–92, 6 pls.
Schill, R. 1978. Palynologische Untersuchungen zur systematischen Stellung der Apostasiaceae. Bot. Jahrb. Syst. 99:353–362.
Schill, R., and W. Pfeiffer. 1977. Untersuchungen an Orchideenpollinien unter besonderer Berucksichtigung ihrer Feinskulpturen. Pollen et Spores 19:5–118.
Schmid, R. 1969. The pollination of *Polycycnis barbata* (Stanhopeinae) by the euglossine bee *Eulaema speciosa*. Orchid Digest 33:220–223.
Smith, G. R., and G. E. Snow. 1976. Pollination ecology of *Platanthera* (*Habenaria*) *ciliaris* and *P. blephariglottis* (Orchidaceae). Bot. Gaz. 137:133–140.
Stallberg-Stenhagen, S., E. Stenhagen, and G. Bergstrom. 1973. Analytical techniques in pheromone studies. Zoon Suppl. 1:77–82.
Stoutamire, W. P. 1967. Flower biology of the lady's-slippers (Orchidaceae: *Cypripedium*). Michigan Botanist 6:159–175.
——. 1968. Mosquito pollination of *Habenaria obtusata* (Orchidaceae). Michigan Botanist 7:203–212.
——. 1974a. Relationships of the purple-fringed orchids *Platanthera psychodes* and *P. grandiflora*. Brittonia 26:42–58.
——. 1974b. Australian terrestrial orchids, thynnid wasps, and pseudocopulation. Amer. Orchid Soc. Bull. 43:13–18.
——. 1975. Pseudocopulation in Australian terrestrial orchids. Amer. Orchid Soc. Bull. 44:226–233.
Sweet, H. R. 1973. Orquideas andinas poco conocidas. VII. *Schlimia* Planchon & Linden ex Lindley. Orquideología 8:3–14.
——. 1974. Orquideas andinas poco conocidas. X. *Lycomormium* Rchb. f. Orquideología 9:183–199.
Takhtajan, A. 1969. Flowering plants: Origin and dispersal. Smithsonian Institution Press, Washington, D.C.
Tengo, J. 1979. Odour-released behavior in *Andrena* male bees (Apoidea, Hymenoptera). Zoon 7:15–48.
Tengo, J., and G. Bergstrom. 1976. Comparative analyses of lemon-smelling secretions from heads of *Andrena* F. (Hymenoptera, Apoidea) bees. Comp. Biochem. Physiol. 55B:179–188.
——. 1977. Comparative analyses of complex secretions from heads of *Andrena* bees (Hym., Apoidea). Comp. Biochem. Physiol. 57B:197–202.
Thien, L. B. 1969a. Mosquitoes and *Habenaria obtusata* (Orchidaceae). Mosquito News 29:252–255.
——. 1969b. Mosquito pollination of *Habenaria obtusata* (Orchidaceae). Amer. J. Bot. 56:232–237.

——. 1971. Orchids viewed with ultraviolet light. Amer. Orchid Soc. Bull. 40:877–880.

Thien, L. B., and B. G. Marcks. 1972. The floral biology of *Arethusa bulbosa, Calopogon tuberosus,* and *Pogonia ophioglossoides* (Orchidaceae). Can. J. Bot. 50:2319–2325.

Thien, L. B., and F. Utech. 1970. The mode of pollination in *Habenaria obtusata* (Orchidaceae). Amer. J. Bot. 57:1031–1035.

Thorne, R. F. 1976. A phylogenetic classification of the Angiospermae. Evol. Biol. 9:35–106.

Vermeulen, P. 1965. The place of *Epipogium* in the system of Orchidales. Acta Bot. Neerl. 14:230–241.

Vogel, E. F. de. 1969. Monograph of the tribe Apostasieae (Orchidaceae). Blumea 17:313–350.

Vogel, S. 1954. Blütenbiologische Typen als Elemente der Sippengliederung. Botanische Studien 1:1–338.

——. 1963a. Duftdrüsen im Dienste der Bestäubung: Über Bau und Funktion der Osmophoren. Akad. Wiss. Lit. (Mainz), Abh. Math.-Naturwiss. Kl., Jahrgang 1962:599–763.

——. 1963b. Das sexuelle Anlockungsprinzip der Catasetinen- und Stanhopeen-Blüten und die wahre Funktion ihres sogenannten Futtergewebes. Oesterr. Bot. Z. 100:308–337.

——. 1966a. Scent organs of orchid flowers and their relation to insect pollination, p. 253–259. *In* L. R. DeGarmo (ed.), Proc. 5th World Orchid Conf., Long Beach, Calif.

——. 1966b. Parfümsammelnde Bienen als Bestäuber von Orchidaceen und *Gloxinia.* Oesterr. Bot. Z. 113:302–361.

——. 1966c. Pollination neotropischer Orchideen durch duftstoffhoselnde Prachtbeinen-Männchen. Naturwiss. 53(7):181–182.

——. 1967. "Parfümblumen" und parfümsammelnde Bienen. Umschau in Wissenschaft und Technik 10(67):327.

——. 1974. Ölblumen und ölsammelnde Bienen. Akad. Wiss. Lit. (Mainz), Math-Naturwiss. Kl., Tropische und subtropische Pflanzenwelt 7:283–547.

Webster, G. L., and W. S. Armbruster. 1979. A new euglossine-pollinated species of *Dalechampia* (Euphorbiaceae) from Mexico. Brittonia 31:352–357.

Wiehler, H. 1976. A report on the classification of *Achimenes, Eucodonia, Gloxinia, Goyazia,* and *Anetanthus* (Gesneriaceae). Selbyana 1:374–404.

——. 1978. The genera *Episcia, Alsobia, Nautilocalyx,* and *Paradrymonia* (Gesneriaceae). Selbyana 5:11–60.

Wille, A. 1963. Behavioral adaptations of bees for pollen collecting from *Cassia* flowers. Rev. Biol. Trop. 11:205–210.

Williams, C. A. 1979. The leaf flavonoids of the Orchidaceae. Phytochem. 18:803–813.

Williams, N. H. 1974. Taxonomy of the genus *Aspasia* Lindley (Orchidaceae: Oncidieae). Brittonia 26:333–346.

——. 1979. Subsidiary cells in the Orchidaceae: Their general distribution with special reference to development in the Oncidieae. Bot. J. Linn. Soc. 78:41–66.

Williams, N. H., J. T. Atwood, and C. H. Dodson. 1981. Floral fragrance analysis in *Anguloa, Lycaste,* and *Mendoncella* (Orchidaceae). Selbyana 5:291–295.

Williams, N. H., and C. R. Broome. 1976. Scanning electron microscope studies of orchid pollen. Amer. Orchid Soc. Bull. 45:699–707.

Williams, N. H., and C. H. Dodson. 1972. Selective attraction of male euglossine bees to orchid floral fragrances and its importance in long distance pollen flow. Evolution 26:84–95.

Williams, N. H., and R. L. Dressler. 1976. Euglossine pollination of *Spathiphyllum* (Araceae). Selbyana 1:349–356.

Wilson, E. O. 1971. The insect societies. Harvard Univ. Press, Cambridge, Mass.

Zucchi, R., B. L. de Oliveira, and J. M. F. Camargo. 1969a. Notas bionômicas sôbre *Euglossa* (*Glossura*) *intersecta* Latreille 1938 e descrição de suas larvas e pupa (Euglossini, Apidae). Bol. Univ. Federal do Paraná, Zool. 3:203–224.

Zucchi, R., S. F. Sakagami, and J. M. F. de Camargo. 1969b. Biological observations on a neotropical parasocial bee, *Eulaema nigrita,* with a review on the biology of Euglossinae (Hymenoptera, Apidae). A comparative study. J. Fac. Sci. Hokkaido Univ., Series VI, Zool., 17(2):271–380.

5

Carbon Fixation in Orchids*

POPURI NAGESWARA AVADHANI, CHONG JIN GOH,
ADISHESHAPPA NAGARAJA RAO, and JOSEPH ARDITTI

*The literature survey pertaining to this chapter was concluded in March 1980; the chapter was submitted in April 1980, and the revised version was received in May 1980.

Introduction

Essentially there are three photosynthetic pathways by which green plants fix carbon dioxide (Fig. 5-1). The most common among these is the Calvin-Benson (C3) pathway. In this sequence ribulose-bisphosphate (RUBP) is the carbon acceptor and the first stable product is 3-phosphoglycerate. This pathway occurs either alone or with other sequences in all green plants (Kelly and Latzko, 1976).

The C4 (Hatch and Slack) pathway was discovered relatively recently (Kortschak *et al.*, 1965; Hatch and Slack, 1970). In this pathway phosphoenolpyruvate (PEP) is the initial carbon acceptor, and the product is oxaloacetate, which is readily converted to malate or aspartate. The malate is then decarboxylated to produce CO_2 which is refixed by RUBP carboxylase. C4 pathway occurs primarily in plants of tropical origin growing under high light intensity and high temperature.

A third pathway, involving crassulacean acid metabolism (CAM), is operative mainly in succulents. The initial reactions in CAM, namely, the fixation of CO_2 to malate, are similar to the C4 pathway except that they occur in the dark. During periods of illumination, malate is decarboxylated to yield CO_2 which is fixed via the C3 pathway.

A comparison of the characters associated with these pathways shows adaptation to specific conditions.

Both C3 and CAM have been shown to operate in orchids (Table 5-1). The possibility that C4 exists in orchids has also been suggested (Arditti, 1979; Avadhani *et al.*, 1978; Kluge and Ting, 1978).

C3 Photosynthesis

All thin-leaved orchids investigated to date fix carbon via the C3 (Calvin-Benson) pathway (Adams, 1970; Arditti, 1979; Arditti and Dueker, 1968; Avadhani and Goh, 1974; Avadhani *et al.*, 1978; Bendrat, 1929; Dueker and Arditti, 1968; Erickson, 1957a, b; Hew, 1976; McWilliams, 1970; Neales and Hew, 1975; Nuernbergk, 1963; Rubenstein *et al.*, 1976; Warburg, 1886–1888; Wong and Hew, 1973). The thin-leaved orchids have fewer layers of smaller mesophyll cells and a larger number of stomata than thick-leaved species (Table 5-2). Their mesophyll cells change in shape gradually from round near the upper epidermis to oblong-rectangular close to the lower epidermis. Neither chloroplast dimorphism nor Kranz anatomy is present (Goh, 1975).

Chlorophyll a/b ratios in the orchids *Arundina graminifolia, Bromheadia finlaysoniana,* and *Spathoglottis plicata* are within the range normally found in other plants which fix carbon via this pathway (Higgs, 1973; Avadhani and Goh, 1974). The CO_2 compensation points of these plants as well as *Coelogyne mayeriana, C. rochussenii, Cymbidium sinense,* and *Eulophia keithii* are close to or greater than 50 ppm (Higgs, 1973; Wong and Hew, 1973). Thin-leaved orchids were found to have a ^{13}C value of approximately −26.6% (Neales and Hew, 1975). These characteristics are indicative of C3 photosynthesis.

Direct confirmation of C3 fixation in orchids was obtained from ^{14}C feeding experiments. In mature leaves of *Arundina graminifolia* 37.8% of ^{14}C fixed during 5-second

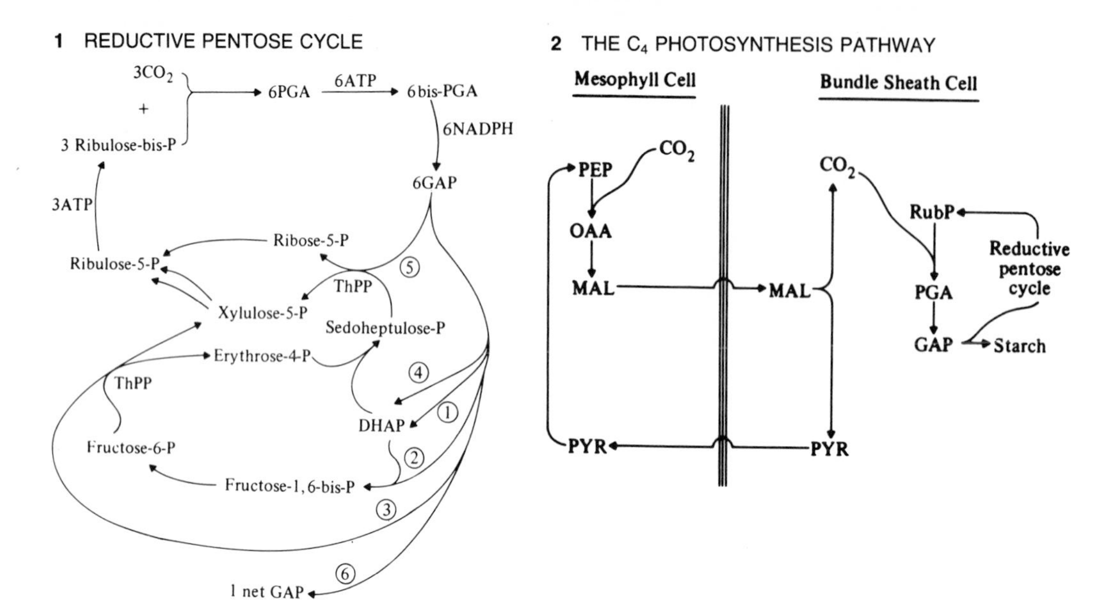

3 THE CAM PATHWAY

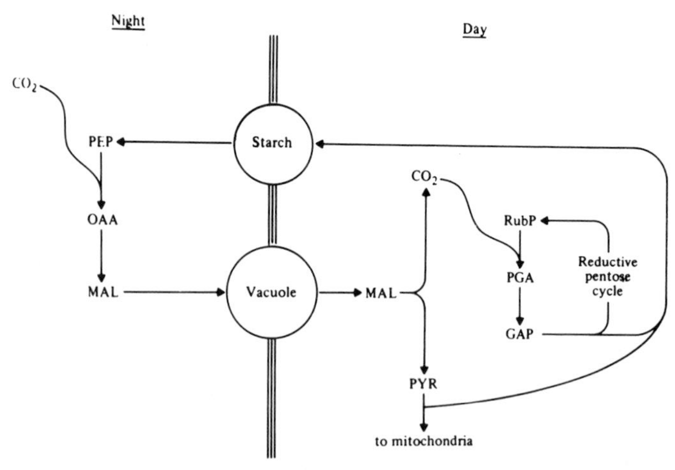

Fig. 5-1. Pathways of carbon fixation. **1:** The reductive pentose cycle, also known as C3 or Calvin-Benson pathway. **2:** The C4 or Hatch and Slack pathway. **3:** The Crassulacean acid metabolism (CAM) pathway of carbon fixation. Abbreviations: ATP, adenosine triphosphate; DHAP, dihydroxyacetone phosphate; GAP, glyceraldehyde phosphate; MAL, malic acid; NADPH, niacinamide dinucleotide phosphate, reduced; OAA, oxaloacetic acid; P, phosphate; PEP, phosphoenolpyruvic acid; PGA, phosphoglyceric acid; PYR, pyruvic acid; RubP, ribulose-bis-phosphate; ThPP, thiamine pyrophosphate. (Illustrations reproduced from Irwin P. Ting, *Plant Physiology,* © 1982. Addison-Wesley, Reading, Mass. Figures 16.6, 16.11, and 16.14. Reprinted with permission.)

period of photosynthesis was present in phosphorylated compounds (Avadhani and Goh, 1974). During a 60-second exposure to light 31.6% of ^{14}C fixed was found in 3-phosphoglycerate, 45.7% in hexose monophosphate, and 7.8% in hexose diphosphate (Higgs, 1973). These figures are in line with those for other C3 plants. Additional radioactivity was found in malate (4–8%), aspartate (0–6%), and citrate and/or isocitrate (5–8%), among others. In *Bromheadia finlaysoniana,* phosphorylated compounds were the main products of ^{14}C fixation in both young and mature leaves (Avadhani and Goh, 1974). Similarly, phosphoglycerate was found to be the primary product of a 20-second $^{14}CO_2$ feeding experiment in *Spathoglottis plicata* (Wong and Hew, 1973).

Stomata of the C3 orchids *Arundina graminifolia, Bromheadia finlaysoniana,* and *Spathoglottis plicata* are open during the day and reach maximum stomatal conductance at noon in Singapore. Variations in the acidity of their guard cells are small (Goh *et al.,* 1977).

C4 Photosynthesis

Although the C4 pathway has not been demonstrated conclusively in orchids (Hew, 1976) there are several indications that it may be operative in some species (Arditti, 1979; Avadhani, 1976). In the young leaves of *Arundina graminifolia,* 24.6% of ^{14}C fixed during a 5-second period of illumination was present as malate. This value decreased to 14.5% following a 180-second fixation. These figures prompted the suggestion that these leaves may photosynthesize "at least in part" via the C4 pathway (Avadhani and Goh, 1974). Also, only malate and aspartate were detected in epidermis-free leaves of *Arachnis* cv Maggie Oei following a 10-second exposure to $^{14}CO_2$ in the light (Avadhani *et al.,* 1978). This tends to suggest that *Arachnis* cv Maggie Oei is capable of both C4 and CAM (see below). A similar suggestion (CAM at night, C4 during the day) has been made regarding a *Schombocattleya* hybrid (Rubenstein *et al.,* 1976).

Many orchids grow in habitats suitable for C4 photosynthesis, but few have been investigated. Therefore, it is not unreasonable to assume that C4 orchids may be discovered. However, the results of such studies (and in fact data from most carbon-fixation experiments) should be interpreted with great caution since what may appear as C4 intermediates may well be the products of residual dark fixation.

Crassulacean Acid Metabolism (CAM)

The earliest report on the occurrence of CAM among orchids was the demonstration of the fluctuation of titratable acidity in *Cattleya* (Warburg, 1886–1888). All succulent leaves of orchids investigated to date exhibit CAM (Table 5-1, 5-2; Arditti, 1979; Goh *et al.,* 1977; Sanders, 1979). As pointed out earlier (Kluge and Ting, 1978), the main criterion for CAM is not the succulence of the leaf as such but that of the mesophyll cells. Among orchids, all those known to exhibit CAM have mesophyll cells 60 m in diameter or larger (Tables 5-1, 5-2).

Characters that point to the existence of CAM in leaves include cell succulence, diurnal fluctuations in titratable acidity, fluctuations of malic acid content, fixation of $^{14}CO_2$ into malate in the dark, C4-like δ ^{13}C values as well as nocturnal stomatal opening (Table 5-3). However, caution is necessary in the evaluation of leaf analyses since most plants synthesize large amounts of malic acid under low light intensities and high CO_2 levels.

Members of the genera *Arachnis, Aerides,* and *Vanda* were shown to exhibit qualitative

Table 5-1. Carbon-fixation patterns in some orchids

Orchid	Organ	Probable pathway	Remarks	Reference
Diandrae				
Cypripedioideae				
Cypripedium acaule	Leaves (0.41 mm thick)	No CAM		McWilliams, 1970
Paphiopedilum barbatum	Leaf	C3	Species photorespires	Hew, 1976; Wong and Hew, 1973
P. insigne	–	CAM		Nuernbergk, 1961
P. venustum	Leaf (0.38 mm thick)	No CAM		McWilliams, 1970
P. villosum	Leaf (soft)	No CAM		Bendrat, 1929
P. cv Mildred Hunter	Leaf (soft, leathery)	CAM at night, C3 during day		Rubenstein *et al.*, 1976
Monandrae				
Basitonae				
Ophrydoideae				
Habenarieae				
Habenaria platyphylla	Whole plant	C3		Raghavendra and Das, 1976
Platanthereae				
Aceras anthropophorus	Labellum	Cyclic photophosphorylation		Schmid *et al.*, 1976
Acrotonae				
Polychondroideae				
Listereae				
Neottia nidus avis	Plastids from labellum; whole plant	Only PSI is operative; no light or dark carbon fixation	Mycotrophic species	Menke and Schmid, 1976; Reznik, 1958
Cephalanthereae				
Limodorum abortivum	All parts of the plant	Minimal fixation in light despite presence of chlorophyll	Probably mycotrophic	Chatin, 1874–1875; Griffon, 1898, 1899
Vanilleae				
Vanilla aromatica	Whole plant	CAM		Nuernbergk, 1963
V. fragrans	Leaf (0.22 mm thick)	CAM		McWilliams, 1970
V. planifolia	Leaf (thick)	CAM		Bendrat, 1929; Warburg, 1886–1888
V. sp.	Leaves	CAM		Coutinho, 1969
Sobralieae				
Arundina graminifolia	Leaf (0.3 mm thick)	No CAM; C3	Species photorespires	Avadhani and Goh, 1974; Goh *et al.*, 1977; Hew, 1976; Neales and Hew, 1975; Wong and Hew, 1975
Thunia marshalliana	–	CAM		Nuernbergk, 1961
Spirantheae				
Spiranthes speciosa	Leaf (0.51 mm thick)	No CAM		McWilliams, 1970
Erythrodeae				
Goodyera pubescens	Leaf (0.43 mm thick)	No CAM		McWilliams, 1970

Kerosphaeroideae				
Pleurothallideae				
Pleurothallis ophiocephalus	–	CAM-like $\delta\ ^{13}C$ values		Szarek and Ting, 1977
Collabieae				
Tainia penangiana	Leaf	C3	Species photorespires	Hew, 1976; Wong and Hew, 1975
Coelogyneae				
Coelogyne cristata	Leaves	CAM		Nuernbergk, 1961, 1963
C. massangeana	Leaves (plicate)	C3		Rubenstein *et al.*, 1976
C. mayeriana	Leaf (0.4 mm thick)	C3	Species photorespires	Hew, 1976; Neales and Hew, 1975; Neales and Hew, 1975
C. rochussenii	Leaf (0.2 mm thick)	C3	Species photorespires	Hew, 1976; Wong and Hew, 1975; Neales and Hew, 1975
Laelieae				
Brassavola perrinii	Leaf	Nocturnal CO_2 uptake		Coutinho, 1964; Szarek and Ting, 1977
Cattleya autumnalis	Leaves	CAM		Coutinho, 1969
C. bicolor	Leaves	CAM		Coutinho, 1969
C. forbesii	Leaf	CAM		Hew, 1976; Neales and Hew, 1975
C. gigas	Roots, stems, leaves	Fixation in the light		Dycus and Knudson, 1957
C. guttata	Leaves	CAM		Coutinho, 1969
C. intermedia	Leaves	CAM		Coutinho, 1969
C. labiata	Whole plant	CAM		Nuernbergk, 1961, 1963
C. loddigesii	Whole plant	CAM		Nuernbergk, 1961, 1963
C. mossiae	Whole plant	CAM		Nuernbergk, 1961, 1963
C. skinneri	Whole plant	CAM		Nuernbergk, 1961, 1963
C. trianae	Whole plant	CAM		Nuernbergk, 1961, 1963
C. walkeriana	Whole plant	CAM		Nuernbergk, 1961, 1963
C. warneri	Whole plant	CAM		Nuernbergk, 1961, 1963
C. sp.	Leaf (thick)	CAM		Warburg, 1886–1888
C. sp.	Roots, leaves	Fixation in the light		Erickson, 1975a,b
C. cv Bow Bells	Leaf (0.5 mm thick)	CAM		Hew, 1976; Neales and Hew, 1975
C. cv White Blossom 'Stardust' × cv Bob Betts 'Glacier'	Whole plant	CAM		Knauft and Arditti, 1969
Encyclia atropurpurea	Leaves	CAM		Nuernbergk, 1961, 1963
E. flabellifera	Leaves	CAM		Coutinho, 1969
E. odoratissima	Leaves	CAM		Coutinho, 1969
Epidendrum alatum	Leaf (1.40 mm thick)	CAM		McWilliams, 1970
E. ciliare	Leaves (succulent)	CAM		Bendrat, 1929
E. ellipticum	Leaves	CAM		Coutinho, 1963, 1964, 1965
E. floribundum	Leaves	CAM		Coutinho, 1969; Nuernbergk, 1961
E. moseni	Leaves	CAM		Coutinho, 1969

(*continued*)

Table 5-1.—Continued

Orchid	Organ	Probable pathway	Remarks	Reference
E. radicans	Leaves (2.16 mm thick)	CAM		McWilliams, 1970
E. schomburgkii	Leaves	CAM		Nuernbergk, 1961
E. xanthinum	Roots, stems, leaves	Fixation in the light		Dycus and Knudson, 1957
Laelia cinnabarina	Leaves	CAM		Coutinho, 1969
L. crispa	Leaves	CAM		Coutinho, 1969
L. flava	Leaves	CAM		Coutinho, 1969
L. millerii	Leaves	CAM		Coutinho, 1969
L. perrinii	Leaves	CAM		Coutinho, 1969
L. purpurella	Leaves	CAM		Coutinho, 1969
L. xanthina	Leaves	CAM		Coutinho, 1969
Lanium avicula	Leaves	CAM		Coutinho, 1969
Schomburgkia crispa	Leaves	CAM		Coutinho, 1969; Nuernbergk, 1961
Sophronitis cernua		Nocturnal CO_2 uptake		Coutinho, 1969
Brassolaeliocattleya cv Maunalani	Leaf	Nocturnal CO_2 uptake and acidification		McWilliams, 1970
Schombocattleya hybrid	Leaves	CAM at night; C3 during day		Rubenstein *et al.*, 1976
Dendrobieae				
Dendrobium taurinum	Leaf (1.5 mm thick)	CAM		Neales and Hew, 1975
Polystachyeae				
Bromheadia finlaysoniana	Leaf (thin)	No CAM; C3		Avadhani and Goh, 1974; Goh *et al.*, 1977
Phajeae				
Aplectrum hyemale	Leaf	No CAM		Adams, 1970
Calanthe vestita	Leaves	CAM		Nuernbergk, 1963
Spathoglottis plicata	Leaf (0.3 mm thick)	No CAM; C3	Species photorespires	Goh *et al.*, 1977; Hew, 1976; Neales and Hew, 1975
Bulbophylleae				
Bulbophyllum gibbosum	Leaf (1.42 mm thick)	CAM		McWilliams, 1970
Cyrtopodieae				
Cyrtopodium paranaensis	Leaves	CAM		Coutinho, 1969
Eulophia keithii	Leaf	C3	Species photorespires	Hew, 1976; Wong and Hew, 1975
Cymbidieae				
Cymbidium chinense	Leaves (thin)	CAM (?)		Warburg, 1886–1888
C. sinense	Leaves	C3	High rate of ^{14}C fixation in the dark	Wong and Hew, 1973
C. lowianum 'Yorktown'	Leaves	CAM		Nuernbergk, 1961
C. cv Chelsea	Sepals, petals, leaves	C3	High rate of fixation in the dark	Arditti and Dueker, 1968; Dueker and Arditti, 1968

C. cv Independence Day	Sepals, petals, leaves	C3	High rates of fixation in the dark	Arditti and Dueker, 1968; Dueker and Arditti, 1968
C. hybrid	Leaves (thin)	C3		Rubenstein *et al.*, 1976
Cataseteae				
Catasetum fimbriatum	Leaves	C3		Nuernbergk, 1963
Gongoreae				
Acropera loddigesii	Pseudobulbs	No CAM		Warburg, 1886–1888
Maxillarieae				
Maxillaria aromatica	Leaves, pseudobulb	No CAM		Warburg, 1886–1888; Coutinho, 1963
Ornithidium densum	Leaf (thick)	CAM		Warburg, 1886–1888
Oncidieae				
Oncidium flexuosum	Leaves (0.33 mm thick)	C3	Species photorespires	Neales and Hew, 1975; Wong and Hew, 1975
O. lanceanum	Leaves (0.33 mm thick)	C3		Hew, 1976
O. pumilum	Leaves	C3		Coutinho, 1969
O. sphacelatum	Leaves (thin)	No CAM; C3	Species photorespires	Bendrat, 1929; Wong and Hew, 1975
O. sp.	Leaves (type unknown)	CAM	Some *Oncidium* species have thick leaves	Warburg, 1886–1888
Sarcantheae				
Aerides odoratum	Leaves (2.0 mm thick)	CAM		Avadhani *et al.*, 1978
Angraecum sesquipedale	Leaves	CAM		Nuernbergk, 1963
Arachnis flos-aeris	Leaf (1.5 mm thick)	CAM		Avadhani *et al.*, 1978
A. hookeriana	Leaf (1.5 mm thick)	CAM		Avadhani *et al.*, 1978
A. cv Maggie Oei	Leaf (1.5 mm thick)	CAM		Goh *et al.*, 1977; Lee, 1970; Neales and Hew, 1975
Ascocentrum ampullaceum	Leaf (1.270 mm thick)	CAM		McWilliams, 1970
Phalaenopsis amabilis	Leaves	CAM		McWilliams, 1970
P. schilleriana	Leaves	Nocturnal CO_2 uptake and acidification		McWilliams, 1970
P. sp.	Leaf (thick)	CAM		Borris, 1967
P. sp.	Root, stem, leaves	Fixation in the light		Dycus and Knudson, 1957
Vanda tessellata	Whole plant	C3		Raghavendra and Das, 1976
V. sp.	Root, stem, leaves	Fixation in the light		Dycus and Knudson, 1957
V. cv Miss Joaquim	Leaves (4.0 mm thick)	CAM		Khan, 1964; Avadhani *et al.*, 1978
Aeridachnis cv Bogor	Leaves (1.5 mm thick)	CAM		Avadhani *et al.*, 1978
Aranda cv Deborah	Leaf (1.5 mm thick)	CAM		Goh *et al.*, 1977
A. cv Wendy Scott	Leaf (1.5 mm thick)	CAM		Hew, 1976; Neales and Hew, 1975
Aranthera cv James Storie	Leaf (1.5 mm thick)	CAM		Neales and Hew, 1975

Table 5-2. Leaf characteristics of some orchids (Goh, 1975)

Orchid	Cuticle thickness (μ)		Stomatal density[a]		Stomatal size (μ)		Mesophyll					Leaf thickness (mm)
	Upper epidermis	Lower epidermis	Upper epidermis	Lower epidermis	Outer ledges	Central pore	No. of cell layers	Cell size (μ)	Palisade layer	Fibers[b]	Cell shape	
Arundina graminifolia	2	2	0	18000	30 × 30	10	11–12	20 × 18 18 × 18 18 × 30 18 × 36	Slightly differentiated	–	Round to oblong	0.3
Bromheadia finlaysoniana	14	9	0	17500	21 × 21	6	10	28 × 40 28 × 50	Slightly differentiated	–	Rectangular	0.4
Bulbophyllum vaginatum	7	7	0	3000	24 × 24	9	12–15	40 × 110 56 × 140 70 × 112 70 × 56 70 × 70	Slightly differentiated	–	Elongated to round	1.8
Dendrobium crumenatum	6	3	0	6500	24 × 24	6	18–20	70 × 56 56 × 42	–	B	Round	1.0
Epidendrum radicans	12	8	0	3300	30 × 30	14	13–15	112 × 140 98 × 140 98 × 84	–	B	Hexagonal	2.0
Phalaenopsis violacea	6	3	0	800	9 × 9	3	12	112 × 140 112 × 84 98 × 98 84 × 84 70 × 70	–	–	Round	2.0
Spathoglottis plicata	2	2	0	14000	24 × 24	9	5	28 × 28	–	–	Round	0.3
Thrixspermum calceolus	14	14	0	800	110 × 70	28	16–18	140 × 140 130 × 140 110 × 110	–	–	Round	1.7
Arachnis												
cv Maggie Oei	9	9	0	4000	42 × 42	14	12–15	70 × 70	–	+	Round, isodiametric	1.2
Aranda												
cv Deborah	14	11	0	3000	42 × 42	14	18–21	60 × 60 70 × 70	–	++	Round, isodiametric	1.6
cv Hilda Galistan	14	14	0	2800	42 × 42	14	18–20	70 × 70 56 × 70 98 × 84	Slightly differentiated	++	Round, polyhedral to slightly elongated	1.6

cv Lucy Laycock	14	11	0	3000	42 × 56	14	15–18	84 × 84 84 × 98 112 × 112 140 × 140	Slightly differentiated	++	Round to slightly elongated	1.7
cv Nancy	14	14	0	3100	42 × 42	14	11–14	140 × 84 126 × 98 84 × 70	Slightly differentiated	++	Slightly elongated near upper epidermis, round	1.6
cv Wendy Scott	14	14	0	3000	42 × 42	14	16–18	112 × 70 112 × 60 84 × 70	–	++	Slightly elongated	1.5
Aranthera												
cv James Storie	14	11–14	0	3400	42 × 42	14	12–13	112 × 112 84 × 84 70 × 70	–	+	Round	1.8
Dendrobium												
cv Caesar	6	6	0	3800	30 × 30	13	15	98 × 98 84 × 84 70 × 70	–	–	Round	1.5
Oncidium												
cv Golden Shower	3	3	0	7500	28 × 28	13	10–12	56 × 40 40 × 40 30 × 30	–	B	Round	0.5
Vanda												
cv Miss Joaquim	14	14	0	2500	50 × 42	14	35	140 × 112 140 × 98 140 × 70 112 × 112 98 × 98 84 × 84	Slightly differentiated	–	Slightly elongated to round	4.0[c]
cv Tan Chay Yan	14	14	0	2400	50 × 42	14	–	168 × 84 168 × 70 154 × 112 154 × 70	–	++	Slightly elongated	2.0
cv TMA	14	14	0	2400	50 × 42	14	22–24	168 × 70 154 × 112 154 × 70	–	++	Slightly elongated	2.0

[a] Stomatal density is expressed as number of stomata/cm^2 of leaf surface area.
[b] This refers to fiber cells not associated with vascular bundles; −, absent; +, present; ++, present in abundance; B, present in the form of bundles.
[c] Diameter.

Table 5-3. Characteristics of C3, C4, and CAM plants

Characteristics	C3	C4	CAM
Structural			
Succulence	–	Some cells are succulent	Cell succulence
Kranz anatomy (presence of chloroplast-containing bundle sheath)	Absent	Present	Absent
Chloroplast dimorphism	Absent	Not always present	Absent
Peripheral reticulum of mesophyll and bundle sheath chloroplasts	Absent	Present	Absent
Biochemical			
Initial atmospheric CO_2-fixing enzyme	RUDP carboxylase	PEP carboxylase	PEP carboxylase
Product	3-PGA	OAA	OAA
Temperature optima for the primary CO_2-fixing enzyme	25°C	35°C	5°C
ATP requirement (moles ATP/moles CO_2 fixed)	3	5	4
Chlorophyll a/b ratios (approximate)	2/1	4.5/1	2/1
δ^{13} mode	−7.4‰ to −89.2‰	−9‰ to −14‰	−11‰ to −33‰
Physiological			
Photorespiration	Present	Reduced or not apparent	Present
CO_2 compensation point	50	5–50	–
Light saturation	2000 lux	Saturates at full sunshine	–
Effect of low O_2 concentration	Increase in fixation rate	No increase in fixation rate	Decrease in fixation rate
Transpiration ratio (water utilization per unit dry matter production)	450–950 g/gDW	250–350 g/gDW	18–20 g/gDW
Maximum rate of net photosynthesis	15–40 $mgCO_2/dm^2/h$	40–80 $mgCO_2/dm^2/h$	1–4 $mgCO_2/dm^2/h$
Maximum growth rate	0.5–2 $gdw/dm^2/d$	4–5 $gdw/dm^2/d$	0.015–0.018 $gdw/dm^2/d$
Ecological			
Most common habitat	Mesic	Tropical, open spaces with seasonal aridity	Arid, epiphytic
Habit	Mesophytic	Halophytic or xerophytic	Mesophytic/xerophytic

and quantitative characteristics of CAM (Avadhani *et al.*, 1978). They exhibit diurnal fluctuation of titratable acidity and dark $^{14}CO_2$ fixation into malate as well as low-temperature enhancement (Figs. 5-2 to 5-4). In *Arachnis* cv Maggie Oei, 25% of ^{14}C fixed by acid-depleted leaves in the dark over a period of 22 hours appeared in malate (Lee, 1971). Thus it seems that this orchid could fix CO_2 by CAM as well as the C4 pathway as pointed out earlier. In *Aranda* and *Cattleya,* malate was either the only or the principal product of dark $^{14}CO_2$ fixation (Avadhani *et al.*, 1978; Knauft and Arditti, 1969).

The δ ^{13}C values of *Aranda* cv Wendy Scott, *Aranthera* cv James Storie, and *Cattleya forbesii,* were in the range of −15% to −16%, indicating their CAM or C4 nature (Neales and Hew, 1975).

The stomatal rhythms of all succulent orchids are consistent with CAM in that they open at night (Goh *et al.*, 1977). However, it is interesting to note that nocturnal opening does not occur unless the leaves have been sufficiently deacidified during the day (Fig. 5-5).

In another CAM orchid, *Vanda* cv Miss Joaquim (Table 5-4), there appears to be some correlation between flowering and acidity fluctuations (Avadhani *et al.*, in press). Interestingly, this plant flowers throughout the year in the equatorial regions where night temperatures are less variable than in temperate areas. Its flowering is relatively profuse during winter months in Northern Thailand where conditions are more conducive for greater acidity fluctuations.

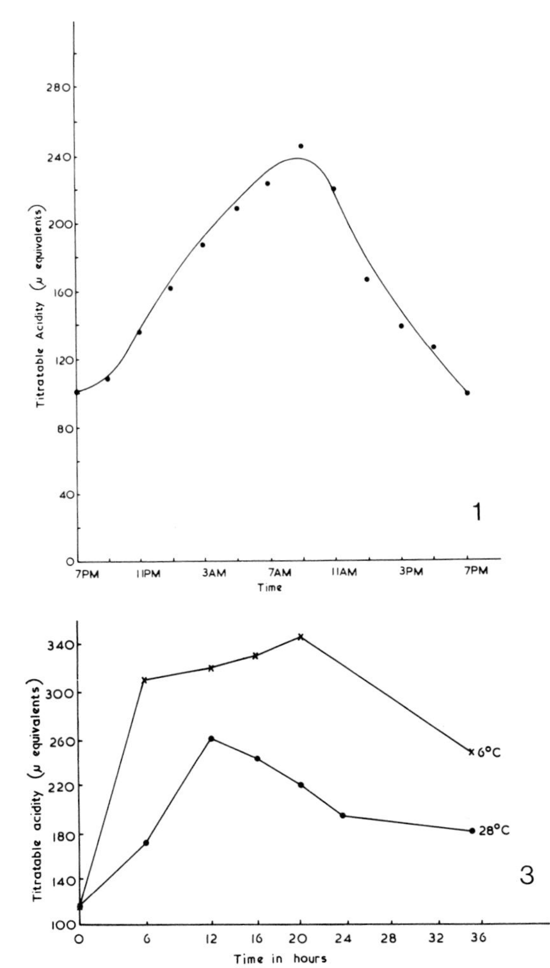

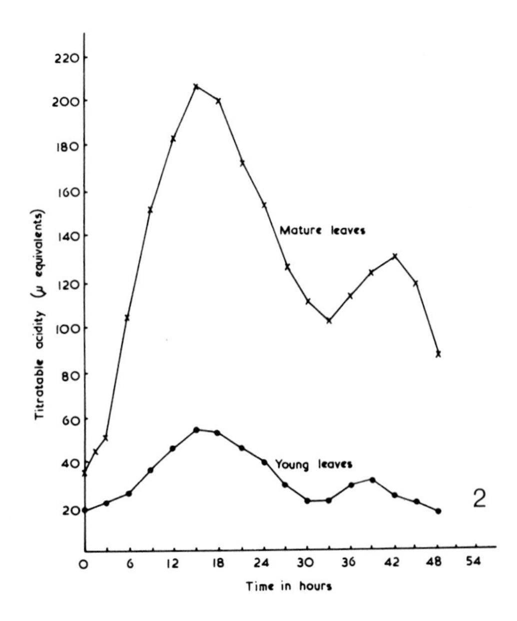

Fig. 5-2. Acidity in orchid leaves (reproduced from Avadhani *et al.*, 1978). **1:** Titratable acidity of leaves of *Vanda* cv Miss Joaquim at different times of the day. **2:** Changes of titratable acidity of young and mature *Vanda* leaves kept in the dark. **3:** Effect of temperature on titratable acidity of leaves of *Vanda* cv Miss Joaquim.

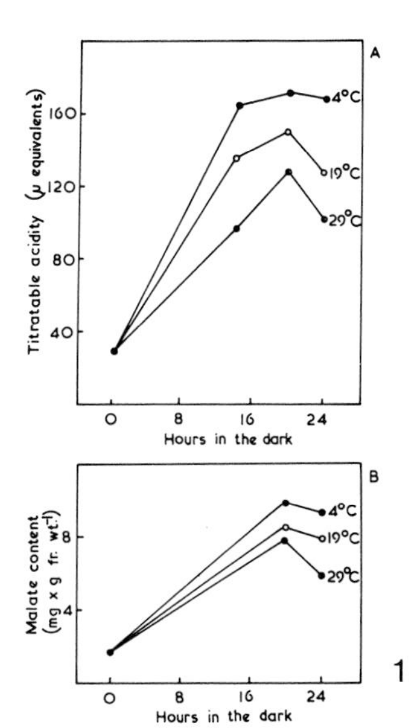

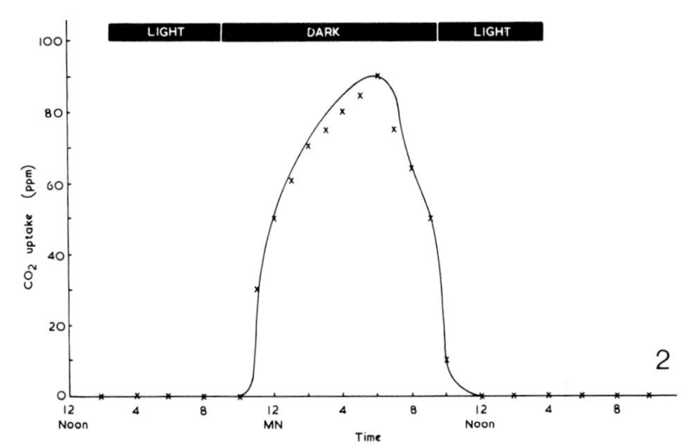

Fig. 5-3. Acidification and CO_2 uptake by *Arachnis* (reproduced from Avadhani *et al.*, 1978). **1:** The effect of temperature on dark acidification of *Arachnis* cv Maggie Oei, showing changes in titratable acidity (A), and changes in malate content (B). **2:** Uptake of carbon dioxide in the dark by leaves of *Arachnis* cv Maggie Oei.

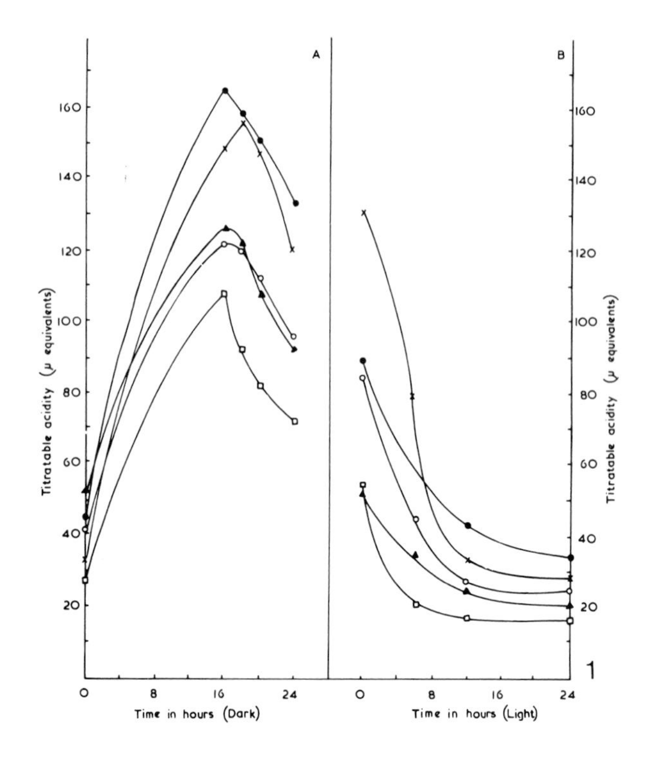

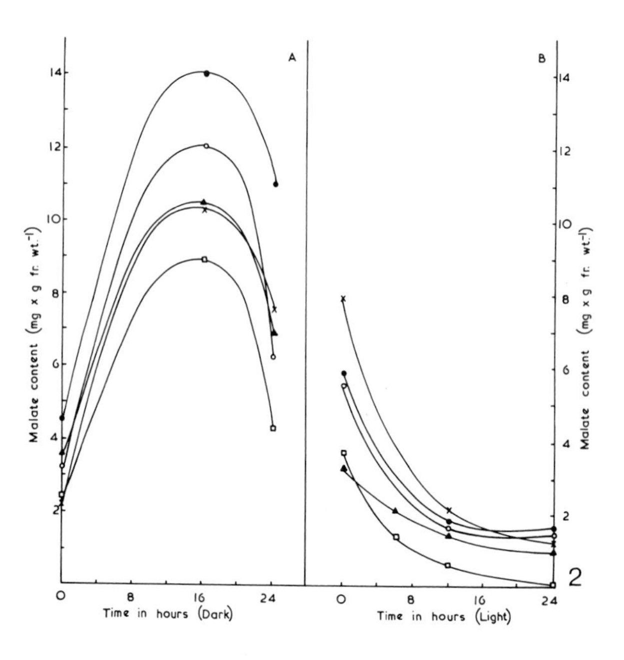

Fig. 5-4. Titratable acidity and malate content in leaves of succulent orchids (reproduced from Avadhani *et al.*, 1978). Explanation of symbols: x, *Arachnis* cv Maggie Oei; closed circle, *A. flos-aeris;* open circle, *A. hookeriana* var *luteola;* open square, *Aerides odoratum;* closed triangle, *Aeridachnis* cv Bogor. **1:** Changes in titratable acidity of leaves of different succulent orchids in the dark (A) and in the light (B). **2:** Changes in malate content of leaves of different succulent orchids in the dark (A) and in the light (B).

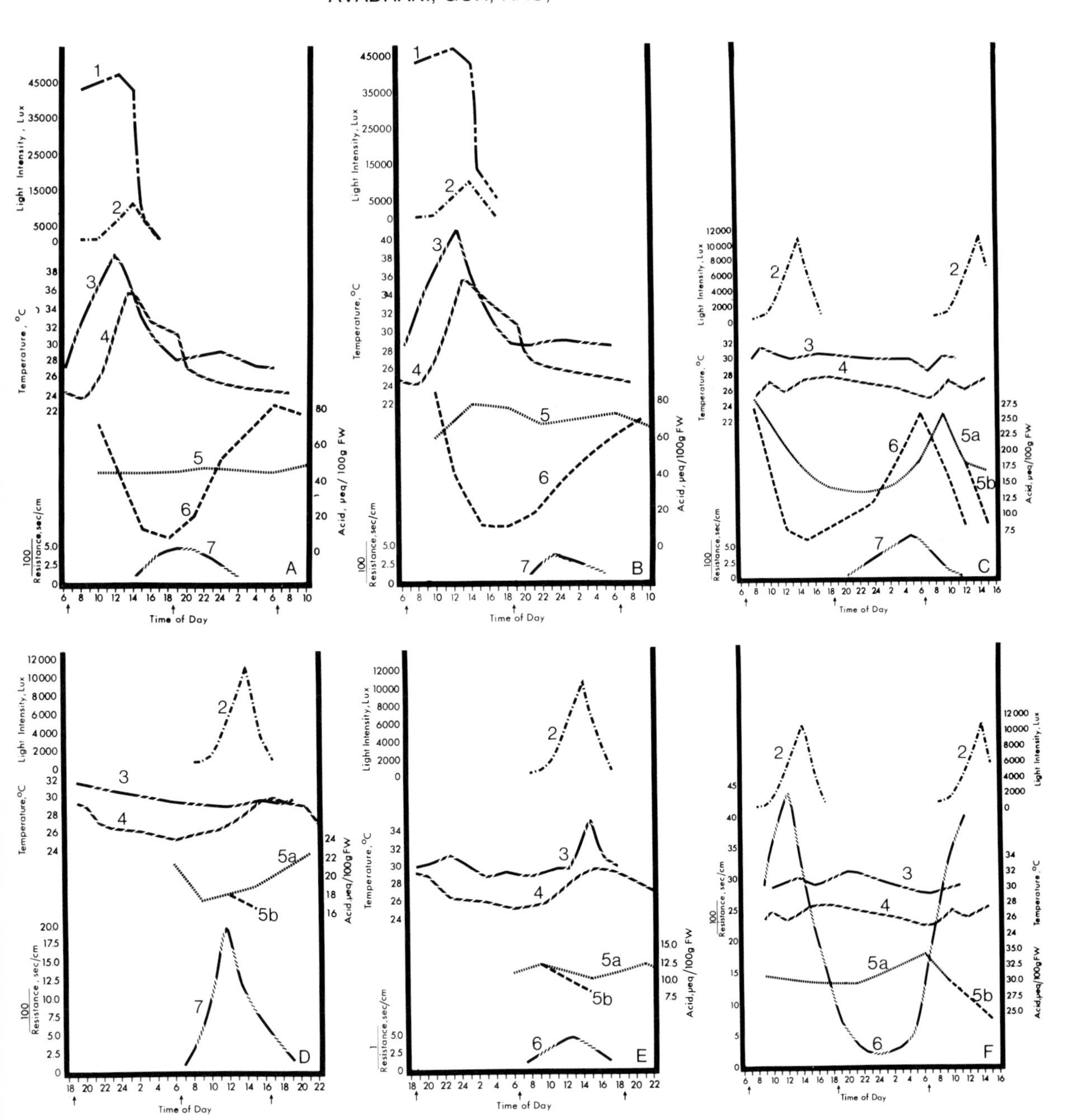

Fig. 5-5. Stomatal and acidity rhythms in orchids (Goh *et al.*, 1977). A: *Arachnis* cv Maggie Oei; B: *Aranda* cv Deborah; C: *Cattleya bowringiana* × *C. forbesii;* D: *Arundina graminifolia;* E: *Bromheadia finlaysoniana;* F: *Spathoglottis plicata.* Explanation of symbols: 1, light intensity in open field (full sun); 2, light intensity in partial shade; 3, leaf temperature; 4, air temperature; 5 or 5a, acidity of leaves in partial shade; 5b, acidity of leaves moved from partial shade to full sun in open; 6, acidity of leaves under full sun; 7, stomatal opening as reciprocal of resistance ($100/r_1$); arrows indicate time of dawn and dusk.

Table 5-4. Titratable acidity of *Vanda* cv Miss Joaquim leaves at different positions on the plant (Avadhani *et al.*, 1978)

Leaf position	Titratable acidity (μequivalents/gFw)[a]
Adjacent to flower buds	190
Adjacent to open flowers	169
Not adjacent to any flower or flower bud	156

[a] Each value is an average of 7 determinations.

The magnitude of malic acid production, changes in titratable acidity (malic and isocitric), and $^{14}CO_2$ fixation may vary with genera and species and their hybrids (Table 5-5). Such variation can be the result of either enzyme activities or vacuolar storage capacity.

Mixed Pathways

It has been suggested that some orchids may fix carbon by more than one pathway (Avadhani *et al.*, 1978; Avadhani and Goh, 1974; Lee, 1971); the predominance of any one pathway depends upon age and/or internal acidity. As in other CAM plants, leaf age significantly affects the amplitude of acidification in *Vanda* cv Miss Joaquim (Khan, 1964).

Short-term products of ^{14}C fixation by succulent leaves of *Sedum* can be either those of C3 or C4 depending upon its internal acidity (Avadhani *et al.*, 1971). In *Bryophyllum* it was shown that only young leaves which have not yet developed their maximum potentials for acidification can fix external CO_2. With increase in age the capacity to fix external CO_2 decreased concomitantly with the elevation of CAM potential (even after removal of the epidermis). The products of this fixation were malate and aspartate. Thus it is possible that the young leaves of some CAM orchids may prove to be C4 (but see cautionary note in the section on CAM).

The report that a *Paphiopedilum* hybrid is C3 during the day and CAM at night (Rubenstein *et al.*, 1976) may be misleading since all CAM plants utilize the C3 pathway during the day for the fixation of internal or external CO_2 (Fig. 5-1). There are also reports that some *Paphiopedilum* plants did not take up CO_2 in the dark and failed to

Table 5-5. Maximum titratable acidity, malate content, and relative ^{14}C incorporation in certain orchids (Avadhani *et al.*, 1978)

Acidity/ CO_2 fixation[a]	*Arachnis hookeriana*	*Arachnis flos-aeris*	*Arachnis* cv Maggie Oei	*Aerides odoratum*	*Aeridachnis* cv Bogor
Titratable acidity (μeq/gFw)	98.5	128.0	127.8	92.4	106.0
Malate (mg/gFw)	10.6	12.4	9.2	8.9	9.4
^{14}C fixation (cpm/gFw)	4×10^4	7×10^4	4.5×10^4	6.3×10^4	3.8×10^4

[a] Total ^{14}C activity after 18 hours fixation in the dark.

acidify (Nuernbergk, 1963; Bendrat, 1929; McWilliams, 1970). Thus, either *Paphiopedilum* lacks CAM or the lack of dark acidification may be due to inadequate deacidification during the preceding day, as in the case of *Aranda* cv Deborah (Goh *et al.*, 1977).

Carbon Fixation by Nonfoliar Organs

Nonfoliar organs of several orchids also fix carbon. In green *Cymbidium* flowers the amounts fixed decrease with flower age (Fig. 5-6). Roots of *Epidendrum xanthinum*, *Cattleya gigas*, and *Phalaenopsis* hybrids fix more carbon than their leaves (Table 5-6). All carbon fixation in autotrophic leafless species occurs in the roots (Arditti, 1979). However, in aerial roots of *Aranda*, although CO_2 fixation occurred, the gas-exchange data indicated that there was no net CO_2 uptake in light (Hew, 1976). Green *Cattleya* roots can also carry out photosynthesis (Erickson, 1957a).

Ecology

Orchidaceae is the largest family of flowering plants and its members occupy a great many ecological niches (Holttum, 1953). As a result a large number of morphological, anatomical, and biochemical (e.g., carbon-fixation pathways) adaptations can be found among the orchids. Epiphytic orchids, for example, exist in xeric habitats. Under such conditions adaptive xerophytic characters are of survival advantage for the orchids.

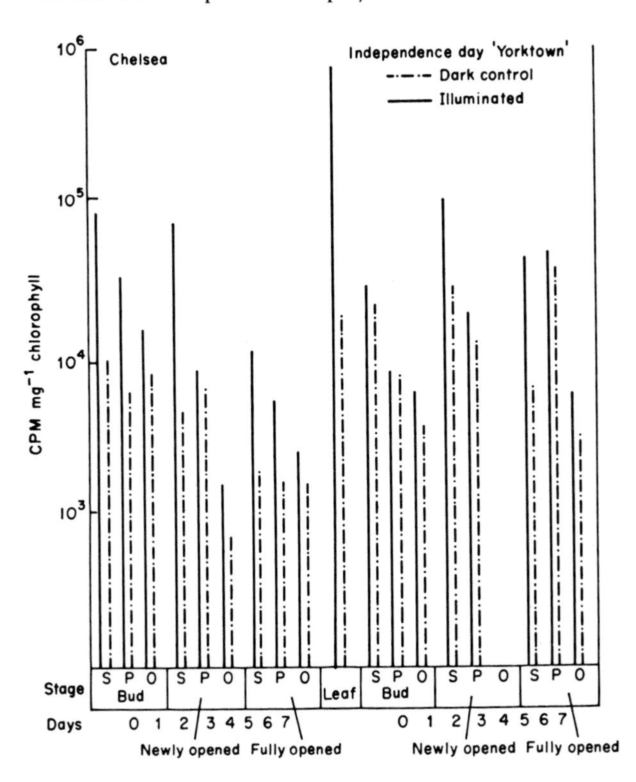

Fig. 5-6. $^{14}CO_2$ fixation by *Cymbidium* buds, leaves, and newly and fully opened flowers. S, sepals; P, petals; O, ovary. (Reproduced from Arditti, 1979, with permission from *Advances in Botanical Research* 7:421–655 [1979]. Copyright: Academic Press Inc. [London] Ltd.)

Table 5-6. Net carbon fixation in the light by orchid plant organs expressed as percent of the amount fixed by leaves (From Arditti, 1979, with permission from *Advances in Botanical Research* 7:421–655 [1979]. Copyright: Academic Press, Inc. [London] Ltd.)

	Cymbidium						
Organ	× Chelsea	× Indep. Day "Yorktown"	*Epidendrum xanthinum*	*Cattleya gigas*	*Phalaenopsis* hybrids	*Vanda suavis*	*Cattleya*
Bud							
Sepals	13.6	1.0					
Petals	5.5	0.1					
Ovary	1.2	0.4					
Newly opened flowers							
Sepals	10.7	11.9					
Petals	0.5	1.0					
Ovary	0.1						
Fully opened flowers							
Sepals	1.4	4.9					
Petals	0.6	1.1					
Ovary	0.2	0.4					
Lower leaves			100			100	
Upper leaves			111			129	
Stems and leaves			80				
Stem				113			
Leaf and stem					100		
Root			1814	645	3689		
Leaf	100	100		100			
1 year old							100
2 years old							103
3 years old							103
4 years old							91.6
5 years old							70
6 years old							50
7 years old							38

CAM results in carbon fixation by the orchids at night when both humidity and CO_2 levels are high. Many orchids, terrestrial and epiphytes alike, exist under mesic conditions. Such orchids can be expected to fix carbon by the C3 pathway, and indeed they do (for a review see Arditti, 1979).

C4 would be beneficial for orchids which grow in tropical areas with elevated temperatures and high light intensities (Goh *et al.*, 1977), but it is not a requirement for survival as indicated by the fact that most plants in these environments are C3.

Phylogeny

The taxonomic range of orchids studied with respect to CO_2 fixation and/or other partial reactions of photosynthesis (Table 5-1) is wide, covering 47 genera and 93 species belonging to 26 tribes of the two subfamilies, Diandrae and Monandrae, plus a number of hybrids. However, even this large number represents only a small fraction of the Orchidaceae (600–800 genera and 20,000–30,000 species). Furthermore, many of the reports are empirical observations or preliminary surveys. Nevertheless, the available data suggest that as in the case of CAM plants (Kluge and Ting, 1978) the carbon-fixation patterns of orchids are, perhaps, more closely related to their habitats and anatomical characters than to their phylogenetic position within the family.

Practical Applications and Potential

There have been several horticultural attempts (for a review see Arditti, 1979) to fertilize orchids with CO_2. For the most part the results have been variable (Anonymous, 1941, 1965; Davidson, 1967; Miwa, 1937; Quis, 1980; Tsuchiya, 1935). *Cymbidium* seedlings seem to be an exception in that they have been reported to respond positively to CO_2 enrichment (Borg, 1965; Wright, 1967). There could be several reasons for the mixed results: (1) *Cymbidium* species have thin leaves and are apparently C3 plants which would respond to additional CO_2 during the day. And, in most reported cases, CO_2 was provided during daylight hours. (2) In some cases CAM orchids were provided with CO_2 during the day when they were incapable of fixing it. Even if these orchids were deacidified and had open stomata, levels of fixation would be small. (3) If the total levels of CO_2 were not adjusted to the compensation points of the individual orchid, maximum response would not have been attained. (4) Unless the added CO_2 was mixed continuously with the air surrounding the orchid, considerable boundary-layer resistance would develop between the plant and the atmosphere, thereby effectively reducing CO_2 availability. (5) Even if additional CO_2 was provided in the dark to CAM orchids, it could have been given following light periods inadequate in intensity or duration for deacidification.

Clearly, then, successful CO_2 enrichment must be based on more precise information about stomatal rhythms, CO_2-fixation pathways, endogenous rhythms of acidification and deacidification, light intensity requirements, and CO_2 compensation points.

Literature Cited

Adams, M. S. 1970. Adaptations of Aplectrum hyemale to the environment: Effects of preconditioning temperature on net photosynthesis. Bull. Torrey Bot. Club 97:219–224.

Anonymous. 1941. Air fertilization of orchids. Orchid Digest 5:189–190.

——. 1965. CO_2 enrichment. Orchid Rev. 73:108.

Arditti, J. 1979. Aspects of orchid physiology. Adv. Bot. Res. 7:421–655, Academic Press, London.

Arditti, J., and J. Dueker. 1968. Photosynthesis by various organs of orchid plants. Amer. Orchid Soc. Bull. 37:862–866.

Avadhani, P. N. 1976. Carbon dioxide fixation in orchids. Proc. 8th World Orchid Conf., Frankfurt (1975), p. 412–413.

Avadhani, P. N., and C. J. Goh. 1974. Carbon dioxide fixation in the leaves of Bromheadia finlaysoniana and Arundina graminifolia (Orchidaceae). J. Singapore Nat. Acad. Sci. 4:1–4.

Avadhani, P. N., I. Khan, and Y. T. Lee. 1978. Pathways of carbon dioxide fixation in orchid leaves, p. 1–12. *In* E. S. Teoh (ed.), Proc. Symp. on Orchidology. Orchid Soc. S.E. Asia, Singapore.

Avadhani, P. N., C. B. Osmond, and K. K. Tan. 1971. Crassulacean acid metabolism and the C4 pathway of photosynthesis in succulent plants, p. 288–293. *In* M. D. Hatch, C. B. Osmond, and R. O. Slayter (eds.), Photosynthesis and photorespiration. Wiley-Interscience, New York.

Bendrat. M. 1929. Zur physiologie der organischen Säuren in grünen pflanzen. VI. Ein beitrag zur kenntnis des säurestoffwechsels sukkulenter pflanzen. Planta 7:508–584.

Benedict, C. R. 1978. The fractionation of stable carbon isotopes in photosynthesis. What's New in Plant Physiol. 9:13–16.

Borg, F. 1965. Some experiments in growing Cymbidium seedlings. Orchid Rev. 73:278–280.

Borriss, H. 1967. Kohlenstoff—Assimilation und diurnaler Saurerythmus epiphytischer Orchideen. Die Orchidee 18:396–406.

Chatin, J. 1874–1875. Sur la presence de la chlorophylle dans le Limodorum abortivum. Rev. Sci. Nat. 3:236–240.

Coutinho, L. M. 1963. Algumas informaçṏes sôbre a ocorrência do "efeito de De Saussure" em epifitas e herbáceas terrestres de mata pluvial. Botânica 20:81–98.

——. 1964. Untersuchungen über die Lage des Lichtkompensations punktes einiger Pflanzen zu verschiedenen Tageszeiten mit besonderer Berücksichtigung des "de Saussure effektes" bei sukkulenten, p. 1–8. *In* K. Kreeb (ed.), Beitrage zur Physiologie. Verlag Eugen Ulmer, Stutgart.

——. 1965. Algumas informaçṏes sôbre a capacidade rítmica diária da fixacao e acumulação de CO_2 no escuro em epifitas e herbáceas te estres da mata pluvial. Bôtanica 21:395–408.

——. 1969. Novas obervaçṏes sôbre a ocorrência do "Efeito de De Saussure" e suas relaçoes com a suculência, a temperatura folhear e os movimentos estomaticos. Bôtanica 24:77–102.

——. 1970. Sôbre a assimilação noturna de CO_2 em orquideas e bromélias. Cíencia e Cultura 22:364–368.

Coutinho, L. M., and C. A. F. Schrage. 1970. Sôbre o efeito da temperatura na ocorrência de fixação noturna de CO_2 em orquideas e bromélias. An. Acad. Brasil. Ciec. 42:843–849.

Davidson, O. W. 1967. Question box. Amer. Orchid Soc. Bull. 36:811.

Dueker, J., and J. Arditti. 1968. Photosynthetic $^{14}CO_2$ fixation by green Cymbidium (Orchidaceae) flowers. Plant Physiol. 43:130–132.

Dycus, A. M., and L. Knudson. 1957. The role of the velamen of the aerial roots of orchids. Bot. Gaz. 119:78–87.

Erickson, L. C. 1957a. Respiration and photosynthesis in Cattleya roots. Amer. Orchid Soc. Bull. 26:401–402.

——. 1957b. Leaf age in Cattleya. Amer. Orchid Soc. Bull. 26:560–563.

Goh, C. J. 1975. The anatomy of orchid leaves. Malayan Orchid Rev. 12:19–23.

Goh, C. J., P. N. Avadhani, C. S. Loh, C. Hanegraaf, and J. Arditti. 1977. Diunal stomatal and acidity rhythms in orchid leaves. New Phytol. 78:365–372.

Griffon, M. (ed.) 1898. L'assimilation chlorophyllienne chez les orchidées terrestres et en particulier chez le Limodorum abortivum. Compt. Rend. Physiol. Veg. 127:973–976.

——. 1899. L'assimilation chlorophyllienne chez les orchidées terrestres et en particulier chez le "Limodorum abortivum." Ann. Sci. Nat., Bot. Ser. 18, 10:71–76.

Hatch, M. D., and C. R. Slack. 1970. Photosynthetic CO_2 fixation pathways. Ann. Rev. Plant Physiol. 21:141–162.

Hew, C. S. 1976. Patterns of CO_2 fixation in tropical orchid species. Proc. 8th World Orchid Conf., Frankfurt (1975), p. 426–430.

Higgs, R. E. A. 1973. Some aspects of photosynthesis among plants of the "Adinandra Belukar." B.Sc. Hons. thesis, Botany Dept., Univ. of Singapore.

Holttum, R. E. 1953. Orchids of Malaya. Government Printing Office, Singapore.

Kelly, G. J., and E. Latzko. 1976. Regulatory aspects of photosynthetic carbon metabolism. Ann. Rev. Plant Physiol. 27:181–205.

Khan, I. 1964. Preliminary investigations into the Crassulacean acid metabolism of Vanda Miss Joaquim. B.Sc. Hons. thesis, Botany Dept., Univ. of Singapore.

Kluge, M., and I. P. Ting. 1978. Crassulacean acid metabolism. Ecological studies 30. Springer Verlag, Berlin.

Knauft, R. L., and J. Arditti. 1969. Partial identification of dark $^{14}CO_2$ fixation products in leaves of Cattleya (Orchidaceae). New Phytol. 68:657–661.

Kortschak, H. P., C. E. Hartt, and G. O. Burr. 1965. Carbon dioxide fixation in sugar cane leaves. Plant Physiol. 40:209–213.

Lee, Y. T. 1970. Carbon dioxide fixation in orchid leaves. B.Sc. Hons. thesis, Botany Dept., Univ. of Singapore.

McWilliams, E. L. 1970. Comparative rates of dark CO_2 uptake and acidification in Bromeliaceae, Orchidaceae and Euphorbiaceae. Bot. Gaz. 131:285–290.

Menke, W., and G. H. Schmid. 1976. Cyclic photophosphorylation in the mykotrophic orchid Neottia nidus-avis. Plant Physiol. 57:716–719.

Miwa, A. 1937. Carbon dioxide content of the atmospheric air of the greenhouse. Orchid Rev. 45:146–152.

Neales, T. F., and C. S. Hew. 1975. Two types of carbon assimilation in tropical orchids. Planta 123:303–306.

Noggle, G. R., and G. J. Fritz. 1976. Introductory plant physiology. Prentice-Hall, Englewood Cliffs, N.J.

Nuernbergk, E. L. 1961. Kunstlicht und Pflanzenkultur. BLV. Verlagsgesellschaft, Munich, W. Germany.
——. 1963. On the CO_2 metabolism of orchids and its ecological aspect. Proc. 4th World Orchid Conf., p. 158–169. Straits Time Press, Singapore.
Quis, P. 1980. CO_2-Düngung bei Orchideen. Die Orchidee 31:62.
Raghavendra, A. S., and V. S. R. Das. 1976. Distribution of the C4 dicarboxylic acid pathway of photosynthesis in local monocotyledonous plants and its taxonomic significance. New Phytol. 76:301–305.
Reznik, H. 1958. Vergleich einer Weissen mutante von Neottia nidus-avis (L.) L. C. Rich. mit der braunen Normalform. Planta 51:694–704.
Rubenstein, R., D. Hunter, R. E. McGowan, and C. L. Withner. 1976. Carbon dioxide metabolism in various orchid leaves. Abstr. Proc. Amer. Soc. Plant Physiol. (NE Regional Meeting).
Salisbury, F. B., and C. W. Ross. 1978. Plant physiology. 2d ed. Wadsworth Publishing Co., Belmont, Calif.
Sanders, D. J. 1979. Crassulacean acid metabolism and its possible occurrence in the plant family Orchidaceae. Amer. Orchid Soc. Bull. 48:796–798.
Schmid, G. H., M. Jankowicz, and W. Menke. 1976. Cyclic photophosphorylation and chloroplast structure in the labellum of the orchid *Aceras anthropophorum.* J. Microscop. Biol. Cell 26:25–28.
Szarek, S. R., and I. P. Ting. 1977. The occurrence of crassulacean acid metabolism among plants. Photosynthetica 11:330–342.
Ting, I. P. 1982. Plant Physiology. Addison-Wesley, Reading, Mass.
Tsuchiya, I. 1935. Air fertilization of orchid seedlings. Orchid Rev. 46:330–346.
Warburg, O. 1886–1888. Uber die Bedeutung der orgaischen Säuren fur den Lebensprocess der Pflanzen (speziell der sog. Fettpflanzen). Unters. Bot. Inst. Tubingen 2:51–150.
Wong, S. C., and C. S. Hew. 1973. Photosynthesis and photorespiration in some thin-leaved orchid species. J. Singapore Nat. Acad. Sci. 3:150–157.
——. 1975. Do orchids photorespire? Amer. Orchid Soc. Bull. 44:902–906.
Wright, D. 1967. Carbon dioxide enrichment for cymbidiums. Orchid Rev. 75:120–122.

PHYSIOLOGY

6

Mineral Nutrition of Orchids*

HUGH A. POOLE and THOMAS J. SHEEHAN

*The survey of the literature pertaining to this chapter was concluded in January 1980; the chapter was submitted in April 1980 and the revised version was received in May 1980.

Introduction

Orchidaceae is a large and very diverse plant family, and any attempt to write a thorough review of mineral nutrition or to make generalizations that would cover all aspects would be foolhardy and impossible. Consequently, the major objectives of this chapter are to review significant research findings concerning mineral nutrition, primarily of mature orchids. The discussion will be limited to the more commonly grown genera, but such a limitation is only slightly restrictive as most research has been directed toward the epiphytic or semiepiphytic orchids.

Nutrition of Plants in the Wild

Our knowledge and understanding of mineral nutrition of epiphytes is often prejudiced by our experiences in a terrestrial (i.e., soil) environment. Nutrients are absorbed by plant roots from exchange sites in the soil. Factors such as cation exchange capacity, pH, soluble salts, buffering capacity, air porosity, and others all contribute to the fertility of the soil and the ultimate growth responses of a specific plant. A terrestrial environment also provides a readily available water supply from which the plant can continue to draw moisture (except under drought conditions) without significant stress to the roots or to the plant.

In fertile soils, nutrients are available for exchange and absorption by the plant whenever water is not limiting. Those nutrients not absorbed will generally be available at a later time as leaching of nutrients is relatively slow. Certainly establishment of mycorrhizal relationships in many plants plays an important role in nutrient absorption. Similarly, the chelation of metals by organic compounds in soil humus prevents precipitation of some minerals and facilitates absorption by roots.

The situation of orchids growing in the crotches of trees in a tropical forest is unique and presents special problems regarding nutrients and water uptake. The plant is situated on the limbs of trees in a manner that permits rainwater to fall on it and its roots. The roots provide support for the plant as well as an extensive network for water absorption as they penetrate the bark or merely attach themselves to it. Tree bark holds water for a considerable length of time, much as a sponge does. Orchid roots are also well suited for water absorption by being enveloped by a fleshy velamen layer that acts like a sponge when wetted. Small hairs on the root reduce surface tension and spread water droplets across its surface. A light rainfall or heavy condensation from fog or dew provides moisture which is available for absorption during several hours, and the moist velamen acts as a water reservoir until it dries; after that it can absorb moisture again. Normally humus collects between the rhizomes, roots, and leaves of orchid plants. This organic matter is easily wetted and acts as another reservoir of moisture and also protects orchid roots from drying rapidly. Many species have phyllotaxy that channels water down toward the base of the plant rather than shedding it away from the root system.

Moisture on leaf surfaces may be a significant source of water for some plants that live in very dry locations and have adapted to those conditions. Water flow and availability are key considerations in a discussion of mineral nutrition. When water is available, nutrients are absorbed and plant growth is optimal. When water is not available for extended periods of time, growth is inhibited. Plant growth is generally continuous when the supply of water and nutrients is satisfactory. The evolutionary adaptations of orchids which provide for an adequate supply of moisture should not be taken lightly since they also affect horticultural practices.

The major sources of nutrients are many but are often poorly documented. Orchid embryos are dependent from the onset on mycorrhizal fungi for sugars and possibly other nutrients. This symbiotic relationship is maintained for the entire life span of an orchid plant and facilitates the close interrelationships between orchids and their niche in the ecosystem.

Rainfall is a valuable source of nutrients for epiphytic plants since it washes dust particles out of the air and onto them. The atmosphere is also an excellent source of nitrates, especially during electrical storms. Water flowing over leaf surfaces leaches mineral and organic nutrients from the leaves. Thus the leaf canopy of the host tree becomes a nutrient source that enriches the water before it reaches the orchid plant. The major source of nutrients, however, is probably the slow decomposition of organic matter (both flora and fauna) that accumulates in tree crotches and among the bark, roots, rhizomes, and leaves of orchid plants. This humus, which also supplies moisture to the roots, is composed of bark, leaves, fruits, carrion, feces, bones, and the like, and contains a wealth of nutrients. Many of these nutrients are parts of organic compounds and are therefore not lost due to oxidation or leaching. Some metals may be chelated by organic molecules; this chelation reduces the possibility of leaching and increases uptake. Thus the ecosystem provides nutrients to epiphytic orchids that would otherwise be available only to terrestrial plants.

Early Beliefs regarding Orchid Nutrition

Epiphytic orchids grown in the early traditional media of osmunda or spagnum moss seldom responded to infrequent applications of nutrients. These media provided many of the required nutrients for acceptable growth. Liquid nutrients, if applied, were probably dilute and therefore dispersed quickly through the porous media and were lost. Not surprisingly, few early growers felt that orchids behaved like other commonly cultivated plants. Orchids were considered to be slow-growing plants that received little nutrition from the growing medium and relied on elements in the air for survival.

One of the classical fertilizer recommendations prior to World War II epitomizes these misconceptions. Cow manure (quality and quantity unknown) was placed in a bucket, and water (quantity unknown) was added. The solution was allowed to stand for several days and then the supernatant was poured off. This liquid was then diluted with water (dilution unknown) and used as a fertilizer solution. It was sprinkled liberally upon the hot steam lines in the greenhouse, whereupon it vaporized and its components were subsequently absorbed by the leaves of the orchid plants.

Burberry (1894) had the following recommendations regarding the fertilization of epiphytic orchids. "Guano, in small quantities sprinkled on the floor occasionally, I never knew to do any harm, and I have thought that plants showed increased vigour and strength when it was used in this way. I also use liquid farmyard manure, by pouring the contents of a small watering can on the floor occasionally, and of an evening when all is quiet. The evaporation from such stimulants being desirable, not so much for inducing moisture as for producing a manurial stimulant in the atmosphere."

Fortunately, many orchids are hardy plants and survive despite such tender loving care.

Burberry's suggestions also illustrate another major problem in the early history of orchid growing. Perhaps because of the snob appeal of orchids or a trait of human nature, early orchid growers were typically very tight-lipped about their cultural practices. Early recommendations in the literature were often incomplete or so confusing and complicated that they were of no practical value. Nutrient sources, dilution rates, frequency of application, and other key data were often deleted and ultimately lost to all. With increased interest in orchids, many new growers, and more periodicals on orchids requiring this type of information, such incomplete recommendations are becoming the exception rather than the rule.

Advice regarding orchid fertilization programs was quite varied and often confusing. The first nutritional studies on mature orchid plants were undertaken by Evers and Laurie (1940) using gravel culture. They concluded that orchids could be successfully grown in such inert media as haydite (a porous lightweight aggregate used in the building trade) and silica gravel with periodic applications of chemical nutrient solutions. Scully (1951) stated that orchid plants required little fertilizer because of their slow growth, whereas Adams (1970) recommended five different fertilizers monthly for best growth and flowering. Davidson (1960) and Beaumont and Bowers (1954) noted that growth was more affected by fertilizers than by the potting medium. Fennell (1951) observed best orchid growth in fir bark when plants were given liquid fertilization weekly. Poole and Sheehan (1970) found that antagonisms among nutrients increased when *Cattleya* plants were fertilized more frequently.

There is little information in the literature on nutrient deficiencies and toxicities of orchids. Cibes *et al.* (1949), studying the importance of nutrition on growth and chemical composition of *Vanilla fragrans* grown in gravel culture with Hoagland solutions, found that nitrogen deficiency caused a reduction of fresh and dry weights of the plants and reduced leaf size and stem diameter. Leaves were dark green but had no marginal or tip burn. Plants receiving a solution deficient in calcium grew well. They had a relatively low Ca content in their leaves, but evidently could grow satisfactorily, at least for the first year, on a relatively low calcium supply. Interestingly, similar plants grown in an organic mulch (forest litter) with distilled water had larger leaves and girth and more shoot growth than those grown in gravel with a full nutrient solution (Cibes and Childers, 1947). This finding indicates that the conditions of the experiments were not optimal for plant growth. Cibes *et al.* (1949) later grew *Vanilla* vines in gravel culture with three levels of nitrogen [0, 10, and 81 parts per million (ppm)] and potassium (0, 7, and 40 ppm) in factorial combinations. An increase in either N or K levels resulted in more vine growth.

Best growth occurred with the highest N-K level. Plants receiving low N were chlorotic and those given high N had a dark green color. There was no apparent effect of K on foliage color.

Davidson (1960) observed that in gravel culture deficiencies of N and P limited growth of *Cattleya* more drastically than did K, Ca, or Mg and that plants were more likely to respond to fertilization with N or P or a combination of the two. He reasoned that plants lacking an external supply of K are capable of translocating this nutrient from old tissues and reutilizing it to meet most of the growth requirements of new organs. Likewise, Ca and Mg present in old tissues are reutilized, but to a lesser extent than K. He stated that small amounts of K, Ca, and Mg often present in rooting media or water supplies may satisfy the requirement for these elements by *Cattleya.* Davidson reported that the microelement requirements of slow-growing plants, such as some orchids, were low. Gravel-culture experiments in which *Cattleya* seedlings were grown in purified quartz and nutrient solutions prepared from purified reagents, but without external supplies of Fe or Mn, failed to demonstrate deficiency symptoms after seven months of growth. Under similar conditions, many rapidly growing horticultural plants would have exhibited symptoms of Fe deficiency in a few days. He stated that he had never seen known symptoms of microelement deficiencies exhibited by orchids. However, he had encountered well-defined toxicity symptoms of Fe, Zn, and B. According to Davidson, water supplies, organic potting media, and commercial-grade fertilizers normally provide amounts of microelements adequate to meet requirements for maximum growth of orchids, thus making special applications of microelements unnecessary.

Chin (1966) found that omission of N, P, K, Ca, or Mg severely affected dry weights of *Dendrobium phalaenopsis* hybrid seedlings, but leaves dropped before deficiency symptoms appeared. He also found that root development of plants growing in the deficient culture media was more vigorous than in the complete nutrient solution.

Sheehan (1961) reported that an increased N fertilization level resulted in higher numbers of flowers on *Cattleya* 'Trimos' grown in Ease (cedar tan) and Ivory (fir) barks. Similar studies with *Phalaenopsis* hybrid seedlings showed a significant interaction between N levels and types of barks. Best growth, however, generally occurred at the higher N levels. Davidson (1957) suggested fertilizing every second watering with a 3–1–1 (N–P_2O_5–K_2O) fertilizer for plants grown in fir bark since this medium is deficient in N due to a high C : N ratio. Lunt and Kofranek (1961) stated that high rates of liquid N fertilizer applied weekly promoted vegetative growth at the expense of flowering in *Cymbidium* grown in two grades of fir bark. They suggested heavy N fertilization for two or three months after flowering, followed by a light N program to achieve good vegetative growth without jeopardizing flowering the following year.

Curtis and Spoerl (1948) evaluated NO_3 versus NH_4NO_3 at constant pH levels and reported that both ions were usable by orchid embryos. NH_4NO_3 was superior to NO_3 for *Cymbidium,* slightly better for *Cattleya,* and inferior for *Vanda.* Lunt and Kofranek (1961) found more N in mature leaves, but no growth differences, among *Cymbidium* plants fertilized with NH_4NO_3 compared to those fertilized with $CaNO_3$. Poole (1971) noted that *Cattleya* plants receiving NH_4 fertilization weekly had greater dry weights of roots, greater fresh weight, and more leaf area of new vegetative growth than those receiving NO_3. When plants were similarly fertilized at two- and three-week intervals, there were no differences in growth between plants receiving NH_4 and those receiving NO_3. However, plants receiving NH_4 fertilization at three-week intervals showed greater chlorosis of the leaves after ten months of growth than did plants given NO_3. This chlorosis was probably due to an NH_4-induced K deficiency, resulting from greater antagonisms between the NH_4 and K ions at less frequent fertilization intervals.

Zakrejs (1976), using N ratios established by Penningsfeld and Fast (1970, 1973), found that a ratio of 1.25 N to 0.4 $P_2O_5 : 0.75\ K_2O : 0.1$ MgO was the optimal level and further increases in N did not affect growth. The most suitable form of N in Zakrejs's experiments was urea. He further noted that using NH_4 as the source of N during the dark winter months had no negative effect.

Sheehan (1961) was unable to affect flowering of *Cattleya* 'Trimos' or growth of *Phalaenopsis* seedlings in three hardwood barks by varying levels of P or K. Poole and Sheehan (1970) increased growth of mericloned *Cattleya* plants by higher levels of K fertilization at low P levels. They also induced Ca deficiency and leaf-tip dieback symptoms of new leaves by increasing P fertilization and noted Ca deficiency among several orchid genera in the absence of Ca in the regular fertilizer program (Poole and Sheehan, 1970, 1973a).

Work by Penningsfeld and his co-workers (1970, 1973, 1980) indicated that *Paphiopedilum* is more sensitive to salt concentration in the medium than either *Cattleya* or *Cymbidium*. The nutrient ratios Penningsfeld and coworkers feel are best for the growth of certain orchids are listed in Table 6-1.

Nutrient-culture studies of *Cattleya, Cymbidium,* and *Phalaenopsis* were undertaken by Poole and Seeley (1978). The growth of these plants was compared in factorial experiments using three levels each of N, K, and Mg. N concentration was the most important factor determining growth of all three genera. For *Phalaenopsis* and *Cymbidium,* 100 ppm N appeared to be adequate or near optimal for best growth. *Cymbidium* plants grown with 50 ppm N had a greater incidence of leaf-tip dieback among the older leaves. Plants receiving 200 ppm N were shorter and had fewer leaves and lower root dry weights than those given 100 ppm N. However, the response of *Cattleya* to N levels was drastically different. Plants grown at 50 ppm N had greater leaf and root dry weights and larger leaves than plants receiving 100 or 200 ppm N. They also had greater dry weights of pseudobulbs and flowers and more growths than plants receiving 200 ppm N. Whereas 50 ppm N was optimal for *Cattleya* in these studies, both *Cymbidium* and *Phalaenopsis* responded favorably to higher N levels.

Table 6-1. Optimal nutrient ratios for several orchids (Penningsfeld and Fast, 1970, 1973; Penningsfeld and Forchthammer, 1980)

Genus	pH	Nutrient concentration (%)	Nutrient ratio $N:P_2O_5:K_2O$
Cattleya	4.0	1.0–2.0	1.0:0.4:0.75
Cymbidium	5.0	2.0	1.0:0.4:0.75
Paphiopedilum	5.5–6.5	0.5	1.0:0.8:1.0
Phalaenopsis	4.7	1.0–1.5	1.0:0.8:1.5

The K levels (50, 100, and 200 ppm) tested had virtually no effect on orchid growth. The 50 ppm K level proved adequate for good vegetative growth of orchids, and substantially higher concentrations did not adversely affect plant performance. It should be noted, however, that K levels in plant tissue appeared to be adequate even at this low concentration.

High Mg levels (100 ppm) adversely affected growth of all three orchid genera. Relatively small plants of *Cymbidium* and *Phalaenopsis* responded best to 25 ppm Mg. However, larger plants of *Cattleya* had greater root and leaf dry weights at 50 ppm Mg. It was not determined whether *Cattleya* had a higher requirement for Mg than either *Cymbidium* or *Phalaenopsis* or if the larger physical size of the *Cattleya* plants placed a greater demand on the available Mg.

A major concern in the work by Poole and Seeley (1978) was the dramatic reduction in growth of *Cattleya* due to increasing N levels. Generally, increments of added N result in increased growth up to a point when the N concentration in the tissues becomes toxic or other factors necessary for good growth become limiting. The possibility that a limiting factor(s) was responsible appeared likely since N increments drastically reduced growth of *Cattleya* with only a slight increase in N concentrations in the leaves and a decrease in total absorption by the plant. With *Cattleya,* increasing N levels decreased growth of roots more severely than that of leaves or pseudobulbs. The reduction in root growth could have resulted from a decrease in synthesis of necessary carbohydrates by the leaves and translocation and/or utilization of these carbohydrates by the roots. Their (Poole and Seeley, 1978) conclusion was that high levels of organic acids (due to NO_3 absorption) and antagonisms between cations resulting in potential deficiencies of the cations could both be occurring simultaneously.

Tissue Analysis

There is a dearth of knowledge concerning the chemical composition of various orchid genera. Yet researchers and growers need this basic information to understand the mineral nutrition of these plants and to recommend fertilization practices. Pertinent information about nutrient and metabolic requirements of orchid plants can be derived from chemical analysis of plant parts, especially in well-designed nutritional experiments.

Cattleya

The work of Withner and Van Camp in 1948, Erickson in 1957, and Davidson in 1961 contributed considerably to our early knowledge of the chemical composition of *Cattleya,* and their findings are summarized in Tables 6-2 and 6-3. Withner and Van Camp (1948) made the first intensive study of the tissue analyses of three *Cattleya* species and seedlings of one hybrid with plants growing in osmunda or haydite gravel. Mature leaves and pseudobulbs analyzed together for N, P, K, Ca, Mg, S, Fe, Mn, Zn, and Cu levels were highest in the haydite-grown plants of *C. percivaliana,* and Mn levels were highest in the hybrid seedlings.

Table 6-2. Tissue analyses of major elements of *Cattleya*

		Percent dry weight					
Tissue	Medium	N	P	K	Ca	Mg	S
Lc. Canhamiana alba[a]	Osmunda	1.18	0.17	2.13	1.74	0.75	1.34
X *C.* Eucharis seedlings[a]	Haydite	1.07	0.16	1.41	1.78	0.65	1.20
Cattleya bowringiana[a]							
mature leaves & pseudobulbs	Osmunda	1.03	0.12	2.46	1.78	0.33	1.33
Cattleya gaskelliana[a]							
mature leaves & pseudobulbs	Osmunda	0.85	0.11	1.80	1.17	0.34	1.22
Cattleya percivaliana[a]							
mature leaves & pseudobulbs	Haydite	0.98	0.15	2.54	1.71	0.27	0.83
Cattleya hybrid[b]							
Leaves							
1 year	Osmunda	0.93	0.13	2.61	0.90	0.35	–
2 year		0.74	0.10	2.76	1.62	0.35	–
3 year		0.68	0.09	2.21	1.84	0.36	–
4 year		0.68	0.10	2.13	2.22	0.34	–
5 year		0.59	0.09	1.65	2.51	0.44	–
6 year		0.57	0.09	1.88	1.41	0.36	–
7 year		0.56	0.08	1.49	1.90	0.41	–
Cattleya trianaei[c]							
Leaves							
1 year	Bark	–	0.08	2.58	1.09	0.26	–
2 year		–	0.06	1.27	2.88	0.24	–
3 year		–	0.07	1.25	2.61	0.26	–
4 year		–	0.05	0.94	2.46	0.57	–
1 year	Osmunda	–	0.07	3.18	0.61	0.27	–
2 year		–	0.03	2.26	1.37	0.55	–
3 year		–	0.03	1.73	1.31	0.45	–
4 year		–	0.03	1.99	1.37	0.64	–
Pseudobulbs							
1 year	Bark	–	0.06	2.32	1.69	0.28	–
2 year		–	0.03	0.85	1.67	0.14	–
3 year		–	0.03	0.81	1.45	0.13	–
4 year		–	0.03	0.20	1.23	0.22	–
1 year	Osmunda	–	0.11	2.45	0.78	0.31	–
2 year		–	0.06	1.01	1.17	0.43	–
3 year		–	0.04	1.08	1.10	0.42	–
4 year		–	0.03	0.66	1.02	0.63	–

[a] Reference: Withner and Van Camp, 1948.
[b] Reference: Erickson, 1957.
[c] Reference: Davidson, 1961.

Table 6-3. Tissue analysis of microelements in *Cattleya* (Withner and Van Camp, 1948)

		Parts per million dry weight				
Tissue	Medium	Fe	Mn	Zn	Cu	B
Lc. Canhamiana alba	Osmunda	150	270	160	50	12
× *C.* Eucharis seedlings	Haydite	170	260	150	50	13
Cattleya bowringiana						
mature leaves & pseudobulbs	Osmunda	350	180	120	18	10
Cattleya gaskelliana						
mature leaves & pseudobulbs	Osmunda	250	180	120	28	12
Cattleya percivaliana						
mature leaves & pseudobulbs	Haydite	850	30	240	85	16

Davidson (1961) compared P, K, Ca, and Mg levels in leaves and pseudobulbs of *C. trianaei* of four different ages grown in a fir-bark medium with regular fertilization and in osmunda without fertilization. His analyses were reported on a fresh-weight basis, but his data in Table 6-2 have been converted to a dry-weight basis, assuming a 90% moisture content, for purposes of comparison. He found no physiologically significant differences in mineral composition of leaves or pseudobulbs due to media or to age of tissue. However, his data showed that K levels decreased and Ca levels increased with age. Leaves of plants grown in osmunda generally had higher levels of K and lower Ca levels than plants maintained in bark regardless of age. Mg levels also appeared to be higher in two- and three-year-old leaves of osmunda-grown plants.

Erickson (1957) analyzed leaves of various ages on six plants of a *Cattleya* hybrid grown in osmunda. Leaves (one to seven years old) were analyzed for moisture content and for N, P, K, Ca, and Mg, and the results were analyzed statistically. The youngest leaf had a greater moisture content than the older leaves. N and K contents of leaves generally decreased with age whereas Ca levels increased. Mg and P levels varied only slightly.

Poole and Sheehan (1973b) analyzed new leads, leaves (by age), pseudobulbs (by age), sheaths, and flowers of *Laeliocattleya* 'Culminant' for dry matter and for N, P, K, Ca, Mg, Fe, Mn, Zn, and Cu contents (Table 6-4). The plants used in this study were grown in a fir-bark medium under a relatively high N, P, K fertilization regime. Foliar N decreased with age, whereas N accumulated in the older pseudobulbs. P levels increased and K levels decreased with increasing age of both leaves and pseudobulbs. Levels of Mg, Fe, Zn, and Cu remained relatively constant regardless of tissue age. There was a direct relationship, however, between age and foliar levels of Ca and Mn. Ca levels in pseudobulbs decreased with age, but Mn concentrations remained relatively constant

Table 6-4. Tissue analyses of plant parts of *Laeliocattleya* 'Culminant' (Poole and Sheehan, 1973b)

		Percent, dry weight					Parts per million, dry weight			
Tissue	Percent dry matter	N	P	K	Ca	Mg	Fe	Mn	Zn	Cu
New leads	7.3	3.1	.21	1.55	0.22	0.34	65	26	210	38
Leaves										
Immature	7.6	2.5	.12	2.41	0.31	0.31	93	42	95	40
1 year	9.3	2.0	.10	1.35	0.71	0.49	124	74	100	38
2 year	10.9	1.9	.09	0.69	1.81	0.79	150	82	121	36
3 year	11.2	2.0	.11	1.00	1.86	0.77	120	359	143	42
4 year	11.4	1.7	.16	0.79	2.22	0.51	132	489	163	47
5 year	11.2	1.9	.20	0.50	2.47	0.51	155	448	152	57
6 year	10.4	1.6	.29	0.54	2.55	0.59	190	450	155	68
7 year	9.9	1.7	.25	0.13	2.20	0.81	153	302	100	55
8 year	8.2	1.9	.41	0.32	2.70	0.70	210	430	108	56
Pseudobulbs										
Immature	7.7	2.1	.09	1.80	0.43	0.35	90	44	100	40
1 year	6.4	1.8	.20	1.29	0.41	0.45	86	23	107	38
2 year	8.1	1.8	.19	0.92	0.64	0.59	82	21	78	35
3 year	8.6	2.4	.26	0.48	0.52	0.66	112	21	163	36
4 year	8.8	3.2	.35	0.30	0.33	0.48	91	20	115	38
5 year	8.8	3.7	.39	0.23	0.12	0.31	92	26	125	55
6 year	7.8	3.8	.38	0.29	0.16	0.40	88	50	175	64
7 year	7.6	4.6	.66	0.33	0.16	0.29	163	59	155	61
Flower sheaths	7.6	1.5	.05	1.45	0.69	0.45	127	73	100	38
Flower										
Petals & sepals	5.7	2.8	.13	2.34	0.30	0.28	117	55	103	41
Column & ovary	5.4	3.3	.14	2.09	0.39	0.46	62	48	98	43

regardless of the age of the pseudobulbs. This study and that by Davidson (1961) suggest that plants grown in bark media have higher Ca levels in older leaves than plants grown in other potting mixtures. The findings of Poole and Sheehan (1973, 1974) indicate that both Ca and Mn are preferentially translocated to and accumulated in mature leaves rather than in roots or pseudobulbs. In general, the older leaves of *Cattleya* plants grown in fir-bark media have higher levels of Mn than of other microelements. The same is true for comparison of Mn levels in these plants and those grown in other media.

Poole and Sheehan (1977), working on the effects of microelement nutrition and media, also found that in a fir-bark medium Mn levels in leaves increased to a greater extent than those of other microelements. They hypothesized that natural chelating agents released during the decomposition of the fir bark particles aided in increasing the availability of Mn and other microelements. Wallace (1962) found that the binding capacities of metals with chelating agents decreased in the following order: Fe^{+++}, Cu^{++}, Zn^{++}, and Mn^{++}. Several factors may account for the preferential absorption of Mn by orchids in this medium (although Mn probably forms the least stable chelate). Brown *et al.* (1960) found that an excess of chelating agents decreased Fe uptake by roots. They also noted that a competition for Fe existed between roots and chelating agents. Hale and Wallace (1962) observed that Fe could increase Mn uptake from soil because Fe (with a higher stability constant than Mn) could displace chelated Mn from soil organic matter and render it more available. Since Fe, Cu, and Zn have higher stability constants than Mn, they can compete successfully with Mn for chelation sites and release it for absorption. Wallace and Bhan (1962) felt that if both quantity and quality of natural chelating agents were involved, then competitive chelation may be an important factor. For instance, a primary product of fir-bark decomposition—capable of chelating microelements of low solubility and high stability—could compete for metal ions more effectively than more decomposed products, such as organic acids, which could form chelates of higher solubility but lower stability (Wallace, 1956). Mn released from more stable chelates would be available for chelation by organic acids and absorption by the plants. The more stable chelates would successfully compete with the less stable ones and orchid roots for Fe, Cu, and Zn and reduce absorption of these ions by the plants. However, Mn levels would be increased in comparison, due to its displacement from the stable chelate and/or those of Fe, Cu, and Zn.

Chemical analyses of leaves and pseudobulbs are summarized in Table 6-5 for the four media tested by Poole and Sheehan (1977). Analyses of plants grown in tree fern and in a combination of tree fern (70%) and redwood bark (30%) were not significantly different. Plants grown in fir bark and in a combination of sphagnum peat moss (50%) and perlite (50%) had higher potassium levels than those maintained in tree fern. The peat-perlite medium provided higher Mg levels than any other medium. Plants grown in fir bark had higher levels of foliar Ca and Mn, as was found in other studies (Poole and Sheehan, 1973b, 1974). Plant growth response in the two tree-fern media and the fir-bark medium were not significantly different. However, plants grown in the peat-perlite medium were superior in all growth responses to those maintained in the other three media (Table 6-6). Poole and Sheehan attributed the good growth to excellent water retention, aeration, high cation exchange capacity, and good buffering capacity against high salts.

Table 6-5. Main effects of media on chemical composition of leaves and pseudobulbs of *Laeliocattleya* 'Aconcagua' (Poole and Sheehan, 1977)

	Percent, dry weight					Parts per million, dry weight			
Tissue and medium	N	P	K	Ca	Mg	Fe	Mn	Zn	Cu
Leaves									
Tree fern	1.85b	0.07a	1.94a	1.05a	1.11a	311a	842a	88a	13a
Tree fern & redwood	1.78ab	0.08a	2.10a	1.63b	1.11a	295a	760a	90a	13a
Fir bark	1.68a	0.06a	2.72b	1.60b	0.99a	405a	1047b	87a	13a
Peat and perlite	1.80ab	0.06a	2.77b	1.18a	1.43b	352a	724a	145b	15b
Pseudobulbs									
Tree fern	1.24a	0.06b	0.77a	1.12a	0.81b	270a	351b	117a	19a
Tree fern & redwood	1.20a	0.07b	0.91b	1.08a	0.77ab	283a	332b	98a	18a
Fir bark	1.11a	0.03a	0.94b	1.38b	0.70a	293a	458c	107a	16a
Peat and perlite	1.29a	0.06b	0.79a	1.31b	1.11c	212a	250a	203b	28b

Note: Means within a vertical column for each tissue followed by the same letter are not statistically different at the 5% level.

Their work indicated that proper air and water relations as well as nutrient retention were important factors in increasing growth and quality of young *Cattleya* plants.

In the studies mentioned above, *Cattleya* and other genera were shown to absorb relatively high levels of K, Ca, Mg, and several of the microelements (especially Mn, in older leaves). However, there is little evidence to date which indicates that the plants (1) require high levels of these elements, (2) utilize these nutrients rather than accumulating them, or (3) benefit (in terms of survival) from these levels. In fact, analyses of upper (acropetal) and lower (basipetal) halves of one- and two-year-old *Cattleya* leaves of plants grown under poor fertilization practices in a fir-bark medium (Poole and Sheehan, 1973b) indicate that at least Ca, Mg, and Mn are preferentially translocated to the upper halves of older leaves and accumulated (Table 6-7). This physiological response may be necessary to reduce nutrient antagonisms or imbalances in the younger and meristematic tissues. Plants in this study exhibited severe chlorosis in the upper halves of two-year-old and older leaves but only slight signs in the lower halves. The leaves could possibly be showing symptoms of K deficiency caused by low K levels coupled with relatively high concentrations of Ca and Mg, especially in the upper half of the leaf. The one-year-old leaves were a pale but acceptable green color (this is possibly a symptom of Fe deficiency due to low Fe concentrations and relatively high Mn levels in the upper halves of the young leaves).

The most extensive nutritional studies of the three major orchid genera (*Cattleya*, *Cymbidium*, and *Phalaenopsis*) were carried out by Poole (1974) and Poole and Seeley

Table 6-6. Main effects of media on growth responses of *Laeliocattleya* 'Aconcagua' (Poole and Sheehan, 1977)

			Dry weight (grams)			
Medium	Number of leaves	Number of new leads	Leaves	Pseudobulbs	Roots	Leaf/root ratio
Tree fern	8.57a	1.10a	2.87a	0.66a	2.68a	1.12a
Tree fern & redwood	7.19a	0.86a	3.01a	0.92a	2.81a	1.08a
Fir bark	6.67a	0.81a	2.36a	0.85a	2.15a	1.09a
Peat & perlite	10.67b	1.95b	5.59b	2.20b	4.03b	1.42b

Note: Means within a vertical column followed by the same letter are not statistically different at the 5% level.

Table 6-7. Chemical composition of upper and lower portions of one- and two-year-old leaves of a *Cattleya* hybrid plant grown in a fir-bark medium and exhibiting chlorotic symptoms on mature leaves (Poole and Sheehan, 1973b)

	Percent, dry weight					Parts per million, dry weight				
Tissue	N	P	K	Ca	Mg	Fe	Mn	Zn	Cu	B
One-year-old leaves										
Upper half	1.15	0.10	2.08	2.69	0.89	37	122	18	12	17
Lower half	1.03	0.10	1.53	2.45	0.75	45	80	19	13	16
Two-year-old leaves										
Upper half	0.75	0.09	0.57	5.41	1.66	50	266	29	19	47
Lower half	0.79	0.10	0.39	4.03	1.24	45	69	24	18	18

Note: Figures within the table are means of 3 samples.

(1978) in solution culture using marbles for the substrate. These studies were accomplished using similar nutritional levels and growing conditions and provide a great deal of information concerning elemental interactions, plant growth response, and tissue concentrations of most nutrients. Chemical composition of various plant parts of all three genera are summarized in Table 6-8. In *Cattleya* plants, N and P levels were nearly the same for leaves, pseudobulbs, and roots. K levels were highest in the leaves and lowest in the roots. The highest levels of Ca and Mn were found in the leaves; the lowest concentrations were in the pseudobulbs. Pseudobulbs also had the lowest levels of Mg and Fe; roots had the highest levels. Roots contained the highest concentrations of Zn and Cu, and very little variation in these elements was found between leaves and pseudobulbs. Levels of B were highest in leaves; roots and pseudobulbs contained similar amounts. The relatively high levels of Mg, Fe, Zn, and Cu in roots may have been due to absorption or to the precipitation of salts onto their surfaces.

Cymbidium

Information concerning tissue analyses of *Cymbidium* is summarized by Poole (1974) and Poole and Seeley (1978) and presented in Table 6-8. The chemical composition of *Cymbidium* leaves and roots is very similar to that of *Cattleya*. Major exceptions are that *Cymbidium* leaves had higher levels of N and lower levels of K than *Cattleya* leaves. In fact, symptoms of N deficiency in *Cymbidium* (blackened tips of older leaves) were very appar-

Table 6-8. Chemical composition of plant parts of *Cattleya, Cymbidium,* and *Phalaenopsis* after experimentation in solution culture (Poole, 1974)

	Percent, dry weight					Parts per million, dry weight						
Orchid	N	P	K	Ca	Mg	Fe	Mn	Zn	Cu	B	Na	Al
Cattleya leaves	1.83	0.20	4.24	1.29	0.47	66	79	28	10	41	11	54
Cymbidium leaves	2.33	0.26	2.93	0.97	0.33	133	54	46	12	48	253	72
Phalaenopsis leaves	1.99	0.25	7.06	2.79	0.53	97	210	23	5	47	192	76
Cattleya pseudobulbs	2.04	0.27	3.25	0.48	0.34	24	–[a]	25	9	16	44	24
Cattleya roots	1.96	0.30	2.18	0.84	0.82	440	28	118	25	19	783	98
Cymbidium roots	2.31	0.66	3.84	0.87	0.83	546	–[a]	116	16	17	880	181
Phalaenopsis roots	3.89	0.27	3.53	1.20	0.67	502	30	86	6	8	365	136

Note: Figures within the table are means of 108 samples.
[a] Concentration is less than 10 ppm.

Table 6-9. Summary of tissue analyses of several parts of *Phalaenopsis* hybrids grown in a fir-bark medium (Poole and Sheehan, 1974)

Tissue	Percent dry matter	Percent, dry weight					Parts per million, dry weight			
		N	P	K	Ca	Mg	Fe	Mn	Zn	Cu
First leaf	5.9	3.4	0.29	5.9	1.6	0.90	120	220	88	23
Second leaf	6.2	2.6	0.18	5.3	1.5	0.82	130	360	76	22
Third leaf	6.7	2.5	0.22	4.9	2.6	0.92	150	520	93	21
Fourth leaf	6.6	2.3	0.17	3.9	2.4	1.07	170	350	94	20
Fifth leaf	6.8	2.2	0.14	3.3	2.4	1.02	150	270	100	20
Sixth leaf	6.5	2.4	0.09	3.2	2.4	0.96	170	250	100	17
Seventh leaf	6.5	2.3	0.08	3.0	2.0	0.81	160	180	120	21
Eighth leaf	6.6	2.3	0.08	2.6	2.2	0.95	170	190	120	22
Plant stem	15.0	1.7	0.18	1.4	1.2	0.40	130	67	93	23
Roots in media	7.8	3.5	0.31	1.4	0.3	0.51	250	28	140	28
Roots in air	9.4	2.8	0.48	2.3	0.4	0.50	170	49	110	24
Flower spike	20.3	1.4	0.11	1.1	0.2	0.11	70	38	70	22
Flowers	6.8	2.5	0.32	5.0	0.5	0.44	110	150	88	21
Fruit	9.6	2.5	0.22	4.2	0.2	0.28	80	55	100	20

ent in the low N treatments which produced the best growth for *Cattleya*. It appears that *Cymbidium* has higher N requirements than *Cattleya* as expressed by foliar analysis, symptoms, and growth response.

Phalaenopsis

Poole and Sheehan (1974) analyzed the mineral content of leaves of several ages, stems, reproduction tissues, and roots of three different *Phalaenopsis* hybrids (Table 6-9). Variations between the three hybrids were minimal; only averages are included for discussion. N, P, and K levels decreased while Ca levels increased with leaf age. Mn levels also increased up to the third or fourth leaf and then decreased with age. The other elements tested (Mg, Fe, Zn, and Cu) remained relatively steady in all leaves which were analyzed. Once again, Ca and Mg levels increased in older leaves of plants being grown in a fir-bark medium.

The findings of Poole (1974) and Poole and Seeley (1978) concerning nutrition and mineral composition of *Phalaenopsis* provide an interesting contrast to those findings for *Cattleya* and *Cymbidium* (Table 6-8). Leaves of *Phalaenopsis* had higher concentrations of K, Ca, and Mn than did those of either *Cattleya* or *Cymbidium*. *Phalaenopsis* roots also had higher levels of Ca and N than did those of the other two genera studied. The researchers were unable to obtain a growth response with increased levels of K, and it seems therefore that *Phalaenopsis* can accumulate a large amount of K in the leaves in apparent "luxury consumption."

Dendrobium

Recently, Yamaguchi (1979) reported from Japan the mineral composition of various plant parts of *Dendrobium nobile* as detected by neutron-activation analysis (naa). His findings are summarized in Table 6-10, although very few elements were determined for all plant parts and they appear to be much lower than would be expected based upon spectrographic analyses of other orchid genera. Among young and old leafless shoots, an increase in content of Fe and Zn were found with age.

Table 6-10. Mineral composition of plant parts of *Dendrobium nobile* obtained by neutron-activation analysis (Yamaguchi, 1979)

	Percent dry weight			Parts per million, dry weight		
	K	Ca	Mg	Fe	Mn	Zn
New shoot						
Top	0.07	0.06	0.06	–	14	–
Bottom	0.10	0.07	0.09	25	8	14
Flowering shoot						
Top	0.04	0.09	0.06	–	31	–
Bottom	0.06	0.05	0.11	66	3	25
Mature shoot						
Top	0.09	0.05	0.08	–	5	–
Bottom	0.06	0.03	0.08	91	8	26
Shoot (top)	0.04	–	0.06	–	31	–
Flowers	0.18	–	0.01	–	3	–
Leaves	0.20	–	0.23	–	176	–
Aerial roots	0.25	–	0.07	–	25	–
Shoot (bottom)	0.06	–	0.11	–	3	–

Odontoglossum

Gething (1977) conducted several experiments on the effects of nutrition on growth of *Odontoglossum* seedlings using factorial combinations of N and K levels. He found that seedlings grown in a compost made up from sedge peat, sphagnum moss, and charcoal responded to N levels as long as P was at a satisfactory level in the compost. Plants grown in composts lacking P showed severe P deficiency. Gething attributed the lack of plant response to K levels to the presence of sufficient K in the compost or in rainwater.

Foliar Application of Nutrients

There has been a long-standing controversy as to whether foliar application of nutrients is effective in orchids. Some workers feel that the heavy cutin layer on many

Table 6-11. Percentages of ^{32}P absorbed by *Cattleya* 'Trimos' orchids as affected by time and method of application (Sheehan *et al.*, 1967)

Stage and treatment	Hours after application ½	2	12	24	120	Least significant difference for means within table
First mature leaf						
Foliar application[a]	0.025	0.036	0.038	0.034	0.077	0.05 = NS
Medium drench[b]	0.035	0.061	0.237	0.123	0.960	0.01 = NS
Second pseudobulb						
Foliar application	6.023	3.892	10.565	34.450	37.730	0.05 = 15.91
Medium drench	0.036	0.023	0.150	0.143	2.440	0.01 = 22.04
Third leaf						
Foliar application	0.018	0.021	0.028	0.052	0.042	0.05 = 00.12
Medium drench	0.024	0.025	0.103	0.138	0.480	0.01 = 00.16
Third pseudobulb						
Foliar application	0.028	0.069	0.265	0.163	0.337	0.05 = 00.35
Medium drench	0.032	0.072	0.065	0.128	2.160	0.01 = 00.49

[a] Foliar application. Foliar spray was applied to second mature leaf.
[b] Medium drench = pot drench.

orchids prevents nutrient penetration and that actually the plants obtain their nutrients from the runoff through their roots. In 1967, Sheehan *et al.* treated mature *Cattleya* 'Trimos' plants with both foliar sprays and root-drench applications of ^{32}P derived from phosphoric acid. Plants were fractionated ½, 2, 12, 24, and 120 hours after application and analyzed for ^{32}P content. ^{32}P was readily absorbed through the leaves (Table 6-11), and almost 35% of the amount supplied was found 24 hours later in the pseudobulb below the leaf that had been sprayed. This efficient absorption of ^{32}P through the leaves suggests that other plant nutrients will also be absorbed and therefore that foliar application is an effective way to apply nutrients to orchid plants.

Root Absorption of Nutrients

The question has often been posed: Since orchids produce a new set of roots annually, are the old ones functional? In some cases growers severely prune back the roots when repotting as they feel the new ones are those that are functional. Sheehan *et al.* (1967) studied the absorptive capabilities of roots one to two years old versus those of three- to four-year-old roots. The rhizomes of mature plants of *Cattleya* 'Trimos' were severed to prevent movement of ^{32}P within the plant between young and old roots. Roots of all ages studied absorbed equal amounts of ^{32}P applied as a medium drench (Table 6-12). This finding indicates that even old roots are functioning and of value to the plants.

Summary

A review of the current literature on orchid nutrition would not indicate any great differences between epiphytic orchids and terrestrial ornamentals, although orchid genera certainly differ somewhat in nutrient requirements and ability to accumulate certain nutrients. We feel that the available data provide sufficient information on the chemical composition of three orchid genera (*Cattleya, Cymbidium,* and *Phalaenopsis*) to serve as a reference point in determining the nutrient status of a plant. Suggested ranges of dry-weight mineral composition of orchid leaves are summarized in Table 6-13 as a guide for future work rather than as a recommendation. Due to the length of time necessary for the appearance of most nutrient deficiencies, chemical analysis of the upper halves of one- or two-year-old leaves can be beneficial in determining the nutritional status and

Table 6-12. Percentages of ^{32}P absorbed by *Cattleya* 'Trimos' orchids as affected by root age (Sheehan *et al.,* 1967)

Organ	Hours after application			Least signficant difference for means within table
	12	24	120	
Pseudobulb				
Roots 1–2 years old	0.071	0.230	0.760	0.05 = .66
Roots 3–4 years old	0.064	1.422	0.362	0.01 = .92

Table 6-13. Suggested ranges of mineral composition (dry-weight basis) of one- or two-year-old leaves of *Cattleya, Cymbidium,* and *Phalaenopsis*

	Element									
	Percent					Parts per million				
	N	P	K	Ca	Mg	Fe	Mn	Zn	Cu	B
Cattleya	1.5–2.5	0.1–0.2	2.0–3.0	0.4–1.0	0.3–0.6	50–100	40–80	25–75	10–30	25–50
Cymbidium	1.5–2.5	0.1–0.3	2.0–3.0	0.4–1.0	0.3–0.6	50–100	40–80	25–75	10–25	25–50
Phalaenopsis	2.0–3.5	0.2–0.3	4.0–6.0	1.5–2.5	0.4–0.8	80–150	100–200	20–60	10–25	25–50

requirements of these orchid genera by researchers and commercial growers before the actual symptoms are evident.

The growth of orchids is influenced by many environmental and physiological factors. Nutrient sources, concentrations, and frequency of applications are important in work on orchid nutrition. However, such factors as water quality and frequency of irrigation; light intensity, quality, and duration; and temperature are of similar importance. The most critical factors affecting orchid nutrition are the medium, degree of decomposition, and the age of the plant.

Much of the present confusion surrounding orchid nutrition is due to nonstandard methods of plant sampling for tissue analyses and insufficient research with plants grown in inert media. The use of organic potting mixes, especially osmunda and fir bark, confounds much of the early nutrition work, although it provides a great deal of information to growers using these media. Fairly good descriptions of symptoms of N, P, and Ca deficiencies exist, and possible deficiencies of K, Mg, and Fe have been reported in the literature. As growers move toward purer forms of higher-analysis fertilizer sources, cleaner water, and inert growing media, deficiency symptoms will become more prevalent.

Understanding the role of the organic medium and its gradual decomposition into inorganic constituents and possible chelating compounds represents a tremendous challenge to our knowledge of the nutrition of orchids, both in their native habitat and in the greenhouse environment. Such an understanding would greatly increase what we now know of orchid nutrition and improve our cultural practices. More research is needed and many challenges must still be met before adequate answers can be provided concerning the mineral nutrition of orchids.

Literature Cited

Adams, J. D. 1970. A tailored plan. Amer. Orchid Soc. Bull. 39:139–142.

Beaumont, J. H., and F. A. K. Bowers. 1954. Interrelationships of fertilization, potting media and shading on growth of seedlings Vanda orchid. Hawaii Agr. Expt. Sta. Tech. Paper 334:16.

Brown, J. C., L. O. Tiffin, and R. S. Holmes. 1960. Competition between chelating agents and root as a factor affecting absorption of iron and other ions by plant species. Plant Physiol. 35:878–886.

Burberry, H. A. 1894. The amateur orchid cultivators' guide book. Blake and MacKenzie, Liverpool.

Chin, T. T. 1966. Effect of major nutrient deficiencies in Dendrobium phalaenopsis hybrids. Amer. Orchid Soc. Bull. 35:549–554.

Cibes, H. R., C. Cernuda, and A. J. Loustalot. 1949. Vanilla physiological studies. Puerto Rico (Mayaguez) Fed. Expt. Sta. Rpt. 1948:22–23.

Cibes, H. R., and N. F. Childers. 1947. Vanilla: Agronomic studies. Puerto Rico (Mayaguez) Fed. Expt. Sta. Rpt. 1946:36–38.
Curtis, J. T., and E. Spoerl. 1948. Studies on the nitrogen nutrition of orchid embroys. II. Comparative utilization of nitrate and ammonium nitrogen. Amer. Orchid Soc. Bull. 17:111–114.
Davidson, O. W. 1957. New orchid potting medium lowers cost of production. Amer. Orchid Soc. Bull. 26:409–411.
——. 1960. Principles of orchid nutrition. Proc. 3d World Orchid Conf., p. 224–233.
——. 1961. Principles of orchid nutrition. Amer. Orchid Soc. Bull. 30:277–285.
Erickson, L. C. 1957. Leaf age in Cattleya. Amer. Orchid Soc. Bull. 26:560–563.
Evers, O. A., and A. Laurie. 1940. Nutritional studies with orchids. Bimo. Bull. Ohio Agr. Exp. Sta. 25:166–173.
Fennell, T. 1951. Food for thought on orchid feeding. Amer. Orchid Soc. Bull. 20:455–459.
Gething, P. A. 1977. The effect of fertilizers on growth of orchid (Odontoglossum) seedlings. Expl. Hort. 29:94–101.
Hale, V. Q., and A. Wallace. 1962. Effect of chelating agents and iron chelates on absorption of Mn by bush beans. *In* A. Wallace (ed.), A decade of synthetic chelating agents in inorganic plant nutrition. Edward Bros., Ann Arbor, Mich.
Lunt, O. R., and A. M. Kofranek. 1961. Exploratory nutritional studies on *Cymbidium* using two textures of fir bark. Amer. Orchid Soc. Bull. 30:297–302.
Penningsfeld, F., and G. Fast. 1970. Ernahrungstragen bei Paphiopedilum callosum. Gartenwelt 9:205–208.
——. 1973. Ernahrungstragen bei Disa uniflora. Die Orchidee 24:10–13.
Penningsfeld, F., and L. Forchthammer. 1980. Ergebnisse neunjahriger Cymbidien-Ernährungsversuche. Die Orchidee 31:11–19.
Poole, H. A. 1971. Effects of nutrition and media on growth and chemical composition of mericloned plants of Cattleya. M.S. thesis, Univ. of Florida, Gainesville.
——. 1974. Nitrogen, potassium and magnesium nutrition of three orchid genera. Ph.D. diss., Cornell Univ., Ithaca, N.Y.
Poole, H. A., and J. G. Seeley. 1978. Nitrogen potassium and magnesium nutrition of three orchid genera. J. Amer. Soc. Hort. Sci. 103:485–488.
Poole, H. A., and T. J. Sheehan. 1970. Effects of levels of phosphorus and potassium on growth, composition and incidence of leaf-tip die-back in Cattleya orchids. Proc. Fla. State Hort. Soc. 83:465–469.
——. 1973a. Leaf-tip die-back—what's the real cause? Amer. Orchid Soc. Bull. 42:227–230.
——. 1973b. Chemical composition of plant parts of Cattleya orchids. Amer. Orchid Soc. Bull. 46:889–895.
——. 1974. Chemical composition of plant parts of Phalaenopsis orchids. Amer. Orchid Soc. Bull. 43:242–246.
——. 1977. Effects of media and supplementary microelement fertilization on growth and chemical composition of Cattleya. Amer. Orchid Soc. Bull. 46:155–160.
Scully, R. M. 1951. Should orchids be fertilized? Amer. Orchid Soc. Bull. 20:137–139.
Sheehan, T. J. 1961. Effects of nutrition and potting media on growth and flowering of certain epiphytic orchids. Amer. Orchid Soc. Bull. 30:289–292.
Sheehan, T. J., J. N. Joiner, and J. K. Cowart. 1967. Absorption of ^{32}P by *Cattleya* 'Trimos' from foliar and root applications. Proc. Fla. State Hort. Soc. 80:400–404.
Wallace, A. 1956. Introduction: Metal chelates in agriculture, p. 4–23. *In* A. Wallace (ed.), Symposium on the use of metal chelates in plant nutrition. National Press, Palo Alto, Calif.
——. 1962. Chelation in heavy metal induced iron chlorosis, p. 25–28. *In* A. Wallace (ed.), A decade of synthetic chelating agents in inorganic plant nutrition. Edward Bros., Ann Arbor, Mich.
——, and K. C. Bhan. 1962. Plants that do poorly in acid soils, p. 36–38. *In* A. Wallace (ed.), A decade of synthetic chelating agents in inorganic plant nutrition. Edward Bros., Ann Arbor, Mich.
Withner, C. L., and J. Van Camp. 1948. Orchid leaf analyses. Amer. Orchid Soc. Bull. 17:662–663.
Yamaguchi, S. 1979. Determination of several elements in orchid plant parts by neutron activation analysis. J. Amer. Soc. Hort. Sci. 104:739–742.
Zakrejs, J. 1976. Nutrition of Cattleya by different sources of nitrogen. Proc. 8th World Orchid Conf., p. 414–419.

7

Flower Induction and Physiology in Orchids*

CHONG JIN GOH, MICHAEL S. STRAUSS,
and JOSEPH ARDITTI

*The literature survey pertaining to this chapter was concluded in January 1980; the chapter was submitted in April 1980, and the revised version was received in May 1980.

History

With a few exceptions, orchids are grown for their flowers. It is, therefore, not surprising that considerable interest has been focused on flower production. However, research on flower induction, development, and physiology was initiated only recently. The first studies on flower induction were probably those concerned with *Dendrobium crumenatum* and other gregariously flowering orchids (Massart, 1895; Treub, 1887; Went, 1898) at the Bogor Botanical Gardens (now known as Hortus Botanicus Bogoriensis or Kebun Raya Indonesia and in those days called the Buitenzorg Botanical Gardens). Subsequent studies were carried out in the United States (Curtis, 1954; Rotor, 1952; Vacin, 1952), West Africa (Sanford, 1971), Venezuela (Dunsterville and Dunsterville, 1967), Philippines (Quisumbing, 1968), Poland (Źotkiewicz, 1961), Singapore (Goh, 1970, 1973, 1975, 1976a,b, 1977a,b,c,d), and the original habitat of *Cymbidium* (Vacin, 1952). Horticulturally oriented studies were undertaken in research institutions (Casamajor and Went, 1953; Montgomery and Laurie, 1944; Wróbel-Stremińska, 1961), and private collections (Hager, 1957; Hamilton, 1977; McDade, 1947; Urmston, 1949). Additionally, several deductions were based on general observations (for reviews see Brieger *et al.*, 1977; Goh and Arditti, in press; Nuernbergk, 1961).

Juvenile Phase

Like other flowering plants, an orchid must reach a certain stage of maturity before it can flower. The period of juvenility varies among species and among hybrids and individual plants derived from seeds of the same capsule. Data concerning this aspect of flower physiology of species generally are scant, but records regarding several hybrids raised at the Singapore Botanic Gardens are detailed and instructive (Table 7-1). In each case the seeds were germinated and the seedlings cultivated in the same manner, but the plants flowered at different ages. For example, seeds of *Aranda* cv Hilda Galistan (*Arachnis hookeriana* × *Vanda tricolor* var. *suavis*), were sown on June 17, 1935, and the plants first flowered on February 25, 1940, indicating a juvenile period of about five years. A closely related hybrid, *Aranda* cv Lucy Laycock (*Arachnis hookeriana* × *Vanda tricolor* var. *purpurea*) required more than 13 years to flower. Generally the period from seed germination to first flowering takes four to seven years, but longer periods are not uncommon. Sometimes seedlings never flower. For instance, of about 30 seedlings of *Aranthera* cv Mohamed Haniff raised at the Singapore Botanic Gardens, "only one ever flowered in Singapore" (Holttum, 1962). On the whole it seems safe to assume that the duration of the juvenile phase is determined genetically.

Ecological Observations

It is common knowledge that some orchids flower only under specific ecological conditions. For example, *Cymbidium roseum,* one of the finest Malayan cymbidiums, flowers in

Table 7-1. Time from seed sowing to flowering of some orchid hybrids (adapted from Wee, 1971)

Hybrid	Juvenile period (years-months-days)	Hybrid	Juvenile period (years-months-days)
Dendrobium		Lily Josephine Wong	8-6-14
Anouk	5-6-10	Storiata	9-3-1
Wee Cheow Beng	5-7-24	Jane McNeill	6-6-22
Tumphal	4-2-1	Joan Mah	9-0-0
Tess Kleinman	4-5-0	Ammani	6-10-15
Ellen Harris	4-10-9	Chia Shui May	6-9-5
Sarie Marijs	3-4-10	Chia Shui Hai	8-2-25
Tanglin	6-8-24	*Vandaenopsis*	
Lee Ewe Boon	7-8-23	Kapden	8-3-17
Lin Yoke Ching	8-2-12	Suavei	4-6-18
Curlylocks	7-7-27	Pang Nyuk Yin	5-3-19
Noor Aishah	6-5-2	Catherine	5-0-6
Michiko	6-7-4	Mary Seal	5-9-6
Vanda		*Phalaenopsis*	
Miss Joaquim	3-1-5	Valentinii	4-11-17
Ruby Prince	3-4-16	Harriettiae	3-8-27
Nellie Corbett	8-10-16	*Arachnopsis*	
Pata	9-6-24	Eric Holttum	7-5-2
Tan Chin Tuan	8-4-2	Lee Siew Chin	9-7-0
Chia Shui May	6-8-4	Adrian Cheok	7-10-17
Sanada Kuma	9-9-27	Khalsom	7-1-7
Mimi Palmer	6-4-22	Napier	5-2-0
Anna Jackson	6-4-28	*Others*	
Joan Warne	10-2-3	*Aeridachnis* Alexandra	5-10-18
Eisenhower	9-7-26	*A.* Lim Theng Hin	5-6-0
Aranda		*Aeridovanda* Chiun Hai Cheok	5-10-23
Hilda Galistan	4-8-8	*Arachnoglottis* Brown Bars	6-4-1
Lucy Laycock	13-3-0	*Brassidium* Tan Lean Bee	6-1-1
Wong Mook Kwi	7-3-23	*B.* Betty H. Shiraki	8-6-13
Anna Braga	7-9-5	*Cymbidium* Faridah Hashim	5-0-20
Wendy Scott	4-7-21	*Laeliocattleya*	6-7-14
Eileen Addison	7-10-6	Cheah Chuan Keat	
Majula	8-9-10	*Paphiopedilum* Shireen	8-5-0
Elizabeth Douglas-Home	6-10-6	*Paphiopedilum*	4-3-16
Freckles	5-2-20	*Renanthoglossum* Ahmad	10-10-11
Wee Huck Lay	9-10-15	*Spathoglottis* Penang Beauty	2-11-13
Grandeur	5-4-6	*S.* Parslee	3
Peng Lee Yeoh	5-6-24	*Vandachnis* Scarlet Runner	11-10-20
Eric Mekie	8-9-28	*V.* Woo Cheng Ee	6-2-6
Tourism Singapura	8-6-20	*Trigeneric hybrids*	
Aranthera		*Burkillara* Henry	5-9-22
Anne Black	5-10-5	*Holttumara* Cochineal	8-0-24
Bloodshot	5-10-16	*H.* Indira Gandhi	6-6-21
Beatrice Ng	6-1-3	*H.* Loh Chin	8-3-19
Star Orange	5-10-11	*Lymanara* Mary Ann	6-4-16
Dainty	6-7-2	*Ridleyara* Fascad	5-3-2
Lilliput	5-7-14	*Sappanara* Ahmad Zahab	3-4-10
Lim Hong Choo	7-5-2		
Renantanda			
Prince Norodom Sihanouk	7-1-0		

the Cameron Highlands, Gunung Tahan, and Gunung Batu Berinchang at altitudes of 1500 to 1800 m above sea level, but "it is unlikely [to] flower under lowland conditions" (Alphonso, 1965). Similarly, plants of *Paphiopedilum barbatum* and *P. bullenianum,* both native to the Malayan highlands at altitudes of 600 m, rarely flower when grown under lowland conditions (Holttum, 1953). At the Singapore Botanic Gardens, however, they bloom readily when transferred to the newly constructed temperate greenhouse (G. Alphonso, Singapore Botanic Gardens, personal communication).

The most extensive and impressive ecological studies were carried out in West Africa (Sanford, 1971, 1974). These studies show that some orchids may be long-day (LD), short-day (SD), or neutral-day (ND) plants,[1] whereas others are influenced by low or fluctuating temperatures (see Table 7-A at the end of the chapter). In some cases it may be "difficult or even impossible to separate temperature effects from those of light, but... at least 35% of the [West African] epiphytes and perhaps at least 27% [of the] terrestrials are directly day length controlled" (Sanford, 1974). This and other reports (Arditti, 1966a, 1967b, 1968, 1979; Rotor, 1952, 1959) seem to substantiate the suggestion that "tropical plants are more sensitive to small differences in day length than are temperate-zone plants" (Sanford, 1974). Such sensitivity would be evolutionarily advantageous since day-length differences are less pronounced in the tropics.

On the other hand, there are reports that tropical orchid hybrids such as *Arachnis* cv Maggie Oei, *Aranda* cv Deborah, *Aranda* cv Wendy Scott, and *Vanda* cv Miss Joaquim as well as some *Dendrobium* hybrids are indifferent to day length. However, *Arachnis* cv Maggie Oei and *Aranda* cv Wendy Scott did initiate buds following a change in photoperiod (Byramji and Goh, 1976; Murashige *et al.,* 1967; Sheehan *et al.,* 1965). With *Vanda* cv Miss Joaquim it has been established that flowering is dependent on the availability of sunlight (Goh and Wan, 1974; Murashige *et al.,* 1967).

Low temperature requirements for flower induction have been documented in tropical orchids. They may vary from the more pronounced elevation-dependent temperature fluctuations required by *Phalaenopsis schilleriana* (de Vries, 1953) and *Polystachya cultriformis* (Sanford, 1971, 1974) to the subtler rain-induced cooling, which initiates flower development and anthesis in *Dendrobium crumenatum* and other *Dendrobium* species as well as in *Bromheadia alticola* and *B. finlaysoniana* (Coster, 1925; Holttum, 1953, 1957, 1962, 1969; Smith, 1926, 1927; for a review see Arditti, 1979). Low temperature requirements have also been reported for *Cattleya* and *Cymbidium* (Rotor, 1952). Altogether, the successful existence in the tropics of orchids that are induced to flower by cooling is an indication that they are sensitive to slight variations in temperature.

Eulophia cucculata, one of "the showiest of Kenya's indigenous orchids,... was only found in flower in areas of grassland recently subjected to burning" (Perkins, 1962). This may be "possibly without real significance" (Perkins, 1962) or could indicate that changes (possibly ethylene gas or increased illumination due to the elimination of other plants) caused by the fire induced blooming or made it possible. However, more extensive observations in Africa suggest that there are "no instances of apparent correlation that are not simply fortuitous" (Sanford, 1974). Still, it is necessary to recall that fire dependence has been reported for some nonorchidaceous plants.

Assertions that the flowering of orchids is affected by the phases of the moon are archaic and not supported by the available evidence (unless, of course, the species in question are photoperiod-dependent and sensitive to very low light intensities).

[1]Long-day plants flower when the days are longer than a certain minimum (in reality they require nights that are shorter than a specific maximum); short-day plants require days shorter than a defined maximum (i.e., nights longer than a minimum); neutral-day (or day-length-neutral) plants are insensitive to day length.

Experimental Evidence

The available experimental evidence on responses to day length, temperature, and hormones is limited to relatively few species and hybrids (Table 7-A). Most of the findings are in line with what is known about other plants. Studies on the control of flower bud initiation conducted with the ND *Aranda* cv Deborah (Goh, 1970, 1975, 1976a,b, 1977a,b,c,d; Goh and Seetoh, 1973) have shown the existence of a flowering gradient which is greatest near the apex and diminishes basipetally. Development of axillary buds is inhibited by apical dominance, and decapitation removes this correlative inhibition. Applications of anti-auxins, growth retardants, or cytokinins have the same effect. Floral buds initiated following decapitation emerged through the leaf sheaths and became visible in seven to 10 days. Cytokinin applications brought about initiation in five to seven days (Goh, 1977a,b,c,d; Goh and Seetoh, 1973). These findings suggest that photoperiodic and low-temperature flower induction in sympodial orchids could have been due to changes in the levels of endogeneous growth regulators. There are several reports that photoperiodic and/or low-temperature treatments can affect the endogenous levels of such substances (Evans, 1971; Zeevart, 1976).

Floral bud initiation by cytokinins has been demonstrated in three *Dendrobium* hybrids (Goh, 1979; Goh and Yang, 1978). Gibberellins have also been used to induce flowers in *Bletilla striata* as well as *Cymbidium* and *Cattleya* hybrids. When plants of *Cattleya* cv Geriant and *C.* cv Los Gatos were treated with gibberellins by injection inside the sheath, flower stems elongated, but the buds aborted or became deformed. Applications of gibberellin-containing lanolin pastes to inflorescences produced results that "were not too spectacular" (Smith, 1958). When gibberellins were applied "more than two months in advance of bloom," "a slight favorable response occurred" (Smith, 1958).

In *Paphiopedilum,* gibberellin treatments "did not hasten... the first bloom [but] later blooms did... flower... earlier"; the "chief effect," however, "was to shorten the time over which the crop was produced" (Smith, 1958). Gibberellin treatments increased the length of the flower stem.

Another substance that can induce flowering in some orchids is salicylic acid (see Table 7-A at the end of the chapter). This compound has also been shown to induce flowering in *Lemna* (Cleland and Ajami, 1974). These findings and (1) the appearance of shoots with both vegetative and reproductive growth (Goh, 1970, 1976a, 1977c, 1979), (2) the reversion of vegetative and reproductive growth in growing apices (Goh, 1976a, 1979), and (3) flower bud initiation by auxins and cytokinins suggest that the regulation of flowering in both monopodial and sympodial orchids by growth substances may be qualitative and quantitative (Goh, 1976a, 1979).

There are a number of reports that treatments with inhibitors, antibiotics, and synthetic growth regulators induce flowers in several orchids. It is quite likely that many of these compounds act by putting plants under physiological stress rather than through direct induction.

As already mentioned, the flowering of many orchids is controlled photoperiodically (Table 7-2). However, sometimes this effect cannot be easily separated from that of temperature. In *Cattleya gaskelliana, C. mossiae, C. mendelii,* and *C. schroederae,* changes in one flower-inducing stimulus require adjustment of another (Table 7-A). High light intensity can also affect flower production (Montgomery and Laurie, 1944).

Perhaps the most interesting and widely studied example of flower induction in orchids is that of *Dendrobium crumenatum,* the Indonesian and Malayan epiphytic *Anggrek merpati* (dove or pigeon orchid). All plants in an area flower at the same time, but the blooms last for only one day. This simultaneous flowering, which seems to ensure pollination, has interested investigators for about a hundred years (Arens, 1923; Beaumée, 1927; Burkill, 1917; Coomans de Ruiter, 1930; Coster, 1925; Kuijper, 1931, 1933; Massart, 1895; Rutgers and Went, 1915; Seifriz, 1923; Smith, 1926, 1927; Treub, 1887; F. A. F. C. Went, 1898, 1917; F. A. F. C. Went and Rutgers, 1915; F. W. Went, 1930; for reviews see Arditti, 1966a, 1967b, 1968, 1979; Holttum, 1953, 1969). Inflorescences of *Dendrobium crumenatum* are terminal and produce several flower buds, each protected by a bract. The flower buds develop until all parts are formed and the anther is "almost fully grown" (Holttum, 1969). Development then ceases and the buds become dormant. Further development is initiated only by a drop in temperature (approximately 5° C). In Malaya, Singapore, and Indonesia sudden temperature drops usually occur following rainstorms and the subsequent evaporative cooling of the buds. Prolonged gradual cooling (ca. 24 hrs) can also induce flowering. Exactly nine days after the cooling [in most cases; one of us (JA) has also heard of eight days in isolated cases and counted 10 for a limited number of flowers in one instance], just before dawn the flowers (pinkish white)

Table 7-2. Dendrobium, Sarcochilus, and *Thrixspermum* species that flower in response to low temperature grouped according to the number of days between stimulus and flowering (modified from Coster, 1925)[a]

8 days	9 days	10 days	11 days
	Dendrobium		
acuminatissimum *bicostatum* *brevicolle* *citrinicostatum* *comatum* *compressicolle* *crenulatum* *flabellum* *flabelloides* *luxurians* *macraei* *maculosum* *papilioniferum* *spathilingue* *validicolle*	*aratriferum* *bancanum* *crumenatum* *feuilletaui* *filiforme* *fugax* *insigne* *nitidicolle*	*dendrocolla* *dilatatocolle* *ecolle* *padangense* *spurium* *tunense*	*angulatum* *pumilum* *xantholeucum*
	Sarcochilus		
appendiculatus[b] *pallidus*[b] *zollingeri*[b]	*compressus*[c]	*teres*[d]	
	Thrixspermum		
arachnites *calceolus* *inquinatum*		*raciborskii* *subulatum*	

[a] Flowers of these species generally last only one day (see also Table 7-4).
[b] The actual number of days may be 8–9.
[c] The actual number of days may be 9–10.
[d] The actual number of days may be 10–11.

open and start emitting a very pleasant aroma. Pollinator bees are attracted by the combination of color, fragrance, and large numbers of flowers and arrive shortly after dawn. [Some of these bees never make it to the flowers since spiderwebs with dead pollinator bees have been observed (by JA in the Bogor Botanical Gardens) in front of *Anggrek merpati* clusters.] Despite the discovery of the stimulus (low temperature) that promotes development of the flower buds, the control mechanism itself is still unknown.

Recent attempts to induce flowering by applications of growth regulators (auxins, cytokinins, and gibberellins) to dormant buds either directly or through the stem by injection have all failed (Goh, unpublished).

An interesting aspect of this dependence on cooling is that it also ensures an adequate supply of moisture before the development of the many blossoms (Holttum, 1953). In Singapore, the absolute minimum and maximum temperatures are 21° C and 34° C, respectively (70–93° F). The normal range during a day may be 23–32° C (sunny days) and 24–27° C (rainy days). In the hottest month (May), the mean temperature is only 1.75° C (3° F) higher than in December, the coolest period (all figures are from Holttum, 1953). Altogether, temperature fluctuations are not a major aspect of the climate. Thus a dependence on cooling in such equatorial climates may appear to be a risky evolutionary strategy, but this is obviously not so. Similar mechanisms have evolved in other orchids (Table 7-2, 7-A). These include *Bromheadia alticola, B. finlaysoniana* (seven days from cooling to flowering), a number of *Thrixspermum* species, and several *Dendrobium* species, some of which flower one or two days after *D. crumenatum* and others which bloom one day before it (Arditti, 1979; Coster, 1925; Holttum, 1953; Smith, 1926). Several non-orchidaceous flowers also respond to cooling including a large deciduous Malayan native tree, *Pterocarpus indicus*, Fabaceae; *Epiphyllum oxypetalum*, Cactaceae, from Central America; and *Zephyranthes rosea*, Amaryllidaceae, an Indian bulbous plant (Holttum, 1953).

Several workers have reported that growth regulators improve flower development and quality. Gibberellic acid, for example, can increase flower size and raceme length and accelerate flowering in *Cymbidium* cv Sicily 'Grandee' and *C.* cv Guelda (Bivins, 1968, 1970). Benzyladenine brings about an increase in the length of mature inflorescences of *Dendrobium* cv Louisae and the number of flowers they produce (Goh, 1979). Thus growth regulators are implicated not only in flower induction but also in the development of inflorescences (Goh, 1977d, 1979; Goh and Yang, 1978; Goh and Wan, 1974). In *Aranda* cv Deborah, development may consist of two processes which are controlled by apical dominance through endogenous auxin levels (Goh, 1977a,b). Initiation itself can be stimulated by a decreased auxin level in stem tissues whereas subsequent development of inflorescences may depend on the availability of the necessary nutrients.

Genetics of Flower Induction

Observations of West African orchids indicate that the blooming time response of some is genetically controlled (Sanford, 1971; see Table 7-A at the end of the chapter). Information on the inheritance of the responses to flower-inducing stimuli can also be adduced from comparisons between hybrids (McDade, 1947) and their parents (Table 7-A). For instance, *Cattleya* cv Harold [a hybrid between *C. gaskelliana* (LD at 13° or 18° C,

SD at 13°) and *C. gigas* (SD at 13°)] flowers in early summer (McDade, 1947), suggesting that it is SD and implying dominance of this characteristic. On the other hand, *C.* cv Enid, which blooms in the autumn (McDade, 1947) and is known to be insensitive to day length (Table 7-A), is a hybrid between *C. gigas* and *C. mossiae,* both of which are SD. This observation seems to indicate a somewhat more complicated inheritance pattern than in *C.* cv Harold. A statement that "the blooming time of the hybrid will vary according to the strength of the influence of the parents" (McDade, 1947) can be interpreted to suggest that photoperiodic responses in orchids may be due to relatively simple inheritance of long- and short-day characteristics.

More complex inheritance patterns can be found among Malayan orchids. Of these "two of the most free-flowering are *Vanda hookeriana* and *Arachnis hookeriana.* But when we crossed these two together and raised a few seedlings, not one of them flowered, though we kept them many years. In the same way, our first attempt to cross *Arachnis flos-aeris* and the common form of *A. hookeriana* gave plants which flowered very rarely.... But when *A. flos-aeris* was crossed with *A. hookeriana* var. *luteola* it gave us Maggie Oei which is astonishingly free-flowering" (Holttum, 1949).

Studies of the flowering gradient in *Aranda* (Goh, 1975, 1977b) have shown that in five *Aranda* hybrids involving *Arachnis hookeriana* as the seed parent, this character is not derived from it. This conclusion is based on the fact that neither *A. hookeriana* (including var. *luteola*) nor its hybrid *Arachnis* cv Maggie Oei exhibit the same character. Therefore, it seems possible that this character was inherited from the *Vanda* parent. However, it could also have been the result of intergeneric hybridization since many hybrids of the *Vanda-Arachnis* alliance flower even more freely than their parents (Holttum, 1949, 1953).

The available information on the inheritance of flowering in orchids is obviously insufficient, and additional research is required. Parentages of orchid hybrids are carefully recorded. Therefore, valuable information could be obtained from analysis of available data on the flowering of hybrids and their parents.

Control of Flower Sex

The genera *Catasetum, Cycnoches,* and *Mormodes* can produce female, male, or hermaphrodite flowers. *Catasetum* and *Cycnoches* plants may bear all three kinds on the same or different inflorescences during one or several flowering seasons. Or, to put it differently, these species may be dioecious or monoecious, and in cases of the latter flowers may be staminate, pistillate, or complete (i.e., perfect). This phenomenon led to considerable taxonomic confusion at first because plants were assigned to different genera or species due to the difference in floral morphology. The confusion was not resolved until the turn of the century when greenhouse plants in England produced all three kinds of flowers. However, the physiological control of sex regulation in orchids is yet to be elucidated despite very interesting recent research (Gregg, 1973, 1975, 1977).

In *Catasetum* or *Cycnoches,* robust plants or those grown under full sun bear female flowers. Smaller plants or those maintained in shade produce male blossoms (Gregg, 1975, 1977). Female flower production is accompanied by increased ethylene evolution, which can be inhibited by placing the plant under shade or covering the raceme tip with a foil cap (Gregg, 1973). These findings suggest that femaleness is induced by ethylene. However, subsequent work seems to indicate that this is not the case since auxin applications and ethylene treatments have failed to induce the production of female flowers (Gregg, 1973, 1975). Hence, it seems that ethylene production may be (1) brought about by strong light (a stress response?), independently of female blossom production, (2) a byproduct of female flower formation, or (3) a combination of the two (Arditti, 1979).

Induction of female flowers in *Catasetum* and *Cycnoches,* by high light intensities, is an adaptive feature because it seems to ensure sufficient photosynthesis to support seed production. Capsules of *Cycnoches* have been reported to contain up to 4,000,000 seeds (Arditti, 1967a), and for *Catasetum* the estimates are as high as 2,000,000 (Gregg, 1975). Such large numbers, even of very small seeds, require considerable amounts of photosynthesis products. *Catasetum* and *Cycnoches* flowers are green. Therefore, it is possible that like green *Cymbidium* blossoms (Arditti and Dueker, 1968; Dueker and Arditti, 1968) they carry out photosynthesis and contribute to their own energy needs. Under reduced light intensities, the level of photosynthesis by the entire plant or the flowers may be insufficient. Thus it appears that a mechanism that makes seed production possible only when there is enough energy to support it is important for the survival of these species. However, other orchid species which usually grow under low light intensities and do not have green flowers also produce millions of seeds (though none have been reported to contain as many per capsule as *Cycnoches*).

Flower Development

Inflorescence primordia may be initiated in the axils of pseudobulb bases, as is the case in *Oncidium sphacelatum.* Floret formation usually is initiated approximately six months after the inflorescence starts to develop. The flowers open two months later (Kosugi, Yokoi, and Yokobori, 1973). In *Domingoa haematochila* and *Nageliella purpurea,* each inflorescence axis flowers periodically "in successive years" (Ebel and Mörchen, 1977).

Once initiated, flower buds do not always develop to a mature flower. In *Aranda* cv Deborah, even among decapitated plants, about 30% of the buds do not develop (Goh and Seetoh, 1973). However, once the buds reach a length of more than 2 mm, they continue to develop and reach maturity (Goh, 1977a,b). Development of inflorescences in the *Vanda-Arachnis* alliance usually requires two months. The growth curve of the inflorescence stalk is typically sigmoid, and during the exponential growth period, floral bud differentiation is very slow. It accelerates when the growth of inflorescence stalk stops (Ede, 1963). The flowers usually open in an acropetal sequence at one-day intervals (Goh, 1977c).

Organogenesis of floral parts in buds of *Arundina graminifolia* and *Bromheadia finlaysoniana* is acropetal (Jeyanayaghy and Rao, 1966; Rao, 1967). In the latter, primordia start as conical outgrowths which eventually flatten. The perianth lobes differentiate on these structures. Following the formation of the outer and inner perianth lobes the column (gynostemium) forms in the center, where it is "parallel to the inner surface of the outer median perianth lobe" (Jeyanayaghy and Rao, 1966). During the early stage the primor-

Table 7-3. Weight increases in flowers of *Dendrobium crumenatum* and *Thrixspermum arachnites* following low-temperature stimulus (modified from Coster, 1925)

	Number of days after flower induction								
Species	0	1	2	3	4	5	6	7	8
Dendrobium crumenatum									
Fresh weight (mg)	3.0	3.2	6.0	7.4	16.6	22.4	41.0	66.6	153.4
Dry weight (mg)	0.7	0.7	1.2	1.2	2.4	3.0	4.4	6.8	12.0
Water (% of fresh weight)	77	78	80	84	86	87	89	90	92
Thrixspermum arachnites									
Fresh weight (mg)	2.0	3.0	7.0	10.4	18.0	23.4	44.2	75.4	138.0
Dry weight (mg)	0.45	0.6	1.15	1.4	2.0	2.2	3.8	5.6	8.3
Water (% of fresh weight)	78	80	84	87	88	90	91	93	94

dium of the column is a dome-shaped structure that elongates and differentiates into a globular head suspended by a stalk. Pollinia differentiate in its head. The flower buds elongate during this period, and the gynoecium appears as an invagination below the floral segments.

As might be expected, buds (and primordia) increase in weight as they develop. For example, during the nine days following low-temperature treatments the fresh weight of *Dendrobium crumenatum* flowers increases 50-fold. Dry weight increases approximately 17-fold (Table 7-3; Coster, 1925). The corresponding increases during an eight-day period in *Thrixspermum arachnites* are 69- and 18-fold, respectively. Obviously, a great deal of dry matter and water must be transported into the developing flower from the pseudobulb and leaves.

Flower Longevity

Orchid flowers vary greatly in their lifespan; some may live for up to several months, whereas others are ephermeral, lasting for only a single day (Table 7-4). The life span of

Table 7-4. Longevity of orchid flowers

Orchid	Longevity (days)	Orchid	Longevity (days)
Aerides multiflorum	14–18	*C. labiata*	30
Angraecum	40	*C. trianei*	60
Arachnopsis cv Eric Holttum	10	*C. trianei* (normal day length)	20
Aranda cv Lucy Laycock	15	*C. trianei* (8-h days)	17
A. punctata cv City of Singapore	5	*Ceratostylis*	1[a]
A. cv Wendy Scott	24–32	*Cirrhopetalum ornatissimum*	15–30
Arundina graminifolia	5	*Coelogyne*	21
		C. cristata	15–30
Brassia brachiata	<15	*C. flaccida*	15–30
Bulbophyllum section Oxysepalum	1[a]	*C. speciosa*	<15
		Cymbidium lowianum	<15–90
Calanthe	>30	*C. virens*	14–25
C. discolor	14	*Cypripedium*	60
C. vestita var regniei	30		
Cattleya	45–60	*Dendrobium*	19
C. bowringiana (normal day length)[c]	14	*D. aggregatum*	16
C. bowringiana (8-h days)	9	*D. agneum*	30
C. hybrids	15–30	*D. angulatum*	1

(*continued*)

Table 7-4.—Continued

Orchid	Longevity (days)
D. auricolor	1
D. bancanum	1
D. barbatulum	19
D. section Bolibidium	1[a]
D. cv Brown Curls	30
D. calceolaria	11
D. cv Caprice	35
D. cv Champagne No. 6	30
D. chrysanthum	8
D. chrysotoxum	13
D. citrinicostatum	1
D. comatum	1
D. compressicolle	1
D. convexum	1
D. crenulatum	1
D. crumenatum	1
D. denigratum	1
D. section Desmotrichum	1[a]
D. dilatatocolle	1
D. section Diplocaulobium	1[a]
D. ecolle	1
D. section Euphlebium	1[a]
D. falconeri	15
D. fimbriatum	15
D. flabellum	1–2
D. gibsonii	10
D. cv Gillian	27
D. section Grastidium	1[a]
D. haemoglossum	8
D. herbaceum	10
D. heterocarpum	12
D. indragiriense	1
D. insigne	1
D. cv Jimmy Leow	12
D. cv Lam Soon	18
D. lawanum	9
D. linearifolium	3–4
D. loddigesii	15–30
D. cv Louis Bleriot	55
D. cv Louisae 'Dark'	44–45
D. cv Louisae 'Elegance'	35
D. luxurians	1
D. macrostachyum	23
D. nanum	54
D. cv Neo Hawaii	27
D. nitidicolle	1
D. nobile	15–30
D. nobile var *pendulum*	20
D. nutans	29
D. papilioniferum	2–3
D. parishii	9
D. cv Parkstance	45
D. cv Phyllis	21
D. pierardii	<15
D. pililobum	1
D. regale	1
D. rhipidolobum	1
D. cv Rose Marie	30
D. superbiens	30
D. superbum	14
D. validicolle	1
D. vanilliodorum	1
D. cv Varsity	40
D. wardianum	29
D. xantholeucum	1
Dendrochilum cobbianum	15–30
Epidendrum ciliare	<15
Epipactis erecta	8–10
E. falcata	8–12
E. papillosa	7
E. thunbergii	7–10
Eria section Cylinolobus	1[a]
Grammatophyllum multiflorum	270
Gymnadenia cucullata	8–10
Laelia anceps	<15
Liparis section Distichon	1[a]
Lycaste skinneri	60
Megaclinium	"several"
Microsaccus	1[a]
Odontoglossum	20–80
O. alexanderae	50
O. cariniferum	60
O. cervantesii	50
O. grande	30
O. hallii	30
O. luteo-purpureum	40
O. rossii var. *majus*	80
O. uro-skinneri	50
Oncidium	26–60
O. cruentum	60
O. serratum	60
O. tigrinum	8–10
Opsisanda cv Singapore	20
Paphiopedilum	90–120
P. hybrids	>30
P. insigne	40
P. villosum	70
Peristeria elata	1–2
Phajus tankervilliae (9-h day)	17
P. tankervilliae (10–13-h day)	18
P. tankervilliae (8-h day)	17
Phalaenopsis	35
P. amabilis	120
P. grandiflora	50
P. violacea	30
Pholidota imbricata	<15
Phragmopedilum sedenii	>30
Platanthera yatabei	7–10
Renanthera marshalliana (10–13-h day)	16
R. marshalliana (long day)	18
R. marshalliana (9-h day)	18
Rhynchostylis retusa	30
R. retusa (10–13-h day)	15
R. retusa (long day)	17
R. retusa (9-h day)	12
Sarcochilus appendiculatus	1
S. compressus	1
S. pallidus	1
S. teres	1

(*continued*)

Table 7-4.—Continued

Orchid	Longevity (days)	Orchid	Longevity (days)
S. teysmanii	1	*Vanda*	60–90
S. zollingerii	1	*V.* cv B. P. Mok 'Dainty'	12
Sarcostoma	1[a]	*V.* cv B. P. Mok 'Old Rose'	10
Sobralia macrantha	1	*V. coerulea*	30
Sophronitis grandiflora	30	*V. flava*	15
Spiranthes australis	10	*V.* cv Jean Kinlock Smith No. 5	20
Stanhopea	3	*V.* cv La Madona	30
S. tigrina	<15	*V. suavis*	60
		V. cv Tan Chay Yan	10, 27–28[b]
Taeniophyllum	1[a]		
Thunia marshalliana	15–30	*V.* cv Triple Cross	10

Sources: Addison, 1957; Arditti, 1979; Bhattacharjee, 1977a, 1977b, 1978a, 1978b, 1979; Bose and Mukhopadhyay, 1977; Coster, 1925; Fitting, 1909; Hew, 1980; H. R., 1895; Kerr, 1972; L. L., 1895a; 1895b; Morita, 1918; Poddubnaya-Arnoldi and Selezneva, 1957; Rao *et al.,* 1979; Smith, 1926; Tchertovitch, 1980.

[a] Some or most species.

[b] Probably full bloom period of inflorescence is 10 days and individual flower lasts 28 days.

blooms or inflorescences of different plants in one species or hybrid is usually the same. However, considerable differences may exist between different (even if related) hybrids or species.

Ethylene can shorten the life of orchid plants (for a review see Arditti, 1979). For example, acetylene (whose effects are similar to those of ethylene) at 100–500 ppm accelerated the senescence of mature, harvested inflorescence of *Arachnis* cv Maggie Oei, but a ten-minute treatment with 100–1200 ppm of $AgNO_3$ (similar to that reported by Beyer, 1976) prolonged the shelf life by three days, an increase of nearly one-fifth of the life span (J. Goh, 1977).

Physiology of Flowers

Most studies of orchid flower physiology are concerned with postpollination phenomena which are beyond the scope of this review. Relatively few studies have been conducted with unpollinated flowers (for a review see Arditti, 1979).

Photosynthesis

Green *Cymbidium* flowers possess chlorophyll (Arditti, 1966b; Matsumoto, 1966) and are capable of photosynthesis. For example, sepals, petals, and ovaries of *Cymbidium* cv Independence Day 'Yorktown' and *Cymbidium* cv Chelsea can fix $^{14}CO_2$. The magnitude of fixation by these floral segments varies with age and is lower than photosynthesis levels in leaves (Fig. 7-1).

Phosphate Uptake and Transport

^{32}P phosphate taken up by intact racemes of *Cymbidium* accumulates in gynostemia (columns) and ovaries (Oertli and Kohl, 1960). In *Arundina* flowers phosphate content decreases with age (Lim *et al.,* 1975). Pollination increases the uptake and intensifies the differences in content between the perianth and the ovary and gynostemium (Hsiang,

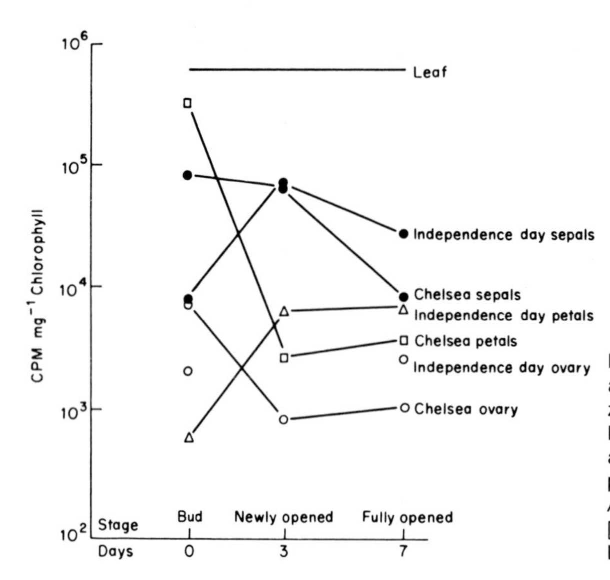

Fig. 7-1. Net $^{14}CO_2$ fixation by flowers in the light as a function of age in *Cymbidium* cv Chelsea and *C.* cv Independence Day 'Yorktown.' The horizontal bar, representing $^{14}CO_2$ fixation by a mature leaf, is included for purposes of comparison. (Arditti and Dueker, 1968; Dueker and Arditti, 1968; reproduced from Arditti, 1979 with permission from *Advances in Botanical Research* 7:421–655 [1977]. Copyright: Academic Press Inc. [London] Ltd.)

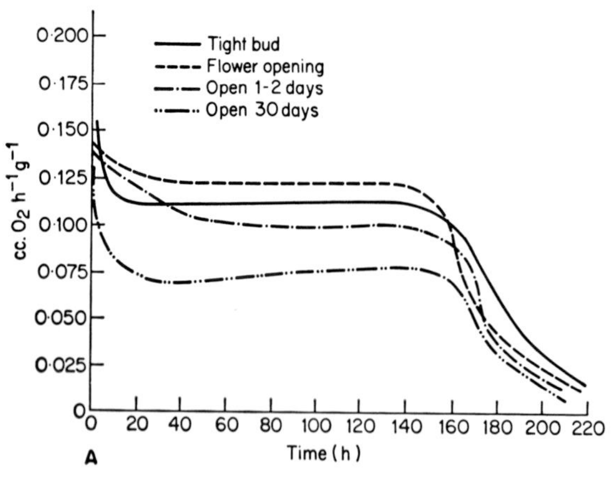

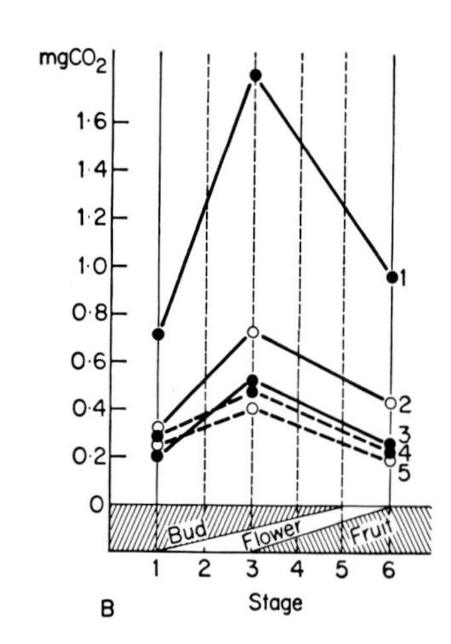

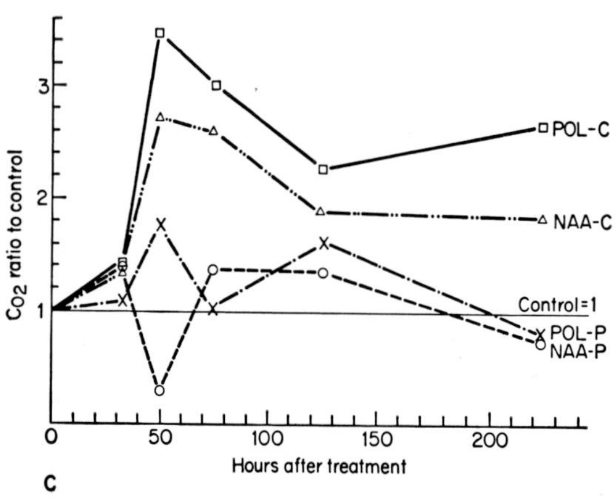

Fig. 7-2. Respiration by orchid flowers. **A:** Respiration of *Cattleya mossiae* flowers in relation to age at 15°C (Sheehan, 1954). **B:** Respiration of *Listera ovata* (1), *Platanthera chlorantha* (2), *Neottia nidus avis* (3), *Epipactis latifolia* (4), and *Goodyera repens* (5). Ordinate: respiration as mg CO_2 released h^{-1} g^{-1} fresh weight (Rosenstock, 1956). **C:** Change in respiratory rate (mm^3 O_2 g^{-1} dry weight h^{-1}) of *Cymbidium lowianum* as ratio to control following pollination or auxin treatment (Hsiang, 1951). (Figures reproduced from Arditti, 1979, with permission from *Advances in Botanical Research* 7:421–655 [1979]. Copyright: Academic Press Inc. [London] Ltd.)

1951; Oertli and Kohl, 1960). These reports indicate that transport of substances occurs within flowers. Experiments involving ^{32}P applications to labella, dorsal sepals, and stigmas of individual flowers have demonstrated that this is indeed the case (Harrison and Arditti, 1972, 1976). Transport levels and directions are changed by pollination.

Respiration

Respiration rates of *Cattleya mossiae* flowers decrease as the flowers age (Fig. 7-2-A; Sheehan, 1952, 1954) and increase very noticeably with fruit set (Fig. 7-2-B; Rosenstock, 1956). Pollination and auxin treatments bring about increases in respiratory levels (Fig. 7-2-C; Hsiang, 1951). Much of the increase in *Vanda* cv Miss Joaquim is centered in the placenta and exhibits three peaks: one immediately after pollination; a second 35 days later; and the third at 12 weeks (Table 7-5; Advanhani *et al.*, 1971). The second peak may be reminiscent of the climacteric reported for several fruits. Thus *Vanda* differs in this respect from *Cattleya mossiae,* which showed "no evidence that a climacteric occurred in the respiration" (Sheehan, 1954; for a review see Arditti, 1979).

A circadian rhythm of CO_2 production was demonstrated in the flowers of *Arachnis* cv Maggie Oei, *Oncidium* cv Goldiana, *Vanda dearei,* and *Vanda* cv Tan Chay Yan, among others. Such a rhythm was not observed in flowers of *Arachnis hookeriana* var. *luteola, Aranda* cv Wendy Scott, and *Phalaenopsis cornu-cervi* (Hew *et al.,* 1978).

Resupination

In the great majority of orchids the labellum points upward in the bud and downward in the open flower (Plate 7-1). This change in position results from a 180° twisting of the pedicel (Ames, 1938, 1945, 1946; Brieger *et al.,* 1977; van der Pijl and Dodson, 1966; Withner *et al.,* 1974). Some orchids do not resupinate at all, and their labella always point upward (Plate 7-1), and at least one orchid, *Malaxis paludosa,* twists 360° and appears to be nonresupinate. In *Catasetum barbatum* (Ames, 1945, 1946; Schomburgk, 1837) the male flowers resupinate, but the female ones do not. The ecological function of resupination is clear; it facilitates pollination (Ames, 1938, 1945, 1946; van der Pijl and Dodson, 1966).

Little is known about the physiology of resupination. The available experimental evidence (from the use of clinostats and the exclusion of gravity effects) suggests that the process is geotropic in nature (for reviews see Ames, 1938, 1946; Brieger *et al.,* 1977; van der Pijl and Dodson, 1966; Withner *et al.,* 1974). However, arguments against gravity

Table 7-5. Respiration rates in placentae of *Vanda* cv Miss Joaquim. (Summarized from data in Avadhani et al., 1971, by Arditti, 1979, with permission from *Advances in Botanical Research* 7:421–655 [1979]. Copyright: Academic Press Inc. [London] Ltd.)

Time	Respiration, oxygen uptake ($\mu l\ g^{-1}$ fresh weight)
Immediately after pollination	750
Three weeks after pollination (when ovule primordia are organized)	400
Five weeks after pollination	575
Nine weeks after pollination (when 8-celled embryo is formed)	150
Twelve weeks after pollination	375
Unspecified time period, later than twelve weeks	225

Plate 7-1. Resupinate (B and b) and nonresupinate (A and a) orchid flowers. **A:** *Cycnoches chlorochilon* Kl. (reproduced from Dunsterville and Garay, 1979). **B:** *Cattleya violacea* (H. B. K.) Rolfe (reproduced from Dunsterville and Garay, 1979). **C:** *Catasetum barbatum* Lindl., female (a) and male (b) flowers (redrawn from Schomburgk, 1837, in Ames, 1945).

effects have also been advanced (Ames, 1946). One possible reason for these diverging views may be the effect of age on resupination: "If plants [of *Goodyera pubescens*] with the lowermost flowers fully expanded are inverted, the... remaining flowers will... complete resupination.... If plants with very young buds are inverted, resupination is checked" (Ames, 1938).

Conclusion

The intricate and varied mechanisms associated with the flowering of orchids (induction, development, and gregariousness) may appear as frivolous "antics" of Nature. This is not the case, however, because the underlying theme is survival of the species. The "antics" are actually mechanisms that bring about flower production during appropriate periods and synchronize flowering with the attraction of pollinators, all of which ensures seed production. To quote Jacob Breynius, who wrote in his *Exoticarum Aliarumque Minus Cognitum Plantarum* (1678):

> If nature ever showed her playfulness in the formation of plants this is visible in the most striking way among the orchids. The manifold shape of these flowers arouses our highest admiration. They take the form of little birds, of lizards, of insects. They look like a man, like a woman, sometimes like an austere sinister fighter, sometimes like a clown who excites our laughter. They represent the image of a lazy tortoise, a melancholy toad, an agile, ever-chattering monkey. Nature has formed orchid flowers in such a way that, unless they make us laugh, they surely excite our greatest admiration. The causes of their marvelous variety are (at least in my opinion) hidden by nature under a sacred veil. [Translated by Ames, 1948]

John Lindley, the British botanist who established the family Orchidaceae, was no less eloquent—

> Orchidaceae are remarkable for the bizarre figure of their multiform flower, which sometimes represents an insect, sometimes an helmet with the visor up, and sometimes a grinning monkey; so various are these forms, so numerous their colours, and so complicated their combination, that there is scarcely a common reptile or insect to which some of them have not been likened.

—but eventually concentrated on scientific fundamentals:

> They all, however, will be found to consist of three outer pieces belonging to the calyx, and three inner belonging to the corolla. [Lindley, 1830].

We venture to add that physiologically orchid blossoms probably resemble more prosaic flowers, and while "they surely excite our greatest admiration... the causes of their marvelous variety [remain] hidden by nature under a... veil" which is anything but sacred.

Table 7-A. Factors controlling flowering of orchids

Species	Factors that control or induce flowering	Reference[a]
Aerangis biloba	Probably long days	Sanford, 1971
A. kotschyana	Probably long days	Sanford, 1971
A. multiflorum	Short days	Bose and Mukhopadhyay, 1977
A. mystacidii	Bright, dry resting period	Koopowitz, 1964
Ancistrochilus rothschildianus	Probably short days	Sanford, 1971
Ancistrorhynchus capitatus	Probably day-length-sensitive; genetic	Sanford, 1971
A. cephalotes	Probably day-length-sensitive	Sanford, 1971
A. clandestinus	Probably day-length-neutral; low or fluctuating temperatures; genetic	Sanford, 1971
A. metteniae	Probably long days	Sanford, 1971
A. recurvus	Probably day-length-sensitive	Sanford, 1971
A. straussii	Probably long days; genetic	Sanford, 1971
Angraecopsis ischnopus	Probably day-length-neutral; low or fluctuating temperatures	Sanford, 1971
A. parviflora	Probably day-length-sensitive	Sanford, 1971
A. tridens	Probably day-length-sensitive	Sanford, 1971
Angraecum angustipetalum	Probably day-length-neutral; low or fluctuating temperatures	Sanford, 1971
A. birrimense	Probably long days	Sanford, 1971
A. chevalieri	Probably day-length-neutral; low or fluctuating temperatures	Sanford, 1971
A. distichum	Probably day-length-neutral; low or fluctuating temperatures	Sanford, 1971
A. multinominatum	Probably day-length-neutral; low or fluctuating temperatures; genetic	Sanford, 1971
A. podochiloides	Probably day-length-neutral; low or fluctuating temperatures	Sanford, 1971
A. pungens	Probably day-length-neutral; low or fluctuating temperatures	Sanford, 1971
A. subulatum	Genetic	Sanford, 1971
Ansellia africana	Probably short days	Sanford, 1971
A. gigantea var *milotica*	Probably day-length-neutral; low or fluctuating temperatures	Sanford, 1971
Arachnis cv Maggie Oei	Plant growth regulators and salicylic acid, triiodobenzoic acid and coumarin	Byramji and Goh, 1976; Goh, 1976a,b
Aranda cv Christine No. 1	Decapitation induces flowering shoots near the apex	Koay and Chua, 1979
A. cv Deborah	Inhibited by actinomycin D and cycloheximide. Apical control; inhibited by auxin; flowering gradient affected by growth regulators; decapitation may induce flowering; indifferent to day length; salicylic acid, triiodobenzoic acid, coumarin, B995, phosphon-D and CCC enhance flowering	Goh and Seetoh, 1973; Goh, 1975; 1976a,b, 1977a,b,c,d
A. cv Hilda Galistan	Similar to *Aranda* cv Deborah	Goh, 1977b
A. cv Lucy Laycock	Similar to *Aranda* cv Deborah	Goh, 1977b
A. cv Mei Ling	Similar to *Aranda* cv Deborah	Goh, 1977b
A. cv Nancy	Similar to *Aranda* cv Deborah	Goh, 1977b
A. cv Peter Ewart	See *Aranda* Christine No. 1	Koay and Chua, 1979
A. cv Wendy Scott	Indifferent to day length. Salicylic acid, triiodobenzoic acid, and coumarin enhance flowering	Byramji and Goh, 1976; Goh, 1976b
Bletilla striata	Gibberellin treatments, 50 ppm, enhance flowering	Sano *et al.*, 1961
Bolusiella talbotii	Probably day-length-sensitive	Sanford, 1971
Brassavola nodosa	Short days (less than 14 h) enhance flowering	Brieger *et al.*, 1977
Brassia verrucosa	Vegetative growth is enhanced by 14-h days	Brieger *et al.*, 1977
Brassocattleya hybrids	More flowers under short days	Sheehan *et al.*, 1965
Bromheadia alticola	Low temperature	Holttum, 1953
B. finlaysoniana	Flowering is stimulated by wet and cool days and retarded by drought. Dry periods check flower bud development.	Holttum, 1953; Jeyanayaghy and Rao, 1966; Sanford, 1971

(*continued*)

Table 7-A.—Continued

Species	Factors that control or induce flowering	Reference[a]
Bulbophyllum barbigerum	Probably day-length neutral; low or fluctuating temperatures	Sanford, 1971
B. bufo	Probably short days	Sanford, 1971
B. buntingii	Probably day-length-sensitive	Sanford, 1971
B. calamarium	Probably day-length-neutral; low or fluctuating temperatures	Sanford, 1971
B. calyptratum	Probably day-length-neutral; low or fluctuating temperatures	Sanford, 1971
B. colubrinum	Probably day-length-neutral; low or fluctuating temperatures	Sanford, 1971
B. congalanum	Probably day-length-neutral; low or fluctuating temperatures	Sanford, 1971
B. distans	Probably day-length-neutral; low or fluctuating temperatures; genetic	Sanford, 1971
B. falcatum	Short days	Sanford, 1971
B. flavidum	Probably day-length-insensitive; low or fluctuating temperatures	Sanford, 1971
B. fuerstenbergianum	Probably day-length-sensitive	Sanford, 1971
B. fuscum	Probably short days	Sanford, 1971
B. intertextum	Probably day-length-neutral; low or fluctuating temperatures	Sanford, 1971
B. josephii	Probably day-length-sensitive	Sanford, 1971
B. kamerunense	Probably short days	Sanford, 1971
B. lobii	Vegetative growth is enhanced by 14-h days	Brieger *et al.*, 1977
B. lupulinum	Probably short days	Sanford, 1971
B. magnibracteatum	Probably day-length-sensitive	Sanford, 1971
B. melanorrhachis	Probably short days	Sanford, 1971
B. nigericum	Probably day-length-sensitive	Sanford, 1971
B. oreonastes	Probably day-length-neutral; low or fluctuating temperatures	Sanford, 1971
B. pavimentatum	Probably day-length-sensitive	Sanford, 1971
B. pipio	Probably day-length-sensitive	Sanford, 1971
B. porphyroglossum	Probably day-length-neutral	Sanford, 1971
B. porphyrostachys	Probably day-length-sensitive	Sanford, 1971
B. rhizophorae	Probably short days	Sanford, 1971
B. schimperanum	Probably day-length-neutral; low or fluctuating temperatures	Sanford, 1971
B. simonii	Probably day-length-sensitive	Sanford, 1971
B. winkleri	Probably day-length-neutral	Sanford, 1971
Calanthe rosea	Vegetative growth is enhanced by 14-h days	Brieger *et al.*, 1977
C. cv Veitchii	Vegetative growth is enhanced by 14-h days	Brieger *et al.*, 1977
C. vestita	Vegetative growth is enhanced by 14-h days	Brieger *et al.*, 1977
Calyptrochilum christyanum	Probably day-length-neutral; low or fluctuating temperatures	Sanford, 1971
C. emarginatum	Probably day-length-sensitive	Sanford, 1971
Catasetum	High light intensities stimulate formation of female flowers; plants in shade produce male flowers. Vegetative growth under 14-h illumination	Brieger, 1957; Gregg, 1975; Brieger *et al.*, 1977
Cattleya	Temperatures of close to 17°C are favorable for abundant flowering	Tran Thanh Van, 1974
C. amabilis	Short days	Urmston, 1949
C. cv Bow Bells	16½-h photoperiods delay flowering	Franklin, 1967
C. bowringiana	14-h days do not prevent flowering; 8-h day brings about earlier flowering	Arditti, 1968; Bhattacharjee, 1979
C. cv Dupreana	GA, 10 μg/sheath advanced flowering by 1–2 days	Arditti, 1966a
C. cv Enid	Not affected by day length and temperature	Arditti, 1968
C. gaskelliana	9-h days at 13°C or 16-h days at 18°C (night temperatures) induce flowering. Long-day plant at 18°C; nonphotoperiodic at 13°C	Arditti, 1966a, 1967b, 1968; Rotor, 1952, 1959; Haber, 1952

(*continued*)

Table 7-A.—Continued

Species	Factors that control or induce flowering	Reference[a]
C. cv Geriant	GA, less than 15 μg/sheath induces deformed earlier flowers	Arditti, 1966a
C. gigas	See *C. warscewiczii*	Arditti, 1966
C. cv Jean Barrow	Short days enhanced flowering, but there was no peak period	Sheehan *et al.*, 1965
C. cv Joyce Hannington	See *C.* cv Bow Bells	Franklin, 1967
C. labiata	Short days (less than 16½ h) induce flowering	Arditti, 1966a, 1967b, 1968; Franklin, 1967; Rotor, 1952, 1959
C. cv Los Gatos	Nights 8–12 h initiate flowering	Arditti, 1966a
C. mendelii	Short days at 13°C and18°C (night temperature), 9-h days at 13°C induce flowering. Will not bloom on 16-h days at 13°C or 18°C night temperature; produces flowers on 9-h days at 18°C.	Urmston, 1949
C. mossiae	Same as *C. mendelii*	Arditti, 1966a, 1967b, 1968; Rotor, 1952, 1959; Haber, 1952
C. cv Oenone	16½- to 17-h light starting in late June, then short days 4 months before desired blooming date	Franklin, 1967
C. percivaliana	Short days (9 h) under 18°C at night	Arditti, 1966a, 1967b, 1968; Haber, 1952
C. cv Pinole	Flowers form only on shoots 12–18 cm long	Rotor, 1952, 1959; Urmston, 1949; Johnson and Laurie, 1943, 1945
C. schroederae	Same as *C. mendelii*	Urmston, 1949
C. skinneri	Could be short-day plant whose photoperiodic response may be modified by temperature	Rotor, 1952
C. trianae	8-h days bring about earlier flowering. Short-day (9 h) plant under both 13°C and 18°C. Short days (9 h) induce flowering whereas long days (16 h) delay it	Bhattarcharjee, 1979; Haber, 1952; Rotor, 1952, 1959; Urmston, 1949; Holdson and Laurie, 1949
C. warscewiczii	Short days at 13°C	Rotor, 1952, 1959; Urmston, 1949
Chamaeangis lanceolata	Probably short days; genetic	Sanford, 1971
C. odoratissima	Probably day-length-sensitive	Sanford, 1971
C. vesicata	Probably long days	Sanford, 1971
Chaulodon buntingii	Probably day-length-neutral; low or fluctuating temperatures	Sanford, 1971
Corymborkis corymbosa	Probably day-length-neutral	Sanford, 1971
Cycnoches	High light intensities stimulate production of female flowers	Gregg, 1975
Cymbidium	Three months of 13°C night temperature. Nights in the range of 7°C to 13°C coupled with bright, sunny days initiate flowering	Davidson, 1977; Arditti, 1966a, 1967b, 1968; Rotor, 1952, 1959; Went, 1946, 1951; Casamajor and Went, 1953
	Night temperatures of 5–8°C are best for flower induction	Richter, 1966
Cymbidium (standard)	14°C nights; 21°C days in summer and 17°C in winter preferable	Leffring, 1972
C. cv Desiree A'Logann	Abscisic acid did not delay flowering	Brewer *et al.*, 1969
C. ensifolium	Formation and development of inflorescences to anthesis are promoted by 30°C days and 25°C nights	Kako and Ohno, 1980
C. cv Guelda	Gibberellic acid causes earlier opening of flowers and larger blossoms	Bivins, 1970
C. insigne	Formation and development of inflorescence to anthesis were not affected by day/night temperatures of 35°C/25°C, 25°C/20°C, and 15°C/10°C	Kako and Ohno, 1980
C. lowianum	Flowers formed when night temperatures were 6–10°C	Arditti, 1979
C. pumilum	Initiation, but not development of inflorescences, favored by day/night temperatures of 25/20°C	Kako and Ohno, 1980

(*continued*)

Table 7-A.—Continued

Species	Factors that control or induce flowering	Reference[a]
C. cv Rozette	Flowers formed when plants were maintained for 14-h days, 1000 ft-cdls, and 22°C and 18°C nights	Arditti, 1968
C. cv Sazanami 'Haru-o-umi'	Formation of inflorescences was not affected by day/night temperatures of 30°C/25°C, 25°C/20°C, and 15°C/10°C; the highest temperature caused abortion of all inflorescences at an early stage of development	Kako and Ohno, 1980
C. cv Sicily 'Grandee'	Gibberellic acid increased flower size and raceme length and accelerated flowering	Bivins, 1968
C. virescens	Flower stalk elongation is enhanced by low night temperatures (5°C)	
Cyrtorchis arcuata subsp. *variabilis*	Probably long days	Sanford, 1971
C. aschersonii	Probably day-length-neutral; low or fluctuating temperatures	Sanford, 1971
C. chailluana	Probably day-length-sensitive	Sanford, 1971
C. hamata	Probably long days	Sanford, 1971
C. monteiroae	Probably day-length-sensitive	Sanford, 1971
C. ringens	Probably day-length-sensitive; genetic	Sanford, 1971
Dendrobium	Not day-length-sensitive. Temperatures close to 17°C are favorable for abundant flowering. Cytokinins can stimulate flowering. Their effect is enhanced by gibberellins	Sheehan *et al.*, 1965; Van der Donk, 1974; Goh, 1979
D. acuminatissimum	Low temperature	Coster, 1925
D. cv Anne Marie	Flowers when grown warm and moist	Coster, 1925
D. angulatum	Low temperature	Coster, 1925
D. aratriferum	Low temperature	Coster, 1925
D. bancanum	Low temperature	Coster, 1925
D. bicostatum	Low temperature	Coster, 1925
D. brevicolle	Low temperature	Coster, 1925
D. cv Buddy Shepler × *D.* cv Peggy Shaw	Cytokinins stimulate flowering; their effect is enhanced by gibberellins	Goh and Yang, 1978
D. citrinicostatum	Low temperature	Coster, 1925
D. comatum	Low temperature	Coster, 1925
D. compressicolle	Low temperature	Coster, 1925
D. crenulatum	Low temperature	Coster, 1925
D. crumenatum	Low temperature	Coster, 1925; Holttum, 1953, 1969
D. dendrocolla	Low temperature	Coster, 1925
D. dilatatocolle	Low temperature	Coster, 1925
D. draconis	Prefers cool climate	Kamemoto and Sagarik, 1965
D. ecolle	Low temperature	Coster, 1925
D. feuilletaui	Low temperature	Coster, 1925
D. filiforme	Low temperature	Coster, 1925
D. findlayanum	May require cold temperatures for flower initiation	Arditti, 1966a
D. flabelloides	Low temperature	Coster, 1925
D. flabellum	Low temperature	Coster, 1925
D. formosum	Plants flower after long dry season	Sanford, 1971
D. fugax	Low temperature	Coster, 1925
D. heterocarpum	See *D. findlayanum*	
D. infundibulum	Low temperature	Kamemoto and Sagarik, 1965
D. insigne	Low temperature	Coster, 1925
D. cv Jaquelyn Thomas	Not induced by light	Sheehan *et al.*, 1965
D. cv Lady Fay	Not induced by light	Sheehan *et al.*, 1965
D. cv Lady Hochoy	Cytokinins stimulate flowering; their effects are enhanced by gibberellins	Goh and Yang, 1978
D. luxurians	Low temperature	Coster, 1925
D. macraei	Low temperature	Coster, 1925
D. maculosum	Low temperature	Coster, 1925
D. cv Merlin	Blooms when grown warm and moist	Arditti, 1966a
D. nitidicolle	Low temperature	Coster, 1925

(*continued*)

Table 7-A.—Continued

Species	Factors that control or induce flowering	Reference[a]
D. nobile	Flowering is induced by low temperatures	Arditti, 1966a; Kosugi, 1952; Rotor, 1952, 1959
D. padangense	Low temperature	Coster, 1925
D. papilioniferum	Low temperature	Coster, 1925
D. phalaenopsis	Short days and 18°C or 13°C night temperature induce flowering	Rotor, 1952, 1959
D. pumilum	Low temperature	Coster, 1925
D. scabrilingue	See *D. infundibulum*	
D. spathilingue	Low temperature	Coster, 1925
D. spurium	Low temperature	Coster, 1925
D. cv Thwaitesiae	Lowering of temperature at night may induce flowering	Arditti, 1966a
D. tunense	Low temperature	Coster, 1925
D. validicolle	Low temperature	Coster, 1925
D. xantholeucum	Low temperature	Coster, 1925
Diaphananthe bidens	Probably day-length-sensitive	Sanford, 1971
D. curvata	Probably long days	Sanford, 1971
D. kamerunensis	Probably long days	Sanford, 1971
D. longicalcar	Probably day-length-sensitive	Sanford, 1971
D. obanensis	Probably day-length-neutral; low or fluctuating temperatures	Sanford, 1971
D. pellucida	Probably long days	Sanford, 1971
D. plehniana	Probably day-length-sensitive; genetic	Sanford, 1971
D. rutila	Probably long days	Sanford, 1971
Encheiridion macrorrhynchium	Probably long days	Sanford, 1971
Epidendrum radicans	Flowers initiated in October 1970 developed to pollen formation in early January 1971 and flowered on 15 January 1971	Kosugi *et al.*, 1973
Eria	Low temperature	Smith, 1926, 1927
Eulophia cucculata	Burning of grasslands may induce flowering	Sanford, 1971
E. euglossa	Probably day-length-neutral; low or fluctuating temperatures	Sanford, 1971
E. gracilis	Probably short days	Sanford, 1971
E. guineensis	Probably long days	Sanford, 1971
E. horsfallii	Probably day-length-neutral; low or fluctuating temperatures	Sanford, 1971
E. quartiniana	Probably day-length-neutral	Sanford, 1971
Eulophidium maculatum	Probably day-length-neutral; low or fluctuating temperatures	Sanford, 1971
E. saundersianum	Probably day-length-neutral; low or fluctuating temperatures	Sanford, 1971
Eurychone rothschildiana	Probably long days	Sanford, 1971
Genyorchis pumila	Probably short days	Sanford, 1971
Grammatophyllum rumphianum	Alternation of wet and dry seasons may regulate flowering	Holttum, 1957; Sanford, 1971
Graphorkis lurida	Probably short days; genetic	Sanford, 1971
Habenaria engleriana	Probably long days	Sanford, 1971
H. macrandra	Probably day-length-sensitive	Sanford, 1971
Hetaeria stammleri	Probably short days	Sanford, 1971
Laelia albida	Short days may enhance flowering	Brieger *et al.*, 1977
L. purpurata	Long days	Arditti, 1968
Laeliocattleya cv Canhamiana	Day length of no less than 16 h induces flowering	Hampton, 1955
Liparis caillei	Probably long days	Sanford, 1971
L. epiphytica	Probably day-length-neutral; low or fluctuating temperatures	Sanford, 1971
L. nervosa	Probably long days	Sanford, 1971
L. platyglossa	Probably day-length-neutral	Sanford, 1971
L. suborbicularis	Probably day-length-neutral	Sanford, 1971
L. tridens	Probably long days	Sanford, 1971
Listrostachys pertusa	Probably day-length-sensitive	Sanford, 1971
Malaxis katangensis	Probably long days	Sanford, 1971
M. maclaudii	Probably day-length-neutral	Sanford, 1971
M. prorepens	Probably day-length-neutral	Sanford, 1971

(*continued*)

Table 7-A.—Continued

Species	Factors that control or induce flowering	Reference[a]
M. weberbauerana	Probably day-length-neutral; low or fluctuating temperatures	Sanford, 1971
Microcoelia caespitosa	Probably day-length-neutral; low or fluctuating temperatures	Sanford, 1971
M. dahomeensis	Probably day-length-neutral; low or fluctuating temperatures	Sanford, 1971
Miltonia	Temperatures cose to 17°C are favorable for abundant flowering	Tran Thanh Van, 1974a
M. anceps	Long days enhance flower formation	Brieger *et al.*, 1977
M. loezlii	Long days enhance flower formation	Brieger *et al.*, 1977
M. spectabilis	Long days enhance flower formation	Brieger *et al.*, 1977
Mormodes	See *Catasetum* and *Cynoches*	Dodson, 1962
Nervilia adolphii	Probably day-length-neutral	Sanford, 1971
N. fuerstenbergiana	Probably day-length-neutral	Sanford, 1971
N. kotschyi	Probably day-length-neutral	Sanford, 1971
N. reniformis	Probably day-length-neutral	Sanford, 1971
N. umbrosa	Probably day-length neutral	Sanford, 1971
Odontoglossum bictonense	Long days enhance flower formation	Brieger *et al.*, 1977
O. citrosmum (*O. pendulum*)	Periods of low temperature initiate flowering	Arditti, 1966a
O. hybrids	"Daylength is an important factor in the production of spikes."	Baker, 1968
Oncidium sphacelatum	Floret initiation in late December in Japan	Kosugi *et al.*, 1973
O. splendidum	Flowers are initiated during short, cool days	Franklin, 1967
Paphiopedilum	No photoperiodic response. Night temperatures of 13°C for 2–3 weeks	Franklin, 1967; Davidson, 1977
P. insigne	Temperatures of 14–18°C are best for flower induction. No photoperiodic response. Night temperatures of 13°C for 2–3 weeks	Richter, 1966
Phajus tankervilliae	Best flowering occurred under 10½ and 13½ hours	Bose and Mukhopadhyay, 1977
Phalaenopsis	No photoperiodic response. Flowering induced by 3 weeks of 14°C nights and 20°C days. Flowering induced by low (10–15°C) temperature. Floral initiation takes place when night temperature varies between 12°C and 17°C and day temperatures do not exceed 27°C. Treatment time: 2–5 weeks	Franklin, 1967; Halperin and Halevy, 1974, 1975; Halevy, 1975; Tran Thanh Van, 1970, 1974; Nishimura *et al.*, 1972, 1976
P. amabilis	Flowering is induced by short days and 18°C	Rotor, 1952, 1959
P. schilleriana	Similar to *P. amabilis.* Flowering is stimulated by 2–3 weeks exposure to night temperatures below 21°C	Rotor, 1952, 1959; De Vries, 1953
Plectrelminthus caudatus	Probably long days	Sanford, 1971
Podangis dactyloceras	Probably short days	Sanford, 1971
Polystachya adansoniae	Probably day-length-neutral	Sanford, 1971
P. affinis	Probably short days	Sanford, 1971
P. albescens subsp. *albescens*	Probably day-length-neutral; low or fluctuating temperatures	Sanford, 1971
P. albescens subsp. *angustifolia*	Probably day-length-neutral; low or fluctuating temperatures	Sanford, 1971
P. calluniflora	Probably day-length-neutral	Sanford, 1971
P. caloglossa	Probably day-length-neutral; low or fluctuating temperatures	Sanford, 1971
P. coriscensis	Probably day-length neutral; low or fluctuating temperatures	Sanford, 1971
P. cultriformis	Same as *Phalaenopsis schilleriana*	Sanford, 1971
P. dolichophylla	Probably day-length-neutral	Sanford, 1971
P. galeata	Probably day-length-neutral; low or fluctuating temperatures	Sanford, 1971
P. golungensis	Probably short days	Sanford, 1971

(*continued*)

Table 7-A.—Continued

Species	Factors that control or induce flowering	Reference[a]
P. laxiflora	Probably day-length-neutral; low or fluctuating temperatures	Sanford, 1971
P. modesta	Probably long days	Sanford, 1971
P. mukandaensis	Probably long days; genetic	Sanford, 1971
P. odorata var. *odorata*	Probably day-length-neutral	Sanford, 1971
P. paniculata	Probably day-length-sensitive	Sanford, 1971
P. polychaete	Probably day-length-neutral; low or fluctuating temperatures	Sanford, 1971
P. ramulosa	Probably day-length neutral; low or fluctuating temperatures	Sanford, 1971
P. rhodoptera	Probably day-length-sensitive	Sanford, 1971
P. saccata	Probably day-length-sensitive	Sanford, 1971
P. subulata	Probably long days	Sanford, 1971
P. supfiana	Probably day-length neutral; low or fluctuating temperatures	Sanford, 1971
P. tessellata	Probably day-length-neutral; low or fluctuating temperatures	Sanford, 1971
Rangaeris muscicola	Probably long days	Sanford, 1971
R. rhipsalisocia	Probably short days	Sanford, 1971
Renanthera imschootiana	Short days (9 h) induced earlier flowering	Bose and Mukhopadhyay, 1977
Rhynchostylis gigantea	Short days (8 h) at low temperatures (10°C for 16 h/day) induce early flowering	Inthuwong and Watcharaphai, 1964
R. retusa	See *Renanthera imschootiana*	
Sarcochilus appendiculatus	Low temperature	Coster, 1925
S. compressus	Low temperature	Coster, 1925
S. pallidus	Low temperature	Coster, 1925
S. teres	Low temperature	Coster, 1925
S. zollingeri	Low temperature	Coster, 1925
Solenangis clavata	Probably long days	Sanford, 1971
S. scandens	Probably day-length-sensitive	Sanford, 1971
Stolzia repens	Probably day-length-neutral; low or fluctuating temperatures	Sanford, 1971
Thrixspermum	A sudden drop in temperature may initiate flowering in some species	Holttum, 1957
T. arachnites	Low temperature	Coster, 1925
T. calceolus	Low temperature	Coster, 1925
T. inquinatum	Low temperature	Coster, 1925
T. raciborskii	Low temperature	Coster, 1925
T. subulatum	Low temperature	Coster, 1925
Tridactyle anthomaniaca	Probably day-length-neutral; low or fluctuating temperatures	Sanford, 1971
T. bicaudata	Probably long days	Sanford, 1971
T. brevicalcarata	Probably long days	Sanford, 1971
T. gentilii	Probably day-length-sensitive	Sanford, 1971
T. lahosens	Probably long days	Sanford, 1971
T. tridactylites	Probably short days	Sanford, 1971
Vanda	Weekly sprays of 10 ppm gibberellins induce flower formation	O'Neill, 1958
V. cv Miss Joaquin	Longer periods in sunlight brought about more profuse flowering; endogenous gibberellins "showed no obvious correlations . . . Auxin levels in the shoot apex may be responsible" (Goh and Wan, 1974)	Goh and Wan, 1974; Murashige *et al.*, 1967
Vanilla	Drought, age, temperature, pruning (decapitation), and training of vines all affect flowering	Childers *et al.*, 1959
Zeuxine elongata	Probably short days	Sanford, 1971
Zygopetalum	See *Vanda*	O'Neill, 1958

[a] For simplicity, several reviews of original literature, rather than the literature itself, are cited as sources.

Literature Cited

Addison, G. H. 1957. Longevity in orchid flowers. Malayan Orchid Rev. 5:27–28.

Alphonso, A. G. 1965. Cymbidium roseum. Malayan Orchid Rev. 8:48.

Ames, O. 1938. Resupination as a diagnostic character in the Orchidaceae with special reference to Malaxis monophyllos. Bot. Mus. Leafl. Harvard Univ. 6:145–183.

——. 1945. The strange case of Catasetum barbatum. Amer. Orchid Soc. Bull. 13:289–294.

——. 1946. Notes on resupination in the Orchidaceae. Amer. Orchid Soc. Bull. 15:18–19.

——. 1948. Orchids in retrospect. Botanical Museum Harvard Univer., Cambridge, Mass.

Arditti, J. 1966a. Flower induction in orchids. Orchid Rev. 74:208–217.

——. 1966b. The green color of Cymbidiums: What is it? Cymbidium Soc. News 20:10–11.

——. 1967a. Factors affecting the germination of orchid seeds. Bot. Rev. 33:1–97.

——. 1967b. Flower induction in orchids. II. Orchid Rev. 75:253–256.

——. 1968. Flower induction in orchids. III. Orchid Rev. 76:191–197.

——. 1979. Aspects of the physiology of orchids. Adv. Bot. Res. 7:421–655. Academic Press, London.

Arditti, J., and J. Dueker. 1968. Photosynthesis by various organs of orchid plants. Amer. Orchid Soc. Bull. 37:862–866.

Arens, P. 1923. Periodische Blütenbildung bei einigen Orchideen. Ann. Jard. Bot. Buitenzorg 32:103–124.

Avadhani, P. N., A. N. Rao, and P. Y. Ong. 1971. Pod development and seed germination of Vanda Miss Joaquim. 58th Indian Sci. Congr., Bangalore, p. 8–10.

Baker, P. G. 1968. The influence of day length on Odontoglossum flowering. Orchid Rev. 76:199–201.

Beaumée, J. 1927. Note. De Tropische Natuur 16:54–55.

Beyer, E. N. 1976. A potent inhibitor of ethylene action in plants. Plant Physiol. 58:268–271.

Bhattacharjee, S. K. 1977a. The native habenarias of India. Orchid Rev. 85:13–16.

——. 1977b. Indian coelogynes. Orchid Rev. 85:168–171.

——. 1978a. Native erias of India. Orchid Rev. 86:246–249.

——. 1978b. Some light on Indian Calanthe. Orchid Rev. 86:332–334.

——. 1979. Regulation of growth and flowering in Cattleya orchids by altered daylengths. Singapore J. Primary Industries 7:90–92.

Bivins, J. L. 1968. Effect of growth regulating substances on the size of flower and bloom date of Cymbidium Sicily 'Grandee.' Amer. Orchid Soc. Bull. 37:385–387.

——. 1970. Effect of gibberellic acid on flower size and bloom date of Cymbidium Guelda. Amer. Orchid Soc. Bull. 39:1005–1006.

Bose, T. K., and T. P. Mukhopadhyay. 1977. Effects of day length on growth and flowering of some tropical orchids. Orchid Rev. 85:245–247.

Brewer, K., C. Gradowski, and M. Meyer. 1969. Effect of abscisic acid on Cymbidium orchid plants. Amer. Orchid Soc. Bull. 38:591–592.

Brieger, F. G. 1957. Research on orchids at Piracicaba, Sao Paulo, Brazil. Amer. Orchid Soc. Bull. 26:546–550.

Brieger, F. G., R. Maatsch, and K. Senghas (eds.). 1977. Schlechter, Die Orchideen, 3d ed. Paul Parey, Berlin.

Burkill, I. H. 1917. The flowering of the Pigeon-orchid, Dendrobium crumenatum. Gardens Bull., Straits Settlements Ser. 3, Vol. 1:400–405.

Byramji, H., and C. J. Goh. 1976. Photoperiodic responses of some local orchid hybrids. J. Singapore Nat. Acad. Sci. 5:15–17.

Casamajor, R., and F. W. Went. 1953. Preliminary report on Cymbidium research. Amer. Orchid Soc. Bull. 22:126.

Childers, N. F., H. R. Cibes, and E. Hernández-Medina. 1959. Vanilla—the orchid of commerce, p. 477–508. *In* C. L. Withner (ed.), The orchids, a scientific survey. Ronald Press, New York.

Cleland, C. F., and A. Ajami. 1974. Identification of a flower inducting factor isolated from aphid honeydew as being salicylic acid. Plant Physiol. 54:904–906.

Coomans de Ruiter, L. 1930. De bloei van Dendrobium crumenatum Sw. gedurende het jaar 1929 te pontianak. Tropische Natuur 19:116–119.

Coster, C. 1925. Periodische Blüteerscheinungen in den tropen. Ann. Jard. Bot. Buitenzorg 35:125–162.

Curtis, J. T. 1954. Annual fluctuation in rate of flower production by native cypripediums during two decades. Bull. Torrey Bot. Club 81:340–352.

Davidson, O. W. 1977. Question box. Amer. Orchid Soc. Bull. 46:798.

De Vries, J. T. 1953. On the flowering of Phalaenopsis schilleriana Rchb. f. Annales Bogoriensis 1:61–76.

Dodson, C. 1962. Variation in the Catasetinae. Ann. Missouri Bot. Gard. 49: 35–36.

Dueker, J., and J. Arditti. 1968. Photosynthetic $^{14}CO_2$ fixation by green Cymbidium (Orchidaceae) flowers. Plant Physiol. 43:130–132.

Dunsterville, G. C. K., and E. Dunsterville. 1967. The flowering season of some Venezuelan orchids. Amer. Orchid Soc. Bull. 36:790–797.

Dunsterville, G. C. K., and L. A. Garay. 1979. Orchids of Venezuela—An illustrated field guide. Botanical Museum, Harvard Univ. Museum Books, New York.

Ebel, F., and G. Mörchen. 1977. Ein Beitrag zu Morphologie und Rhythmik der Orchidee Domingoa haematochila (Rchb. f.) Carabia. Flora 166:35–44.

Ede, J. 1963. Some observations on the flowering characteristics of Arachnis Maggie Oei. Malayan Orchid Rev. 7:76–78.

Evans, L. T. 1971. Flower induction and the florigen concept. Ann. Rev. Plant Physiol. 22:365–394.

Ex-Cantab. 1871. Cacti versus orchids. Gard. Chron. 31:516.

Fitting, H. 1909. Die Beinflussung der Orchideenblüten durch die Bestäubung und durch andere Umstande. Zeit. fur Bot. 1:1–86.

Franklin, L. W. 1967. Initiation of flowers in several orchid species. Orchid Rev. 75:213–215.

Goh, C. J. 1970. Unusual flowering habit in Aranda hybrids. Malayan Orchid Rev. 2:112–114.

——. 1973. On the flowering of some local monopodial orchids. Malayan Orchid Rev. 11:16–18.

——. 1975. Flowering gradient along the stem axis in an orchid hybrid Aranda Deborah. Ann. Bot. 39:931–934.

——. 1976a. Reversion of vegetative and reproductive growth in monopodial orchids. Ann. Bot. 40:645–646.

——. 1976b. Effects of salicylic acid on the flowering of some monopodial orchid hybrids. Plant Physiol. 57 (Ann. Meeting Suppl.):64.

——. 1977a. The nature of the flowering young axillary shoots in Aranda Deborah. Ann. Bot. 41:1065–1067.

——. 1977b. Further studies on the flowering gradient in Aranda orchid hybrids. Ann. Bot. 41:1061–1063.

——. 1977c. Studies on the control of flowering in orchids. Report of the Second ASEAN Orchid Seminar. Jakarta, Indonesia.

——. 1977d. Regulation of floral initiation and development in an orchid hybrid Aranda Deborah. Ann. Bot. 41:763–769.

——. 1979. Hormonal regulation of flowering in a sympodial orchid hybrid Dendrobium Louisae. New Phytol. 82:375–380.

Goh, C., and J. Arditti. In press. Flowering in orchids. *In* A. H. Halevy (ed.), Handbook on flowering. CRC Press, Boca Raton, Fla.

Goh, C. J., and H. C. Seetoh. 1973. Apical control of flowering in an orchid hybrid, Aranda Deborah. Ann. Bot. 37:113–119.

Goh, C. J., and H. Y. Wan. 1974. The role of auxins in the flowering of a tropical orchid hybrid Vanda Miss Joaquim, p. 945–952. *In* Y. Sumiki (ed.), Plant growth substances, 1973. Hirotaka Publishing Co., Tokyo.

Goh, C. J., and A. L. Yang. 1978. Effects of actinomycin D and cycloheximide on the flowering response of Aranda Deborah following decapitation. Plant Physiol. 59 (Suppl.):121.

Goh, J. 1977. Preliminary studies on the postharvest longevity and bud opening of orchid inflorescences. B.Sc. (Hons.) thesis, Botany Dept., University of Singapore.

Gregg, K. B. 1973. Studies on the control of sex expression in the genera Cycnoches and Catasetum, subtribe Catasetinae, Orchidaceae. Ph.D. Diss., Univ. Miami, Coral Gables, Fla.

——. 1975. The effect of light intensity on sex expression in species of Cycnoches and Catasetum (Orchidaceae) Selbyana 1:101–113.

——. 1977. Ethylene physiology, sunlight, intensity, and sex in the orchid genera Catasetum and Cycnoches. West Virginia Acad. Sci. 49:10–11.

——. 1978. The effects of light intensity and sex on ethylene production in developing racemes of Cycnoches and Catasetum (Orchidaceae). Plant Physiol. 61 (Suppl.):50.
H. R. 1895. La durée des fleurs d'orchidées. J. Orchidées 6:213.
Haber, W. 1952. Über orchideenkultur und künstliche Beleuchtung. Die Orchidee 3:14–18.
Hager, H. 1957. Control of flowering and Cattleya. Proc. 2d World Orchid Conf., p. 130–132.
Halevy, A. H. 1975. Light intensity as a factor in flower formation and development. Proc. 3d MPP Meeting, Ege Univ., Bornova, Izmir, Turkey, p. 84–88.
Halperin, M., and A. Halevy. 1974/1975. Regulation of flowering in Phalaenopsis. Ann. Rept. Dept. Ornamental Hort., Hebrew Univ., Rehovoth, p. 49–52.
Hamilton, R. M. 1977. When does it flower? R. M. Hamilton Publisher, Richmond, B.C., Canada.
Hampton, J. 1955. Control of Cattleya flowering by light. Orchid Digest 19:53–58.
Harrison, C. R., and J. Arditti. 1972. Phosphate movement, water relations, and dry weight variation in pollinated Cymbidium (Orchidaceae) flowers. Amer. J. Bot. (Suppl.) 59:699 (abst.).
——. 1976. Post-pollination phenomena in orchid flowers. VII. Phosphate movement among floral segments. Amer. J. Bot. 63:911–918.
Hew, C. S. 1980. Respiration of tropical orchid flowers. Proc. 9th World Orchid Conf., Bangkok (1978), p. 191–195.
Hew, C. S., Y. C. Thio, S. C. Wong, and T. Y. Chin. 1978. Rhythmic production of CO_2 by tropical orchid flowers. Physiologia Plantarum 42:226–230.
Holdson, J., and A. Laurie. 1949. The effect of supplementary illumination on the flowering of the orchid Cattleya trianae. Proc. Amer. Soc. Hort. Sci. 57:379–380.
Holttum, R. E. 1949. Freedom of flowering in orchids in Singapore. Malayan Orchid Rev. 4:15–17.
——. 1953. Evolutionary trends in an equatorial climate. Symp. Soc. Exp. Biol. 7:159–173.
——. 1957. A revised flora of Malaya. Vol. I, Orchids of Malaya. 2d ed. Government Printing Office, Singapore.
——. 1962. Aranthera Mohamed Haniff. Malayan Orchid Rev. 7:21.
——. 1969. Plant life in Malaya. Longman, London.
Hsiang, T-H. T. 1951. Physiological and biochemical changes accompanying pollination in orchid flowers. II. Respiration, catalase activity, and chemical constituents. Plant Physiol. 26:708–721.
Inthuwong, O., and T. Watcharaphai. 1964. Daylength and temperature in relation to flowering of Rhynchostylis gigantea. Kasetsart J. 4:1–6.
Jeyanayaghy, S., and A. N. Rao. 1966. Flower and seed development in Bromheadia finlaysoniana. Bull. Torrey Bot. Club 93:97–103.
Johnson, E., and A. Laurie. 1943. Flower bud differentiation in Cattleya Pinole. Proc. Amer. Soc. Hort. Sci. 42:607–608.
——. 1945. Flower initiation and development in the orchid Cattleya Pinole. Proc. Amer. Soc. Hort. Sci. 46:388.
Kako, S. and H. Ohno. 1980. The growth and flowering physiology of Cymbidium plants. Proc. 9th World Orchid Conf. Bangkok (1978), p. 233–241.
Kamemoto, H., and R. Sagarik. 1965. The Nigrohirsutae of Thailand. Florida Orchidist 8:161.
Kerr, A. D. 1972. Ephemeral means "Don't turn your head." Amer. Orchid Soc. Bull. 41:208–211.
Koay, S. H. and S. E. Chua. 1979. Evaluation and commercial applications of flowering potential in Aranda Peter Ewart and Aranda Christine No. 1. Singapore J. Primary Industries 7:51–61.
Koopowitz, H. 1964. Aerangis—an extrapolation from ecology to cultivation. Amer. Orchid Soc. Bull. 33:1026–1028.
Kosugi, K. 1952. Effects of soil moisture and low temperature upon the flower bud differentiation in the Dendrobium nobile. J. Hort. Assoc. Japan 21:179–182.
Kosugi, K., M. Yokoi, and S. Maekawa. 1973. Flower bud formation in orchids. III. Floral initiation and development in Epidendrum radicans. Tech. Bull. Fac. Hort. Chiba Univ. 21:25–30.
Kosugi, K., M. Yokoi, and H. Yokobori. 1973. Flower bud formation in orchids. IV. Floral initiation and development in Oncidum sphacelatum. Tech. Bull. Fac. Hort. Chiba Univ. 21:31–35.
Kuijper, J. 1931. Het in bloei trekken van duiffes-orchideeën. Tropische Natuur 20:90–94.
——. 1933. Zur Frage der periodischen Blüte von Dendrobium crumenatum Lindl. Rec. Trav. Bot. Neerl. 30:1–22.
L. L. 1895a. La durée de la floraison. J. Orchidées 6:75–77.
——. 1895b. La durée des fleurs d'orchidées. J. Orchidées 6:231–233.

Leffring, L. 1972. Cymbidium bloeibeinvloeding. Aalsmeer Jahrbuch, Netherlands, p. 59–60.
Lim, S. L., T. Y. Chin, and C. S. Hew. 1975. Biochemical changes accompanying the senescence of Arundina flowers. Proc. Seminar Singapore Inst. Biol. and Singapore Nat. Acad. Sci., p. 18–26.
Lindley, J. 1830. An introduction to the natural system of botany, p. 263. Longman, London.
McDade, E. 1947. Flowering season of Cattleya hybrids. Orchid Digest 11:141–146.
Massart, J. 1895. Un botaniste en Malaisie. Bull. Soc. Roy. Bot. Belgique 34:151–343.
Matsumoto, K. 1966. Determination of the chorophyll content of Cymbidium blooms. Cymbidium Soc. News 20:11–14.
Montgomery, J., and A. Laurie. 1944. The effect of certain environmental factors on the growth of Cattleya orchids. Ohio Expt. Sta. Bimonthly Bull. 29:48–55.
Morita, K. 1918. Influences de la pollinisation et d'autres actions extérieures sur la fleur du Cymbidium virens, Lindl. Bot. Mag. (Tokyo) 32:39–52.
Murashige, T., H. Kamemoto, and T. J. Sheehan. 1967. Experiments on the seasonal flowering behavior of Vanda Miss Joaquim. Proc. Amer. Soc. Hort. Sci. 91:672–679.
Nishimura, G., K. Koshugi, and J. Furukawa. 1972. Flower bud formation in orchids. II. On the floral initiation and development in Phalaenopsis. J. Japan Soc. Hort. Sci. 41:297–300.
——. 1976. Flower bud formation in Phalaenopsis. Orchid Rev. 84:175–179.
Nuernbergk, E. L. 1961. Kunzlicht und Pflanzenkultur. BLV Verlagsgesellschaft, Munich.
Oertli, J. J., and H. C. Kohl. 1960. Der Einfluss der Bestäubung auf die Stoffbewegung in Cymbidiumblüten. Gartenbauwissenschaft 25:107–114.
O'Neill, M. W. 1958. Use of gibberellin for growth promotion of orchid seedlings and breaking dormancy of mature plants. Amer. Orchid Soc. Bull. 27:537–540.
Perkins, B. L. 1962. Eulophia cucullata and other orchids in Kenya. Orchid Rev. 70:281–282.
Pijl, L. van der, and C. H. Dodson. 1966. Orchid flowers: Their pollination and evolution. Univ. Miami Press, Coral Gables, Fla.
Poddubnaya-Arnoldi, V. A., and V. A. Selezneva. 1957. Orchidea i ih koultoura (Orchids and their culture). USSR Acad. Sci., Moscow.
Quisumbing, E. 1968. The flowering seasons of Philippine orchids. Araneta. J. Agric. 15:195–212.
Rao, A. N. 1967. Flower and seed formation in Arundina graminifolia. Phytomorphology 17:291–300.
Rao, A. V. N., S. N. Hedge, and A. K. Banerjee. 1979. Cultivation and flowering behaviour of orchids. Orchid Rev. 87:195–201.
Richter, W. 1966. Die Temperatur als ausschlaggebender Faktor für die Blüteninduktion bei Cymbidium und Paphiopedilum insigne. Deutsche Gartenbau 12:316–317.
Rosenstock, G. 1956. Die Atmung von Orchideeninfloreszenzen in Verlaufe ihner Vegetationsperiode. Zeit. fur Bot. 44:77–87.
Rotor, G. B., Jr. 1952. Daylength and temperature in relation to growth and flowering of orchids. Cornell Univ. Agric. Expt. Sta. Bull. 885.
——. 1959. The photoperiodic and temperature response of orchids, p. 397–417. *In* C. L. Withner (ed.), The orchids: A scientific survey. Ronald Press, New York.
Rutgers, A. A. L., and F. A. F. C. Went. 1915. Periodische Erscheinungen bei den Blüten des Dendrobium crumenatum Lindl. Ann. Jard. Bot. Buitenzorg 16:129–160.
Sanford, W. W. 1971. The flowering time of West African orchids. Bot. J. Linn. Soc. 64:163–181.
——. 1974. The ecology of orchids, p. 1–100. *In* C. L. Withner (ed.), The orchids: Scientific studies. Wiley-Interscience, New York.
Sano, Y., K. Kataoka, and K. Kosugi. 1961. On the flower bud differentiation and the effect of gibberellin on the forcing in Bletilla striata Reich. J. Japanese Soc. Hort. Sci. 30:178–182.
Sawa, Y., M. Shisa, and M. Torikata. 1967. Effects of temperature and growth substances on flowering of Cybidium virescens. Japan Orchid Soc. Bull. 13:3–7.
Schomburgk, R. 1837. On the identity of three supposed genera of Orchidaceous epiphytes. In a letter to A. B. Lambert. Trans. Linn. Soc. London 17:551–552, pl. 29.
Seifriz, W. 1923. The gregarious flowering of the orchid Dendrobium crumenatum. Amer. J. Bot. 10:32–37.
Sheehan, T. J. 1952. A study of respiration and storage of flowers of the family Orchidaceae. Ph.D. diss., Cornell Univ., Ithaca, N.Y.
——. 1954. Respiration of cut Cattleya flowers. Amer. Orchid Soc. Bull. 23:241–246.
Sheehan, T. J., T. Murashige, and H. Kamemoto. 1965. Photoperiodism effects on growth and flowering of Cattleya and Dendrobium orchids. Amer. Orchid Soc. Bull. 34:228–232.

Smith, D. E. 1958. Effect of gibberellins on certain orchids. Amer. Orchid Soc. Bull. 27:742–747.
Smith, J. J. 1926. Ephemeral orchids. Ann. Jard. Bot. Buitenzorg 35:55–70.
——. 1927. Ephemeral orchids. Orchid Rev. 35:13.
Tchertovitch, V. N. 1980. Introduktsya orchidei, p. 89–90. *In* K. Kaazik, M. Lyik, U. Martin, T. Relve, and V. Reest (eds.), Observation and cultivation of orchids: Abstracts of the All-Union Symposium, March 18–20, 1980. Academy Sci. Estonian SSR, Tallinn Nat. Bot. Gard., Tallinn.
Tran Thanh Van, M. 1970. Phytotronic techniques applied to growth and development studies of certain species of ornamental orchids: Control of floral induction in Odontonia, Phalaenopsis, and Vanda; optimal conditions for Cymbidium growth. Proc. Internat. Hort. Conf. Tel Aviv, p. 35–36.
——. 1974. Methods of acceleration of growth and flowering in a few species of orchids. Amer. Orchid Soc. Bull. 43:699–707.
Treub, M. 1887. Quelques observations sur la végétation dans l'ile de Java. C.R. Seances Soc. Roy. Bot. Belgique 26:182–186.
Urmston, J. W. 1949. Controllable Cattleya hybrids. Amer. Orchid Soc. Bull. 18:652–655.
Vacin, E. F. 1952. Growth and flowering of cymbidiums in their original habitat. Amer. Orchid Soc. Bull. 21:601–613.
Van der Donk, J. A. W. M. 1974. Differential synthesis of RNA in self- and cross-pollinated styles of Petunia hybrida L. Molec. Gen. Genet. 131:1–8.
Wee, S. H. 1971. Maturation period of pods and time taken for plant to flower. Malayan Orchid Rev. 10:42–46.
Went, F. A. F. C. 1898. Die Periodicitaet des Bluehens von Dendrobium crumenatum Lind. Ann. Jard. Bot. Buitenzorg Suppl. II:73–77.
——. 1917. Periodische Erscheinungen beim Blühen tropischer Gewächse. Die Naturwissenschaften 5:72–76.
Went, F. A. F. C., and A. A. L. Rutgers. 1915. Over den invloed van Uitwendige omstandigheden op den bloei van Dendrobium crumentatum Lindl. Koninklijke Akad. Wetenschappen Amsterdam, Afdeeling Natuurkunde 24:513–517.
Went, F. W. 1930. Note. Tropische Natuur 19:119.
——. 1940. Soziologie der Epiphyten eines tropischen Urwaldes. Ann. Jard. Bot. Buitenzorg 50:1–198.
——. 1946. The control of external conditions in the growing of orchids. Orchid Digest 10:86–89.
——. 1951. Cymbidium research. Cymbidium Soc. News. 6:10–12.
Withner, C. L., P. K. Nelson, and P. J. Wejksnora. 1974. The anatomy of orchids, p. 267–347. *In* C. L. Withner (ed.), The orchids: Scientific studies. Wiley-Interscience, New York.
Wróbel-Stermińska, W. 1961. O nowych storczykach otrzymanych z Brazylic. Biul. Ogorod. Bot. 5:163–165.
Zeevart, J. A. D. 1976. Physiology of flower formation. Ann. Rev. Plant Physiol. 27:321–348.
Żotkiewicz, R. 1961. Kwitnienie storczyków szklarniowych w ogrodzie botanicznym Uniwersytetu Warszawskiego. Biul. Ogorod. Bot. 5:101–104.

APPENDIX

Orchid Seed Germination and Seedling Culture—A Manual

Introduction

Joseph Arditti

With the discovery that orchid seeds germinate only following fungal infection, a practical seed-germination method utilizing fungi was developed. This is the so-called symbiotic method, which was used by orchid breeders, primarily in England, for many years. After Knudson discovered that orchid seeds can germinate on a relatively simple sugar-containing medium, the symbiotic procedure was largely abandoned and his asymbiotic method gained wide acceptance. With time this method was perfected and in some cases adapted to the special requirements of selected genera and species. However, some orchids germinate poorly or not at all under asymbiotic conditions and require a fungal symbiont for best germination.

Altogether a variety of procedures and media are used and presented in this appendix. To ensure accurate and thorough coverage I have asked five experts to write on the germination of selected groups. I wrote the rest and edited the entire appendix to ensure uniformity of style and format.

As with the tissue-culture appendix in the first volume of *Orchid Biology*, we cannot undertake to supply reprints of any of the papers mentioned in this section, or provide additional details in private correspondence. Some of the basic information required for the preparation of orchid culture is provided in the General Outline of Techniques and Procedures; the rest can be obtained from the original literature or from elementary textbooks. Included with this information is a list of supplies needed for orchid seed germination (List 1) and of manufacturers of these supplies (List 2). For manufacturers mentioned in the tables, addresses are given only for companies not in List 2. The addresses provided are those known to the authors. It may be possible to obtain supplies from other sources such as local suppliers or sales representatives.

Following these lists is a list of conversion factors for length, area, volume, capacity, mass, power, and illumination.

The appendix is arranged alphabetically according to species or geographic region to allow both broad coverage and attention to individual orchids. European orchids, like most North Temperate terrestrial species, present special problems. Some species can be germinated on several media or a number of modifications of one medium. Other species have very specific requirements and can be germinated on only one medium. Requirements may also vary within a species or genus, depending on region. To cover these variations, I have included two sections on European orchids.

General Outline of Techniques and Procedures

Joseph Arditti

Orchid-seed germination *in vitro* is not a difficult or complex procedure, but it does require the acquisition of certain skills and knowledge. The general outline of these skills as well as the lists of methods, media, and apparatus which follow are taken (with slight modifications) from "Taro Tissue Culture Manual" (Arditti and Strauss, 1979) and the orchid literature.

Culture Media

The media used for orchid-seed germination may reflect preferences of the investigators who carried out the initial research or the special requirements of each species. When one prepares media it is important to follow instructions strictly as given in the recipes and to measure or weigh all compounds accurately. Every effort should be made to use exactly the substances indicated in the tables of this appendix; if modifications are required because some components are unavailable, they should be made judiciously and carefully. For example, it is possible to substitute anhydrous salts for ones that contain water of crystallization, but in such cases the amounts to be added must be recalculated. In some instances it may be possible to substitute one hormone for another within a particular group, but in such cases allowances must be made for the relative activity of each compound (for example, among the auxins, NAA and IAA may be interchangeable qualitatively, but not quantitatively). Thus it is best to prepare for orchid-seed germination well in advance by ordering the necessary substances from appropriate sources (List 2). On receipt each substance should be stored as advised on the package. If in doubt store organics in a freezer or a refrigerator at approximately 4° C.

Stock Solutions

To save work and increase accuracy it is advisable to prepare stock solutions of most media components. These are concentrated solutions (10, 100, or even 1000 times) of each compound. Stock solutions save work because only one weighing is necessary to prepare enough concentrate for 10, 100, or even 1000 liters. They increase accuracy since larger amounts are weighed and because it is easier (and faster) to measure large or small volumes of solution accurately than it is to weigh solids.

To prepare a stock solution weigh the required amount as indicated in each recipe (given in the tables), and add distilled water to the desired final volume. Label the bottle with the information listed below and store in a refrigerator.

1. Name of compound
2. Formula of compound
3. Concentration of stock solution (10×, 100×, or 1000×)
4. Amount to use per liter of culture medium
5. Date
6. Name of the person who made the solution

Individual stock solutions should be prepared for each macroelement, vitamin, amino acid or hormone. All microelements should be combined into one stock solution. Stock solutions containing nitrogen (NO_3^-, NH_4^+, urea) tend to become contaminated with time. Therefore, these substances should be weighed every time, or if stock solutions are prepared, they must be kept frozen between uses.

Stock solutions of hormones and vitamins should be in 70% ethanol (ethyl alcohol) in distilled water (70% by volume; 70 ml of 100% ethanol or 74 ml of 95% ethanol made up to 100 ml with distilled water). If necessary a few drops of sodium or potassium hydroxide or hydrochloric, sulfuric, nitric, or acetic acid can be added to the alcohol to increase solubility of some substances. The use of ethanol not only prevents contamination of the stock solution, but also eliminates the need for sterilization since the alcohol is a sterilant. However, under conditions of very high humidity (as in Fiji, for example, according to M. Krishnamurthi of the Fiji Sugar Corporation Experimental Station in Lautoka) solutions in 70% ethanol may become contaminated. To prevent such contamination, stock solutions in 70% ethanol should be stored in a freezer or made in 95% ethanol.

Hormones, vitamins, and amino acids may not be stable for prolonged periods. It is best, therefore, to prepare only small volumes (10–15 ml) of stock solutions. For a 10-ml stock solution, weigh carefully the required amount of substance and place it in a volumetric flask (see List 1 for descriptions, and List 2 for sources of glassware). Then add 5 ml of 100% (absolute) or 5.2 ml of 95% ethanol (do not use methylated spirits or any other form of denatured ethanol) and shake the flask gently. If the substance fails to dissolve, add a drop or two of acid and shake again (for kinetin, add sodium hydroxide). Should it be necessary, one or two additional drops of acid may be added. After the substance has dissolved completely, add another 2 ml absolute ethanol (or 2.1 ml of 95%) and then make up the volume to 10 ml with distilled water for 70% ethanolic solution. When a 70% solution is undesirable, make up the volume to 10 ml with 95% ethanol. For a 20-ml stock solution double the amount of substance and volumes of ethanol and water. Use 2.5 times as much for a 25-ml stock solution, and multiply by 5 for 50 ml. When larger volumes are made, the number of acid or sodium hydroxide drops used to increase solubility should always be kept to a minimum.

Stock solutions of organic substances should be stored in a freezer or refrigerator. Do not make stock solutions of inositol (this polyol is variously known as *myo*-inositol, *meso*-inositol, *i*-inositol, and cyclohexitol, which may be confusing), sugar (sucrose or good-quality pure white refined kitchen sugar), or agar.

pH

The term is indicative of the alkalinity or acidity (i.e., hydrogen ion concentration) of a medium: pH 7 is neutral; values below 7 indicate acidity, and those above 7 are alkaline. The pH of culture media used for orchid-seed germination should be between 4.8 and 5.5. Solid media may not solidify if the pH is much below 4.0. Seedling growth may be inhibited when the pH is lower than 4.0 or higher than 8.0. To measure the pH of a medium accurately it is best to use a pH meter. If one is not available, however, pH indicator paper may be used (List 1). If the pH of a medium is above the desired value, it is too alkaline and must be adjusted down with a few drops of acid (hydrochloric, nitric, phosphoric, or sulfuric). Should the pH be lower than required, the medium is too acid and must be adjusted up by a base (alkali) such as ammonium hydroxide, potassium hydroxide, or sodium hydroxide. Concentrated acids or bases change the pH very rapidly and should be added slowly and very carefully or not used at all. For reasons of safety, convenience, and accuracy they should be diluted 10–20 times before use.

State of the Medium

In the great majority of cases, solid media are used for orchid-seed germination. Agar is used to solidify media and can be added and dissolved in several ways. One is simply to pour the required amount into the solution and sterilize by autoclaving. The heat of sterilization dissolves the agar, and the medium (which remains liquid while warm) is poured into preautoclaved culture vessels. These are then allowed to cool and the medium to solidify. A second method is to bring the solution to a slow and gentle boil and then add the agar slowly while stirring vigorously to prevent the formation of clumps. When the agar is completely dissolved, the medium turns a clear golden color and is then poured into culture vessels and sterilized by autoclaving.

Sterilization

Media and tissues must be sterilized before starting cultures or they will be overrun by fungi and bacteria that will smother or otherwise kill the seedlings. Depending on components and available equipment, several methods can be used to sterilize seed-germination media.

Heat Sterilization. Inorganic components (macroelements, microelements), sugars, agar, and some complex organic additives (coconut water, casein hydrolysate, peptone, yeast extract, banana homogenate, etc.) can be heat-sterilized in an autoclave or pressure cooker. The standard conditions for sterilization, 121° C and 1 atmosphere pressure for 15–20 min, can be obtained automatically in autoclaves or very easily in pressure cookers, and the entire sterilization process is very simple.

Filtration. Some media components are destroyed by elevated temperatures and cannot be heat-sterilized. Solutions containing these substances may be sterilized by passing them through very fine sterilizing filters (see List 1; one example is the Millipore brand filter) which permit the passage of liquids but not particles larger than 0.22 μ or 0.45 μ and thereby retain all contaminants. This procedure is cumbersome and expensive and not recommended for smaller laboratories.

Solvents. A simple way to sterilize heat-labile substances is to prepare their stock solutions in 70% or 95% ethanol (ethyl alcohol) in distilled water [70 ml absolute (100%) ethanol or 74 ml of 95% ethanol made up to 100 ml with distilled water] since this solvent is also an excellent sterilant. Our experience is that the addition of up to 5 ml of 70% ethanol per liter of medium does not have a deleterious effect on cultures. If the stock solutions are prepared properly, it is not necessary to add more than that. Methylated spirits or other forms of denatured ethanol should not be used for this purpose.

Preparation of a Medium

Preparing a medium may appear complex to those who have not done it before. The step-by-step sequence described below and illustrated in Fig. A-1 is intended to simplify the procedure.

1. Add the correct volume of each of the several macroelement stock solutions to 250 ml of distilled water. To measure volumes of 10 ml or more use a volumetric cylinder. Smaller volumes should be measured with pipettes. In each case the smallest suitable volumetric glassware should be used. For example, 1 ml should be dispensed with a 1-ml pipette, not a 5-ml or a 10-ml one. For 3 ml use a 5-ml pipette and not a larger one or a 1-ml pipette three times. For 7 ml use a 10-ml pipette. A 0.1-ml pipette should be used for dispensing 0.1 ml. If one is not available, a 0.5-ml or even a 1-ml pipette may be used provided they have the proper graduations (for sources of volumetric glassware see List 1).
2. Dispense the proper amount of microelement stock solution.
3. If inositol is included in the medium, add it.
4. Incorporate into the medium whatever complex additives may be part of the recipe (note that some media may not require such additives).
5. Bring total volume to approximately 900 ml.
6. Adjust pH.
7. Weigh and add sugar (sugar may also be added before pH adjustment).
8. Pour medium into a volumetric flask and adjust the total volume to one liter with distilled water. If distilled water is not available, rainwater (preferably fresh) collected in a glass container may be used. Transfer solution to an Erlenmeyer flask or bottle.
9. For solid media add agar.
10. Next sterilize the medium in an autoclave or pressure cooker. Sterilize the medium in an Erlenmeyer flask or bottle with a capacity twice the total volume of the solution being sterilized (for example: one liter of medium should be sterilized in a 2-liter flask). Never use a volumetric flask as a container for sterilization because the heat may reduce its accuracy. A flask may be adequately covered for autoclaving by inverting a beaker over the neck.
11. Culture vessels must be sterilized, either before the medium is sterilized or at the same time.
12. While the medium is being sterilized combine appropriate volumes of all hormone, vitamin, amino acid, and any other necessary stock solutions (all of which may be dissolved in 95% or 70% ethanol) in a vessel just large enough to contain the total volume (which will usually not exceed 5 ml). Suitable containers for this purpose are 5–10-ml Erlenmeyer flasks, 5–10-ml volumetric flasks, 5–10-ml bottles from the local pharmacy (drugstore, chemist), or small test tubes. After introducing each of the required solutions into the small container, stopper it and shake a few times to sterilize all inner surfaces. Then place the stoppered container in the working area and sterilize its external surfaces by spraying with 70% ethanol or hypochlorite solution (described later).
13. After the medium has been sterilized and while it is still hot (and therefore still liquid if it contains agar), pour the hormones, vitamins, etc. (see item 12 above) into it,

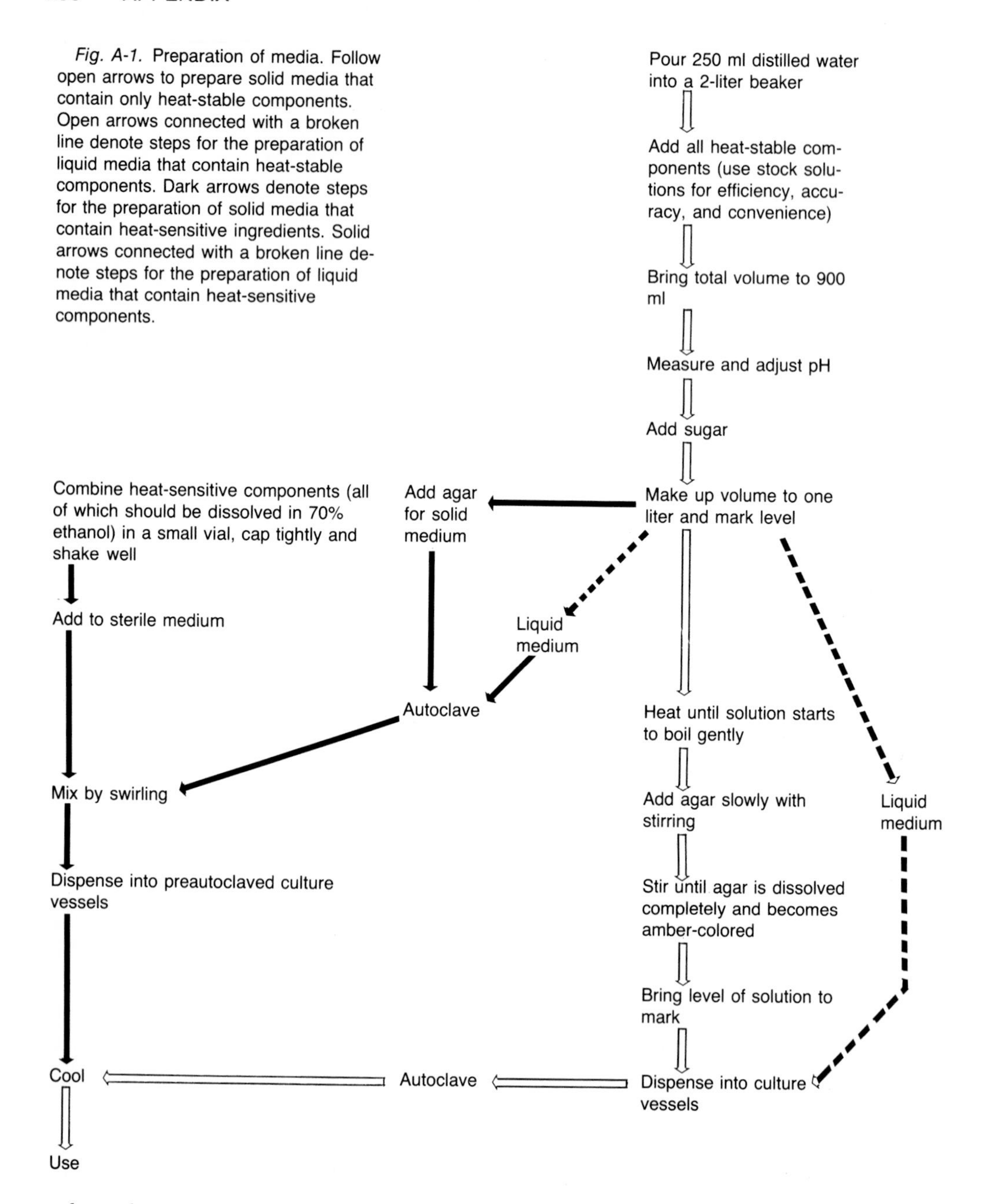

Fig. A-1. Preparation of media. Follow open arrows to prepare solid media that contain only heat-stable components. Open arrows connected with a broken line denote steps for the preparation of liquid media that contain heat-stable components. Dark arrows denote steps for the preparation of solid media that contain heat-sensitive ingredients. Solid arrows connected with a broken line denote steps for the preparation of liquid media that contain heat-sensitive components.

replace the stopper or cover on the flask, and swirl vigorously to ensure quick and thorough mixing.

14. Dispense the complete medium into preautoclaved culture vessels. It is important not to wet the necks of these containers with culture medium since this may lead to contamination later. Therefore, it is best to pour the medium through a sterile funnel.

15. Stopper the vessels.

16. Place the stoppered culture vessels in a clean dry place to cool (and solidify if they contain agar).

17. Start cultures. It is best not to allow media to stand for long periods before using them. Ideally a medium should be used within a week.

Culture Vessels

Test tubes (which in this case are referred to as "culture tubes") and Erlenmeyer flasks (List 1) are considered to be ideal vessels for seed germination. This may well be so, but prescription bottles, used liquor bottles (square scotch and gin bottles are especially good, but any can be used), soft-drink bottles, plasma bottles from hospitals, and a variety of other glass containers can be used. Containers with very wide necks (jars, for example) should not be used since cultures in such vessels are easily contaminated.

Before a container can be used as a culture vessel it must be washed thoroughly. A new culture vessel, especially if it is a container or bottle intended or used for purposes other than tissue culture or seed germination, whether it has been used previously or not, must be washed with aqua regia (List 1), which must be used with great care since it is very dangerous. Rinse the inside of each bottle with the acid (be extremely careful: this mixture of acids is especially dangerous, but each acid alone can damage clothing and skin and cause blindness if splattered into eyes; therefore it is best to wear protective plastic clothing and glasses or goggles while washing glassware with acid). Drain the acid (which can be reused many times if employed only for washing glassware) and rinse the bottles three times with clean tap water. After that, rinse them three additional times with distilled water or rainwater freshly collected in a glass container.

Allow the clean culture vessels to drain and dry by placing them with their openings down on a rack. When the bottles are completely dry push a bun made of nonabsorbent cotton into their necks (the cotton bun should be tight, but still loose enough to allow air movement; if the empty bottle can be lifted by holding the cotton then the bun is just tight enough). Cover that with aluminum foil (or paper if foil is not available; if paper is used it must be tied below the rim on the neck) and heat-sterilize in an autoclave or in a pressure cooker. After removing the bottles from the sterilizer place them in the clean working area (see *Sterile box* in List 1). In high-humidity areas moisture may condense on the cotton, enabling fungi to grow on it and contaminate the medium. In these areas paper, rather than aluminum foil, should be used to cover the cotton. One- or two-hole rubber stoppers (the holes must be stuffed with cotton) may be used instead of cotton buns.

Culture Conditions

The conditions under which seed and seedling cultures are maintained can determine the success or failure of the attempt and also affect growth and development.

Temperature

When adjusting the temperature for orchid seed germination, one should follow instructions carefully. If this proves unsuitable for a particular new cross or a species being germinated for the first time the appropriate temperature can be determined only by trial and error. In some cases reduction of the temperature to as low as 4° C for a short period may induce or modify germination and/or seedling growth. Very high temperatures (35–40° C) may also have an effect. The temperatures given in the specific procedures in this appendix are those recommended by the original workers and should be used, at least initially, to ensure best results.

Light

Illumination is important, of course, because light is necessary for photosynthesis, but its duration (photoperiods), quality (color), and intensity (amount) affect development.

Duration. Orchids may be grown under light or dark periods of up to 24 hr each (i.e., continuous light or continuous dark). They may also be grown under light periods (photoperiods) of varying durations. The appropriate photoperiod must be determined experimentally for genera and species that have not been germinated before. When using established methods it is best to use the photoperiod recommended by the original workers. Seedlings grown under continuous darkness do not produce chlorophyll. Illuminated cultures turn green and may develop shoots and roots.

Intensity. The proper intensity (strength, brightness) of the light is important because a very bright light may bleach the seedlings and a dim source would not be sufficient. We have found that a combination of two 40-watt cool or warm white fluorescent tubes and two to four 25- to 50-watt incandescent bulbs placed 30–50 cm above the cultures is sufficient. To keep cultures in the dark (zero light intensity) place them in a light-tight enclosure and/or carefully wrap them in aluminum foil and/or black plastic.

Agitation

Liquid media must be agitated to allow for gas exchange, to improve contact between the seeds or seedlings and liquid, and to influence growth and development. Agitation may be gyrorotatory (wrist action), oscillatory (back and forth), or rotatory (rotating on a wheel with its axis parallel to the ground). Shakers can be purchased (List 1) or constructed locally. We have found that operators of machine shops and most people who are handy can easily design and construct a shaker. The oscillating shakers (back and forth) are the easiest to build and most suitable for the widest variety of applications.

The speed of shaking is important. If shaken too fast or too slow seedlings may die. Rotatory shakers should rotate at the rate of 1–3 rotations per minute (rpm), oscillating ones should move back and forth 60 times a minute (60 oscillations/min), and gyrorotatory units should shake approximately 30–40 times per minute.

On the whole, liquid media are used very seldom for orchid-seed germination and seedling culture. Therefore, agitation is seldom an important factor.

Sterilization

Culture media, tools, working space, and tissues must be sterilized and work must be carried out under aseptic conditions to ensure success; otherwise cultures will become contaminated and fail. A number of methods can be used to ensure sterility.

Heat

Elevated temperatures are a convenient and simple means of destroying potential contaminants.

Autoclaves. Spores of microorganisms in liquid (i.e., culture media) can survive elevated temperatures under normal (that is, atmospheric) pressure. Therefore, culture media must be sterilized under high temperature and pressure (usually 15 lb/sq in. or 6.8 kg/6.5 cm^2). These conditions can be easily generated in autoclaves. A large number of autoclave models and sizes are available (Lists 1 and 2). Some require electricity, others do not. Several models are very complex and fully automatic, but there are also types which differ very little or not at all from a kitchen-type pressure cooker. Indeed, if an autoclave is not available, media, tools, culture vessels, and other items can be sterilized in a kitchen pressure cooker. Some medium components are not heat-stable and therefore may not be autoclaved. These must be sterilized by alternative methods.

Open Flame. Burning can be used to sterilize tools and the necks of bottles while making cultures. A natural-gas burner is best because it produces a clean, nonsmoking, high-temperature flame. If one is not available an alcohol (methylated spirits or denatured ethanol can be used as fuel) flame can be used, but it may not be hot enough. Another possibility is to dip the tools to be sterilized in alcohol (methyl, ethyl, or isopropyl) and ignite the liquid with an alcohol flame to sterilize their surfaces. A simple lamp can be prepared by filling a bottle with alcohol and inserting a wick (cotton or a piece of cloth are satisfactory). Kerosene or automotive gasoline (petrol, benzin) should not be used as fuel since they produce a lot of smoke and soot and may be explosive.

Filters

Sterilizing filters (List 1) with pore size of 0.22 or 0.45 microns (micrometers, μm) can be used to sterilize liquids since they retain contaminants. Because some of these filters can be used with a number of solvents while others are limited to water, it is necessary to read instructions carefully. Sterilizing filters are expensive and require a vacuum pump and other sophisticated and expensive equipment. For this reason they are not advisable for small laboratories that are not fully equipped.

Liquids

Work areas, tools, tissues, and culture-medium components can be sterilized with liquid sterilants.

Hypochlorite Solutions. Preparations such as Clorox, Purex, Domestos, Milton's, Snow-White, and other household bleaches contain between 4.75% and 5.25% sodium hypochlorite and are, therefore, excellent sterilants. Undiluted, they can be employed to wash tools, working areas, and the outside of culture bottles. If used to sterilize seeds or capsules these bleaches should be diluted according to instructions in specific procedures. To determine the correct dilution it is necessary to consider the sodium hypochlorite content of each preparation. For example, if a procedure calls for a 50% dilution

(50 ml household bleach made up to 100 ml with water) of a brand that contains 5.25% sodium hypochlorite, the diluted solution has 2.625% active agent. Therefore a brand that contains only 4.75% sodium hypochlorite should be used at a lower dilution (55 ml bleach made up to 100 ml with distilled water). Household bleaches cannot be used to sterilize media components.

A saturated solution of calcium hypochlorite is used to surface-sterilize orchid seeds. This solution is prepared by dissolving 10 g calcium hypochlorite in 140 ml water (7 g/100 ml), stirring vigorously, and allowing the solution to stand for 3–5 min. After that the solution is stirred again, allowed to stand until the precipitate has settled, and filtered or decanted. The clear, yellowish liquid is used as the sterilant. It should be used within 12 hr.

Alcohol. Ethyl alcohol (ethanol, drinking alcohol) pure or denatured (methylated spirits), methyl alcohol (methanol), and isopropyl alcohol (isopropanol) can be used to sterilize work areas, tools, and outside surfaces of culture vessels by swabbing. These alcohols can be used in concentrated form or as 70% aqueous solutions (70 ml alcohol made up to 100 ml with water). If at all possible the use of methanol should be avoided since it is toxic to people and may cause blindness.

Some culture-medium components can be sterilized by preparing their stock solutions in 95% or 70% ethyl alcohol (95 ml or 70 ml pure ethyl alcohol made up to 100 ml with distilled water; do not use methyl alcohol, methylated spirits, or other forms of denatured alcohol), as described in the section on stock solutions, above.

Surface Decontamination

A number of methods are used for surface decontamination. In each case it is best to follow the specific procedure outlined in this appendix. Generally, capsules ("pods") are first cleaned by a gentle washing and scrubbing to remove dirt and soil. After that they are dipped briefly in either 95% or 70% ethanol (2–3 sec) and then immersed (5–20 min) in a diluted household bleach or calcium hypochlorite solution (see specific procedures for concentrations and times). The sterilant is then removed from the tissues with several washings in a sterile box or other suitable area (List 1). Seeds are usually sterilized by soaking them in calcium hypochlorite for 5–20 min. The sterilant is then removed by washing with sterile distilled water.

Inoculation

When seeds are placed on a solid medium it is important to establish good contact between them and the medium and to distribute them evenly. They should not be totally buried in the agar to prevent death from improper gas exchange. All inoculations should be carried out in a sterile work area, except as otherwise noted.

Work Areas

Mixing the heat-sterilized solution with the components in the 95% or 70% ethanol stock solutions, pouring the medium into culture vessels, splitting the capsules, and introducing the seeds must be carried out under aseptic conditions. Such conditions can be achieved in several ways.

Laminar-Flow Hoods

At present the best and most efficient, but unfortunately the most expensive, method of assuring sterility in the working area is to use a laminar-flow hood (List 1). Air coming into these hoods is driven through filters that remove all particles. The sterile air is blown gently across the working area toward the operator, and this generally prevents contamination of cultures. Tools must, of course, be sterilized, even when used in such a hood, and care must be taken to prevent the introduction of contaminants from unsterile surfaces into vessels. The work space in these hoods is large enough for comfortable, fast, and efficient movement. The price of smaller hoods is now low enough to justify their purchase by most laboratories where seed germination, seedling culture, and micropropagation may become routine activities.

Sterile Rooms

These are usually small rooms fitted with hard-surface benches which are kept clean by swabbing with alcohol or hypochlorite solution and irradiation with sterilizing ultraviolet lamps (which must be allowed to stay on for at least 30 min to ensure sterility but have to be turned off when the operator enters the room). All culture vessels and tools are placed in these rooms, sterilized by washing or spraying with alcohol or hypochlorite, and irradiated with ultraviolet light.

When everything is sterile the operator enters, having washed his or her hands carefully (short clean nails are important if no gloves are worn). Alternatively, the operator should wear surgical gloves (List 1), which are kept sterile by periodic swabbing with alcohol or hypochlorite. Tools must be kept sterile by flaming and/or dipping them in alcohol or hypochlorite before and after every use. The necks of culture vessels must be flamed after removing the cotton buns and following the introduction of tissues.

Sterile rooms have been and are still being used in some tissue-culture laboratories, but they tend to be expensive (in the range of small laminar-flow hoods) and are not very comfortable, efficient, or desirable working areas for tissue culture or seed germination.

Sterile Boxes

An enclosure made of plastic, glass, stainless steel, wood painted with hard polyurethane or plastic, or a cardboard box lined with aluminum foil (see List 1) can prove to be a most satisfactory aseptic working area for tissue culture. The box is kept sterile by washing it with alcohol or hypochlorite. Irradiating it with sterilizing ultraviolet lamps is very desirable, but not strictly necessary. Tools, a burner, culture vessels containing medium, and containers with sterilizing solutions are placed in the box and sterilized by spraying or swabbing them with alcohol or hypochlorite. Approximately 20 min later the operator (preferably wearing gloves) can insert his or her hands into the box through the front openings and start to work. All other procedures are as in a sterile room.

Sterile boxes are very suitable for a small laboratory or one which is just initiating a tissue-culture program. With a minimum of training, dexterity, and experience most operators can use such a box successfully.

A Clean Laboratory Bench

In some places (clean laboratories, areas of low atmospheric humidity) experienced operators can simply use a clean laboratory bench as an appropriate work area, but this is not generally advisable.

Storing Seeds

To protect the viability of orchid seeds it is best to store them over a dessicant at 4° C (39° F). In my laboratory we place a small amount of anhydrous calcium chloride ($CaCl_2$) in the bottom of a screw-cap test tube and cover it with a layer of cotton. The seeds are placed on the cotton and the cap is screwed on tightly before placing the tube in the vegetable crisper of a household refrigerator. Before sowing seeds we remove the tubes from the refrigerator and place them on a laboratory bench for 2–3 hr before opening to allow for temperature equilibriation. If this is not done water from the air may condense on the seeds, causing them to mold.

Washing Glassware

All glassware used in orchid seed germination (culture vessels, volumetric flasks or cylinders, beakers, test tubes, etc.) must be chemically clean. This is especially true for containers that have been previously used for other purposes (to hold ketchup, liquor, soft drinks, or medicines, for example) and for new vessels that are not made of heat-resistant glass.

One method for washing such bottles is to first rinse them with aqua regia (List 1), a volatile, yellow, fuming, corrosive, and extremely dangerous liquid which must be handled with utmost care. Workers using it must wear plastic face guards, plastic or rubber gloves, and aprons, and work in a well-aerated area. Bottles containing aqua regia must be clearly labeled and capped tightly (preferably glass-stoppered). They must be kept in a cool, dark, safe, well-protected place. To rinse a container with aqua regia a small amount of the liquid should be poured into it. The container is then rotated to ensure that all inner surfaces come into contact with the acid. Then the acid is poured into the next container to be washed or back into the bottle (aqua regia can be reused many times). Next the container being washed is rinsed at least three times with tap water before three distilled-water rinses.

A second method is to carefully scrub the glassware inside and out with a lukewarm soap solution and a brush before rinsing with tap and distilled water as described above.

Incorporation of Anticontaminants in Culture Media

Efforts to formulate orchid-seed and seedling-culture media which do not require sterilization or can reduce contamination started shortly after the Knudson C medium was developed (for a short review see Thurston *et al.*, 1979). Vanillin derivatives (Knudson, 1947; McAlpine, 1947 and personal communication) and several antibiotics (Schaffner, 1954) have been tested as additives for this purpose, but were found to be phytotoxic and unsuitable. More recently several useful combinations were formulated (Table A-1) for use in seedling culture media following the screening of a number of substances (Thurston *et al.*, 1979).

Medium is prepared as for tropical orchids (p. 250) through the step of dissolving the agar. The anticontaminants (Table A-1), all dissolved or suspended and mixed in a total of 6 ml of 70% ethanol, are added to the medium at this point and mixed well by vigorous stirring. To facilitate preparation of media, stock solutions can be prepared (Table A-2). After mixing, the medium is poured into culture vessels which have been washed with 70% ethanol or rubbing (i.e., isopropyl) alcohol.

All culture vessels, funnels, and other glassware used with unsterilized anticontaminant-containing media must be washed with 70% ethanol or rubbing (i.e., isopropyl) alcohol and allowed to dry upside down in clean, dust-free areas. Tools must be washed similarly and flamed before use. Work surfaces must be first washed with soap and water and then with 70% ethanol or rubbing alcohol. Water used for the preparation of media must be boiled for 5 min, allowed to stand in a covered vessel for 24 hr, and boiled again for another 5 min.

All work areas must be clean and dust-free. Work must be carried out quickly and efficiently.

The formulations described here are not suitable for use in seed-germination media and should not be employed for this purpose. Also, it is advisable to test each combination with a few cultures prior to large-scale use because as this is being written the formulations have only been tested with *Cattleya* and *Stanhopea* seedlings (Thurston *et al.*, 1979) and *Phalaenopsis* flower-stalk-node cultures (Spencer *et al.*, 1979/1980).

Table A-1. Formulations of anticontaminants for use in culture media for orchid seedlings (Thurston *et al.*, 1979)

Number	Formulation[a,b]
1	Benlate + nystatin + penicillin G + gentamycin + graphite[c]
2	Benlate + nystatin + penicillin G + gentamycin + sodium omadine + graphite[c]
3	Benlate + nystatin + penicillin G + gentamycin + amphotericin B + vancomycin + graphite[c]

[a] *Concentrations:* amphotericin B, 10 mg/liter; benlate, 50 mg/liter; gentamycin, 50 mg/liter; nystatin, 25 mg/liter (100,500 units/liter); penicillin G, 100 mg/liter (159,500 units/liter); sodium omadine, 5 mg/liter; vancomycin, 50 mg/liter; graphite, 2 g/liter.

[b] *Suppliers:* amphotericin B, gentamycin, nystatin, penicillin G, and vancomycin can be obtained from Sigma Chemical Co.; sodium omadine is available from the Olin Corporation, Agricultural Division, 700 N. Buckeye St., Little Rock, AR 72114 USA; benlate formulations are sold by retail nurseries and plant shops. Graphite may be purchased from the J. T. Baker Chemical Co.

[c] Not an anticontaminant, but used as a darkening agent to prevent photodestruction of light-sensitive compounds.

Table A-2. Stock solutions of anticontaminants[a]

Compound	Amount per liter of culture medium (final concentration in culture medium)	Stock solution (a concentrate prepared for repeated and convenient use)	Volume of stock solution per liter of culture medium	Remarks
Amphotericin B	10 mg	100 mg/10 ml 70% ethanol[b]	1 ml	keep frozen between uses
Benlate	50 mg	500 mg/10 ml distilled water[c] or 70% ethanol	1 ml	keep frozen between uses
Gentamycin	50 mg	sterile injectable liquid prepared according to instructions in package[d]	depending on instructions in package	keep frozen between uses
Nystatin	25 mg	250 mg/10 ml absolute ethanol	1 ml	keep frozen between uses
Penicillin G	100 mg	1 g/10 ml 70% ethanol	1 ml	keep frozen between uses
Sodium omadine	5 mg	50 mg/10 ml 70% ethanol	1 ml	keep frozen between uses
Vancomycin	50 mg	500 mg/10 ml 70% ethanol	1 ml	keep frozen between uses
Graphite[e]	2000 mg	no stock	no stock	weigh

[a] To prepare a mixture for use, mix the required compounds in a small vial approximately 1–2 hr before needed, add the graphite, and shake well. Add this mixture to the medium after agar has been dissolved.
[b] The 70% ethanol solution is prepared by bringing 737 ml of 95% ethanol to 1000 ml with distilled water. Ethanol (95%) can be purchased in drugstores with prescription.
[c] A precipitate will form. Shake well before use.
[d] This step requires a sterile syringe-and-needle combination which can be purchased in drugstores with prescription.
[e] Not an anticontaminant, but used to darken media to prevent photodestruction of light-sensitive compounds.

Incorporation of Charcoal in Culture Media

The first attempt to darken a culture medium used for orchid-seed germination seems to have been made in an effort to germinate native American *Cypripedium* species (Curtis, 1943). Lampblack (obtained by burning various fats, oils, resins, etc., under suitable conditions) was added to the medium at the rate of 3 g per liter. Germination of *Cypripedium reginae* was very low on this medium, but this species always germinates poorly on asymbiotic media and therefore the effect of the darkening is difficult to evaluate. Darkening with charcoal had a clearly positive effect on the growth of *Cymbidium* plantlets *in vitro* (Werkmeister, 1970a,b, 1971) and the same is true for *Phalaenopsis* and *Paphiopedilum* seedlings (Ernst, 1974, 1975, 1976). These observations led to the development of charcoal-containing media (Ernst, 1974, 1975, 1976), which have gained wide acceptance in a relatively short time.

There is no simple explanation for the growth effects of charcoal or of darkened media. The available experimental evidence does not fully support suggestions that the charcoal adds microelements, establishes polarity by darkening the medium, affects substrate temperature, or absorbs toxic substances produced by the orchids. At least for the present the most tenable explanation seems to be that the charcoal improves aeration (Arditti, 1979). Addition of charcoal to culture media is a simple process. Vegetable charcoal [Nuchar C was used in the original research (Ernst, 1974) and will probably prove suitable for most, even all, media] should be used and added to the medium at a concentration of 2 g per liter. The charcoal tends to settle and must therefore be dispersed well with a blender or by vigorous stirring or swirling the flasks before the medium has solidified. There is no reason to believe that the addition of charcoal may prove detrimental to any orchid species. In the absence of published information on a particular orchid, however, growers would be wise to perform preliminary tests before large-scale use.

Charcoal has the capacity of irreversibly adsorbing certain compounds which might be added to orchid-culture media (including certain vitamins and hormones). Therefore, it cannot be used in media that contain such additives. If darkening of media that do contain them becomes necessary, 2 g graphite per liter (Thurston *et al.*, 1979) should be used instead of charcoal.

Vegetable charcoal may be listed under several headings (which are not always clear) in catalogs. To ensure purchase of the appropriate charcoal, it is best to contact the suppliers and inquire.

Literature Cited

Arditti, J. 1979. Aspects of the physiology of orchids. Adv. Bot. Res. 7:421–655. Academic Press, London.

Arditti, J., and M. S. Strauss. 1979. Taro tissue culture manual. Information document no. 44-1979. South Pacific Commission, Noumea, New Caledonia.

Curtis, J. T. 1943. Germination and seedling development in five species of Cypripedium L. Amer. J. Bot. 30:199–206.

Ernst, R. 1974. The use of activated charcoal in asymbiotic seedling culture of Phaphiopedilum. Amer. Orchid Soc. Bull. 43:35–38.

——. 1975. Studies in asymbiotic culture of orchids. Amer. Orchid Soc. Bull. 44:12–18.

——. 1976. Charcoal or glasswool in asymbiotic culture of orchids. Proc. 8th World Orchid Conf., Frankfurt (1974), p. 379–383.

Knudson, L. 1947. Orchid seed germination by the use of fungicidal or fungistatic agents. Amer. Orchid Soc. Bull. 16:443–445.

McAlpine, K. L. 1947. Germination of orchid seeds. Orchid Rev. 55:8–10, 21–22.

Schaffner, C. P. 1954. Studies on orchid media inhibitory to fungi. Amer. Orchid Soc. Bull. 23:798–802.

Spencer, S. J., K. C. Thurston, J. A. Johnson, R. G. Perera, and J. Arditti. 1979/1980. Media which do not require sterilization for the culture of orchid seedlings and Phalaenopsis flower stalk nodes. Malayan Orchid Rev. 14:44–53.

Thurston, K. C., S. J. Spencer, and J. Arditti. 1979. Phytotoxicity of fungicides and bactericides in orchid culture media. Amer. J. Bot. 66:825–835.

Werkmeister, P. 1970a. Die Steuerung von Vermehrung (Proliferation) und Wachstum in der Meristemkultur von Cymbidium und die Verwendung eins Kohle-Nährmediums. Die Orchidee 21:126–131.

——. 1970b. Über die Lichtinduktion der geotropen Orientirung von Luft- und Bodenwurzeln in Gewebekulturen von Cymbidium. Ber. Deutsch. Bot. Gaz. 83:19–26.

——. 1971. Light induction of geotropism, and the control of proliferation and growth of Cymbidium in tissue culture. Bot. Gaz. 132:346–350.

List 1

General Information on Supplies, Equipment, Terms, and Reagents

This appendix may be used by experienced workers who are familiar with seed-germination techniques. However, it may well be a major source of information for hobbyists, commercial growers, and investigators who have little or no experience. In more isolated laboratories it could be the first or only source of information. Therefore, every effort has been made to include in it all necessary information. This list describes equipment and chemicals, in alphabetical order. The numbers in parentheses refer to the sources of supply given in List 2.

Acid, washing. Mixtures of equal volumes of nitric and hydrochloric, or hydrochloric and sulfuric, or nine parts sulfuric and one part nitric acids can be used for washing glassware. Alternatively, any one of these acids or aqua regia (see below) can be employed alone. It is important to keep in mind that all acids, and especially mixtures, are dangerous and must be used with extreme caution. When acids are mixed heat may be liberated and explosions can occur. Therefore, acids must be mixed very slowly and carefully in a fume hood and the containers should be kept in a safe, well-ventilated place. Workers who prepare such mixtures should wear protective clothing and glasses or goggles. If aqueous solutions are needed, always pour *acid* into *water* (to remember this fact just recall that in the alphabet *A* comes before *W*).

Agar may vary in purity, mineral content, organic substances, or the amount required to solidify a medium, depending on its source or batch. Bacto agar (Difco Laboratories, Source 7) is a commonly used brand in the U.S., but we have used other brands with equal success. Agar we have purchased in food stores has been suitable or even preferable in some instances. In general, it may be wise to test agar from a new source before using it in critical cases (7, 8, 9).

Alcohol is a term that describes a large number of compounds. The alcohol recommended for use in this manual is the one most familiar to people—ethanol or drinking alcohol. Ethanol must be used to make all alcohol stock solutions and should be employed as a sterilant since methyl alcohol (methanol), methylated spirits, and other forms of denatured ethanol may be toxic. If pure ethanol is not available, use an inexpensive brand of vodka (10).

Amino acids are components of proteins and may be required in some culture media (1, 5, 7, 15).

Aqua regia is used to wash glassware. It consists of 18 ml nitric acid (HNO_3) and 82 ml hydrochloric acid (HCl) and should be handled with extreme care because it is corrosive and dangerous and can cause burns and blindness.

Ascorbic acid is another name for vitamin C, which is sometimes used as an antioxidant (1, 5, 7, 15).

Autoclaves are sterilizers that develop high temperature and pressure. Many types are available, the simplest being a pressure cooker (14, 16).

Auxins are plant hormones. Indoleacetic acid (IAA) occurs naturally, but a number of synthetic auxins are also available and often used in culture media. These include naphthaleneacetic acid (NAA), *para*-chlorophenoxyacetic acid (p-CPA), 2,4-dichlorophenoxyacetic acid (2,4-D), and 2,4,5-trichlorophenoxyacetic acid (2,4,5-T), among others (1, 5, 9, 15).

Balances are required for accurate weighing of most medium components. Two types are necessary. The first is a simple swing balance for weighing several grams at a time. For quantities down to 1 mg an analytical balance, though expensive, is necessary. If costs are prohibitive a balance accurate to 10 mg may be used but will require making larger amounts of stock solutions (14, 16).

Banana. Banana homogenate can enhance the growth of orchid seedlings or plantlets. No information is available on the reason(s) for this effect. Depending on the medium, the homogenate may be of fruits that are green, ripe, or intermediate. Growth enhancement may vary with the variety of banana used, but the differences are not large enough to justify concern. We generally use bananas (bought at the nearest food store) that are ripe or nearly so, but not ones that are very soft, overripe, or have large black patches on the skin.

A simple way to prepare the homogenate is to place the required weight of fruit in a blender with an equal volume of distilled water. Homogenize until the tissue is completely disrupted. Add the homogenate to the medium and adjust the final volume as required with distilled water. Then check the pH and adjust it if necessary.

Banana-containing media can be autoclaved without loss of activity. They are slightly darker in color than media that do not contain banana. If the tissue is not sufficiently homogenized, it may settle to the bottom of the culture vessels. Should a precipitate form after autoclaving, its distribution may be improved by swirling the flasks when the agar has cooled but not yet solidified (10).

Beakers are made of glass or plastic and used in the preparation of solutions. If unavailable use glass jars, cups, or glasses. To withstand heating, the beakers must be made of Pyrex, Kimax, or other heat-resistant glass (4, 14, 16).

Benomyl (Benlate) is a systemic fungicide with minimal phytotoxic effects (10).

Bleach, household. See Household bleach.

Calcium hypochlorite is used in a filtered solution of 7 g/100 ml water to surface-sterilize seeds or capsules. The solution must be used within a few hours. A few drops of mild household detergent can be added as a wetting agent (1, 2, 11).

Casamino acids are acid-hydrolized casein. Hydrolysis is continued until all nitrogenous substances in the casein are converted to amino acids or other relatively simple substances. The result is a complex mixture of a composition which may vary. Therefore, only "typical analyses" are available (Table A-3). Casamino acids can be autoclaved unless otherwise specified (7).

Table A-3. Typical analysis of Difco peptones and hydrolysates[a]

Component	Peptone	Proteose peptone	Proteose peptone No. 3	Tryptone	Tryptose	Neopeptone	Protone	Casitone	Casamino acids (technical grade)	Casamino acids	Yeast extract
Percent											
Ash	3.53	9.61	4.90	7.28	8.44	3.90	2.50	6.66	30.8	3.64	10.1
Ether soluble extract	0.37	0.32		0.30	0.31	0.30	0.31				
Total nitrogen	16.16	14.37	13.06	13.14	13.76	14.33	15.41	13.00	7.85	11.15	9.18
Primary proteose nitrogen	0.06	0.60		0.20	0.40	0.46	5.36				
Secondary proteose nitrogen	0.68	4.03		1.63	2.83	3.03	7.60				
Peptone nitrogen	15.38	9.74		11.29	10.52	10.72	2.40				
Ammonia nitrogen	0.04	0.00		0.02	0.01	0.12	0.05				
Free amino nitrogen (Van Slyke)	3.20	2.66		4.73	3.70	2.82	1.86				
Amide nitrogen	0.49	0.94		1.11	1.03	1.23					
Mono amino nitrogen	9.42	7.61		7.31	7.46	7.56					
Di-amino nitrogen	4.07	4.51		3.45	3.98	4.43					
Arginine	8.0	6.8	5.9	3.3	5.05	4.7	3.9	3.2	1.9	3.8	0.78
Aspartic acid	5.9	7.4	6.6	6.4	6.9	6.7	10.8	6.5	4.0	0.49	5.1
Cystine (Sullivan)	0.22	0.56		0.19	0.38	0.39	0.27				
Glutamic acid	11.0	12.0	11.2	18.9	15.4	15.2	8.1	20.0	12.6	5.1	6.5
Glycine	23.0	11.6	8.9	2.4	7.0	6.3	5.0	2.5	1.3	1.1	2.4
Histidine	0.96	1.7	1.7	2.0	1.8	2.3	5.9	2.1	1.4	2.3	0.94
Isoleucine	2.0	3.3	3.3	4.8	4.0	4.3	0.71	5.0	2.9	4.6	2.9
Leucine	3.5	6.4	6.0	3.5	7.4	8.4	13.6	8.2	4.0	9.9	3.6
Lysine	4.3	5.3	5.1	6.8	6.0	6.4	10.3	7.0	4.4	6.7	4.0
Methionine	0.83	2.0	1.8	2.4	2.2	2.4	1.9	2.6	1.08	2.2	0.79
Phenylalanine	2.3	3.3	3.1	4.1	3.7	4.3	6.8	4.3	2.0	4.0	2.2
Threonine	1.6	3.5	3.2	3.1	3.3	3.7	4.6	4.2	2.2	3.9	3.4

Tryptophan	0.42	0.72	0.85	1.45	1.08	1.01	1.65	1.38	Nil	0.8	0.88
Tyrosine	2.3	3.4	0.36	7.1	5.2	5.3	3.0	2.8	0.52	1.9	0.60
Valine	3.2	4.4	4.0	6.3	5.3	6.0	10.1	6.3	3.8	7.2	3.4
Organic sulfur	0.33	0.60		0.53	0.57	0.63	0.45				
Inorganic sulfur	0.29	0.04		0.04	0.04	0.09	0.16				
Phosphorus	0.079	0.24	0.46	0.75	0.49	0.112	0.15	0.72	0.29	0.35	9.89
Iron	0.0023	0.0038	0.0044	0.0071	0.0054	0.0021	0.0099	0.0039	0.0101	0.0006	0.028
SiO_2	0.042	0.078	0.019	0.090	0.084	0.18	0.52	0.073	0.022	0.053	0.052
Potassium	0.22	0.70	0.21	0.30	0.50	0.85	0.06	0.12	0.16	0.88	0.042
Sodium	1.08	2.84	0.033	2.69	2.76	0.45	0.30	0.24	1.05	0.77	0.32
Magnesium	0.056	0.118	0.00048	0.045	0.081	0.051	0.057	0.00060	0.0039	0.0032	0.030
Calcium	0.058	0.137	0.0396	0.096	0.116	0.198	0.263	0.0913	0.0538	0.0025	0.040
Chlorine	0.27	3.95		0.29	2.77	0.84	0.38				
Chloride	0.27	3.95	4.15	0.29	2.12	0.84	0.38	0.425	21.34	11.2	0.190
Parts per million											
Manganese	8.6	5.3	7.8	13.2	9.2	5.8	6.0	9.7	5.7	7.6	7.8
Lead	15.00	5.00	3.00	6.00	5.50	5.00	9.00	5.00	3.00	4.00	16.00
Arsenic	0.09	0.25	0.00	0.07	0.16	0.37	0.46	0.32	0.00	0.50	0.11
Copper	17.00	31.00	9.00	16.00	23.50	19.00		10.00	8.00	10.00	19.00
Zinc	18.00	44.00	37.00	30.00	37.00	2.00	13.00	10.00	14.00	8.00	88.00
Micrograms per gram											
Pyridoxine	2.5	3.0	4.1	2.6	2.8	5.0	0.24	1.1	0.025	0.073	20.0
Biotin	0.32	0.43	0.24	0.36	0.39	0.73	0.0021	0.34	0.050	0.102	1.4
Thiamine	0.50	3.0	2.7	0.33	1.66	3.4	0.17	0.48	0.02	0.12	3.2
Nicotinic acid	35.00	131.00	169.00	11.00	71.00	134.00	2.1	24.00	2.5	2.7	279.00
Riboflavin	4.00	11.00	13.00	0.18	5.59	11.4	0.046	0.68	0.019	0.03	19.00
Reaction, pH[b]	7.0	6.8		7.2	7.3						

[a] *Sources: Difco Manual,* 9th ed., Difco Laboratories, Detroit, Mich., 1953; H. W. Schoenlein, Difco Laboratories, personal communication, 1957. Courtesy of E. McDonald, Technical Services, Difco Laboratories, Detroit, Mich.

[b] pH of a 1% solution in distilled water after autoclaving 15 min at 121°C.

Cellulose, the substance cell walls are made of, is sometimes used in culture media for symbiotic fungi (17).

Charcoal for use in culture media must be pure and of vegetable origin. One possible source is J. T. Baker Chemical Co. (2).

Clorox. See Household bleach.

Coconut water (sometimes called coconut milk) is a liquid endosperm that can enhance the growth of cells, tissues, organs, or seedlings *in vitro.* Its exact composition and the reasons for its effects have not been determined. Water from green coconut is preferable, but we have had reasonable success with liquid removed from relatively mature nuts purchased in local food stores. Unless otherwise specified, coconut water can be autoclaved.

Complex additives are mixtures whose exact composition is unknown, including coconut water, casein hydrolysate, yeast extract, peptone, and tryptone. They are often added to media (7, 8, 9, 10).

Cylinder, volumetric. See Volumetric cylinders.

Cytokinins (formerly called plant kinins or phytokinins) are a group of hormones which regulate cell division and other plant functions. They may be used in culture media and include compounds such as benzyl adenine (BA), 6-dimethyl-aminopurine (6-DMAP), kinetin, and N-benzyl-9-(tetrahydro-2H-pyran-2-yl), an experimental compound (SD 8339) which may no longer be available (1, 5, 8, 9, 15).

Electric equipment. Proper precautions should be exercised in the purchase of any electrical equipment, and a laboratory's current, cycles, voltage levels, and fluctuations should be considered. If fluctuations are common and excessive, voltage regulators are necessary.

Erlenmeyer flasks have a relatively narrow neck and a broader base and are very useful in preparing solutions. To withstand heating they must be made of Pyrex, Kimax, or other heat-resistant glass (4, 14, 16).

Ethanol. See Alcohol.

Filters. See Sterilizing filters.

Flask, Erlenmeyer. See Erlenmeyer flasks.

Flask, volumetric. See Volumetric flasks.

Gibberellins are plant hormones which may be used in culture media. One possible source is Sigma Chemical Co. (15).

Graduated cylinder. See Volumetric cylinders.

Hormones (auxins, cytokinins, gibberellins) are sparingly soluble or insoluble in water. This problem can be solved in several ways. One possibility is to use their soluble salts (potassium salts of auxins, for example). Another is to dissolve them in ethyl alcohol (ethanol plus a few drops of dilute acid, if necessary—hydrochloric acid, for instance) or base (potassium hydroxide is suitable). We prefer ethanolic solutions because they also serve to sterilize these substances.

Many organic molecules, including plant hormones, are destroyed by autoclaving. Therefore, they must be sterilized by filtration, a process that is somewhat complicated and may require relatively sophisticated or expensive equipment. A simpler approach is to dissolve them in ethyl alcohol (at least 70% in distilled water), which is a good sterilant as well as a suitable solvent. Stock solutions prepared in this manner must be concentrated enough to allow for the addition of each hormone in 1 ml or less per liter of medium. When such stock solutions are used, media are prepared as usual, but the hormones are omitted. They are added to the hot solution following autoclaving and mixed by swirling. The complete medium is then distributed into autoclaved culture vessels (1, 5, 15).

Household bleach is usually a solution containing 4.25–5.25% sodium hypochlorite. It is sold under a large number of brand names all over the world. Simply read the label to determine the active agent(s) and concentration(s) of such bleaches to determine if they are suitable (10).

Inorganic salts are used to provide macro- and microelements in culture media (1, 2, 11).

Inositol is a polyol often added to culture media with beneficial effects. It is also known as *myo*-inositol, *meso*-inositol, *i*-inositol, and a number of other names (1, 2, 5, 8, 9, 11, 15).

Iron presents a problem because it tends to precipitate in media. The addition of organic acids (citric, for example) alleviates the problem, but does not solve it, in media formulated before the advent of chelating agents such as EDTA. In this appendix, all media are presented as originally reported since I did not wish to modify them without prior testing. Therefore, EDTA is not included in several tables. Those who wish to do so may modify these media by using an equal (i.e., equimolar) amount of iron as a chelate, which can be prepared by mixing sodium EDTA with $FeSO_4$ as listed in other tables, or by purchasing one of several chelated iron preparations now available on the market (1, 2, 5, 7, 8, 9, 10, 11, 15).

Laminar-flow hoods (or cabinets) provide a sterile atmosphere for seed- and tissue-culture work. They should be considered for laboratories that plan an intensive program. For a small-scale or one-time operation the sterile box described below is suitable (3, 6, 14, 16).

Lux is a measure of light intensity defined as the direct illumination on a surface that is everywhere one meter from a uniform source of one international candle. It equals 0.092903 foot-candles (one ft-candle = 10.76 lux).

Methanol. See Alcohol.

Microfilters. See Sterilizing filters.

Microscalpels may be used for splitting fruits or excising tissues. They can be purchased or can be made from a razor blade as described in Fig. A-2. Small hypodermic needles from disposable syringes are used by some workers as a substitute for microscalpels.

Niacin, also known as nicotinic acid, a vitamin, may be added to culture media (1, 2, 5, 7, 8, 9, 11, 15).

Neopeptone is an enzymatically digested protein preparation. Its exact composition and the basis for its positive activity are not known. The typical analysis of neopeptone (Table A-3) differs from that of peptone (see below) and may vary with batches or manufacturers (7).

Pasteur pipettes are disposable glass pipettes that work like an eyedropper (4, 14, 16).

Peptone is a water-soluble protein hydrolysate with a high amino-acid content. Its composition may vary and only typical analyses are available (Table A-3). Differences may exist between peptones from various sources or batches (7).

pH paper is impregnated with an indicator dye and can be used to measure the acidity or alkalinity of a solution (14, 16).

pH meters are an accurate way of measuring the acidity or alkalinity of culture media. They are available in a variety of sizes, including portable ones, and should be considered for laboratories planning extensive culture work (14, 16).

Pipettes are made of glass and sometimes plastic (though the former are better) and should be used to measure volumes smaller than 10 ml. They vary in capacity from less than 1 ml to more than 25 ml. The most useful for orchid-culture work are the 0.1-, 1-, 2-, 5-, and 10-ml "to deliver" (TD) pipettes. Always use the nearest larger size to measure a volume (for example, use a 5-ml pipette to measure 3 ml, do not use 10-ml or more or a 1-ml pipette three times). TD pipettes are graduated to their tips and the last drop of solution must be blown out (4, 14, 16).

Scalpels are used to split open capsules (fruits) or to excise tissues. Commercial scalpels with pointed blades are available (14, 16), but it is also possible to construct an even better one from a sewing needle and a section of a razor blade (see Microscalpels).

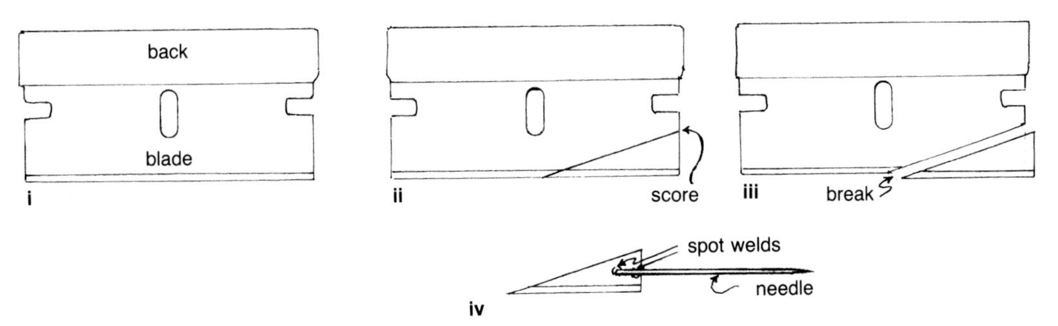

Fig. A-2. Preparation of a microscalpel from a razor blade. To protect eyes, wear goggles during this procedure. **i.** Use an unused, rust-free, single-edge razor blade. **ii.** Score blade on one corner. **iii.** Break off the scored corner by holding the smaller section with one pair of pliers or a small vice. Bend the larger portion (above the scoring line) with another pair of pliers until it breaks. **iv.** Spot-weld the smaller piece of blade to a sewing needle whose eye has been removed. Welds should be made quickly to prevent undue heating of the blade surface. Grasp the needle with a pair of pliers and insert into a wire loop holder (or a piece of wood, pencil, or twig). Trim the blade on a fine grinding wheel. Care should be taken, particularly at the tip, to prevent color changes in the metal, which indicate loss of temper in the blade. Scalpels that have lost their temper can be used, but they do not remain sharp for long.

Shakers are available in a large variety of sizes and types from several suppliers (13, 14, 16).

Sodium omadine is a systemic fungicide with minimal phytotoxic effects. It is available from: Olin Corporation, Agricultural Division, 700 N. Buckeye St., Little Rock, AR 72114 USA.

Sterile box. A very adequate flasking box may be constructed from a cardboard box, Saran Wrap, and aluminum foil. Obtain a 90- × 90- × 90-cm (approx. 3- × 3- × 3-ft, or similar) cardboard box and remove the top flaps. Draw a diagonal line along each side of the box, as shown in Fig. A-3, step i. Cut along these lines to obtain sloping surfaces (step ii). Line the inside of the box with aluminum foil and cover the top (sloping surfaces)

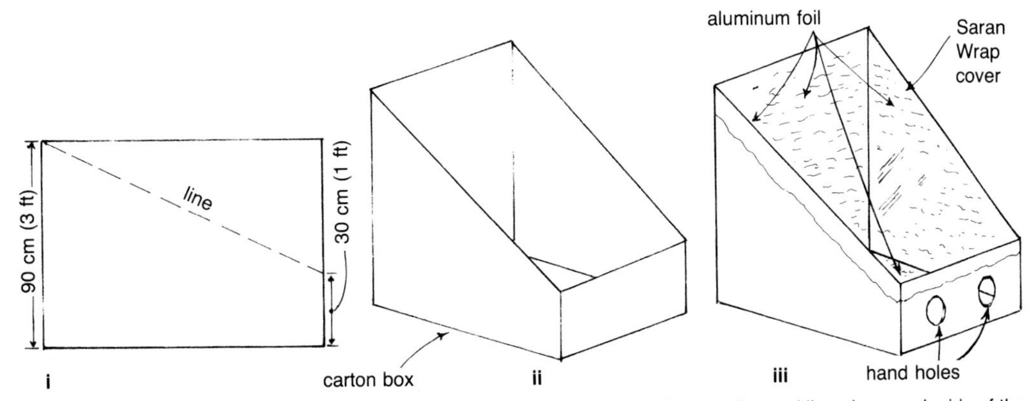

Fig. A-3. Preparation of a sterile work enclosure from a cardboard box: **i:** Draw a diagonal line along each side of the box. **ii:** Cut along diagonal line. **iii:** Line inside of box with aluminum foil, and cover top (sloping surfaces) with Saran Wrap. Cut two circular hand holes in shorter side (front) of box. Such work enclosures, although primitive and not very long lasting, are entirely satisfactory.

with Saran Wrap. Cut two circular holes in the front (shorter) side so that a worker's hands may be inserted for work inside the box (step iii).

The box should be sterilized prior to any flasking operation by swabbing its inside with household bleach. Flasks, jars, pans, and tools should also be sterilized in a similar fashion. Hands should be washed and decontaminated with a mild disinfectant of the type available in drugstores. Chances for contamination may be reduced further by attaching an open-ended plastic bag to each hand hole. Hands can then be inserted into the box through the bags and secured to them with rubber bands.

Sterilization of tissues (actually surface decontamination), media, and working areas is very important since it eliminates organisms that contaminate cultures. Work areas can be sterilized by swabbing with alcohol or a hypochlorite solution (in the U.S., Clorox, Purex, or other commercial preparations of sodium hypochlorite are used). In addition, a sterilizing (UV) lamp can be employed.

Care must be taken in decontaminating tissues because the agent used may kill them. Since tissues may differ in their sensitivity, it is best to use recommended or proven procedures.

Media can be sterilized by autoclaving (in an autoclave or pressure cooker), but some substances may be destroyed in the process (see Hormones). Therefore, media should not be sterilized by autoclaving unless it is known that they will not be affected deleteriously. Filter sterilization (or stock solutions in alcohol, as indicated for hormones) should be used for substances or media that are damaged by autoclaving.

Sterilizing filters are made of either glass (Morton type) or membranes (for example, Millipore). They are expensive and require complex equipment and are more suited for larger and better equipped laboratories (12, 14, 16).

Stock solutions. Weighing components of a medium can be time-consuming and subject to errors. Therefore, more concentrated solutions of some media components are prepared in many laboratories. These are called stock solutions. For example, if the recipe calls for 1 gram per liter (g/liter) of magnesium sulfate, a stock solution containing 10 g/liter can be prepared. Then, in making the medium, 100 ml (one-tenth of a liter and therefore one-tenth of the magnesium sulfate) of the stock solution is used. This is faster than weighing 1 g every time and also more accurate.

Storage. Stock solutions should be stored under appropriate conditions. Solutions of some inorganic salts become contaminated with time. Therefore, it is best to prepare them fresh every few months or when they are no longer completely clear. Storage in a refrigerator or freezer can prolong the life of such solutions.

Solutions of organic substances in water must be kept frozen. Alcoholic solutions should be refrigerated. Since some organic substances may be affected by light, it is wise to keep these stock solutions in the dark. Autoclaved media should not be stored for excessively long periods.

Surfactants are used to decrease the surface tension of liquids or solutions and increase their wetting power. Most mild household detergents can be used for this purpose (10).

Surgical gloves are lightweight rubber gloves for sterile work (14, 16).

Vials and tubes of various kinds can be used as culture vessels (4, 14, 16).

Vitamins should generally be handled like hormones (even if their mode of action is different) in the preparation of stock solutions or the sterilization of media (see Sterilization, Stock solutions). They are added to culture media because some seeds may require them (1, 2, 5, 7, 8, 9, 15).

Volumetric cylinders are made of glass or plastic and graduated to contain specific volumes. Their size designation is based on the maximum volume they can contain. A seed-germination laboratory should have volumetric cylinders in capacities of 10, 25, 50, 100, 250, 500, and 1000 ml. Always use the next largest cylinder to measure a particular volume (for example, use a 50-ml cylinder to measure 30 ml and not a 100-ml cylinder or a 10 ml and/or a 20 ml). Plastic cylinders are less apt to break (4, 14, 16). Cylinders with stoppers can be used like volumetric flasks whereas those with lips are for measuring and pouring; both are graduated.

Volumetric flasks are made of glass or plastic and marked to contain specific volumes. They are designated by the volume they contain. A seed-culture laboratory should have volumetric flasks in capacities of 5, 10, 25, 50, 100, 250, 500, and 1000 ml. To ensure accuracy volumetric flasks rather than beakers or flasks should be used to make dilutions and prepare media.

When using a volumetric flask to make dilutions or prepare solutions pour solvent into it (approximately 20% of its total volume), then introduce the compound to be dissolved or solution to be diluted. Shake and add solvent (water, alcohol, etc.) up to the mark. Then shake again and pour into a bottle. Never use volumetric flasks to store solution or media or for autoclaving. Follow the instructions in the tables and Fig. A-1 when using a volumetric flask to prepare media (4, 14, 16).

Yeast extract is the water extract of autolyzed yeast and contains vitamins and other substances which may be required by tissues. It may vary depending on its mode of preparation or batch (7).

List 2

Sources of Supplies and Equipment

All chemicals and apparatus mentioned in this appendix are available from many suppliers in the U.S. and other countries, but because an occasional item may be difficult to obtain locally, a list of suppliers is provided below. They are coded by numbers that are listed in parentheses at the end of the description of each List 1 entry. In addition, I have summarized the general categories of laboratory equipment, reagents, chemicals, and other supplies offered by each source. There are, no doubt, other equally adequate sources in the U.S. and abroad; the mention of some and omission of others should not be interpreted as an endorsement or lack of it.

It is advisable to request catalogs from suppliers well in advance of the time supplies will be needed. Order by airmail from outside the U.S., and request delivery by air for small items. In some countries the importation of equipment, apparatus, and chemicals requires special permits, which should be secured well in advance.

Some chemicals should be stored under refrigeration, and allowing them to linger in a postal warehouse may cause damage. Therefore, it is advisable to alert postal authorities to the arrival of such an item and make arrangements to pick it up promptly. And a final word: we can neither undertake to supply equipment, chemicals, and reagents nor act as intermediaries between suppliers and buyers.

1 Aldrich Chemical Co., Inc., 940 W. St. Paul Avenue, Milwaukee, WI 53233 USA

Inorganic chemicals, organic chemicals, biochemicals, hormones, amino acids, and vitamins. This company carries one of the largest selections of chemicals in the world.

2 J. T. Baker Chemical Co., 995 Zephyr Avenue, Hayward, CA 94544 USA

Mostly inorganics and some organic chemicals, hormones, amino acids, and vitamins.

3 The Baker Company, Inc., P.O. Drawer E, Sanford Airport, Sanford, ME 04073 USA

Laminar-flow hoods.

4 Bellco Glass, Inc., 340 Edrudo Road, Vineland, NJ 08360 USA

Glassware, culture tubes, laminar-flow cabinets.

5 Cal Biochem, P.O. Box 12087, San Diego, CA 92112 USA

Mostly biochemicals and some organic chemicals, hormones, amino acids, and vitamins.

6 Contamination Control Inc., Forty Foot and Tomlinson Roads, Kulpsville, PA 19443 USA

Laminar-flow cabinets.

7 Difco Laboratories, P.O. Box 1058A, Detroit, MI 48201 USA

Agar and media components.

8 Flow Laboratories, Inc., 936 W. Hyde Park Blvd., Inglewood, CA 90302 USA

Tissue-culture media, hormones.

9 Grand Island Biological Company, P.O. Box 200, Chagrin Falls, OH 44022 USA

Tissue-culture media, hormones.

10 Local stores

Bleach, razor blades, alcohol, and other items.

11 Mallinkrodt Chemicals, Second and Mallinkrodt Streets, P.O. Box 5439, St. Louis, MO 63147 USA

Inorganic and some organic chemicals.

12 Millipore Filter Corporation, Bedford, MA 01730 USA

Sterilizing filters.

13 New Brunswick Scientific Co., Inc., 1130 Somerset Street, New Brunswick, NJ 08003 USA

Shakers.

14 Scientific Products, 17111 Red Hill Avenue, Irvine, CA 92744 USA

Apparatus, glassware, filters, pH paper, plasticware, and laminar-flow cabinets.

15 Sigma Chemical Co., P.O. Box 14508, St. Louis, MO 63178 USA

Biochemicals and organic chemicals, hormones, amino acids, and vitamins.

16 Van Waters and Rogers, P.O. Box 1004, Norwalk, CA 90650 USA

Apparatus, filters, glassware, pH paper, plasticware, and laminar-flow cabinets.

17 Whatman, Inc., Chemical Separation Division, 9 Biedewell Pl., Clifton, NJ 07014 USA

Springfield Mill, Maidstone, Kent, ME1H 2LE England

P.O. Box 900, 41 6th Ave., Singapore 1027

Room 302 Ware Valley Mansion, 11-26 Akasaka 8-Chome, Mineto Ku, Tokyo 107 JAPAN

101 Mittal Chambers, 228 Nariman Point, Bombay 400 0211 INDIA

Filter papers.

Conversion Factors

Length

1 centimeter	0.394 inch
1 inch	2.540 centimeters
1 meter	3.2808 feet
1 foot	0.305 meter
1 meter	1.0936 yards
1 yard	0.9144 meter
1 kilometer	0.62137 mile
1 mile	1.60935 kilometers

Area

1 square centimeter	0.1550 square inch
1 square inch	6.452 square centimeters
1 square meter	10.764 square feet
1 square foot	0.09290 square meter
1 square meter	1.1960 square yards
1 square yard	0.8361 square meter
1 square kilometer	0.3861 square mile
1 square mile	2.590 square kilometers
1 acre U.S.	4840 square yards

Volume

1 cubic centimeter	0.0610 cubic inch
1 cubic inch	16.3872 cubic centimeters
1 cubic meter	35.314 cubic feet
1 cubic foot	0.02832 cubic meter
1 cubic meter	1.3079 cubic yards
1 cubic yard	0.7646 cubic meter

Capacity

1 milliliter	0.03382 ounce (U.S. liquid)
1 ounce (U.S. liquid)	29.573 milliliters
1 milliliter	0.2705 dram (U.S. Apothecaries)
1 dram (U.S. Apothecaries)	3.6967 milliliters
1 liter	1.05671 quarts (U.S. liquid)
1 quart (U.S. liquid)	0.94633 liter
1 liter	0.26418 gallon (U.S. liquid)
1 gallon (U.S. liquid)	3.78533 liters

Mass

1 gram	15.4324 grains
1 grain	0.0648 gram
1 gram	0.03527 ounce (Avoirdupois)
1 ounce (Avoirdupois)	28.3495 grams
1 gram	0.03215 ounce (Troy)
1 ounce (Troy)	31.10348 grams
1 kilogram	2.20462 pounds (Avoirdupois)
1 pound (Avoirdupois)	0.45359 kilogram

Power

1 watt	0.73756 foot pound per second
1 foot pound per second	1.35582 watts
1 watt	0.056884 BTU per minute
1 BTU per minute	17.580 watts
1 watt	0.001341 Horsepower (U.S.)
1 Horsepower (U.S.)	745.7 watts
1 watt	0.01433 kilogram-calorie per minute
1 kilogram-calorie per minute	69.767 watts
1 watt	1×10^7 ergs per second
1 lumen	0.001496 watt

Illumination

1 foot-candle	10.7639 lux
1 lumen	0.079577 candle power (about 1,700 lumens are produced by a standard 100-watt light bulb)
1 lumen/square meter	10.76391 foot-candles (10,000 phots)
1 lux	0.092903 foot-candles

Tropical Orchids (Epiphytic and Terrestrial)

Joseph Arditti

The great majority of orchids (epiphytic and terrestrial) grown by hobbyists or commercial growers are of tropical origin. Most of them (all epiphytes and many terrestrials) are easy to germinate on the most commonly used medium for orchid seed germination—Knudson C (either in the original formulation or with some modifications). Some of the popular genera which fall into this category are *Cattleya, Cymbidium, Oncidium, Odontoglossum, Coelogyne, Vanda* and its allies, *Dendrobium, Brassia, Brassavola, Laelia, Zygopetalum, Phaius,* and *Stanhopea.* In general, this medium is suitable for most orchids except north temperate terrestrials and some others, which are listed separately in this appendix (Arditti, 1967, 1968, 1979; Ernst, 1974, 1975, 1976; Harrison, 1968; Knudson, 1946; Werkmeister, 1970).

Plant Material. Mature or immature seeds [from ripe and dehisced or green capsules ("pods"), respectively] can be germinated using this method. Immature seeds are sometimes incorrectly referred to as "ovules," but what is being cultured are actually embryos in various stages of development. The earliest stage at which embryos can be cultured successfully varies with the genus, species, hybrid, and local conditions and if not specified in the literature (Table A-4) should be determined experimentally. When estimates must be made, a useful but not universal generalization is to culture seeds from fruits (in orchids these are always capsules and never pods) that have been on the plant for one-third to one-half of the time they normally require to reach maturity. For example, if a species requires 6 months for its fruit to mature, chances are good that its immature embryos will grow after 9–14 weeks. Some can start growing much earlier, but this time schedule provides an ample safety margin. Others may not be able to develop *in vitro* until later.

Table A-4. Interval between pollination and growth of immature embryos *in vitro*

Species	Interval (days)	Species	Interval (days)
Aerides odoratum	150–180	*Oncidium altissimum*	110–140
Ansellia species and hybrids	150–180	*O. ampliatum*	45–50
Ascocenda	120–150	*O. bahamense*	65–70
A. hybrids	120–190	*O. baueri*	110–140
A. cv Mem. Jim Wilkins	90–150	*O. carthagenense*	180–240
× *Vanda* cv Patricia Lee		*O. cavendishianum*	180–240
Ascocentrum	110–180	*O. cebolleta*	110–130
Brassavola	120–150	*O. flexuosum*	110–140
B. cucullata	75–80	*O. jonesianum*	110–130
B. nodosa	70–75	*O. kramerianum*	90–120
Brassocattleya	130–180	*O. lanceanum*	180–240
Brassolaeliocattleya	130–180	*O. leochilum*	110–140
Broughtonia	60–75	*O. limminghei*	90–120
B. sanguinea	32–34	*O. lucayanum*	65–70
Cattleya bifoliate group	110–150	*O. luridum*	75–80, 100–180
C. bowringiana	75–80	*O. luridum* × *O. papilio*	150–180
C. elongata	55–60	*O.* cv M. P. de Restrepo	95–100
C. labiata group	130–180	*O. maculatum*	110–140
C. loddigesii	80–85	*O. microchilum*	130–170
C. skinneri	85–90	*O. papilio*	90–120
C. violacea	80–85	*O. papilio* × *O. luridum*	90–120
C. unifoliate group	120–150	*O. pulchellum*	65–70
Chysis	140–180	*O. retemeyerianum*	180–240
Cirrhopetalum	140–180	*O. sanderae*	90–120
Cymbidium	280–360	*O. sphacelatum*	110–140
Cypripedium	30	*O. splendidum*	130–170
Cyrtopodium	150–270	*O. stipitatum*	30–35, 110–130
Dendrobium cv Bali	60–65	*O. teres*	110–130
D. cv D. A. Spalding × *D. striotes*	55–60	*O. tetrapetalum*	65–70
D. devonianum	160–250	*O. urophyllum*	65–70
D. cv J. Thomas	55–60	*O. variegatum*	65–70
D. lituiflorum	150–180	*Orchis morio*	35–40
D. nobile and its hybrids	150–180	*Paphiopedilum*	240–300
D. phalaenopsis and its hybrids	120–140	*Phajus*	120–150
D. cv E. Kawamoto × *D. gouldii*	55–60	*Phalaenopsis* cv Chieftain	70–75
D. stratiotes and its hybrids	150–200	*P.* cv Doris	75–80
D. superbiens and its hybrids	150–200	*P.* cv Elizabethae	75–80
D. superbum and its hybrids	160–250	*P.* hybrids	110–120
Doritaenopsis	90	*P.* species	110–120
Doritis pulcherrima	65–70	*P. stuartiana* × *Phalaenopsis amabilis*	75–80
Encyclia	130–180	*Potinara*	130–150
Epidendrum	100–120	*Renanthera*	150–180
E. atropurpureum	150–160	*R.* cv R. B. Chandler	70–75
E. tampense	70–75	*Rhyncholaelia*	120–180
Laelia anceps	120–150	*Rhynchostylis*	150–250
L. cinnabarina	110–120	*Rodriguezia*	110–130
L. flava	110–120	*Sophrocattleya*	110–150
L. harpophylla	110–120	*Sophrolaelia*	110–150
L. perrinii	120–180	*Sophronitis*	75–100
L. purpurata	120–180	*Vanda* cv Burgeffi	70–75
L. rubescens	120–150	*V.* cv E. Noa × *V. suavis*	70–75
L. xanthina	120–180	*V.* cv M. Foster	70–75
Laeliocattleya hybrids	120–180	*V.* hybrids	150–195
Maxillaria	120–140	*V.* cv H. Paoa	70–75
Miltonia candida	120–140	*V. luzonica* × *V. sanderiana*	70–75
M. clowesii	120–140	*V.* cv Patricia Lee × Self	120–150
M. flavescens	120–140	*V.* cv Patricia Lee	90–150
M. spectabilis	120–140	× *Ascocenda* cv Mem. Jim Wilkins	
Odontoglossum	80–90	*V.* species	150–195

Sources: Fast, 1980; Frosch, 1980; Jones and Scully, 1977; Sagawa and Valmayor, 1966; Sauleda, 1976.

Surface Sterilization. Green capsules should be washed with soap and water and, if muddy or especially dirty, also scrubbed with a soft brush. After being washed they should be rinsed with water and submerged in 50% household bleach (50 ml bleach and 50 ml water) for 10–15 min. Following removal from the bleach and before being split the capsules should be rinsed with distilled water. The interior of intact green capsules is sterile, and sterilization of immature seeds is therefore unnecessary. In fact, such sterilization is not advisable since the sterilant may cause harm.

Mature seeds must be sterilized, and the most commonly used agent for this purpose is saturated calcium hypochlorite. The sterilant is prepared by dissolving 7 grams of dry calcium hypochlorite in 100 ml of water (or 10 g/140 ml) and stirring well several times before filtering or decanting the liquid (which should be used within 12 hr). To improve wetting of the seed, 2–3 drops of a mild household detergent should be added to every 100 ml of sterilizing solution. Seeds are sterilized in vials by being suspended and shaken in approximately 50 times their volume of sterilant (i.e., a batch of seeds which occupies a volume of 1 ml should be suspended in 50 ml of saturated calcium hypochlorite) for 15–20 min. Or a glass tube can be used (Harrison, 1968). Opinions vary on whether seeds can be transferred directly from the sterilant to culture flasks or must be washed with sterile distilled water first. Success has been reported with both procedures, but in case of doubt it may be wise to wash the seeds.

Culture Vessels. Erlenmeyer flasks (125, 250, 500, or even 1000 ml), a variety of bottles, and several kinds of jars or test tubes can be used for orchid seed germination and seedling culture. The vessels must be transparent and clean (even acid-washed in some cases). They should be filled with medium to one-fifth of their capacity. The bottles should be covered with a cotton bun, a plastic cap, or a two-hole rubber stopper, with the holes filled with cotton. Rubber stoppers or cotton buns should be covered with aluminum foil or paper before autoclaving.

Culture Conditions. The best conditions for orchid seed germination and seedling development are 20–25° C, moderate light intensity, and 10- to 14-hr photoperiods. Gro-Lux or other plant growth lights, mixtures of cool or warm white fluorescent tubes and incandescent lamps, and diffuse daylight (but no direct sun) are all suitable illumination.

Culture Medium. Knudson C medium (Table A-5) is used for seed germination and seedling culture of tropical epiphytic orchids. To ensure a steady pH, phosphate buffer can be used instead of KH_2PO_4 (Table A-5).

Seedling growth on media containing vegetable charcoal (Nuchar C) is improved often and well enough to warrant the routine use of this ingredient. However, it is necessary to keep in mind that charcoal may irreversibly bind additives like vitamins or hormones and render them unavailable. In cases where such adsorption must be avoided, graphite should be used.

Banana is an additive which stimulates seedling growth but may hinder germination. Therefore it should only be added to media used to transplant seedlings. A combination

Table A-5. Modified Knudson C medium (Knudson, 1946) for orchid seed germination and seedling culture

Item number	Component	Amount per liter of culture medium (final concentration in culture medium)	Stock solution (a concentrate prepared for repeated and convenient use)	Volume of stock solution per liter of culture medium	Remarks
	Macroelements				
1	Calcium nitrate, $Ca(NO_3)_2 \cdot 4H_2O$[a]	1 g	100 g/liter[a]	10 ml[a]	or weigh[a]
2	Potassium phosphate, KH_2PO_4[b]	250 mg	25 g/liter[b]	10 ml[b]	
3	Magnesium sulfate, $MgSO_4 \cdot 7H_2O$	250 mg	25 g/liter	10 ml	
4	Ammonium sulfate, $(NH_4)_2SO_4$[a]	500 mg	50 g/liter[a]	10 ml[a]	or weigh[a]
5	Ferrous sulfate, $FeSO_4 \cdot 7H_2O$[c]	25 mg	2.5 g/liter[c]	10 ml[c]	
6	Manganese sulfate, $MnSO_4 \cdot 4H_2O$	7.5 mg	750 mg/liter	10 ml	
7	*Microelements*[d]				
a	Boric acid, H_3BO_3	0.056 mg	56 mg/liter	1 ml[d]	one solution[d]
b	Molybdic acid, MoO_3	0.016 mg	16 mg/liter		
c	Cupric sulfate (anh.), $CuSO_4$	0.040 mg	40 mg/liter		
d	Zinc sulfate, $ZnSO_4 \cdot 7H_2O$	0.331 mg	331 mg/liter		
	Sugar[e]				
8	Sucrose	20 g	no stock	no stock	weigh
	Complex additives				
9	Ripe banana[e,f]	100–150 g	no stock	no stock	weigh[f]
10	Vegetable charcoal (Nuchar C)[g]	2 g	no stock	no stock	weigh[g]
11	Water, distilled[e]	to 1000 ml			
12	Agar	12–15 g	no stock	no stock	weigh

[a] Solutions containing ammonium and/or nitrate may become contaminated on standing. Therefore, stock solutions should not be prepared. If made, they must be kept frozen between uses.

[b] A phosphate buffer which will keep the pH constant may be substituted. Prepare buffer by mixing 975 ml of 0.1 M KH_2PO_4 (monopotassium phosphate) solution (13 g/liter) with 25 ml of a 0.1 M K_2HPO_4 (dipotassium phosphate) solution (17.4 g/liter); measure the pH to be certain it is correct (pH 5.1–5.4), adjust if necessary, and use 18 ml/liter of culture medium.

[c] A rust-colored precipitate may form on standing. This solution must, therefore, be shaken vigorously before use.

[d] Add all microelements to the same one liter of distilled water; stir and/or heat until they are dissolved, and add 1 ml per liter of culture medium.

[e] Add items 1–7 and 10 or 1–7 and 9–10 to 650 ml distilled water (item 11) in a blender, homogenize for 1 min, let stand for 2 min and homogenize again for another min, and repeat if necessary to completely disperse the banana; adjust the pH to 5.3; add sugar (item 8) and adjust volume to 1000 ml with more distilled water (item 11). Bring solution to a gentle boil and add agar (item 12) slowly while stirring. When the agar has dissolved completely (and the solution turns a clear amber color), distribute the medium into culture vessels, cover, and autoclave.

[f] Use ripe banana and add only to medium used for transflasking seedlings. Do not add banana to seed or embryo germination media.

[g] Add to both germination and transflasking media.

of vegetable charcoal (Nuchar C) and banana may enhance the growth of many orchid seedlings.

Knudson C medium with or without banana and/or charcoal should be heat-sterilized in an autoclave or pressure cooker at 121° C (250° F) and one atmosphere (1 kg/cm^2 or 15 lb/in^2 of pressure) for 20 min.

Procedure. All transfers of seeds from immature fruits or sterilizing solutions and seedlings between flasks should be carried out under aseptic conditions in a laminar-flow hood or sterile box.

To remove immature seeds, cut open green fruits with a sterile knife, scalpel, razor blade, or spatula. Scrape out contents with a sterile spatula and transfer a small amount to each culture vessel. Distribute the material evenly over the surface of the medium using the spatula and the small amount of water which normally condenses on the agar.

If the amount of condensate is insufficient, add a few drops of autoclaved distilled water with a sterile eyedropper or syringe.

Mature seeds can be placed on the medium in several ways. One is to decant the sterilant and transfer small amounts of seed to the medium with a sterile wire loop. The seeds should be distributed evenly on the agar surface as described above. When this method is used only very small amounts of sterilant adhere to the seeds. It is therefore not strictly necessary to wash them following their removal from the sterilant and before placing them on the culture medium (except for very sensitive seeds or in experiments).

A second method is to decant the sterilant and resuspend the seeds in sterile distilled water. Shake well to ensure an even suspension in the water and use a sterile eyedropper to transfer seeds into culture vessels. Use the water to distribute the seed evenly over the surface of the medium.

The simplest and best method involves the use of a glass tube and a rubber bulb. Cut the top of a Pasteur pipette (List 1) to a length of 15–20 mm above the constriction and 15–20 mm below it. Stuff cotton above the constriction and place enough seed for one culture vessel below it. Insert cotton below the seeds and attach a rubber bulb to the tube. Using the bulb, fill the tube with sterilizing solution and allow the seeds to remain in contact with the sterilant for 20 min. At the end of this period expel the sterilant and replace it with sterile distilled water. Expel the water and replace it several times to wash the seeds. Then take up water again. Next, remove the cotton from below the seeds and expel the seeds into a culture vessel. Use the water to distribute the seeds evenly over the surface of the agar. The tubes can be used and reused many times.

After placing seeds on the medium, plug the flask; move it to a culture room and observe at regular intervals. If proper procedures are followed, few if any cultures become contaminated. Remove contaminated cultures from the culture room and autoclave them before discarding their contents and reusing the vessel.

Cultures will become crowded after various periods of time depending on the number of seeds placed in a vessel, the hybrid, and the species. When crowding occurs the seedlings should be transferred to banana-containing medium. The distance between seedlings should be about 1 cm. One method is to move seedlings directly from one vessel to another with a sterile loop. Another procedure is to place all seedlings in a sterile container (plate, petri dish) and transfer them from there into the next flask. Seedlings may be transflasked two or three times.

Developmental Sequence. Several weeks after seeds are placed on the culture medium their embryos turn green and start to grow. Protocorm formation and seedling development follow. Growth and development will continue after transflasking.

General Comments. With some (mostly minor) modifications these are the most widely used methods for orchid seed germination and seedling culture. Whenever possible, the culture of immature seeds from green capsules is preferable to that of mature seeds from ripe fruits because it saves time and is simpler (especially the surface sterilization), and in some cases this is the only way to obtain germination.

Literature Cited

Arditti, J. 1967. Factors affecting the germination of orchid seeds. Bot. Rev. 33:1–97.

——. 1968. Germination and growth of orchids on banana fruit tissue and some of its extracts. Amer. Orchid Soc. Bull. 33:112–116.

——. 1979. Aspects of the physiology of orchids. Adv. Bot. Res. 7:421–655. Academic Press, London.

Ernst, R. 1974. The use of activated charcoal in asymbiotic seedling culture of Paphiopedilum. Amer. Orchid Soc. Bull. 43:35–38.

——. 1975. Studies in asymbiotic culture of orchids. Amer. Orchid Soc. Bull. 44:12–18.

——. 1976. Charcoal or glass wool in asymbiotic culture of orchids. Proc. 8th World Orchid Conf. Frankfurt (1975), p. 379–383.

Fast, G. (ed.). 1980. Orchideen kultur. Verlag Eugen Ulmer, Stuttgart.

Frosch, W. 1980. Asymbiotische Ausaat von Orchis morio. Die Orchidee 31:123–124.

Harrison, C. R. 1968. A simple method for flasking orchid seeds. Amer. Orchid Soc. Bull. 39:715–716.

Jones and Scully. 1977. Catalog. Miami, FL 33142 USA.

Knudson, L. 1946. A new nutrient solution for the germination of orchid seed. Amer. Orchid Soc. Bull. 15:214–217.

Sagawa, Y., and H. L. Valmayor. 1966. Embryo culture of orchids. Proc. 5th World Orchid Conf., Long Beach, p. 99–101.

Sauleda, R. P. 1976. Harvesting times of orchid seed capsules for the green pod culture process. Amer. Orchid Soc. Bull. 45:305–309.

Werkmeister, P. 1970. Uber die Lichtinduction der geotropen Orientirung von Luft- und Bodenwurzeln in Gewebekulturen von Cymbidium. Ber. Deutsch. Bot. Gaz. 83:19–26.

North American Terrestrial Orchids

Joseph Arditti

More than 125 terrestrial orchids are found in North America north of Florida (Luer, 1975). All are terrestrial and, as far as is known, difficult to germinate. The methods outlined here are taken from the literature and are practical though not always as effective as one might wish. Species other than the ones mentioned here have been germinated, but some of the methods have not been published. Several procedures are not given in sufficient detail in the literature and could not be adapted. A number are based on work from my laboratory (Arditti, Michaud, and Oliva, 1982).

Calopogon pulchellus

The grass pink, *Calopogon pulchellus,* renamed *Calopogon tuberosus,* is a terrestrial orchid found in the eastern United States, Canada, and the Caribbean (Luer, 1975). Attempts to germinate its seeds asymbiotically were made before publication of the Knudson C medium (Liddell, 1944). The method presented here is the last to be formulated (Liddell, 1944).

Plant Material. Mature seeds were probably used in the original research.

Surface Sterilization. The procedures used for tropical orchids (p. 275) are appropriate.

Culture Vessels. Flasks stoppered with one-hole rubber stoppers were used in the original research. The holes in the stoppers were filled with cotton, or a glass tube stuffed

Table A-6. Modified Knudson B medium for the germination of *Calopogon pulchellus* seeds and seedling culture

Item number	Component	Amount per liter of culture medium (final concentration in culture medium)	Stock solution (a concentrate prepared for repeated and convenient use)	Volume of stock solution per liter of culture medium	Remarks
	Macroelements				
1	Calcium nitrate, $Ca(NO_3)_2 \cdot 4H_2O$[a]	1 g	100 g/liter[a]	10 ml[a]	or weigh[a]
2	Monopotassium phosphate, KH_2PO_4[b]	250 mg	25 g/liter[b]	10 ml[b]	
3	Magnesium sulfate, $MgSO_4 \cdot 7H_2O$	250 mg	25 g/liter	10 ml	
4	Ammonium sulfate, $(NH_4)_2SO_4$[a]	500 mg	50 g/liter[a]	10 ml[a]	or weigh[a]
5	Ferric phosphate, $FePO_4$[c]	25 mg	2.5 g/liter[c]	10 ml[c]	
6	Manganese sulfate, $MnSO_4 \cdot FH_2O$	7.5 mg	750 mg/liter	10 ml	
	Sugar[d]				
7	Sucrose	20 g	no stock	no stock	weigh
8	Water, distilled[d]	to 1000 ml			
	Solidifier[d]				
9	Agar	17.5 g	no stock	no stock	weigh

[a] Solutions containing ammonium and nitrate may become contaminated on standing. Therefore, stock solutions should not be prepared. If made, they must be kept frozen between uses.

[b] A phosphate buffer which will keep the pH constant may be substituted. Prepare buffer by mixing 975 ml of 0.1 M KH_2PO_4 (monopotassium phosphate) solution (13.6 g/liter) with 25 ml of a 0.1 M K_2HPO_4 (dipotassium phosphate) solution (17.4 g/liter); measure the pH to be certain it is correct (pH 5.1–5.4), adjust if necessary, and use 18 ml/liter of culture medium.

[c] This solution must be shaken vigorously before use.

[d] Add items 1–6 to 650 ml distilled water (item 8); adjust the pH to 5.3; add sugar (item 7) and adjust volume to 1000 ml with more distilled water (item 8). Bring solution to a gentle boil and add agar (item 9) slowly while stirring. When the agar has dissolved completely (and the solution turns a clear amber color) distribute the medium into culture vessels, cover, and autoclave.

with cotton was inserted through them. Vessels of the kind recommended for tropical species (p. 275) can also be used.

Culture Conditions. The conditions employed for tropical orchids are suitable.

Culture Medium. Knudson B medium (Table A-6).

Procedure. The procedures utilized for tropical species (p. 276) should be employed. Initial germination should be carried out in agar-free (i.e., liquid) solution. When the seeds swell to a diameter of 1.6 mm (1/16 in.) they should be transferred to solid medium.

Developmental Sequence. Germination begins rapidly and some embryos turn green after a few days in liquid medium. The viable seeds swell within a month, reaching a diameter of 16 mm (5/8 in.). Within 3 to 4 months after transfer to solid medium seedlings reach a height of 5 cm (2 in.) and have protocorms that are 3.2 mm (1/8 in.) in diameter.

General Comments. It is interesting to note that *Calopogon puchellus* is not limited to north temperate climates but extends into the Caribbean and tropical conditions. Perhaps this is the reason why its seeds germinate with relative ease.

Literature Cited

Arditti, J., J. D. Michaud, and A. P. Oliva. 1982. Germination of seeds of native California orchids. I. Native California and related species of *Calypso, Epipactis, Goodyera, Piperia,* and *Platanthera.* Bot. Gaz. 142:442–453.

Curtis, J. T. 1936. The germination of native orchid seeds. Amer. Orchid Soc. Bull. 5:42–47.

——. 1937. Some phases of symbiotic and non-symbiotic orchid seed germination. Ph.D. diss., Univ. Wisconsin, Madison.

Liddell, R. W. 1944. Germinating native orchid seed. Amer. Orchid Soc. Bull. 12:344–345.

Luer, C. A. 1975. The native orchids of the United States and Canada, excluding Florida. New York Botanical Garden, New York.

Calypso bulbosa

Like most orchids which are native to California, *Calypso bulbosa* is a terrestrial species threatened with extinction. We are, therefore, trying to develop a practical seed-germination method which can be used to propagate it and increase its numbers (J. Arditti, J. Michaud, and A. Oliva, unpublished).

Plant Material. Immature seeds removed from green capsules collected in Northern California (Humboldt County) were used in the original research.

Table A-7. Modified Curtis (1937) medium as used for seed germination and seedling culture of *Calypso bulbosa*

Item number	Component	Amount per liter of culture medium (final concentration in culture medium)	Stock solution (a concentrate prepared for repeated and convenient use)	Volume of stock solution per liter of culture medium	Remarks
	Macroelements				
1	Potassium phosphate, KH_2PO_4	120 mg	12 g/liter	10 ml	
2	Magnesium sulfate, $MgSO_4 \cdot 7H_2O$	260 mg	26 g/liter	10 ml	
3	Ammonium nitrate, NH_4NO_3[a]	220 mg	22 g/liter[a]	10 ml[a]	or weigh[a]
4	Calcium nitrate, $Ca(NO_3)_2 \cdot 4H_2O$[a]	350 mg	35 g/liter[a]	10 ml[a]	or weigh[a]
	Iron				
5	Ferrous sulfate, $FeSO_4 \cdot 7H_2O$[b]	5.53 mg	553 mg/liter[b]	10 ml[b]	
	Complex additive				
6	Coconut water[c,e]	50 ml	no stock	no stock	
	Special additive				
7	W66$_f$[d]	0.1 ml	commercial preparation[d]		
	Sugar[e]				
8	Glucose (dextrose)	10 g	no stock	no stock	weigh
	Solidifier[e]				
9	Agar	14 g	no stock	no stock	weigh
10	Water, distilled[e]	to 1000 ml			

[a] Solutions containing ammonium and/or nitrate may become contaminated on standing. Therefore, stock solutions should not be prepared. If made, they must be kept frozen between uses.

[b] A rust-colored precipitate may form on standing. This solution must, therefore, be shaken vigorously before use.

[c] Keep frozen between uses.

[d] Available from Edward Gerlack GMBH, Chemische Fabric, 4990 Lubbecke 1, Postfach 1165, West Germany.

[e] Add items 1–5 and 7 to 190 ml distilled water (item 10) and put aside. Filter coconut water (item 6) through Whitman filter No. 1 and add to items 1–5. Adjust pH to 5.5, bring total volume to 250 ml, and sterilize by filtering through a 0.45 μ Millipore filter (p. 268) and set aside. Add sugar and agar (items 8 and 9) to 750 ml distilled water (item 10). Autoclave, and while still hot combine with filter-sterilized solution; mix well by swirling and distribute into culture vessels.

Table A-8. Preparation of potato-dextrose agar for seed germination of *Calypso bulbosa* (Harvais, 1973)

Step	Operation
1	Peel 200 g of potatoes.
2	Slice peeled potatoes into 1 liter of boiling water.
3	Steam for 1 hr in a double boiler.
4	Strain; save the liquid and discard particulate matter.
5	Bring volume of liquid to 1 liter, pour into a 2-liter vessel and mark level.
6	Add 10 g of glucose (dextrose).
7	Add 1% agar.
8	Boil gently until agar has been dissolved. Adjust volume to mark (see step 5).
9	Dispense into culture vessels and cap.
10	Autoclave.
11	Cool.
12	Use.

Surface Sterilization. Capsules should be sterilized by immersing them in filtered saturated calcium hypochlorite (7 g/100 ml water) solution (pp. 254, 261, 275) for 10 minutes.

Culture Vessels. The culture vessels recommended for tropical orchids (p. 275) can be used.

Culture Conditions. Cultures can be maintained from the onset under 18-hr photoperiods provided by two 40-watt Gro-Lux lamps 30 cm (1 ft) above the vessels and 22° ± 2° C until white protocorms appear (4.5 to 5.5 months) and then moved to the light. Alternatively, they can be kept in the dark and transferred to 12-hr photoperiods of 200 lm/0.09 m^2 (ft^2) provided by Gro-Lux lamps and 25° C (Harvais, 1973).

Culture Media. Modified Curtis medium (Table A-7) or potato-dextrose agar (Table A-8; Harvais, 1973) can be used. It is best to attempt the germination of each batch of seeds on both media.

Procedure. Sterilized capsules are sliced open and the seeds are removed with a sterile spatula and placed on the culture medium. It is important to place only very few seeds in each flask to avoid crowding because the seedlings cannot be moved; they die if transflasked.

Developmental Sequence. White protocorms develop both in the dark and under illumination after 4.5 to 5.5 months of culture. Within 3 months some of these protocorms turn green if maintained under light and develop into plantlets. Seedlings may reach a height of 3 cm (1.25 in.) after 15 months of culture.

General Comments. Seeds from mature capsules germinate very poorly. Germination of seeds from immature fruits is slightly better. In any case, only a few plantlets can be obtained, and they are very sensitive. They die if transflasked and it is not clear at this point whether it would be possible to grow asymbiotically produced seedlings in a greenhouse or outdoors.

Germination is slow and uneven. Seeds continue to germinate and form white and green protocorms as late as 18 months after inoculation. The protocorms develop into plantlets. Thus in some cases the interval between inoculation and plantlet production may be 2 or even 3 years and perhaps longer. Media dry during such prolonged periods and the cultures must be irrigated every 3 months, at first with double-distilled sterile water, and later, after plantlet formation, with autoclaved 1% (10 g/liter) sucrose solution. The volume of irrigation should be 1–1.5 ml per 30 ml of medium.

Literature Cited

Curtis, J. T. 1937. Some phases of symbiotic and nonsymbiotic orchid seed germination. Ph.D. diss., Univ. Wisconsin, Madison.

Harvais, G. 1973. Growth requirements and development of Cypripedium reginae in axenic culture. Can. J. Bot. 51:327–332.

Corallorhiza maculata

The coral-root orchid, *Corallorhiza maculata,* a terrestrial, is achlorophyllous and leafless. Therefore it has been assumed to be saprophytic; but this is probably not the case. *C. maculata,* like other chlorophyll-free orchids, depends on its fungus for nutrition (i.e., is parasitic). In the United States it flowers from April to September. Its range extends into Canada where seed germination was studied (Harvais, 1974).

Plant Material. Seeds for the original research were collected in Canada when the capsules began to dehisce, air-dried at 25° C (77° F) for at least 48 hr, and stored dry at 8° C (46° F).

Surface Sterilization. The procedures outlined for tropical orchids (p. 275) are suitable.

Culture Vessels. Tubes 20 × 150 mm containing 10 ml medium were used in the original research, but the culture vessels recommended for tropical species (p. 275) are also suitable.

Culture Conditions. In the original research, "the cultures were incubated at 25° C [77° F] and watered at intervals [as often as needed to compensate for drying without waterlogging the cultures].... Initially the cultures were kept in the dark, but were later transferred to a diurnal 12-[hr] regime in artificial light (Gro-Lux fluorescent lamps, Sylvania Co.) at an intensity of 200 lm/ft^2 when the developed shoots started to etiolate."

Culture Medium. Germination occurs on a relatively simple medium (Table A-9).

Procedure. Seeds are placed on medium following sterilization and maintained under the conditions described above.

Table A-9. Medium for seed germination of *Corallorhiza maculata* (Harvais, 1974)

Item number	Component	Amount per liter of culture medium (final concentration in culture medium)	Stock solution (a concentrate prepared for repeated and convenient use)	Volume of stock solution per liter of culture medium	Remarks
	Macroelements				
1	Calcium nitrate, $Ca(NO_3)_2 \cdot 4H_2O$[a]	800 mg	80 g/liter[a]	10 ml[a]	or weigh[a]
2	Potassium phosphate, KH_2PO_4	200 mg	20 g/liter	10 ml	
3	Magnesium sulfate, $MgSO_4 \cdot 7H_2O$	200 mg	20 g/liter	10 ml	
4	Potassium nitrate, KNO_3[a]	200 mg	20 g/liter[a]	10 ml[a]	or weigh[a]
5	Potassium chloride, KCl	100 mg	10 g/liter	10 ml	
	Iron				
6	Ammonium ferric citrate (about 0.2 mg Fe)	1 mg	100 mg/liter	10 ml	
7	*Microelements*[b]				
a	Manganese sulfate, $MnSO_4 \cdot 4H_2O$	0.5 mg	500 mg/liter	1 ml[b]	one solution[b]
b	Boric acid, H_3BO_3	0.5 mg	500 mg/liter		
c	Zinc sulfate, $ZnSO_4 \cdot 7H_2O$	0.5 mg	500 mg/liter		
d	Copper sulfate, $CuSO_4 \cdot 5H_2O$	0.5 mg	500 mg/liter		
	Sugar[c]				
8	Glucose	10 g	no stock	no stock	weigh
	Solidifier[c]				
9	Agar	10 g	no stock	no stock	weigh
10	Water, distilled[c]	to 1000 ml			

[a] Solutions containing ammonium and/or nitrate may become contaminated on standing. Therefore, stock solutions should not be made. If made, they must be kept frozen between uses.

[b] Add all microelements to the same one liter of distilled water and stir or heat until they are dissolved. Add 1 ml of the stock solution to each liter of medium.

[c] Add items 1–8 to 1000 ml of distilled water (item 10). Bring to a slow boil and add agar with stirring. When agar has dissolved, dispense medium into culture vessels, autoclave, cool, and use.

Developmental Sequence. Germinating seeds accumulate moderately high levels of starch.

General Comments. The germination rate as measured by starch accumulation is low and growth does not occur. Still, this is the best available published method which could serve as a starting point for improved procedures. When other methods are used, seeds of *Corallorhiza maculata* "swell but do not germinate" (Stoutamire, 1974).

Literature Cited

Harvais, G. 1974. Notes on the biology of some native orchids of Thunder Bay, their endophytes and symbionts. Can. J. Bot. 52:455–460.

Stoutamire, W. 1974. Terrestrial orchid seedlings, p. 101–128. *In* C. L. Withner (ed.), The orchids: Scientific studies. Wiley-Interscience, New York.

Cypripedium

Since species now known as *Paphiopedilum* were at one time called *Cypripedium,* care is necessary in deciding which seed-germination procedures to use. *Paphiopedilum* seeds are relatively easy to germinate (pp. 351–353), whereas those of *Cypripedium,* a north temperate terrestrial, present considerable difficulties (Curtis, 1936, 1937, 1943; Harvais, 1973; Muick, 1978; Withner, 1953). Still, some have been germinated.

Cypripedium acaule

Cypripedium acaule differs from other American species of the genus in habitat, flower shape, and seed morphology (Arditti *et al.*, 1979; Luer, 1975; Wherry, 1918). Attempts to germinate its seeds have been only partially successful (Curtis, 1936, 1937).

Plant Material. In the original research, "sowings were made . . . with seeds which varied in age from fresh to four years old. In some instances . . . seeds were . . . removed aseptically" (Curtis, 1936). However, several species were involved and there is no way to determine the age of the *C. acaule* seeds used.

Surface Sterilization. Seeds should be sterilized by immersion in saturated (7 g/100 ml) calcium hypochlorite solution for 15–20 min.

Culture Vessels. Pyrex test tubes 20 cm long and 15–20 cm in diameter, Erlenmeyer flasks, and other containers can be used.

Culture Conditions. Low light intensity ("a shaded place in the greenhouse" [Curtis, 1936]) and a temperature of 20–24° C (68–75° F) are suitable.

Culture Medium. Curtis solution 5 should be used (Table A-10).

Procedure. The procedures suggested for tropical orchids (p. 276) should be used.

Table A-10. Curtis solution 5 for *Cypripedium acaule* seed germination (Curtis, 1936)

Item number	Component	Amount per liter of culture medium (final concentration in culture medium)	Stock solution (a concentrate prepared for repeated and convenient use)	Volume of stock solution per liter of culture medium	Remarks
	Macroelements				
1	Potassium phosphate, KH_2PO_4	120 mg	12 g/liter	10 ml	
2	Magnesium sulfate, $MgSO_4 \cdot 7H_2O$	260 mg	26 g/liter	10 ml	
3	Ammonium nitrate, NH_4NO_3[a]	220 mg	22 g/liter[a]	10 ml[a]	or weigh[a]
4	Calcium nitrate, $Ca(NO_3)_2 \cdot 4H_2O$[a]	350 mg	35 g/liter[a]	10 ml[a]	or weigh[a]
	Iron				
5	Ferric phosphate, $FePO_4$[b]	3 mg	300 mg/liter[b]	10 ml	
	Sugar[c]				
6	Glucose	10 g	no stock	no stock	weigh
	Solidifier[c]				
7	Agar	14 g	no stock	no stock	weigh
8	Water, distilled[c]	to 1000 ml			

[a] Solution containing ammonium and/or nitrate may become contaminated on standing. Therefore, stock solutions should not be made. If made, they must be kept frozen between uses.
[b] Shake well before using.
[c] Add items 1–5 to 800 ml of distilled water (item 8), adjust pH to 4.7, add sugar (item 6), bring volume to 1000 ml with distilled water (item 8), and dissolve sugar. Heat solution to a gentle boil and add agar with stirring. When agar has dissolved distribute medium into culture vessels, autoclave, cool, and use.

Developmental Sequence. Colorless protocorms, up to 3 mm long, form after 4 months in culture.

General Comments. Germination levels are low, and seedling development is very poor. However, this is the only published asymbiotic method for the germination of *C. acaule* seeds. It may serve as a basis for other, improved, procedures. Soaking the seeds in sterilized nutrient solution for 45 days may enhance germination (Kano, 1968).

Cypripedium calceolus

A method for the germination of *Cypripedium calceolus,* a species found in Europe and North America, was developed in Austria (Muick, 1978).

Plant Material. Seeds from ripe capsules should be used.

Surface Sterilization. There is no need to sterilize the seeds.

Culture Vessels. No details are given, but from the context of the article it would appear that pots or nursery beds should be used.

Culture Conditions. Again, no details are given; a protected area outdoors may be suitable.

Culture Medium. Humus, to which were "added such things as coconut milk, banana... yeast and sugar" to "bring about a sufficient amount of mycelia which caused the seed embryos to germinate" (Muick, 1978). This is an important point, and the author suggests that "further questions on details" should be directed to him (F. Muick, Roseggerweg 76, A-8044, Graz, Austria). The seedling medium used was "mycelium-enriched humus without clay at a pH [of] 5.6."

Procedure. The seeds are sown after ripening of the capsules. It is best not to transplant the seedlings (which is a good argument for using pots to hold the humus).

Developmental Sequence. Seedlings reach a length of 2–3 cm (5–7.5 in.) 18 months after sowing. If transplanted to well-rotted humus without clay (pH 5.6), few seedlings survive, and those that do survive develop slowly. Untransplanted seedlings grow much faster and within approximately 2 years have well-developed buds. These develop leaves 3 years after sowing and do not require 10–12 years to flower.

General Comments. This method is successful and its developer has used it to raise seedlings of *C. calceolus* and a cross between it and *C. cordigerum.*

Cypripedium macranthum

Use the procedures recommended for *C. calceolus*.

Cypripedium parviflorum

The procedure suggested for *C. acaule* should be used.

Cypripedium reginae

Seeds of *Cypripedium reginae* can be germinated symbiotically like those of *C. calceolus* and *C. macranthum* (Muick, 1978) or asymbiotically (i.e., axenically) *in vitro* (Harvais, 1973).

Plant Material. Seeds from mature capsules should be used.

Surface Sterilization. The method recommended for tropical species (p. 275) is suitable.

Culture Vessels. Vessels of the types used for *Corallorhiza maculata* (p. 282) or tropical species can be employed.

Culture Conditions. The most suitable temperature is 25° C (77° F). Germination in the dark is much better than under illumination. Seedlings should be moved to light (12-hr photoperiods of 70 lm/0.0929 m^2 provided by Gro-Lux fluorescent lamps) only after they are at least 12 months old. Germination may be improved by covering the seeds with 10–12 mm of agar which contains a darkening agent (Curtis, 1943). A suitable darkening agent may be 0.2% (w/v) graphite.

Culture Medium. A suitable medium has been formulated on the basis of factorial experiments (Table A-11).

Procedure. Soaking the seeds for 45 days in sterile culture medium may improve germination (Kano, 1968). Once the seeds are placed on the medium, the culture vessels should be maintained in the dark for at least 12 months. They may be transferred to light after that, but some seedlings will be lost.

Developmental Sequence. Germination can be observed after 12–14 weeks. Protocorms are formed and roots may appear at the end of 5 months. Mortality may be high during the following 7 months, but the seedlings that survive may produce leaves and roots.

General Comments. This process may produce "uneven populations of protocorms and plantlets," (Harvais, 1973) but it is still a very significant advance in the germination of *Cypripedium* seeds.

Table A-11. Medium for seed germination and seedling culture of *Cypripedium reginae* (Harvais, 1973)

Item number	Component	Amount per liter of culture medium (final concentration in culture medium)	Stock solution (a concentrate prepared for repeated and convenient use)	Volume of stock solution per liter of culture medium	Remarks
	Macroelements				
1	Calcium nitrate, $Ca(NO_3)_2 \cdot 4H_2O$[a]	400 mg	40 g/liter[a]	10 ml[a]	or weigh[a]
2	Potassium phosphate, KH_2PO_4	200 mg	20 g/liter	10 ml	
3	Magnesium sulfate, $MgSO_4 \cdot 7H_2O$	200 mg	20 g/liter	10 ml	
4	Potassium nitrate, KNO_3[a]	200 mg	20 g/liter[a]	10 ml[a]	or weigh[a]
5	Potassium chloride, KCl	100 mg	10 g/liter	10 ml	
6	Ammonium nitrate, NH_4NO_3[a]	400 mg	40 g/liter[a]	10 ml[a]	or weigh[a]
	Iron				
7	Ammonium feric citrate[b] (5 mg/liter Fe)	25 mg	2.5 g/liter[b]	10 ml	
8	*Microelements*[c]				
a	Manganese sulfate, $MnSO_4 \cdot 4H_2O$	0.5 mg	500 mg/liter	1 ml[c]	one solution[c]
b	Boric acid, H_3BO_3	0.5 mg	500 mg/liter		
c	Zinc sulfate, $ZnSO_4 \cdot 7H_2O$	0.5 mg	500 mg/liter		
d	Copper sulfate, $CuSO_4 \cdot 5H_2O$	0.5 mg	500 mg/liter		
	Vitamins[d]				
9	Niacine[e]	5 mg	100 mg/100 ml distilled water[e]	5 ml	
10	Pantothenic acid[e]	0.5 mg	100 mg/100 ml distilled water[e]	0.5 ml	
11	Pyridoxine[e]	0.5 mg	100 mg/100 ml distilled water[e]	0.5 ml	
12	Thiamine[e]	5 mg	100 mg/100 ml distilled water[e]	5 ml	
	Complex additive				
13	Potato extract[e,f]	100 ml	no stock	no stock	
	Sugar[g]				
14	Sucrose	10 g	no stock	no stock	weigh
	Solidifier[g]				
15	Agar	10 g	no stock	no stock	weigh
16	Water, distilled[g]	to 1000 ml			

[a] Solutions containing ammonium and/or nitrate may become contaminated on standing. Therefore, stock solutions should not be prepared. If made, they must be kept frozen between uses.
[b] Shake well before use.
[c] Add all microelements to the same one liter of distilled water, stir or heat until dissolved, and add 1 ml of the combined stock solution to each liter of medium.
[d] May be omitted or can be added to 12-month cultures maintained under light.
[e] Keep frozen between uses.
[f] Prepared as in Table A-8.
[g] Add items 1–13 to 700 ml distilled water (item 16), adjust pH to 5.2, dissolve sugar (item 14) and bring volume to 1000 ml with distilled water (item 16). Heat solution to light boil and add agar while stirring. When agar has dissolved, distribute medium into culture vessels, autoclave, cool, and use.

Literature Cited

Arditti, J., J. D. Michaud, and P. L. Healey. 1979. Morphometry of orchid seeds. I. Paphiopedilum and native California and related species of Cypripedium. Amer. J. Bot. 66:1128–1137.

Curtis, J. T. 1936. The germination of native orchid seeds. Amer. Orchid Soc. Bull. 5:42–47.

——. 1937. Some phases of symbiotic and nonsymbiotic orchid seed germination. Ph.D. diss., Univ. Wisconsin, Madison.

——. 1943. Germination and seedling development in five species of Cypripedium L. Amer. J. Bot. 30:199–206.

Harvais, G. 1973. Growth requirements and development of Cypripedium reginae in axenic culture. Can. J. Bot. 51:327–332.

Kano, K. 1968. Acceleration of the germination of so-called "hard-to-germinate" orchid seeds. Amer. Orchid Soc. Bull. 37:690–698.
Luer, C. A. 1975. The native orchids of the United States and Canada, excluding Florida. New York Botanical Garden, New York.
Muick, F. 1978. Propagation of Cypripedium species from seeds. Amer. Orchid Soc. Bull. 47:306–308.
Wherry, E. T. 1918. The reaction of soils supporting the growth of certain native orchids. J. Wash. Acad. Sci. 8:589–590.
Withner, C. L. 1953. Germination of "Cyps." Orchid J. 2:473–477.

Epipactis gigantea

Several growers in California have succeeded in cultivating *Epipactis gigantea* plants that were collected in the field. Therefore it seems that native terrestrial orchid could be introduced into culture if a seed-germination method were available (collecting the plants is now illegal in California). To meet this need we are trying to develop such a method (J. Arditti, J. D. Michaud, and A. Oliva, unpublished).

Plant Material. Mature and immature seeds from dehisced and green capsules, respectively, can be used.

Surface Sterilization. Immature capsules require sterilization by the procedure outlined for *Calypso bulbosa*. Seeds from mature capsules should be sterilized by soaking them in filtered saturated calcium hypochlorite solution (7 g/100 ml) for 10 min and then rinsed twice with sterile distilled water.

Culture Vessels. The culture vessels used for *Calypso bulbosa* (p. 281) are suitable.

Culture Conditions. Cultures should be maintained under the conditions recommended for *Calypso bulbosa* (p. 281).

Culture Medium. Modified Curtis medium (Table A-7) is suitable.

Procedure. The procedure for immature seeds is the same as for *Calypso bulbosa*. Mature seeds are placed on the medium following sterilization and after being rinsed twice with sterile distilled water.

Developmental Sequence. White protocorms covered with hairs develop within 2 months, and at that time all cultures should be moved to the light. If germination rates are high and cultures become crowded, the protocorms must be transflasked at this time. Rhizomes, first (mostly white) leaves, and some green protocorms appear 3 months later. After an additional 2 months many white and some green plantlets are formed. The green plantlets can be transflasked again and reach a height 3–4 cm by the end of a year.

The white plantlets form big masses, most of which fail to turn green regardless of culture medium or conditions. A few produce small green shoots, which develop into green plantlets. The development of these plantlets (which have separate, green roots) does not affect the white mass of seedlings (whose roots are white) to which they are attached.

Germination rates of mature seeds were lower than those of immature ones. If kept in the dark they produced white, hairy protocorms. When moved to light 5 months after inoculation, they turned green and eventually produced individual seedlings of the same color. These reach a height of 7.5 cm (3 in.) at end of a year.

Like those of *Calypso bulbosa* (p. 282), cultures of *Epipactis gigantea* must be irrigated every 3 months.

General Comments. Seedlings produced by this method have been moved successfully to pots in a greenhouse.

Goodyera

The genus *Goodyera* consists of approximately 20 species and is distributed throughout the world. Of these species, four (all of them terrestrial) are native to the United States and Canada. The seeds of two have been germinated asymbiotically.

Goodyera oblongifolia

A procedure for the germination of *Goodyera oblongifolia* was developed during research on the biology of orchids native to Thunder Bay, Ontario, Canada (Harvais, 1974).

Plant Material. Mature seeds should be used.

Surface Sterilization. The procedures recommended for tropical orchids (p. 275) can be used.

Culture Vessels. Vessels of the kind used for *Corallorhiza maculata* (p. 282) or tropical species (p. 275) are suitable.

Culture Conditions. The cultures should be maintained under the conditions used for *Corallorhiza maculata* (p. 282).

Culture Medium. A modification of the medium used for *Corallorhiza maculata* should be used (Table A-12).

Procedure. Cultures were kept in the dark for 5 months and then transferred to 12-hr photoperiods of 800 lm/0.0929 m^2.

Developmental Sequence. Significant germination occurs within a month, and after 5 months a large proportion of the protocorms develop leaves and roots.

Table A-12. Medium used for seed germination and seedling culture of *Goodyera oblongifolia* (Harvais, 1974)

Item number	Component	Amount per liter of culture medium (final concentration in culture medium)	Stock solution (a concentrate prepared for repeated and convenient use)	Volume of stock solution per liter of culture medium	Remarks
	Macroelements				
1	Calcium nitrate, $Ca(NO_3)_2 \cdot 4H_2O$[a]	400 mg	40 g/liter[a]	10 ml[a]	or weigh[a]
2	Ammonium nitrate, NH_4NO_3[a]	400 mg	40 g/liter[a]	10 ml[a]	or weigh[a]
3	Potassium phosphate, KH_2PO_4	200 mg	20 g/liter	10 ml	
4	Magnesium sulfate, $MgSO_4 \cdot 7H_2O$	200 mg	20 g/liter	10 ml	
5	Potassium nitrate, KNO_3[a]	200 mg	20 g/liter[a]	10 ml[a]	or weigh[a]
6	Potassium chloride, KCl	100 mg	10 g/liter	10 ml	
	Iron				
7	Ammonium ferric citrate[b] (5 mg/literFe)	25 mg	25 g/liter	10 ml	
8	*Microelements*[c]				
a	Manganese sulfate, $MnSO_4 \cdot 4H_2O$	2.03 mg	2.03 g/liter	1 ml[c]	one solution[c]
b	Boric acid, H_3BO_3	0.5 mg	500 mg/liter		
c	Zinc sulfate, $ZnSO_4 \cdot 7H_2O$	0.5 mg	500 mg/liter		
d	Copper sulfate, $CuSO_4 \cdot 5H_2O$	0.5 mg	500 mg/liter		
e	Ammonium molybdate, $Na_2MoO_4 \cdot 2H_2O$	0.02 mg	20 mg/liter		
f	Cobaltous nitrate, $Co(NO_3)_2 \cdot 6H_2O$	0.025 mg	25 mg/liter		
g	Potassium iodide, KI	0.01 mg	10 mg/liter		
	Complex additive				
9	Potato extract[d]	100 ml	no stock	no stock	
	Sugar[e]				
10	Glucose	10 g	no stock	no stock	weigh
	Solidifier[e]				
11	Agar	10 g			
12	Water, distilled[e]	to 1000 ml			

[a] Solutions containing ammonium and/or nitrate may become contaminated on standing. Therefore, stock solutions should not be prepared. If made, they must be kept frozen between uses.

[b] Shake well before use.

[c] Add all microelements to the same one liter of distilled water. Stir or heat until dissolved, and add 1 ml of the stock solution to each liter of culture medium.

[d] Prepare as in Table A-8; keep frozen between uses.

[e] Add items 1–9 to 700 ml of distilled water (item 12), adjust pH to 5.2; add sugar (item 10) and bring volume to 1 liter with distilled water (item 12). Heat to a gentle boil and add agar with stirring. When agar has dissolved, distribute into culture vessels, autoclave, cool, and use.

General Comments. At the time of publication, the cultures were only 7 months old and therefore information is not available on later stages of development.

Literature Cited

Harvais, G. 1974. Notes on the biology of some native orchids of Thunder Bay, their endophytes and symbionts. Can. J. Bot. 52:451–460.

Goodyera pubescens

Under natural conditions seeds of *Goodyera pubescens* do not germinate immediately on being released from the capsule. They germinate during the following year and develop a small protocorm that lives for a year in the humus. The seeds can also be germinated asymbiotically *in vitro* (Knudson, 1941).

Plant Material. Seeds collected in the fall and stored at 4.5° C (40° F) for about 3 months were used in the original research.

Surface Sterilization. The procedures recommended for tropical orchids (p. 275) are suitable.

Culture Vessels. Erlenmeyer flasks, culture tubes, and vessels of the kind used for tropical orchids (p. 275) can be used.

Culture Conditions. About 20–24° C (68–75° F) and 600 ft-candles constitute appropriate culture conditions.

Culture Medium. Knudson B medium adjusted to pH 5.0 and with 20 g glucose instead of sucrose (Table A-6, item 7) should be used.

Procedure. The procedures employed for tropical species are appropriate.

Developmental Sequence. White protocorms, 1 mm long and 0.5 mm wide, are formed 4 months after sowing. The first leaf primordia appear at the apical end. Rhizoids are formed early in the germination process. Ten months after sowing the seeds, protocorms have well-developed, long rhizoids but are still white except at the tips, which are turning green.

General Comments. The original article does not present information on development past the 10-month stage.

Literature Cited

Knudson, L. 1941. Germination of seed of Goodyera pubescens. Amer. Orchid Soc. Bull. 9:199–201.

Goodyera repens

This species has been germinated by the method used for *Cypripedium acaule* (p. 284).

Habenaria

The species germinated as belonging to the genus *Habenaria* have been transferred to *Platanthera* (Luer, 1975). They are *Platanthera (Habenaria) dilatata, P.* (*H.*) *hyperborea, P.* (*H.*) *obtusata, and P.* (*H.*) *psychodes* (Harvais, 1974).

Plant Material. Mature seeds were used in the original research.

Surface Sterilization. The method used for *Corallorhiza maculata* (p. 282) should be employed.

Culture Vessels. Vessels of the kind used for *Corallorhiza maculata* (p. 282) can be used.

Culture Conditions. The conditions used for the culture of *Corallorhiza maculata* seeds (p. 282) are suitable.

Culture Medium. Platanthera dilatata, P. obtusata, and *P. psychodes* can be germinated on the medium used for *Corallorhiza maculata* (Table A-9). For *Platanthera hyperborea* the medium must be modified by the addition of 5 g/liter of Bacto Casamino Acids Certified (Difco).

Procedure. The procedures used for *Corallorhiza maculata* (p. 282) can be used.

Developmental Sequence. After 10 months of culture, 32% of *Platanthera dilatata* seeds germinated and starch levels were very high. *P. hyperborea* had 54% germination and very high starch levels in 8 months. The figures for *P. obtusata* (very high starch levels) and *P. psychodes* (moderately high starch) after 9 months were 21% and 10%, respectively. A direct correlation was noted between percent germination and subsequent growth. Seedlings of *P. dilatata* and *P. obtusata* develop many rhizoids.

General Comments. It is interesting to note that of the four species tested, three germinate on one medium and the fourth has different requirements.

Literature Cited

Harvais, G. 1974. Notes on the biology of some native orchids of Thunder Bay, their endophytes and symbionts. Can. J. Bot. 52:451–460.

Luer, C. A. 1975. The native orchids of the United States and Canada, excluding Florida. New York Botanical Garden, New York.

Malaxis unifolia

A terrestrial species distributed from Newfoundland to Mexico, Honduras, and the Caribbean, *Malaxis unifolia* has been germinated on the medium used for *Cypripedium acaule* (Table A-10).

Pogonia ophioglossoides

The methods used for *Cypripedium acaule* (p. 284; Table A-10) can be used for this species.

Spiranthes

Three species, *Spiranthes cernua, S. gracilis,* and *S. romanzoffiana,* have been germinated by the methods employed for *Cypripedium acaule* (p. 284; Table A-10).

Other North American Terrestrial Species

Photographs depicting the germination of other species have been published (Stoutamire, 1974).

Literature Cited

Stoutamire, W. 1974. Terrestrial orchid seedlings, p. 101–128. *In* C. L. Withner (ed.), The orchids: Scientific studies. Wiley-Interscience, New York.

Arundina bambusifolia

Joseph Arditti

Arundina is a monotypic terrestrial genus found in Ceylon, northern India, southern China, Malaysia, and several Pacific islands. A number of taxa have been described but on closer examination all appear to be hybrids of or to belong to one highly variable species which extends from Sikkim to Java. The oldest name applied to it, *Bletia graminifolia* (1825), was changed a few years later to *Arundina bambusifolia.* Consequently, the most appropriate name at present would appear to be *Arundina graminifolia* (Holttum, 1957). It probably includes *A. speciosa* described from Java by Blume; *A. chinensis* also by Blume; *A. densa* described from Singapore by Lindley; *A. minor* from Sri Lanka (then Ceylon); *A. revoluta* from Taiping; a white form, *A. sanderiana,* from Sumatra; and additional forms from Annam, Carolina Islands, Celebes, China, and Tahiti (Holttum, 1957). Thus the procedures described here could probably be used with all *Arundina* seeds (Mitra, 1971).

Plant Material. Seeds from dehisced and undehisced (i.e., immature or green) capsules were used in the original research.

Surface Sterilization. Undehisced capsules should be washed with soap and water, dipped in absolute (100%) or 95% ethyl alcohol (ethanol) for 12 sec, soaked in saturated (7 g/100 ml) calcium hypochlorite (pp. 254, 261, 275) for 45 min, and rinsed three times with sterile distilled water.

Seeds from dehisced (i.e., mature) capsules can be sterilized by soaking them in saturated calcium hypochlorite for 10 min and then rinsing three times with sterile distilled water.

Culture Vessels. Wide-mouth 150-ml Erlenmeyer flasks containing 50 ml medium and 180 × 25 mm culture tubes with 25 ml solution were used in the original research. Other culture vessels can also be used.

Culture Conditions. During the original research, seed and seedling cultures were maintained under a temperature of 26° ± 1° C and 12-hr photoperiods of 3000 lux provided by Philips "Natural" fluorescent tubes. Gro-Lux or warm white fluorescent tubes or combinations of other fluorescent lights and incandescent bulbs could also be used to provide similar illumination.

Culture Medium. Knudson C, Vacin and Went, and Raghavan and Torrey media were used in the original research. The Raghavan and Torrey medium (Table A-13) is the most suitable for seeds from both dehisced and undehisced capsules.

Procedure. Seeds are placed on the culture medium following surface sterilization. The seedlings should be transflasked when they develop leaves and roots and the cultures appear crowded.

Developmental Sequence. Seeds from mature and immature capsules germinate after 18 and 41 days, respectively, and form protocorms. Leaves and roots form 20 days following germination. After 20 weeks the seedlings may develop 6 to 8 leaves and 4 or 5 roots.

Table A-13. Raghavan and Torrey (1964) medium as used for the germination of mature and immature seeds of *Arundina bambusifolia* (Mitra, 1971)

Item number	Component	Amount per liter of culture medium (final concentration in culture medium)	Stock solution (a concentrate prepared for repeated and convenient use)	Volume of stock solution per liter of culture medium	Remarks
	Macroelements				
1	Potassium phosphate, KH_2PO_4	270 mg	27 g/liter	10 ml	
2	Magnesium sulfate, $MgSO_4 \cdot 7H_2O$	240 mg	24 g/liter	10 ml	
3	Calcium sulfate, $CaSO_4 \cdot 2H_2O$	80 mg	8 g/liter	10 ml	
4	Calcium phosphate, $CaH_4(PO_4)_2$	100 mg	10 g/liter	10 ml	
5	Ammonium nitrate, NH_4NO_3[a]	80 mg	8 g/liter[a]	10 ml[a]	or weigh[a]
6	*Microelements*[b]				
a	Boric acid, H_3BO_3	0.60 mg	600 mg/liter	1 ml[b]	one solution[b]
b	Manganese chloride, $MnCl_2 \cdot 4H_2O$	0.40 mg	400 mg/liter		
c	Zinc sulfate, $ZnSO_4 \cdot H_2O$	0.05 mg	50 mg/liter		
d	Copper sulfate, $CuSO_4 \cdot H_2O$	0.05 mg	50 mg/liter		
e	Sodium molybdate $Na_2MoO_4 \cdot 2H_2O$	0.05 mg	50 mg/liter		
f	Potassium iodide, KI	0.03 mg	30 mg/liter		
g	Cobaltous nitrate, $Co(NO_3)_2$	0.05 mg	50 mg/liter		
	Iron				
7	Ferric tartrate, $Fe_2(C_4H_4O_6)_3$[c]	3 mg	300 mg[c]	10 ml[c]	
	Sugar[d]				
8	Sucrose	20 g	no stock	no stock	weigh
9	Water, distilled[d]	to 1000 ml			
	Solidifier[d]				
10	Agar	9 g	no stock	no stock	weigh

[a] Solutions of nitrogenous salts or organic substances may become contaminated. Therefore, stock solutions should not be prepared. If made, they must be kept frozen between uses.

[b] Add all microelements to the same one liter of distilled water; stir and/or heat until dissolved. Add 1 ml to culture medium.

[c] Ferric tartrate is relatively insoluble. Grinding it with a mortar and pestle before dissolving helps. The addition of a pellet or two of KOH to the solution will increase solubility, but a precipitate may form nevertheless. To ensure equal distribution, shake stock solution well before dispensing.

[d] Dissolve items 1–7 in 500 ml distilled water (item 9); adjust the pH to 5.2–5.5; add the agar (item 10) slowly, while stirring, to the gently boiling solution. When fully dissolved, dispense into culture vessels, autoclave, cool, and use.

General Comments. It is interesting to note that *A. bambusifolia* seed do not germinate well on Knudson C. One reason for this may be the fact that this is a terrestrial species.

Literature Cited

Holttum, R. E. 1957. A revised flora of Malaya. Vol. I. Orchids of Malaya, 2d ed. Government Printing Office, Singapore.
Mitra, G. C. 1971. Studies on seeds, shoot tips, and stem discs of an orchid grown in aseptic culture. Indian J. Exp. Biol. 9:79–85.
Raghavan, V., and J. G. Torrey. 1964. Inorganic nitrogen nutrition of the seedlings of the orchid Cattleya. Amer. J. Bot. 51:264–274.

Australian Native Orchids (Epiphytic and Terrestrial)

Mark A. Clements

Epiphytic Species

Apparently the first published reference to the germination of Australian epiphytic species by asymbiotic methods is to *Phalaenopsis amabilis* (Foote, 1957). While there is a decided lack of published information, unpublished work (Clements and Ellyard) has shown that a large number of Australian epiphytic species (Table A-14) can be successfully germinated on Knudson C. However, some species, particularly those belonging to the subtribe Sarcanthinae, do best when germinated and grown on Vacin and Went medium (Table A-15).

Table A-14. Epiphytic Australian species germinated asymbiotically on Knudson C medium

Species	*Reference*
Arthrochilus irritabilis (syn. *Spiculaea irritabilis*)	Stoutamire, 1964
Caladenia catenata (syn. *C. carnea*)	Stoutamire, 1964; McIntyre *et al.*, 1972
C. patersonii	Stoutamire, 1964
Calanthe triplicata	Unpub.
Caleana major	Stoutamire, 1964
Calochilus robertsonii	Stoutamire, 1964
Chiloglottis gunnii	Unpub.
Diuris sheaffiana	Unpub.
Geodorum densiflorum	Unpub.
Glossodia major	Unpub.
Habenaria ochroleuca	Unpub.
Microtis media	Stoutamire, 1964
M. parviflora	Unpub.
M. unifolia	Unpub.
Phaius australis var. *bernaysii*	Unpub.
P. tancarvilliae	Unpub.
Prasophyllum archeri	Unpub.
P. aureoviride	Unpub.
P. australe	Stoutamire, 1974
P. macrostachyum var. *ringens*	Unpub.

continued

Table A-14.—Continued

Pterostylis concinna	McIntyre *et al., 1972*
P. curta	Unpub. Stoutamire, 1974
P. falcata	Stoutamire, 1974
P. nutans	Stoutamire, 1964
P. parviflora	McIntyre *et al.*, 1972
P. rufa	Stoutamire, 1974
Spathoglottis paulinae	Unpub.
Spathoglottis sp.	Unpub.
Thelymitra carnea	McIntyre *et al.*, 1972
T. chasmogama	McIntyre *et al.*, 1972
T. flexuosa	Stoutamire, 1964
T. fuscolutea	Stoutamire, 1964
T. ixioides	Stoutamire, 1964
T. mucida	Unpub.
T. nuda (as *T. aristata* McIn.)	McIntyre *et al.*, 1972
T. pauciflora	McIntyre *et al.*, 1972 Unpub. Stoutamire, 1964
T. rubra (as *Calochilus robertsonii* McIn.)	McIntyre *et al.*, 1972 Stoutamire, 1964
T. venosa	Unpub.

Terrestrial Species

Asymbiotic Germination

Approximately 400 species of terrestrial orchids are native to Australia. In general they have proved particularly difficult to germinate *in vitro* using asymbiotic methods. A number of species, however, have been germinated on Knudson C (Table A-16; Stoutamire, 1963, 1964, 1974; McIntyre *et al.*, 1972; Clements, unpublished data). *Phaius, Calanthe,* and *Habenaria* germinate readily, and some success can be expected with

Table A-15. Vacin and Went (1949) medium for the culture of members of the Australian Sarcanthinae

Item number	Component	Amount per liter of culture medium (final concentration in culture medium)	Stock solution (a concentrate prepared for repeated and convenient use)	Volume of stock solution per liter of culture medium
	Macroelements			
1	Tricalcium phosphate, $Ca_3(PO_4)_2$	200 mg	no stock	no stock
2	Potassium nitrate, KNO_3	525 mg	no stock	no stock
3	Potassium phosphate, KH_2PO_4	250 mg	no stock	no stock
4	Magnesium sulfate, $MgSO_4 \cdot 7H_2O$	250 mg	no stock	no stock
5	Ammonium sulfate $(NH_4)_2\ SO_4$	500 mg	no stock	no stock
6	Ferric tartrate, $Fe_2(C_4H_4O_6)_3$[a]	28 mg	2.8 g/liter[a]	10 ml[a]
	Microelement			
7	Manganese sulfate, $MnSO_4 \cdot 4H_2O$[b]	7.5 mg	750 mg/liter[b]	10 ml[b]
	Sugar			
8	Sucrose	20 g	no stock	no stock
9	Water, distilled[c]	to 1000 ml	no stock	no stock
	Solidifier[c]			
10	Agar	80 g	no stock	no stock

[a] Stock solution kept refrigerated between use.
[b] Stock solution kept frozen between use.
[c] Mix items 1–8 in 950 ml of distilled water (item 9); adjust pH to 5.2 and then adjust volume to 1500 ml with distilled water (item 9), bring solution to a slow boil, add agar (item 10) while stirring, dispense into culture vessels, and autoclave.

Table A-16. Terrestrial Australian species germinated asymbiotically on Knudson C medium

Acriopsis javanica
var. *nelsoniana*

Bulbophyllum aurantiacum
B. baileyi
B. bowkettae
B. bracteatum
B. elisae
B. johnsonii
B. macphersonii
B. newportii
B. wadsworthii
B. weinthalii

Cadetia maideniana
C. taylori
Chiloschista phyllorhiza
Cymbidium canaliculatum
C. madidum
C. suave

Dendrobium aemulum
D. antennatum
D. bairdianum
D. bifalce
D. bigibbum
var. *bigibbum*
D. bigibbum
var. *superbum*
D. canaliculatum
var. *canaliculatum*
D. cancroides
D. cucumerinum
D. × *delicatum*
D. discolor
D. falcorostrum
D. fleckeri
D. × *foederatum*
D. gracilicaule
var. *gracilicaule*
D. gracilicaule
var. *howeanum*
D. × *grimesii*
D. johannis
D. kingianum
D. lichenastrum
var. *lichenastrum*
D. linguiforme
var. *linguiforme*
D. linguiforme
var. *nugentii*
D. malbrownii
D. monophyllum
D. moorei
D. mortii
D. nindii
D. rigidum
D. ruppianum
var. *ruppianum*
D. schneiderae
D. smilliae
D. speciosum
var. *speciosum*
D. stuartii
D. tenuissimum
D. teretifolium
var. *teretifolium*
D. tetragonum
var. *tetragonum*
D. tetragonum
var. *giganteum*
D. wassellii
D. wilkianum
D. tozerensis
Diplocaulobium glabrum
Drymoanthus minutus

Ephemerantha comata
E. convexa
Eria eriaeoides
E. fitzalanii
E. inornata
E. irukandjiana
E. queenslandica

Liparis nugentae
L. reflexa
Luisia teretifolia

Micropera fasciculata
Mobilabium hamatum

Oberonia palmicola

Phalaenopsis amabilis
var. *papuana*
Pholidota pallida
Phreatia robusta
Plectorrhiza tridentata
Pomatocalpa macphersonii
P. marsupiale
Pteroceras hirticalcar
P. spathulatus
Rhinerrhiza divitiflora

Sarcochilus ceciliae
S. falcatus
S. fitzgeraldii
S. hartmannii
S. moorei

Taeniophyllum glandulosum
T. malianum
Thrixspermum congestum
T. platystachys
Trichoglottis australiensis

Vanda tricolor
V. whiteana

Thelymitra, Diuris, Prasophyllum, and *Microtis.* With these latter four genera Knudson C (Table A-5) at half strength or Zak medium (Table A-17) is recommended.

Plant Material. Preferably, capsules should be collected just prior to splitting, when they begin to turn yellow. Mature seed may also be used.

Surface Sterilization. Green capsules should be sterilized as outlined for tropical orchids (p. 275) or in 1% sodium hypochlorite for 10 min. Mature seed may be sterilized with

Table A-17. Zak medium for the symbiotic germination of orchid seeds[a]

Item number	Component	Amount per liter of culture medium (final concentration in culture medium)
	Sugar	
1	Sucrose	10 g
	Complex additive	
2	Coconut milk (ripe)	20 g
3	Water, distilled[b]	to 1000 ml
	Solidifier	
4	Agar	8 g

[a] Personal communications, G. Mundey, England. Medium formerly used for asymbiotic germination of European terrestrial orchids.

[b] To prepare medium, mix items 1 and 2, bring volume to 900 ml with distilled water (item 3), adjust pH to 5.5, adjust to 1000 ml with more distilled water (item 3), heat to 94°C, then add agar (item 4) slowly; distribute into culture vessels and autoclave.

0.5% sodium hypochlorite for 2 to 5 min and washed with sterile distilled water prior to sowing on the agar medium.

Culture Vessels. Use 105 × 28 mm flat-bottom test tubes containing 20 ml medium.

Culture Conditions. With *Phaius, Calanthe,* and *Habenaria* the culture conditions should be as outlined for tropical orchids (p. 275). Cultures of other genera should be incubated in the dark at 20° C until protocorms are formed. The cultures should then be shifted into diffused daylight or artificial light environment of moderate intensity (150–300 ft candles).

Culture Medium. Knudson C (Table A-5) is suitable. The addition of banana (150 g/liter) and charcoal (1 g/liter) to media used for the transfer of seedlings has proved beneficial.

Procedure. The general procedures outlined for tropical orchids (p. 276) are suitable.

Developmental Sequence. Germination is first evidenced by enlargement of the embryo and splitting of the seed coat. The embryo swells, forming a protocorm with rhizoids protruding. These rhizoids are the nutrient-absorbing organs. When large enough, leaves begin to develop from the growing apex of the protocorm. Roots subsequently emerge from near the base of the leaves. When both leaves and roots are present the plant can best be described as a seedling although the protocorm may persist for many months before it finally senesces.

General Comments. Seedlings grown by this method may produce abnormally shaped tubers and leaves. However, these abnormalities disappear soon after the seedling is transferred to potting mix. Plants large enough for transfer to potting mix have been produced in less than 12 months. This propagation method can be used with northern Australian terrestrial species whose origin appears to be from New Guinea and Indonesia. In contrast, many species of southern Australian terrestrials grown by this method may take more than 2 years before they are large enough for transfer to potting mix.

Symbiotic Germination

Excellent germination of a number of Australian terrestrial species (Table A-18) has been achieved using symbiotic methods (Warcup 1971, 1973; Clements and Ellyard, 1979, unpublished data). This technique produces much quicker germination and in many cases results in the germination of species which fail to germinate under asymbiotic conditions.

Isolation of Fungus. Root, rhizome, or underground stem tissue is selected from orchid plants collected in the wild. Experience has shown that the fungus normally occurs in underground tissue at or near the ground surface (Figs. A-4, A-5). The optimal time for isolation from most of these plants is during the normal rapid vegetative growth period prior to flowering. The tissue containing the fungus is washed in tap water to remove all traces of soil and a thin sliver is removed and placed under a ×12 or ×25 binocular

Table A-18. Australian terrestrial species germinated symbiotically

Species	Medium used	Source of fungus/isolator	Reference
Caladenia alba	War MI	/Warcup	Unpub.
C. barbarossa	Oat	/Warcup	Unpub.
C. catenata (syn. *carnea*)	Czapek-Dox mins/ cellulose	/Warcup	Warcup, 1971
Diuris aequalis	Oat, Zak	*Caladenia reticulata*/Warcup	Unpub.
D. brevifolia	Oat	*Caladenia reticulata*/Warcup	Unpub.
D. laxiflora	Oat	*Caladenia reticulata*/Warcup	Unpub.
D. longifolia	War MI	*Caladenia reticulata*/Warcup	Warcup, 1973
		Thelymitra aristata/Warcup	
D. maculata	Oat	*Caladenia reticulata*/Warcup	Unpub.
D. pedunculata	Oat	*Caladenia reticulata*/Warcup	Unpub.
D. punctata var. *punctata*	Oat, Zak	*Caladenia reticulata*/Warcup	Unpub.
D. punctata var. *albo-violacea*	Oat, Zak, War MI	*Caladenia reticulata*/Warcup	Clements and Ellyard, 1979
D. punctata var. *longissima*	Oat, Zak	*Caladenia reticulata*/Warcup	Clements and Ellyard, 1979
D. sheaffiana	Oat	*Caladenia reticulata*/Warcup	Unpub.
D. venosa	War MI	*Caladenia reticulata*/Warcup	Unpub.
Malaxis latifolia	Oat	*Malaxis latifolia*/Clements	Unpub.
Pterostylis allantoidea	Oat	*Pterostylis allantoidea*/Clements	Unpub.
P. biseta	War MI	*Pterostylis hamata*/Clements	Unpub.
P. boormanii	War MI	*Pterostylis hamata*/Clements	Clements and Ellyard, 1979
P. ceriflora	War MI	*Pterostylis hamata*/Clements	Unpub.
P. cucullata	War MI	*Pterostylis vittata*/Warcup	Unpub.
P. curta	War MI	*Pterostylis vittata*/Warcup	Warcup, 1973
		Soil, wheat field/Warcup	Clements and Ellyard, 1979
P. fischii	Zak	*Pterostylis revoluta*/Clements	Unpub.
P. hamata	War MI	*Pterostylis hamata*/Clements	Clements and Ellyard, 1979
P. × *ingens*	War MI	*Pterostylis vittata*/Warcup	Clements and Ellyard, 1979
P. laxa	Zak	*Pterostylis hamata*/Warcup	Unpub.
P. longifolia	War MI	*Pterostylis vittata*/Warcup	Clements and Ellyard, 1979
P. mutica	War MI	*Pterostylis vittata*/Warcup	Unpub.
		Pterostylis hamata/Clements	Unpub.
P. nutans	War MI	*Pterostylis vittata*/Warcup	Warcup, 1973
		Soil, wheat field/Warcup	Warcup, 1973
P. pulchella	Oat, Zak	*Pterostylis vittata*/Warcup	Unpub.
P. recurva	Oat	*Pterostylis recurva*/Clements	Unpub.
P. rufa subsp. *aciculiformis*	War MI	*Pterostylis hamata*/Clements	Unpub.
P. scabra var. *robusta*	War MI	*Pterostylis* sp. nov./Clements	Unpub.
		Pterostylis hamata/Clements	Unpub.
P. vittata var. *vittata*	War MI, Oat	*Pterostylis vittata*/Warcup	Clements and Ellyard, 1979
		Pterostylis hamata/Warcup	Clements and Ellyard, 1979
		Pterostylis vittata/Warcup	Unpub.
		Pterostylis rufa/Clements	Unpub.
Spathoglottis sp.	Oat	*Malaxis latifolia*/Clements	Unpub.
Thelymitra aristata (syn. *T. grandiflora*)[a]	War MI	*Thelymitra epipactoides*/Warcup	Warcup, 1973
		Thelymitra epipactoides/Warcup	Warcup, 1973
		Dendrobium dicuphum/Warcup	Warcup, 1973
T. flexuousa	War MI	*Caladenia reticulata*/Warcup	Unpub.
T. fusco-lutea	War MI	*Caladenia reticulata*/Warcup	Warcup, 1973
		Thelymitra nuda/Warcup	Warcup, 1973
		Dendrobium dicuphum/Warcup	Warcup, 1973
T. longifolia	War MI	*Thelymitra nuda*/Warcup	Warcup, 1973
		Caladenia reticulata/Warcup	Warcup, 1973
T. luteocilium	War MI, Zak	*Caladenia reticulata*/Warcup	Warcup, 1973; Unpub.
	War MI	*Thelymitra nuda*/Warcup	Warcup, 1973
	War MI	*Paphiopedilum* sp. (cult.)/ Warcup	Warcup, 1973
	War MI	*Dendrobium dicuphum*/Warcup	Warcup, 1973
T. macmillanii	Zak	*Caladenia reticulata*/Warcup	Unpub.
T. pauciflora	Oat, Zak	*Caladenia reticulata*/Warcup	Unpub.
		Diuris maculata/Clements	Unpub.
T. rubra	Oat	*Caladenia reticulata*/Warcup	Unpub.
T. venosa	Oat, Zak	*Caladenia reticulata*/Warcup	Unpub.

[a] The name *Thelymitra aristata* Lindl. now applies to the species formerly known as *T. grandiflora* R. Fitz., and all plants formerly listed under this name (*T. aristata*) belong to the *T. nuda* R.Br. complex.

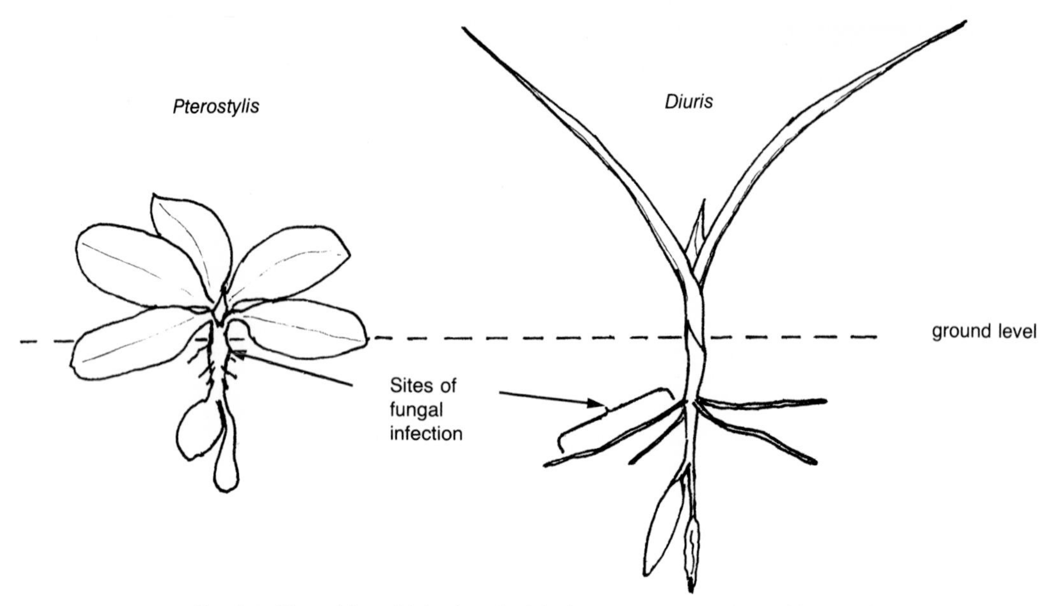

Fig. A-4. Sites of fungal infection of adult plants. (Drawing by Mark Clements.)

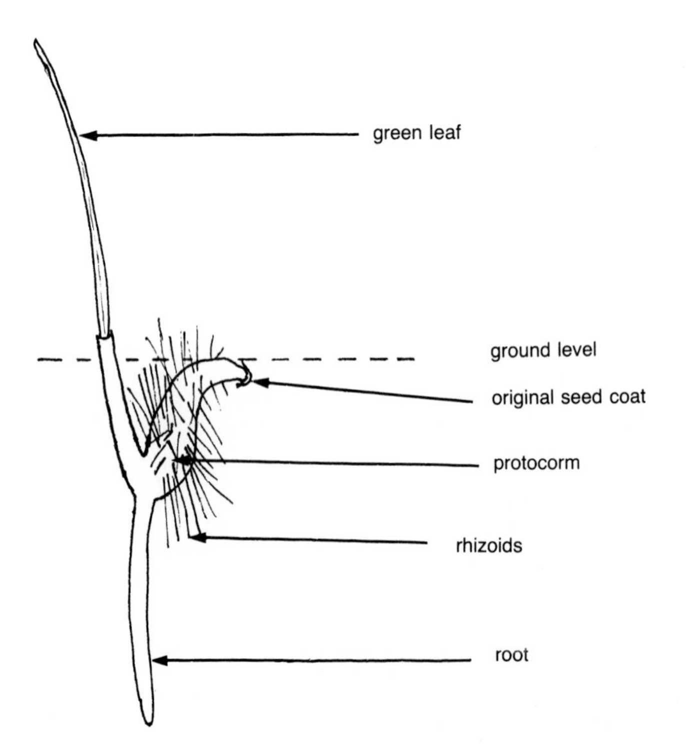

Fig. A-5. Diuris seedling structure; approximately ×10 natural size. (Drawing by Mark Clements.)

microscope. Using a sterile probe and surgical blade tip, one to several fungal hyphae are dissected out and placed in a petri dish containing solidified or partially solidified isolation medium (Table A-19) and cultured at 20° ± 3° C. Subcultures are usually necessary to obtain pure cultures. Pure cultures are stored on the same medium under mineral oil in a cool cupboard.

Surface Sterilization. Green capsules and mature seed are sterilized as outlined above for asymbiotic germination. The sterilizing solution containing the seed is then diluted ×5 with sterile distilled water and the seed collected on a sterile disc of filter paper fitted to a vacuum filter. A further 5–10 ml of sterile distilled water is passed through the filter to remove all traces of sterilizing agent.

Culture Vessels. Test tubes 200 × 30 mm are suitable, containing 50 ml medium.

Culture Conditions. Incubation is carried out at 20° ± 3° C in the dark. Once the protocorm has differentiated to form a leaf the tube is transferred to a 16-hr light/8-hr dark regime and grown until of a size suitable for transfer to a soil medium. Seedlings generally do not require transflasking.

Culture Media. The best germination and growth are obtained on an oat medium (Table A-20). The tubes containing medium are autoclaved at 121° C at one atmosphere for 20 min prior to use and allowed to cool and set as agar slopes. War MI medium

Table A-19. Isolation medium (Clements and Ellyard, 1979) for fungal cultures

Item number	Component	Amount per liter of culture medium (final concentration in culture medium)
	Microelements	
1	Sodium nitrate, $NaNO_3$	0.3 g
2	Potassium phospate, KH_2PO_4	0.2 g
3	Magnesium sulfate, $MgSO_4 \cdot 7H_2O$	0.1 g
4	Potassium chloride, KCl	0.1 g
	Complex additive	
5	Yeast extract (Difco)	0.1 g
	Sugar	
6	Sucrose	5 g
	Antibiotic	
7	Streptomycin sulfate	0.05 g
8	Water, distilled[a]	to 1000 ml
	Solidifier	
9	Agar	10–11 g

[a] To prepare medium mix components 1–8, adjust pH to 4.0–5.0, heat to 94°C, and add agar (item 9) slowly; distribute into culture vessels and autoclave.

Table A-20. Oat medium (Clements and Ellyard, 1979) for the symbiotic germination of orchid seeds

Item number	Component	Amount per liter of culture medium (final concentration in culture medium)
	Complex additive	
1	Whole oats (rolled) blended to powder	2.5 g
2	Water, distilled[a]	to 1000 ml
	Solidifier	
3	Agar	7 g

[a] To prepare medium, mix items 1 and 2, adjust pH to 5.5, heat to 94°C, and add agar (item 3) slowly; distribute into culture vessels and autoclave.

(Table A-21; Warcup, 1973) is also suitable for some species, as is a combination of Czapek-Dox minerals and other additives (Table A-22).

Procedure. The sterile filter paper containing the seed is placed on the agar slope; a small (25 × 25 mm) inoculum of the appropriate fungus culture is added to the edge of the filter paper and the tube plugged with sterile cotton wool. Germination usually occurs within 10–20 days.

Developmental Sequence. The fungus takes several days to reach and inoculate the seeds. Approximately 48 hr or more after inoculation the embryo starts to swell and split

Table A-21. War MI medium (Warcup, 1973) for the symbiotic germination of orchid seeds

Item number	Component	Amount per liter of culture medium (final concentration in culture medium)
	Macroelements	
1	Sodium nitrate, $NaNO_3$	0.3 g
2	Potassium phosphate, KH_2PO_4	0.2 g
3	Magnesium sulfate, $MgSO_4 \cdot 7H_2O$	0.1 g
4	Potassium chloride, KCl	0.1 g
	Complex additive	
5	Yeast extract (Difco)	0.1 g
	Carbon source	
6	Cellulose powder (Whatman Column Chromedia CF11)[a]	20 g
7	Water, distilled[b]	to 1000 ml
	Solidifier[b]	
8	Agar	10 g

[a] Obtainable from Whatman, Inc.; chromatography grade.
[b] To prepare medium, mix items 1–6 in 950 ml distilled water (item 7), adjust pH to 5.5, bring water volume to 1000 ml, heat to 94°C, and add agar (item 8) slowly; distribute into culture vessels and autoclave.

Table A-22. Czapek-Dox minerals (Warcup, 1971) plus yeast extract and powdered cellulose, for the symbiotic germination of orchid seeds

Item number	Component	Amount per liter of culture medium (final concentration in culture medium)	Remarks
	Complex additives		
1	Czapek-Dox minerals[a]	—[a]	quantity not given
2	Yeast extract (Difco)	0.5 g	0.05%
	Carbon source		
3	Cellulose powder (Whatman Column Chromedia CF11)[b]	20 g	2.0%
4	Water, distilled[c]	to 1000 ml	
	Solidifier[c]		
5	Agar	10–12 g	

[a] Dr. J. H. Warcup, Waite Agric. Res. Inst., Univ. Adelaide, Glen Osmond, So. Australia.
[b] Obtainable from Whatman, Inc.; chromatography grade.
[c] To prepare medium, mix items 1–3, adjust pH to 5.5, heat to 94°C and add agar (item 5) slowly, distribute into culture vessels and autoclave.

the seed coat, indicating that germination has occurred. The embryo continues to enlarge, forming the colorless protocorm from which protrude many rhizoids covered in fungus. Differentiation is then observed at the growing tip, resulting in the formation of the first green leaf. Several leaves may emerge prior to the emergence of a root. A single seedling may produce several roots, the largest of these eventually forming a tuber at its apex (Figs. A-3, A-4).

The sequence of events in the development of the protocorm and subsequent leaf and tuber formations is thought to assist in the establishment of these plants in the wild.

General Comments. The presence of both leaves and roots on a seedling indicates that the plant is sufficiently large for transfer to a potting mix. Seedlings normally take from 3 to 6 months to develop using the symbiotic method. The production of a tuber while the seedlings are still in the culture media is not desirable owing to difficulties in controlling this aspect of the plants' growth sequence. It can sometimes be induced, however, by placing the plants under mild stress such as drying the media out slowly.

Literature Cited

Clements, M. A., and R. K. Ellyard. 1979. The symbiotic germination of Australian orchids. Amer. Orchid Soc. Bull. 48:810–816.

Foote, M. 1957. Raising orchids from seed. Aust. Orchid Review 22:26–27.

Knudson, L. 1946. A new nutrient for the germination of orchid seeds. Amer. Orchid Soc. Bull. 15:214–217.

McIntyre, D. K., G. I. Veitch, and J. W. Wrigley. 1972. Australian terrestrial orchids from seed. Aust. Plants 6:250–255.

Stoutamire, W. P. 1963. Terrestrial orchid seedlings. Aust. Plants 2:119–122.

——. 1964. Terrestrial orchid seedlings II. Aust. Plants 2:264–266.

——. 1974. Terrestrial orchid seedlings, p. 101–128. *In* C. L. Withner (ed.), The orchids: Scientific studies. Wiley-Interscience, New York.

Vacin, F., and F. W. Went. 1949. Some pH changes in nutrient solutions. Bot. Gaz. 110:605–613.

Warcup, J. H. 1971. Specificity of mycorrhizal association in some Australian terrestrial orchids. New Phytol. 70:41–46.

——. 1973. Symbiotic germination of some Australian terrestrial orchids. New Phytol. 72:387–392.

Bletilla striata

Joseph Arditti

Bletilla striata, sometimes called *Bletilla hyacinthina,* is a terrestrial species native to Taiwan, China, and Japan. It has been germinated *in vitro* in England (Smith, 1973) and Japan (Ichihashi and Yamashita, 1977).

Plant Material. Mature seeds should be used.

Surface Sterilization. The seeds can be sterilized by soaking in either calcium hypochlorite solution (7 g/100 ml water) for 15–20 min or 0.5–0.8% sodium hypochlorite (1 part household bleach containing 5.25% sodium hypochlorite to 5.5 or 9 parts water) for 5 min. After sterilization the seeds must be washed several times with sterile distilled water.

Culture Vessels. Erlenmeyer flasks, 50-ml capacity, containing 20 ml medium or the vessels used for tropical orchids (p. 275) can be used for *Bletilla striata.*

Culture Conditions. Cultures may be incubated at 22–26° C in the dark, under 400 lux illumination, or under the conditions employed for *Paphiopedilum* (p. 352).

Culture Media. Modified Knudson B medium (Table A-23) or a newly devised medium (Table A-24) may be used. In view of the discovery that *Bletilla striata* seed do not grow

Table A-23. Modified Knudson B medium (Knudson, 1946) for seed germination and seedling culture of *Bletilla striata* (Smith, 1973)

Item number	Component	Amount per liter of culture medium (final concentration in culture medium)	Stock solution (a concentrate prepared for repeated and convenient use)	Volume of stock solution per liter of culture medium	Remarks
	Macroelements				
1	Calcium nitrate, $Ca(NO_3)_2 \cdot 4H_2O$[a]	1 g	100 g/liter[a]	10 ml[a]	or weigh[a]
2	Potassium phosphate, KH_2PO_4[b]	250 mg	25 g/liter[b]	10 ml[b]	
3	Magnesium sulfate, $MgSO_4 \cdot 7H_2O$	250 mg	25 g/liter	10 ml	
4	Ammonium sulfate $(NH_4)_2\ SO_4$[a]	500 mg	50 g/liter[a]	10 ml[a]	or weigh[a]
5	Ferric phosphate, $FePO_4 \cdot 4H_2O$[c]	25 mg	2.5 g/liter[c]	10 ml[c]	
6	Manganese sulfate, $MnSO_4 \cdot 4H_2O$	7.5 mg	750 mg/liter	10 ml	
	Sugar[d]				
7	Glucose	15 g	no stock	no stock	weigh
	Solidifier[d]				
8	Agar	15 g	no stock	no stock	weigh
9	Water, distilled[d]	to 1000 ml			

[a] Solutions containing ammonium and/or nitrate may become contaminated on standing. Therefore, stock solutions should not be prepared. If made, they must be kept frozen between uses.

[b] A phosphate buffer which will keep the pH constant may be substituted. Prepare buffer by mixing 975 ml of 0.1 M KH_2PO_4 (monopotassium phosphate) solution (13.6 g/liter) with 25 ml of 0.1 M K_2HPO_4 (dipotassium phosphate) solution (17.4 g/liter); measure the pH to be certain it is correct (pH 5.1–5.4), adjust if necessary, and use 18 ml/liter of culture medium.

[c] Shake well before use.

[d] Add items 1–6 to 650 ml distilled water (item 9); adjust the pH to 5.2; add sugar (item 7) and adjust volume to 1000 ml with more distilled water (item 9). Bring solution to a gentle boil and add agar (item 8) slowly while stirring. When the agar has dissolved completely (and the solution turns a clear amber color) distribute the medium into culture vessels, cover, and autoclave.

Table A-24. Ichihashi and Yamashita (1977) medium for seed germination and seedling culture of *Bletilla striata*

Item number	Component	Amount per liter of culture medium (final concentration in culture medium)	Stock solution (a concentrate prepared for repeated and convenient use)	Volume of stock solution per liter of culture medium	Remarks
	Macroelements				
1	Calcium nitrate, $Ca(NO_3)_2 \cdot 4H_2O$[a]	826 mg	82.6 g/liter[a]	10 ml[a]	or weigh[a]
2	Ammonium phosphate, $NH_4H_2PO_4$[a]	391 mg	39.1 g/liter[a]	10 ml[a]	or weigh[a]
3	Magnesium sulfate, $MgSO_4 \cdot 7H_2O$	172 mg	17.2 g/liter	10 ml	
4	Potassium nitrate, KNO_3[a]	747 mg	74.7 g/liter[a]	10 ml[a]	or weigh[a]
5	*Microelements*[b]				
a	Boric acid, H_3BO_3	1 mg	1 g/liter	1 ml[b]	one solution[b]
b	Manganese sulfate, $MnSO_4 \cdot 4H_2O$	0.01 mg	10 mg liter		
c	Sodium molybdate, $Na_2MoO_4 \cdot 2H_2O$	0.25 mg	25 mg/liter		
d	Nickel chloride, $NiCl_2$	0.03 mg	30 mg/liter		
e	Cupric sulfate, $CuSO_4 \cdot 5H_2O$	0.03 mg	30 mg/liter		
f	Potssium iodide, KI	0.01 mg	10 mg/liter		
g	Zinc sulfate, $ZnSO_4 \cdot 7H_2O$	7 mg	7 g/liter		
	Iron				
6	Iron chelate, Fe_2EDTA	25 mg	25 g/liter	1 ml	
	Sugar[c]				
7	Sucrose	20 g	no stock	no stock	weigh
8	Water, distilled[c]	to 1000 ml			
	Solidifier				
9	Agar[c]	10 g	no stock	no stock	weigh

[a] Solutions containing ammonium and/or nitrate may become contaminated on standing. Therefore, stock solutions should not be prepared. If made, they must be kept frozen between uses.
[b] Add all microelements to the same one liter of distilled water; stir and/or heat until they are dissolved, and add 1 ml per liter of culture medium.
[c] Add items 1–6 to 650 ml distilled water (item 8), adjust the pH to 5.2; add sugar (item 7) and adjust volume to 1000 ml with more distilled water (item 8). Bring solution to a gentle boil and add agar (item 9) slowly while stirring. When the agar has dissolved completely (and the solution turns a clear amber color), distribute the medium into culture vessels, cover, autoclave, cool, and use.

well on sucrose, it may be advisable to replace this sugar in the Ichihashi and Yamashita medium (Table A-24) with 15 g glucose or trehalose.

Procedure. The procedures recommended for tropical orchids (p. 276) can be used.

Developmental Sequence. When cultured for 9 weeks in the dark at 22° C the seedlings reached a length of 2.575 ± 0.192 mm and a width of 0.68 ± 20 mm. After 12 weeks the protocorms developed leaves.

General Comments. It is interesting to note that *Bletilla striata* does not grow well on sucrose, a sugar which is suitable for many other orchids, or mannitol, a carbohydrate of fungal origin.

Literature Cited

Ichihashi, S., and M. Yamashita. 1977. Studies on the media for orchid seed germination. I. The effects of balances inside each cation and anion group for the germination and seedling development of Bletilla striata seeds. J. Japan Soc. Hort. Sci. 45:407–413.

Knudson, L. 1946. A new nutrient solution for the germination of orchid seed. Amer. Orchid Soc. Bull. 15:214–217.

Smith, S. E. 1973. Asymbiotic germination of orchid seeds on carbohydrates of fungal origin. New Phytol. 72:497–499.

Dactylorhiza (Orchis) purpurella
Joseph Arditti

Work on the germination requirements of *Dactylorhiza purpurella* (T. and T. A. Steph) Soo (*Orchis purpurella*) was initiated in the laboratory of Dr. Geoffrey Hadley at the University of Aberdeen, Scotland. It was continued at the Biology Department, Lakehead University, Thunder Bay, Ontario, Canada (Hadley and Harvais, 1968; Harvais, 1963, 1965, 1972; Harvais and Hadley, 1967).

Plant Material. Seeds collected in Scotland were used in the initial investigations.

Surface Sterilization. The seeds can be sterilized in a solution of sodium hypochlorite containing 0.5% available chlorine, (5% Domestos, a British household bleach) or saturated calcium hypochlorite (7 g/100 ml water) for 15 min or until the seeds are bleached (120 min). In the initial experiments germination rates of bleached seeds were generally higher.

Culture Vessels. Test tubes 20 × 150 mm containing 10 ml medium were used in the original research. In addition, the culture vessels recommended for use with tropical species (p. 275) can also be employed.

Culture Conditions. Cultures should be incubated at 25° C (77° F) in the dark until the first signs of etiolation. After that it is necessary to transfer the cultures to 12-hr photoperiods and a light intensity of 70 lm/ft^2 provided by Gro-Lux fluorescent lamps.

Culture Medium. A medium enriched with casamino acids and yeast extract is suitable (Table A-25).

Procedure. In addition to maintaining the cultures under appropriate conditions, it is also necessary to irrigate them with sterile distilled water to compensate for drying. Cultures that become too crowded should be thinned by transflasking.

Developmental Sequence. Up to 45% germination of bleached seeds can be noted in 3 weeks (vs. 26% for those sterilized for only 15 min). Germination percentages continue to increase and by the end of 30 weeks may reach 85%. As the seeds germinate, the embryos form absorbing epidermal hairs and accumulate starch. A protocorm is formed next and develops a shoot primordium which eventually produces leaves. Roots are formed later.

General Comments. The research which led to the formulation of this procedure has also elucidated some of the germination requirements of *Dactylorhiza purpurella.* The production of "large, healthy protocorms with active roots and green shoots should be an interesting study" (Harvais, 1972).

Table A-25. Harvais medium (1972) for the germination of *Dactylorhiza* (*Orchis*) *purpurella* seeds

Item number	Component	Amount per liter of culture medium (final concentration in culture medium)	Stock solution (a concentrate prepared for repeated and convenient use)	Volume of stock solution per liter of culture medium	Remarks
	Macroelements				
1	Calcium nitrate, $Ca(NO_3)_2 \cdot 4H_2O$[a]	800 mg	80 g/liter[a]	10 ml[a]	or weigh[a]
2	Potassium phosphate, KH_2PO_4	200 mg	20 g/liter	10 ml	
3	Magnesium sulfate, $MgSO_4 \cdot 7H_2O$	200 mg	20 g/liter	10 ml	
4	Potassium nitrate, KNO_3[a]	200 mg	20 g/liter[a]	10 ml[a]	or weigh[a]
5	Potassium chloride, KCl	100 mg	10 g/liter	10 ml	
6	Ammonium ferric citrate[b]	1 mg	100 mg/liter[b]	10 ml[b]	
7	*Microelements*[c]				
a	Boric acid, H_3BO_3	0.5 mg	500 mg/liter	1 ml[c]	one solution[c]
b	Manganese sulfate, $MnSO_4 \cdot 4H_2O$	0.5 mg	500 mg/liter		
c	Copper sulfate, $CuSO_4 \cdot 5H_2O$	0.5 mg	500 mg/liter		
d	Zinc sulfate, $ZnSO_4 \cdot 2H_2O$	0.5 mg	500 mg/liter		
	Complex Additives				
8	Casamino acids, vitamin-free[d]	5 g	no stock	no stock	weigh
9	Yeast extract	1 g	no stock	no stock	weigh
	Sugar[e]				
10	Sucrose or glucose (but not both)	10 g	no stock	no stock	weigh
11	Water, distilled[e]	to 1000 ml			
	Solidifier[e]				
12	Agar	12–15 g	no stock	no stock	weigh

[a] Solutions containing ammonium and/or nitrate may become contaminated on standing. Therefore, stock solutions should not be prepared. If made, they must be kept frozen between uses.

[b] Shake well before dispensing.

[c] Add all microelements to the same one liter of distilled water; stir and/or heat until they are dissolved, and add 1 ml per liter of culture medium.

[d] Available from Difco, Detroit, Michigan, USA.

[e] Add items 1–9 to 650 ml distilled water (item 11); adjust the pH to 5.3–5.6; add sugar (item 10) and adjust volume to 1000 ml with more distilled water (item 11). Bring solution to a gentle boil and add agar (item 12) slowly while stirring. When the agar has dissolved completely (and the solution turns a clear amber color), distribute the medium into culture vessels, cover, and autoclave.

Literature Cited

Hadley, G., and G. Harvais. 1968. The effects of certain growth substances on asymbiotic germination and development of Orchis purpurella. New Phytol. 67:441–445.

Harvais, G. 1963. Investigations of the mycorrhiza of Orchis purpurella and other British orchids. M.Sc. thesis. Univ. Aberdeen, Scotland.

——. 1965. Some aspects of symbiosis in Orchis purpurella. Ph.D. diss., Univ. Aberdeen, Scotland.

——. 1972. The development and growth requirements of Dactylorhiza purpurella in asymbiotic cultures. Can. J. Bot. 50:1223–1229.

Harvais, G., and G. Hadley. 1967. The development of Orchis purpurella in asymbiotic and inoculated cultures. New Phytol. 66:217–230.

Disa

Joseph Arditti

The genus *Disa* contains approximately 80 species of which 70 have been recorded in South Africa (O'Connor, 1975). Among these *Disa uniflora,* the Red Disa or Pride of the Table Mountain, is best known and, in the view of some, the most beautiful (Johnson, 1969). This species can be grown in greenhouses (Lindquist, 1958; Stoutamire, 1977). Its seeds are difficult to germinate but a number of workers have developed usable methods (Collett, 1971; Harbeck, 1968; Johnson, 1969; Lindquist, 1958, 1965; O'Connor, 1975; Stoutamire, 1974, 1977).

Plant Material. It is generally agreed that *Disa* seed must be sown at once (soon after ripening) because "it loses its germination power within a month after the harvest" (Lindquist, 1965).

Surface Sterilization. There is no need to surface sterilize the seeds.

Culture Vessels. Petri dishes, shallow trays, and 12.5-cm (5 in.) plastic pots with drainage holes may be used.

Culture Conditions. Seeds germinate well at temperatures ranging from 14–16° C (57–60° F; Lindquist, 1965) to 21–27° C (70–80° F; Johnson, 1969) in greenhouses or shady locations outdoors in the Cape province of South Africa.

Culture Media. According to one report, *Disa uniflora* seeds germinate "on standard agar media" and seedlings can be cultivated "on agar in a growth chamber" (Stoutamire, 1977). Unfortunately, however, the composition of these standard agar media was not described (Stoutamire, 1977), and since there are many "standard media" one can only wish that the procedures and "standard media" were given in more detail.

More details are available regarding what amounts to symbiotic germination of *Disa* seeds. The English method (Collett, 1971) is to cut the tops of fresh sphagnum moss 2.5 cm (1 in.) from the crown (discard the lower portion), wash them in fresh water and shake several times in a wire strainer or colander. A final few rinses should be with fresh rainwater (or distilled water) to remove "all undesirable algae, liverworts and moss spores." After the final rinses the moss is placed in a cambric or linen kerchief (preferably an old, thin, worn one without holes) whose corners are tied" in a tramp's lunch pack fashion." The resulting ball is placed in a 12.5-cm (5 in.) pot and pressed down until the surface of the ball is at least 2.5 cm (1 in.) below the rim. If the result is a flabby and creased pad it must be removed and tightened (or more moss can be added). The seeds are spread on the wet surface of the handkerchief. The pot is then placed in a box of sphagnum or in a watering tray and covered with clear plastic or glass. If glass is used it must be tilted at a 10° angle to allow condensation to run off without dripping on the seeds.

The Swedish method (Lindquist, 1965) is somewhat different. Short-cut sphagnum is placed vertically in petri dishes, "forming beds which are intermixed with rhizomes from *Disa* plants." A dilute fertilizer and glucose solution may be added to the sphagnum before sowing and the surface of the bed is covered with clean, fine sand.

Procedure. To allow for better distribution the seeds may be mixed with clean, fine sand and spread over the substrate. The sphagnum beds must never become dry and should be watered regularly by adding water at the bottom of the dish with a pipette. It is very important not to disturb the seeds or move them while watering.

Developmental Sequence. Seeds begin to swell in 15–20 days, forming small, pale green seedlings. The first leaves appear in 30 days and the second ones develop after 3.5–4 months. First roots develop after 16–17 weeks. A slow, continuous development follows. At the end of 5 months, when they have two small leaves, the seedlings can be repotted into a mixture of ½ sphagnum moss, ⅙ sphagnum peat, ⅙ leaf compost, and ⅙ sterilized sand. The best sphagnum to use is *Sphagna palustre* or *S. magellanicum.* At this time the temperature may be allowed to reach 20° C (68° F) and several months later 25° C (77° F). During the next winter temperatures should be kept at 6–10° C (43–50° F).

Disa plants must be kept moist or even wet under conditions that resemble a bog. Parasitic fungi may attack the seedlings and can be inhibited by additional illumination and by spreading sterilized sand on top of the cultures. Mature plants require cool, airy, well-illuminated conditions (but no direct sun). Night temperatures should be cool (about 10° C or 50° F). Humidity should be at least 60–65%. The leaves should be misted on hot summer days.

General Comments. The procedures described above were developed in northern latitudes. They may or may not be directly applicable to other areas without additional experimentation.

Literature Cited

Collett, A. 1971. Notes on the growing of Disa uniflora. J. Roy. Hort. Soc. 96:358–361.
Harbeck, M. 1968. Aussaatuntersuche mit Disa uniflora. Die Orchidee 19:1–5.
Johnson, K. C. 1969. Disa uniflora and its hybrids. Amer. Orchid Soc. Bull. 38:135–146.
Lindquist, B. 1958. A greenhouse culture of Disa uniflora Berg in Gothenburg. Amer. Orchid Soc. Bull. 27:652–657.
——. 1965. The raising of Disa uniflora seedlings in Gothenburg. Amer. Orchid Soc. Bull. 34:317–319.
O'Connor, M. J. 1975. Disas of Natal's veld and vlei. Amer. Orchid Soc. Bull. 44:4–11.
Stoutamire, W. 1974. Terrestrial orchid seedlings, p. 101–128. *In* C. L. Withner (ed.), The orchids: Scientific studies. Wiley-Interscience, New York.
——. 1977. Disa uniflora and Disa veitchii. Amer. Orchid Soc. Bull. 46:438–444.

European Terrestrial Orchids (Symbiotic and Asymbiotic Methods)

Gertrud Fast

In comparison to most tropical orchids, European terrestrial species are generally more difficult to germinate under laboratory conditions. Germination often takes place slowly and irregularly; sometimes it fails entirely, or the seedlings may die during early stages of growth. The reasons for these difficulties are not known, but it is possible to surmise that perhaps the germination of European orchids is (1) more highly influenced by fluctuations in climate which are absent in culture, (2) dependent on still unknown growth-regulating substances, supplied in nature by the symbiotic fungi, or (3) inhibited by hypothetical growth inhibitors that are leached out under natural conditions.

After World War I, the pharmaceutical industry became interested in the production of European orchids, because the salep, obtainable from the tubers of some species, was an important remedy against *cholera infantum.* At that time, supported by a U.S. life insurance company,[1] the well-known botanist and orchidologist Prof. Hans Burgeff in Würzburg intensified his earlier studies on the culture of terrestrial orchids by means of a symbiotic method which he had developed some years before. His findings led to the conclusion that germination and cultivation of European orchids would be technically possible, but would not pay off commercially because of the high losses of plants after transferring them to open areas (Burgeff, 1936, 1954).

Recently, Burgeff's experiments were resumed by Seitz (1976) with the goal of finding methods for the preservation and protection of orchids by replanting them in areas from which they have been lost. He suggests that plantlets grown with a symbiont are more resistant and adaptable than those without fungus. This is probably true.

In addition to these efforts to preserve orchids in the wild, there are many hobbyists whose interest in terrestrial orchids has increased during recent years. They want to grow these species, but can no longer purchase them easily because imports from countries where orchids were once found in abundance have diminished due to stronger conservation laws. As a result, botanists and hobbyists have become more concerned with the development of methods for the germination of seeds of European orchids. Unfortunately this is a time-consuming and troublesome process, and these efforts may remain limited.

Reports of previous investigations have demonstrated that, in addition to symbiotically, seeds of terrestrial species can be germinated asymbiotically if composition of the medium, temperature, and light conditions are suitable. But information on the subject is very scattered and often imprecise. Therefore, our knowledge at present is insufficient. The instructions given here are based primarily on the work of several investigators (Borriss, 1969; Burgeff, 1936, 1954; Downie, 1940, 1943; Eiberg, 1969; Fast, 1974, 1976, 1978; Hadley, 1970; Harbeck, 1961, 1963, 1964; Harvais, 1972; Lucke, 1971, 1976; Seitz, 1976; Sprau, 1939; Stoutamire, 1974; Vöth, 1975, 1976).

Plant Material. An important factor in the germination of European terrestrial orchids is the age of seeds. As has often been shown, mature seeds from dehisced capsules do not germinate as well as those from immature fruits in which the seeds are just beginning to turn brown. This is especially true for the north temperate species, and to a lesser extent for those of Mediterranean climate. In most cases seed viability decreases several weeks after they have been collected. Seeds will seldom remain viable over a period of several years.

Treatment of seeds for 2–3 months with pure water to leach out inhibitory substances is based on the assumption that such compounds are formed during ripening (Burgeff, 1936). This can be done either under unsterile or sterile conditions. Unsterile: place the

[1]Northwestern Mutual Life Insurance Company, Pittsburgh, Pa.

seed between two pieces of cotton, fastened to the rim of a pot, and spray regularly with rain- or distilled water. Sterile: disinfected seeds are inserted into a special sterilized glass container with an inner cone-shaped surface (Fig. A-6) where they are washed with water which condenses on the slope and drops on them.

Immature seeds usually need no pretreatment. With two *Cypripedium* species from North America (*C. reginae* and *C. acaule*) we made the interesting observation that

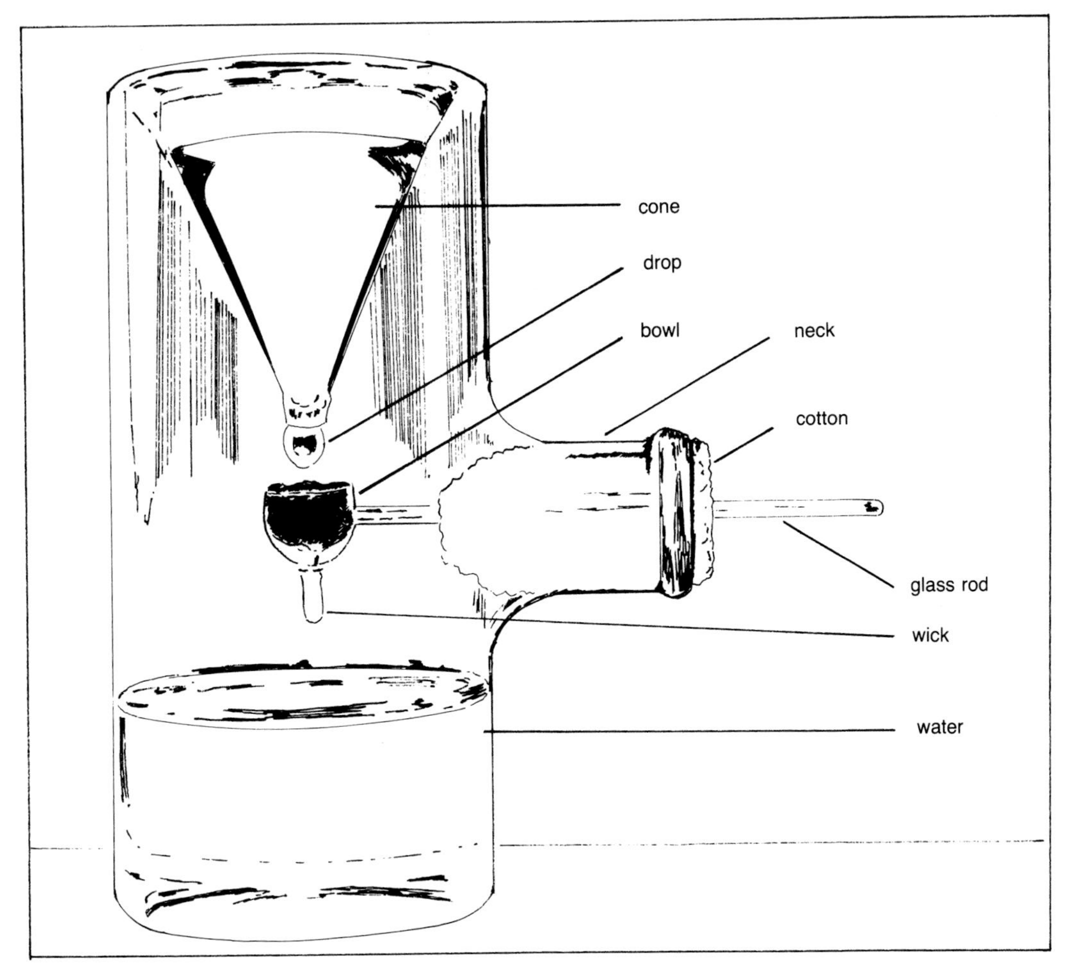

Fig. A-6. Flask for washing orchid seeds (drawn from photo in Seeman, 1953). The flask consists of a closed glass cylinder (ca. 45 mm in diameter and 100 mm in height) with a downward-sloping cone which is approximately 40 mm long. A short neck (20 mm long and 20 mm in diameter) allows for the insertion of a glass rod which is attached to a small bowl that holds the seeds. A wick is inserted through a hole in the bottom of the small bowl to allow for drainage while holding the seeds in place. To start the washing process water is poured into the vessels to a height of 25 mm. The seeds, sterilized with calcium hypochlorite, are placed in the bowl (a flamed spatula may be used in doing so) and the glass rod, wrapped in a cotton bun which must be large enough to fit the neck tightly [or inserted through a rubber stopper—Ed.], is placed in the flask. It is important to position the bowl directly underneath the tip of the cone. When ready the flask is placed in the dark near a north window. Water which evaporates from the reservoir below condenses on the cone, drops on the seeds, washes them, and drains from the bowl through the wick. Drops form at the rate of 10 per 24 hours, but the process can be accelerated by filling the outside of the cone with cold water.

slightly fermented seeds from capsules that apparently were stored in a warm and humid place during their long sea voyage to Germany germinated extraordinarily well (Fast, unpublished).

Surface Sterilization. Undehisced capsules are sterilized best by flaming them after a short dip in isopropyl alcohol. Mature seeds from open capsules are sterilized with calcium hypochlorite like those of tropical orchids (p. 275). To improve wetting of the seed coat, 2–3 drops of alcohol or Tween 80 can be added to the sterilizing solution.

Culture Vessels. All autoclavable, transparent, clean vessels of the type used for tropical orchids (p. 275) are suitable. Larger, wide-neck Erlenmeyer flasks are especially well suited for symbiotic cultures. For methods of filling and covering the vessels see p. 251. Airtight (hermetic) sealing is not advisable, because germination and particularly development of protocorms are greatly inhibited by a lack of oxygen.

Culture Media. Investigators differ considerably as to the composition of a suitable culture medium. Unfortunately, however, accurate and comparative data on the specific macroelements and microelements or organic additives required during the various stages of seed germination and seedling development do not exist at present. In many cases, seeds swell and embryo cells divide in pure water (Eiberg, 1969) or in organic media that are extremely poor in nutrient content (Borriss, 1969; Vöth, 1975). After protocorms appear, their requirements generally increase and a somewhat richer medium is preferable. However, European orchids are considerably more sensitive to salt concentrations than tropical species. Therefore, if media such as Knudson C are used for seed germination, the quantities of salts and sugar should be reduced to approximately one-fourth. Orchids raised symbiotically on polypodium substrate need neither sugar nor nitrogen.

Cultures without fungus on agar medium utilize sugar such as sucrose or glucose plus fructose, nitrogen in organic forms (e.g., fish emulsion) or inorganic (ammonium and/or nitrate ions). In an investigation of the effect of single-salt solutions on the rate of germination in *Dactylorhiza maculata,* nitrate salts were found to be unsuitable when the solution was not supplemented with peptone or other organic substances (Eiberg, 1969). In addition to peptone, fresh or hydrolysated beer yeast (from a brewery), amino-acid combinations, B-group vitamins, and nucleinates seem to stimulate seed germination and seedling development.

Plant hormones or coconut water have a stimulating effect on germination only rarely (Borriss, 1969; Hadley, 1970). Vegetable charcoal is a useful additive for species whose seedlings exude substances that darken the nutrient medium. These exudates can cause a gradual withering of the plantlets. The addition of trace elements is not always necessary, particularly when unpurified agar is being used.

Procedures: Symbiotic Method. The symbiotic method (Burgeff, 1911, 1936, 1954), is based on the natural association of orchid embryos with mycorrhizal fungi. This procedure requires first the isolation and culture of several fungal strains which are then

Table A-26. MN + N medium for the isolation of mycorrhizal fungi (Burgeff, 1936)

Item number	Component	Amount per liter of culture medium (final concentration in culture medium)	Stock solution (a concentrate prepared for repeated and convenient use)	Volume of stock solution per liter of culture medium	Remarks
	Macroelements				
1	Ammonium sulfate, $(NH_4)_2SO_4$ [a]	500 mg	50 g/liter[a]	10 ml[a]	or weigh[a]
2	Dipotassium phosphate, K_2HPO_4	300 mg	30 g/liter	10 ml	
3	Monopotassium phosphate, KH_2PO_4	700 mg	70 g/liter	10 ml	
4	Calcium chloride, $CaCl_2$	100 mg	10 g/liter	10 ml	
5	Sodium chloride, NaCl	100 mg	10 g/liter	10 ml	
6	Magnesium sulfate, $MgSO_4 \cdot 7H_2O$	300 mg	30 g/liter	10 ml	
7	Ferrous sulfate, $FeSO_4 \cdot 7H_2O$ [b]	10 mg	1 g/liter[b]	10 ml[b]	
	Carbohydrate				
8	Potato starch[c]	3 g	no stock	no stock	weigh
9	Water, distilled[d]	to 1000 ml			
	Solidifier[d]				
10	Agar	15 g	no stock	no stock	weigh

[a] Stock solutions containing nitrogen may become contaminated on standing. Therefore, stock solutions should not be prepared. If made, they must be kept frozen between uses.
[b] A rust-colored precipitate may form on standing. This solution must, therefore, be shaken vigorously before use.
[c] Suspend the starch in 100 ml of distilled water (item 9) and stir vigorously.
[d] Add components 1–7 to 750 ml of distilled water (item 9), bring volume to 900 ml, heat to a gentle boil, add the 100 ml which contain the starch (item 8) and stir well; then add the agar (item 10) while stirring. When the agar has dissolved completely, distribute the solution into culture vessels, cover, and autoclave.

tested as to their compatibility with the desired orchid species. One disadvantage of this method is the difficulty in determining the correct partnership and establishing a stable balance between orchid and fungus. On the other hand, growth of seedlings is enhanced remarkably when this balance is achieved. The following four steps constitute Burgeff's procedure.

1. Isolation of the symbiotic fungi from orchid roots and establishment of pure fungal cultures: Fresh, healthy roots of various orchid species from different locations are collected and cleaned thoroughly. After an external disinfection with calcium hypochlorite (7g $Ca0Cl_2$ in 100 ml distilled water) and rinsing in pure sterile water, sections 8–10 cm long are removed a short distance (5–8 mm) from the root tip. These sections are sliced lengthwise (not too thin, ca. 1–2 mm thick) and examined under the microscope. Small pieces of tissue with living hyphae are placed on MN + N medium (Table A-26) in petri dishes, slightly below the surface of the agar. From 3 to 5 sections can be placed in a dish. When the fungi, the strains of which must be kept separate, have penetrated the agar, they should be subcultured in several flasks on Of/N-free medium (Table A-27) in order to establish stock cultures ("Stammkulturen"). These cultures are kept cool and dark and should be subcultured monthly onto new medium.

2. Inoculation of hyphae on polypodium substrate: A mixture of chopped fern roots (*Polypodium, Osmunda,* or tree fern) and rye straw (cut short) is boiled with rain- or distilled water and then rinsed with fresh water. After that the pH is raised to 7.0 with a 1 : 1 mixture of distilled water and lime water (lime water = 1 part CaO to 104 parts distilled water[2]) because it usually drops to pH 6.0 after autoclaving. Following pH

[2]Lime water has 0.15–0.17% $Ca(OH)_2$; its molarity is 74.09.

Table A-27. Of/N-free medium for stock cultures of mycorrhizal fungi (Burgeff, 1936)

Item number	Component	Amount per liter of culture medium (final concentration in culture medium)	Stock solution (a concentrate prepared for repeated and convenient use)	Volume of stock solution per liter of culture medium	Remarks
	Macroelements				
1	Monopotassium phosphate, KH_2PO_4	150 mg	15 g/liter	10 ml	
2	Dipotassium phosphate, K_2HPO_4	350 mg	35 g/liter	10 ml	
3	Magnesium sulfate, $MgSO_4 \cdot 7H_2O$	250 mg	25 g/liter	10 ml	
4	Calcium chloride, $CaCl_2$	100 mg	10 g/liter	10 ml	
5	Ferrous sulfate, $FeSO_4 \cdot 7H_2O$[a]	20 mg	2 g/liter[a]	10 ml[a]	
	Carbohydrate				
6	Potato starch[b]	10 g	no stock	no stock	weigh
7	Water, distilled[c]	to 1000 ml			
	Solidifier[c]				
8	Agar	10 g	no stock	no stock	weigh

[a] A rust-colored precipitate may form on standing. This solution must, therefore, be shaken vigorously before use.
[b] Suspend the starch in 100 ml of distilled water (item 7) and stir vigorously.
[c] Add components 1–5 to 750 ml of distilled water (item 7), bring the volume to 900 ml, heat to a gentle boil, add the 100 ml of distilled water that contain the starch (item 6), and stir well. Then add the agar (item 8) while stirring. When the agar has dissolved completely, distribute the solution into culture vessels, cover, and autoclave.

adjustment the fern-straw mixture is placed in larger flasks over a 10-mm layer of washed quartz sand. Layers of the mixture should be 3–5 times thicker than the sand and should be placed horizontally or at a slight slant. The substrate is then thoroughly soaked with a nutrient solution (Table A-28) and the surplus is siphoned off with a pipette. After autoclaving or vapor sterilization,[3] the medium is inoculated with the various fungal strains and the bottles are placed in the dark.

3. Sowing of orchid seed on the fungus-polypodium substrate: About 18 days later, when the mycelium has grown through the substrate, the seeds (which have been leached with water 2–3 months in advance) can be added. The cultures are kept dark until protocorms have formed and begin to differentiate. Temperature should be maintained between 10° and 20° C. It is advisable to test the association of every species with each fungal strain in advance, to determine their compatibility and the most effective orchid-fungus combination. This can be done with a small quantity of seed in petri dishes.

4. Transfer of seedlings to new substrate ("Net-connection"): After six months at most, seedling development usually becomes slower, and they should be transplanted onto newly inoculated polypodium. In order to ensure better contact between the rhizoids of protocorms and the fungal hyphae, the fern roots should be ground in this case rather than cut.

Procedures: Asymbiotic Method. There is no difference between tropical epiphytic orchids (p. 276) and European terrestrials in the method of asymbiotic germination, except for the choice of medium and culture conditions. Because germination is often poor, seeds must be sown somewhat more densely.

[3]Burgeff recommends in this case sterilization without pressure in steaming vapor twice for 30–40 min; first time 24 hr after filling the vessels, a second time 24 hr later.

Table A-28. Nutrient solution for orchid seed germination and seedling culture with symbiotic fungi on *Polypodium* substrate (Burgeff, 1936)

Item number	Component	Amount per liter of culture medium (final concentration in culture medium)	Stock solution (a concentrate prepared for repeated and convenient use)	Volume of stock solution per liter of culture medium	Remarks
	Macroelements[a]				
1	Dipotassium phosphate, K_2HPO_4	250 mg	25 g/liter	10 ml	
2	Magnesium sulfate, $MgSO_4 \cdot 7H_2O$	100 mg	10 g/liter	10 ml	
3	Ferrous phosphate, $Fe_3(PO_4)_2$	50 mg	no stock	no stock	weigh
4	Water, distilled	to 1000 ml			

[a] Add items 1 and 2 to 750 ml of distilled water (item 4) followed by item 3. Bring volume to 1000 ml with more distilled water (item 4) and stir vigorously. The solution must be shaken before use.

Culture Conditions. Most of the European orchids need darkness during the first stages of growth. After sowing, the flasks must be kept in a dark, well-aerated room or cupboard. Before germination proper the seeds swell, taking up water, regardless of environmental conditions. But further development (splitting of the testa and division of embryonic cells) is influenced to a high degree by culture conditions. Most European terrestrials germinate under normal room temperature, but there are some exceptions. For example, *Epipactis palustris* and probably also *Cypripedium calceolus* require a low temperature of 2–4° C (Borriss, 1969).

After germination, when protocorms have formed and begin to differentiate, the culture flasks must be removed from the dark and placed under diffuse light to prevent etiolation. Under light, the seedlings grow and develop green leaves rather quickly, but in some cases this depends not only on light, but also on temperature. *Dactylorhiza* species, for example, require a chilling period of ten or twelve weeks at 2–3° C for normal shoot development (Borriss, 1969). The chilling can be replaced by three weekly applications of gibberellic acid (GA_3) in which the shoot tips each receive 5 μm GA_3 in an alcoholic solution (Borriss, 1969). For *Orchis mascula* and *O. militaris* the temperature should be lowered to 13–16° C after germination (Burgeff, 1954). Shoots of *Coeloglossum viride* and *Dactylorhiza purpurella* will turn green only at 10° C (Hadley, 1970). Also the subsequent development of orchids with subterranean overwintering organs depends on seasonal thermoperiods. Therefore an annual cooling phase must be included in the culture procedure.

Developmental Sequence. Germination often occurs over a long period of time, sometimes several years. As a result, protocorms appear in succession. In only a few species, which germinate extremely well, do seedlings emerge nearly at the same time. The morphological, anatomical, and developmental characteristics of orchid protocorms were described recently (Veyret, 1974). Some genera, especially *Dactylorhiza,* often show proliferation of protocorms when the nutrient medium contains coconut water, cytokinins, or B-vitamins.

The duration of protocorm stage differs with the species and parentage, but usually differentiation occurs rapidly. Occasionally seedlings become brown and die, either during the early protocorm stage or later on, during formation of the first root. Death of protocorms may be mostly due to an unsuitable medium, whereas plantlets are primarily damaged by an improper climate. Moreover, partial necrosis doubtless occurs during dormant periods in nature as well. Mediterranean orchids cultured on agar often manifest a clearly defined cycle of growth: following temporary wilting, shoots appear again after several weeks.

Procedure of Transplanting and Dividing. It is advisable to transplant seedlings which are growing poorly or too densely as early as possible, because a fresh medium always enhances growth and development. If the germination medium had an extremely low nutrient content, a somewhat richer one, supplemented with special additives, should be used for transflasking (Table A-29). For a method of transferring the plantlets into new vessels see p. 277.

In some cases, transplanting can be combined with division, especially when the rate of germination was very low. With *Dactylorhiza,* in particular, it is easy to section off the new growth of proliferating protocorms and to distribute it into several flasks. *Cypripedium,* which often germinates poorly, can be forced to form adventitious shoots by slicing the young, rooted plantlets. After repeated divisions and further culture in darkness, masses of shoots form, and allow an almost infinite vegetative propagation *in vitro* (Fast, 1974, 1976). When divisions are terminated, the flasks should be placed under daylight, and the hitherto inhibited shoots will soon elongate and become green. This method, first tested with *Cypripedium calceolus,* is also suitable for *Cypripedium guttatum, C. macranthum, C. cordigerum,* and *C. acaule.*

General Comments. Until now, only 50–60 of about 150 orchid species found in Europe have been investigated as to their germination under laboratory conditions. The alpine species and rarer moorland orchids have scarcely been examined. On the other hand, we have a number of reports on germination experiments with species which are more frequently found in woods and meadows. From the survey presented here it is possible to conclude that most of the terrestrial orchids which have been studied will grow on suitable nutrient media, with or without symbiont. But the number of emerging seedlings is variable. Some species germinate remarkably well in most experiments, e.g., *Barlia longibracteata, Dactylorhiza maculata, D. incarnata, D. majalis, Goodyera repens, Gymnadenia conopsea, Ophrys insectifera, O. fusca, O. sphegodes, Orchis militaris, O. morio, Serapias parviflora,* and *S. orientalis.* Others cannot be germinated as easily, for example the genera *Anacamptis, Cephalanthera, Epipactis,* and holosaprophytic orchids like *Neottia nidus-avis.* Here, we need more information on the requirements for germination, be they of biochemical, thermoperiodical or atmospheric nature. Some of these hard-to-grow orchids can form adventitious buds on their roots in the wild (Stoutamire, 1974), and it has been possible to induce the development of protocorms from isolated root tips in *Neottia* (Champagnat, 1971).

In spite of many unsolved problems we can at present summarize the points in which the germination of European terrestrials differs fundamentally from that of tropical epiphytes. The most important are: (a) *Maturity of the seed:* Satisfactory germination *in vitro* can only be expected from seeds which were not completely mature when removed

Table A-29. Germination and seedling culture of European terrestrial orchids on different media

Species	Medium for germination			Result[a]	Medium for transflasking			Reference
	Name	Table	Additives (per liter)		Name	Table	Additives (per liter)	
Anacamptis pyramidalis	FN	A-35	–	0	–	–	–	Fast, 1978
Barlia longibracteata	FN	A-35	–	+++	as for germination			Fast, 1978
	N_3f	A-31	1 g peptone[b]	g	as for germination			Stoutamire, 1974
Cephalanthera damasonium	FN	A-35	–	0	–	–	–	Fast, 1978
	Chang	A-37	–	0	–	–	–	Harbeck, 1963
C. rubra	FN	A-35	–	0	–	–	–	Fast, 1978
Coeloglossum viride	Pfeffer	A-32	–	g	Pfeffer	A-32	1 mg kinetin[c] 1 mg IAA[c]	Hadley, 1970
	N_3f	A-31	1 g peptone[b]	g	as for germination			Stoutamire, 1974
Corallorhiza trifida	symb.[d]	A-32	potato extract[e]	g	as for germination			Downie, 1943
Cypripedium calceolus	Zak	A-39	1 mg kinetin[c]	++	Eg1 mod.	A-36	5 mg GA_3[f]	Borriss, 1969
	F or FN	A-34,35	–	+	FN	A-35	100 ml coconut water[g] or 0.5 mg BAP,[h] 100 g banana[i]	Fast, 1974, 1976, 1978
	symb.[d]	A-28	–	g	as for germination			Seitz, 1976
C. guttatum	FN	A-35	–	+	FN	A-35	as for *C. calceolus*	Fast, 1978
C. macranthum	FN	A-35	–	++	FN	A-35	as for *C. calceolus*	Fast, 1978
Dactylorhiza foliosa	dist. water	–	0.05% Tween 80	+++	no information given			Eiberg, 1969
	FN	A-35	10% coconut water[g]	+++	FN	A-35	–	Fast, 1978
D. incarnata	dist. water	–	0.05% Tween 80	+++	no information given			Eiberg, 1969
	FN	A-35	–	+++	as for germination			Fast, 1978
	Chang	A-37	–	+	as for germination			Harbeck, 1963
	symb.[d]	A-28	–	g	as for germination			Seitz, 1976
D. maculata	Zak	A-39	–	+++	Eg1 mod.	A-30	–	Borriss, 1969
	dist. water	–	0.05% Tween 80	+++	no information given			Eiberg, 1969
	FN	A-35	–	+++	as for germination			Fast, 1978
	Chang	A-37	–	+++	as for germination			Harbeck, 1963
D. majalis	Zak	A-39	–	+++	Eg1 mod.	A-30	–	Borriss, 1969
	dist. water	–	0.05% Tween 80	+++	no information given			Eiberg, 1969
	FN	A-35	10% coconut water[g]	+++	FN	A-35	–	Fast, 1978
	Chang	A-37	–	+	as for germination			Harbeck, 1963
	symb.[d]	A-28	–	g	as for germination			Seitz, 1976
D. papilionacea	Knudson C	A-5	1 g peptone[b]	g	as for germination			Stoutamire, 1974
D. praetermissa	Chang	A-37	–	+	as for germination			
D. purpurella	dist. water	–	0.05% Tween 80	++	no information given			Eiberg, 1969
	Pfeffer	A-32	–	g	Pfeffer	A-32	10 ppm kinetin[c] 1 ppm IAA[c]	Hadley, 1970
	Pfeffer	A-32	5 g casein hydrolysate,[j] 1 g yeast extract[k]	+++	Pfeffer	A-32	5 g casein hydrolysate,[j] 2 g yeast extract,[k] 5 mg Mn, and 10 mg Fe	Harvais, 1972

(*continued*)

Table A-29.—Continued

Species	Medium for germination			Result[a]	Medium for transflasking			Reference
	Name	Table	Additives (per liter)		Name	Table	Additives (per liter)	
D. sambucina	dist. water	–	0.05% Tween 80	++	no information given			Eiberg, 1969
	Chang	A-37	–	+++	as for germination			Harbeck, 1963
D. traunsteineri	dist. water	–	0.05% Tween 80	++	no information given			Eiberg, 1969
	symb.[d]	A-28	–	++	as for germination			Seitz, 1976
Epipactis helleborine	dist. water	–	0.05% Tween 80	0	–	–	–	Eiberg, 1969
	FN	A-35	–	0	–	–	–	Fast, 1978
E. palustris	Zak	A-39	–	++	Eg1 mod.	A-30	–	Borriss, 1969
	FN	A-35	–	0	–	–	–	Fast, 1978
Goodyera repens	Zak	A-39	–	+++	Eg1 mod.	A-30	–	Borriss, 1969
	Pfeffer	A-32	potato extract[e]	+++	Pfeffer	A-32	endophytic fungus	Downie, 1940
	F	A-34	–	+++	FN	A-35	–	Fast, 1976
Gymnadenia conopsea	Zak	A-39	–	++	Eg1 mod.	A-30	–	Borriss, 1969
	symb.[d]	A-28	–	++	as for germination			Burgeff, 1936
	FN	A-35	–	++	FN	A-35	1 g charcoal[l]	Fast, 1978
	Chang	A-37	–	+	as for germination			Harbeck, 1963
G. odoratissima	FN	A-35	5 mg niacin[m]	+	FN	A-35	1 g charcoal[l]	Fast, 1978
Leucorchis albida	Chang	A-37	50 mg GA_3[f]	+++	Chang	A-37	–	Harbeck, 1963
Listera ovata	Zak	A-39	–	++	Eg1 mod.	A-30	–	Borriss, 1969
Himantoglossum hircinum	CT-1	A-36	–	++	CT-1	A-36	33% coconut water[n]	Lucke, 1976
Malaxis monophyllos	FN	A-35	10% coconut water[g]	++	as for germination			Fast, 1978
Neottia nidus-avis	symb.[d]	A-28	–	0	–	–	–	Burgeff, 1936
	FN	A-35	–	0	–	–	–	Fast, 1978
Ophrys apifera	FN	A-35	–	++	FN	A-35	1 g charcoal[l]	Fast, 1978
O. bertolonii	Eg1 mod.	A-30	–	++	as for germination			Borriss, 1969
O. fuciflora	FN	A-35	–	++	FN	A-35	1 g charcoal[l]	Fast, 1978
O. fusca	FN	A-35	5 ml yeast[o]	+++	FN	A-35	1 g charcoal[l]	Fast, 1978
O. insectifera	Zak	A-39	–	+++	Eg1 mod.	A-30	–	Borriss, 1969
	FN	A-35	–	+++	FN	A-35	1 g charcoal[l]	Fast, 1978
	Chang	A-37	–	+	as for germination			Harbeck, 1963
O. speculum	Chang	A-37	–	+	as for germination			Harbeck, 1963
O. sphegodes	FN	A-35	10% coconut water[g]	++	FN	A-35	1 g charcoal[l]	Fast, 1978
	GD	A-33	5% coconut water,[g] 1 mg thiamine, 1 mg pyridoxine, 10 mg niacinamide, 0.1 mg biotin	+++	GD	A-33	5% coconut water,[g] 1 mg thiamine, 1 mg pyridoxine, 5 mg niacinamide, 0.2 mg biotin, 1 g charcoal[l]	Lucke, 1971
Orchis coriophora	Eg1 mod.	A-30	–	+++	as for germination			Borriss, 1969
	NB-1	A-38	–	+++	as for germination			Vöth, 1975
	symb.[d]	A-28	–	+++	as for germination			Seitz, 1976
O. laxiflora	dist. water	–	0.05% Tween 80	+++	no information given			Eiberg, 1969
	Chang	A-37	–	+++	as for germination			Harbeck, 1963
O. longicornu	FN	A-35	8% grass extract[p]	+	as for germination			Fast, 1978

O. mascula	symb.[d]	A-28	0.4 mg thiamine, 50 mg aspargine, 3 g agar, pH 7.0	+++	as for germination			Burgeff, 1954
O. militaris	symb.[d]	A-28	0.4 mg thiamine 50 mg asparagine 3 g agar, pH 7.0	+++	as for germination			Burgeff, 1954
	FN	A-35	8% coconut water[g]	+++	FN	A-35	1 g charcoal[l]	Fast, 1978
	N_3f	A-31	1 g peptone[b]	+++	as for germination			Stoutamire, 1974
O. morio	Zak	A-39	–	+++	Eg1 mod.	A-30	–	Borriss, 1969
	FN	A-35	–	+++	as for germination			Fast, 1978
	symb.[d]	A-28	–	+++	as for germination			Seitz, 1976
								Sprau, 1939
O. patens	FN	A-35	8% grass extract[p]	+	as for germination			Fast, 1978
O. purpurea	Zak	A-39	–	++	Eg1 mod.	A-30	–	Borriss, 1969
	FN	A-35	–	+	as for germination			Fast, 1978
O. sancta	GD or NB-1	A-33,38	–	+++	as for germination			Vöth, 1975
Platanthera bifolia	Zak	A-39	–	++	Eg1 mod.	A-30	–	Borriss, 1969
	symb.[d]	A-28	–	++	as for germination			Burgeff, 1936
	FN	A-35	–	++	as for germination			Fast, 1978
	Pfeffer	A-32	–	g	Pfeffer	A-32	1 ppm kinetin[c] 0.1 ppm IAA[c]	Hadley, 1970
P. chlorantha	Zak	A-39	–	++	Eg1 mod.	A-30	–	Borriss, 1969
	symb.[d]	A-28	–	++	as for germination			Burgeff, 1936
	Chang	A-37	–	+	as for germination			Harbeck, 1963
Serapias parviflora	Eg1 mod.	A-30	–	+++	as for germination			Borriss, 1969
	FN	A-35	–	+++	as for germination			Fast, 1978
	Chang	A-37	–	+++	as for germination			Harbeck, 1963
	NB-1	A-38	–	+++	as for germination			Vöth, 1976
S. orientalis	NB-1	A-38	–	+++	as for germination			Vöth, 1976
Spiranthes spiralis	Eg1 mod.	A-30	–	++	as for germination			Borriss, 1969
	symb.[d]	A-28	–	+++	as for germination			Seitz, 1976
Traunsteinera globosa	dist. water	–	0.05% Tween 80	g	no information given			Eiberg, 1969

[a] Symbols: 0, no germination; g, germination (degree not stated in the original paper); +, moderate germination; ++, good germination; +++, very good germination.
[b] Source of peptone not mentioned in the original paper.
[c] Dissolve kinetin in several drops of NaOH, indoleacetic acid (IAA) in HCl.
[d] Abbreviation for sowing with symbiotic fungi on polypodium or agar medium.
[e] Preparation of potato extract not dealt with in the original paper.
[f] Dissolve gibberellic acid (GA_3) with a few drops of HCl.
[g] Water from green coconuts, stored frozen.
[h] Dissolve benzylamino purine (BAP) with a few drops of NaOH.
[i] Ripe banana, peeled and homogenized with the nutrient solution.
[j] Acid-hydrolized casein (Difco Bacto Casamino Acids, vitamin-free).
[k] Difco Bacto yeast extract.
[l] Activated charcoal, vegetable (E. Merck, Chemicals, 61 Darmstadt, West Germany).
[m] Nicotinic acid, dissolved in 95% ethanol.
[n] Water from ripe coconuts.
[o] Fresh beer yeast from a nearby brewery.
[p] Grass extract: 25 g fresh grass (lawn mixture) boiled for 3–5 min with 250 ml distilled water, and filtered. Add extract to the medium at the rate of 8% (v/v).

Table A-30. Eg1 medium modified, for orchid seed germination and seedling culture (Burgeff, 1936; mod. Borriss, 1969)

Item number	Component	Amount per liter of culture medium (final concentration in culture medium)	Stock solution (a concentrate prepared for repeated and convenient use)	Volume of stock solution per liter of culture medium	Remarks
	Macroelements				
1	Calcium nitrate, $Ca(NO_3)_2 \cdot 4H_2O$[a]	1000 mg	100 g/liter[a]	10 ml[a]	or weigh[a]
2	Ammonium sulfate, $(NH_4)_2SO_4$[a]	250 mg	25 g/liter[a]	10 ml[a]	or weigh[a]
3	Magnesium sulfate, $MgSO_4 \cdot 7H_2O$	250 mg	25 g/liter	10 ml	
4	Ferrous sulfate, $FeSO_4 \cdot 7H_2O$[b]	20 mg	2 g/liter[b]	10 ml[b]	
5	Dipotassium phosphate, K_2HPO_4	250 mg	25 g/liter		
6	Monopotassium phosphate, KH_2PO_4	250 mg	25 g/liter		
	Vitamins[c]				
7	Thiamine (vitamin B_1)	2 mg	200 mg/100 ml 95% ethanol	1 ml[c]	one solution[c]
8	Pyridoxine (vitamin B_6)	2 mg	200 mg/100 ml 95% ethanol		
9	Niacinamide	2 mg	200 mg/100 ml 95% ethanol		
	Complex additive[d]				
10	Coconut water	20 ml			
	Sugar[e]				
11	Sucrose	10 g	no stock	no stock	weigh
12	Water, distilled[e]	to 1000 ml			
	Solidifier[e]				
13	Agar	10 g	no stock	no stock	weigh

[a] Solutions containing ammonium and nitrate may become contaminated on standing. Therefore, stock solutions should not be prepared. If made, they must be kept frozen between uses.

[b] A rust-colored precipitate may form on standing. This solution must, therefore, be shaken vigorously before use.

[c] The vitamins can be combined into one stock solution. Add 1 ml of this solution per liter of culture medium. Keep frozen between uses.

[d] Keep frozen or autoclaved between uses.

[e] Dissolve items 1–4 and 5–6 separately, each in 400 ml of distilled water (item 12), mix the two solutions together, add vitamins (items 7–9) and coconut water (item 10), adjust the pH to 5.3, add sugar (item 11), and bring volume to 1000 ml with more distilled water (item 12). Bring solution to a gentle boil and add agar (item 13) slowly while stirring. When the agar has dissolved completely, distribute the medium into culture vessels, cover, and autoclave.

Table A-31. N_3f medium for orchid seed germination and seedling culture (Burgeff, 1936)

Item number	Component	Amount per liter of culture medium (final concentration in culture medium)	Stock solution (a concentrate prepared for repeated and convenient use)	Volume of stock solution per liter of culture medium	Remarks
	Macroelements				
1	Calcium nitrate, $Ca(NO_3)_2 \cdot 4H_2O$[a]	1000 mg	100 g/liter[a]	10 ml[a]	or weigh[a]
2	Ammonium sulfate, $(NH_4)_2SO_4$[a]	250 mg	25 g/liter[a]	10 ml[a]	or weigh[a]
3	Potassium chloride, KCl	250 mg	25 g/liter	10 ml	
4	Magnesium sulfate, $MgSO_4 \cdot 7H_2O$	250 mg	25 g/liter	10 ml	
5	Ferrous sulfate, $FeSO_4 \cdot 7H_2O$[b]	20 mg	2 g/liter[b]	10 ml[b]	
6	Citric acid	90 mg	no stock	no stock	weigh
7	Dipotassium phosphate, K_2HPO_4	250 mg	25 g/liter	10 ml	
	Sugar[c]				
8	Glucose	10 g	no stock	no stock	weigh
9	Fructose	10 g	no stock	no stock	weigh
10	Water, distilled[c]	to 1000 ml			
	Solidifier[c]				
11	Agar	12 g	no stock	no stock	weigh

[a] Solutions containing ammonium and/or nitrate may become contaminated on standing. Therefore, stock solutions should not be prepared. If made, they must be kept frozen between uses.

[b] A rust-colored precipitate may form on standing. The solution must, therefore, be shaken vigorously before use.

[c] Dissolve items 1–5 and 6–7 separately, each in 400 ml of distilled water (item 10), mix the two solutions together, adjust the pH to 5.3, add sugar, and bring volume to 1000 ml with more distilled water (item 10). Bring the solution to a gentle boil and add agar (item 11) slowly while stirring. When the agar has dissolved completely, distribute the medium into culture vessels, cover, and autoclave.

Table A-32. Pfeffer medium for orchid seed germination and seedling culture (Harvais, 1972)

Item number	Component	Amount per liter of culture medium (final concentration in culture medium)	Stock solution (a concentrate prepared for repeated and convenient use)	Volume of stock solution per liter of culture medium	Remarks
	Macroelements				
1	Calcium nitrate, $Ca(NO_3)_2 \cdot 4H_2O$[a]	800 mg	80 g/liter[a]	10 ml[a]	or weigh[a]
2	Potassium nitrate, KNO_3[a]	200 mg	20 g/liter[a]	10 ml[a]	or weigh[a]
3	Potassium chloride, KCl	100 mg	10 g/liter	10 ml	
4	Potassium phosphate, KH_2PO_4	200 mg	20 g/liter	10 ml	
5	Magnesium sulfate, $MgSO_4 \cdot 7H_2O$	200 mg	20 g/liter	10 ml	
6	Ammonium ferric citrate[a]	1 mg	no stock	no stock	weigh[a]
7	*Microelements*[b]				
a	Manganese sulfate, $MnSO_4$	0.5 mg	0.5 g/liter	1 ml[b]	one solution[b]
b	Boric acid, H_3BO_3	0.5 mg	0.5 g/liter		
c	Zinc sulfate, $ZnSO_4 \cdot 2H_2O$	0.5 mg	0.5 g/liter		
d	Copper sulfate, $CuSO_4 \cdot 5H_2O$	0.5 mg	0.5 g/liter		
	Sugar[c]				
8	Glucose	10 g	no stock	no stock	weigh
9	Water, distilled[c]	to 1000 ml			
	Solidifier[c]				
10	Agar	10 g	no stock	no stock	weigh

[a] Solutions containing ammonium and/or nitrate may become contaminated on standing. Therefore, stock solutions should not be prepared. If made, they must be kept frozen between uses.

[b] Add all microelements to the same liter of distilled water; stir and/or heat until fully dissolved. Add 1 ml per liter of culture medium.

[c] Add items 1–7 to 800 ml of distilled water (item 9), add sugar (item 8) and adjust volume to 1000 ml with more distilled water (item 9). Bring solution to a gentle boil and add agar (item 10) slowly while stirring. When the agar has dissolved completely, distribute the medium into culture vessels, cover, and autoclave.

Table A-33. GD medium for orchid seed germination and seedling culture (Thomale, 1954)

Item number	Component	Amount per liter of culture medium (final concentration in culture medium)	Stock solution (a concentrate prepared for repeated and convenient use)	Volume of stock solution per liter of culture medium	Remarks
	Macroelements				
1	Ammonium sulfate, $(NH_4)_2SO_4$[a]	60 mg	6 g/liter[a]	10 ml[a]	or weigh[a]
2	Potassium nitrate, KNO_3[a]	400 mg	40 g/liter[a]	10 ml[a]	or weigh[a]
3	Ammonium nitrate, NH_4NO_3[a]	370 mg	37 g/liter[a]	10 ml[a]	or weigh[a]
4	Potassium phosphate, KH_2PO_4	300 mg	30 g/liter	10 ml	
5	Magnesium nitrate, $Mg(NO_3)_2 \cdot 6H_2O$	110 mg	11 g/liter	10 ml	or weigh
6	Ferrous sulfate, $FeSO_4 \cdot 7H_2O$[b]	20 mg	2 g/liter[b]	10 ml[b]	
	Sugar[c]				
7	Glucose	10 g	no stock	no stock	weigh
8	Fructose	10 g	no stock	no stock	weigh
9	Water, distilled[c]	to 1000 ml			
	Solidifier[c]				
10	Agar	13 g	no stock	no stock	weigh

[a] Solutions containing ammonium and/or nitrate may become contaminated on standing. Therefore, stock solutions should not be prepared. If made, they must be kept frozen between uses.

[b] A rust-colored precipitate may form on standing. This solution must, therefore, be shaken vigorously before use.

[c] Add items 1–6 to 800 ml of distilled water (item 9), adjust the pH to 5.1, add sugar (items 7, 8) and adjust volume with more distilled water (item 9) to 1000 ml. Bring solution to a gentle boil and add agar (item 10) slowly while stirring. When the agar has dissolved completely, distribute the medium into culture vessels, cover, and autoclave.

Table A-34. F medium for seed germination and seedling culture of European orchids (Fast, 1976)

Item number	Component	Amount per liter of culture medium (final concentration in culture medium)	Stock solution (a concentrate prepared for repeated and convenient use)	Volume of stock solution per liter of culture medium	Remarks
	Macroelements				
1	Calcium nitrate, $Ca(NO_3)_2 \cdot 4H_2O$[a]	80 mg	8 g/liter[a]	10 ml[a]	or weigh[a]
2	Ammonium nitrate, NH_4NO_3[a]	160 mg	16 g/liter[a]	10 ml[a]	or weigh[a]
3	Potassium phosphate, KH_2PO_4	80 mg	8 g/liter	10 ml	
4	Potassium chloride, KCl	160 mg	16 g/liter	10 ml	
5	Magnesium sulfate, $MgSO_4 \cdot 7H_2O$	80 mg	8 g/liter	10 ml	
6	Chelated iron "Fetrilon"[b]	16 mg	no stock	no stock	weigh
7	*Microelements*[c] *(Heller's)*				
a	Zinc sulfate, $ZnSO_4 \cdot 7H_2O$	1 mg	1000 mg/liter		
b	Boric acid, H_3BO_3	1 mg	1000 mg/liter		
c	Manganese sulfate, $MnSO_4 \cdot 4H_2O$	0.1 mg	100 mg/liter		
d	Copper sulfate, $CuSO_4 \cdot 5H_2O$	0.03 mg	30 mg/liter	0.8 ml[c]	one solution[c]
e	Aluminum chloride, $AlCl_3$	0.03 mg	30 mg/liter		
f	Nickel chloride, $NiCl_2 \cdot 6H_2O$	0.03 mg	30 mg/liter		
g	Potassium iodide, KI	0.01 mg	10 mg/liter		
	Complex additives				
8	Peptone[d]	1.5 mg	no stock	no stock	weigh
9	Autolyzed beer yeast[e]	8 ml	no stock	no stock	
	Sugar[f]				
10	Sucrose	12 g	no stock	no stock	weigh
11	Fructose	5 g	no stock	no stock	weigh
12	Water, distilled[f]	to 1000 ml			
	Solidifier[f]				
13	Agar	8 g			

[a] Solutions containing ammonium and/or nitrate may become contaminated on standing. Therefore, stock solutions should not be prepared. If made, they must be kept frozen between uses.
[b] Fetrilon (5% Fe) is a product of Badische Anilin und Sodafabrik (BASF), 67 Ludwigshafen, West Germany.
[c] Add all microelements to the same liter of distilled water; stir and/or heat until dissolved. Add 0.8 ml per liter of culture medium.
[d] Peptone from meat, dried: a product of E. Merck, Chemicals, 61 Darmstadt, West Germany. Dissolve it in several ml of distilled water (item 12).
[e] Incubate fresh beer yeast from a brewery for 48 hr at 30°C, then filter through a paper filter. The filtrate must be kept refrigerated.
[f] Dissolve items 1–7 in 800 ml of distilled water (item 12), add peptone (item 8) and yeast filtrate (item 9). Adjust the pH to 5.5, add sugar (items 10, 11) and bring the volume to 1000 ml with more distilled water (item 12). Add agar (item 13) slowly, while stirring, to the gently boiling solution. When fully dissolved, dispense the medium into culture vessels, cover, and autoclave.

from the undehisced capsules. (b) *Medium:* The agar medium should be (at least until the protocorm stage) of relatively low nutrient content. The addition of peptone, yeast, or B-vitamins at that stage has a growth-promoting effect. (c) *Light:* Nearly all European orchids germinate better in the dark; cultures should be exposed to light only after shoots begin to appear. (d) *Temperature:* For normal development some species require a periodic lowering of temperature to correspond with their annual growth cycle.

By observing these fundamentals, it is possible to germinate and grow European orchids under laboratory conditions in a manner similar to that of tropical epiphytes. However, a difference is that considerable difficulties arise when the seedlings are transferred from sterile media to natural substrates. At this stage growth usually stops and a large number of plants become parasitized and die. The task of future research should be to explore ways in which these obstacles can be overcome, either through use of other,

Table A-35. FN medium for seed germination and seedling culture of European orchids (Fast, 1978)

Item number	Component	Amount per liter of culture medium (final concentration in culture medium)	Stock solution (a concentrate prepared for repeated and convenient use)	Volume of stock solution per liter of culture medium	Remarks
	Macroelements				
1	Calcium nitrate, $Ca(NO_3)_2 \cdot 4H_2O$[a]	65 mg	6.5 g/liter[a]	10 ml[a]	or weigh[a]
2	Ammonium nitrate, NH_4NO_3[a]	65 mg	6.5 g/liter[a]	10 ml[a]	or weigh[a]
3	Potassium phosphate, KH_2PO_4	33 mg	3.3 g/liter	10 ml	
4	Potassium chloride, KCl	33 mg	3.3 g/liter	10 ml	
5	Magnesium sulfate, $MgSO_4 \cdot 7H_2O$	33 mg	3.3 g/liter	10 ml	
6	Chelated iron "Fetrilon"[b]	13 mg	no stock	no stock	weigh
	Complex additives				
7	Sodium nucleinate[c]	0.13 g	no stock	no stock	weigh
8	Peptone[d]	1.3 g	no stock	no stock	weigh
	Sugar[e]				
9	Sucrose	7 g	no stock	no stock	weigh
10	Water, distilled[e]	to 1000 ml			
	Solidifier[e]				
11	Agar	8 g	no stock	no stock	weigh

[a] Stock solutions containing ammonium and/or nitrate may become contaminated on standing. Therefore, stock solutions should not be prepared. If made, they must be kept frozen between uses.

[b] Fetrilon (5% Fe) is a product of the Badische Anilin und Sodafabrik (BASF), 67 Ludwigshafen, West Germany.

[c] Sodium nucleinate from yeast is a product of E. Merck, Chemicals, 61 Darmstadt, West Germany.

[d] Peptone from meat, dried, is a product of E. Merck, Chemicals, 61 Darmstadt, West Germany. Dissolve it in several ml of distilled water (item 10) before adding it to the other components.

[e] Add components 1–6 to 800 ml of distilled water (item 10), add the dissolved sodium nucleinate and peptone (items 7, 8), adjust the pH to 5.5, add sugar (item 9), and bring the volume to 1000 ml with more distilled water (item 10). Add the agar (item 11) slowly, while stirring, to the gently boiling solution. When fully dissolved, distribute the medium into culture vessels, cover, and autoclave.

Table A-36. CT-1 medium for seed germination and seedling culture of European orchids (Lucke, 1976)

Item number	Component	Amount per liter of culture medium (final concentration in culture medium)	Stock solution (a concentrate prepared for repeated and convenient use)	Volume of stock solution per liter of culture medium	Remarks
	Macroelements				
1	Ammonium phosphate, $(NH_4)_2HPO_4$[a]	100 mg	10 g/liter[a]	10 ml[a]	or weigh[a]
2	Potassium phosphate, KH_2PO_4	100 mg	10 g/liter	10 ml	
3	Magnesium sulfate, $MgSO_4 \cdot 7H_2O$	50 mg	5 g/liter	10 ml	
	Vitamins[b]				
4	Thiamine (vitamin B_1)	0.1 mg	10 mg/100 ml 95% ethanol	1 ml[b] (items 4–5)	one solution[b] (items 4–5)
5	D(+)Biotin	0.2 mg	20 mg/100 ml 95% ethanol		
	Complex additive				
6	Dried grass extract[c]	2 g	no stock	no stock	weigh
	Sugar[d]				
7	Sucrose	10 g	no stock	no stock	weigh
8	Water, distilled[d]	to 1000 ml			
	Solidifier[d]				
9	Agar	10 g	no stock	no stock	weigh

[a] Solutions containing ammonium may become contaminated on standing. Therefore, stock solutions should not be prepared. If made, they must be kept frozen between uses.

[b] The vitamins may be combined into one stock solution. Refrigerate between uses.

[c] To prepare dried grass extract: cut fresh grass (lawn mixture), make an extract with water, and dry the filtered liquid under reduced pressure ("Unterdruckverfahren").

[d] Dissolve components 1–6 in 800 ml of distilled water (item 8), adjust the pH to 5.5–6.0, add sugar (item 7), and adjust volume to 1000ml with more distilled water (item 8). Bring the solution to a gentle boil and add agar (item 9) slowly while stirring. When the agar has dissolved completely, distribute the medium into culture vessels, cover, and autoclave.

Table A-37. Chang medium, modified, for seed germination and seedling culture of European orchids (Harbeck, 1963)

Item number	Component	Amount per liter of culture medium (final concentration in culture medium)	Stock solution (a concentrate prepared for repeated and convenient use)	Volume of stock solution per liter of culture medium	Remarks
	Complex substances				
1	Atlas fish emulsion[a]	6 ml	no stock	no stock	measure volume
2	Peptone[b]	2 g	no stock	no stock	weigh
	Sugar[c]				
3	Sucrose	20 g	no stock	no stock	weigh
4	Water, distilled[c]	to 1000 ml			
	Solidifier[c]				
5	Agar	12 g	no stock	no stock	weigh

[a] Atlas fish emulsion: a product of Atlas Fish Fertilizer Co., 1 Drumm Street, San Francisco, CA 94111 USA.
[b] Peptone from meat, dried: a product of E. Merck, Chemicals, 61 Darmstadt, West Germany.
[c] Add items 1 and 2 to 900 ml of distilled water (item 4), adjust the pH to 5.5, add sugar (item 3), and adjust the volume to 1000 ml with more distilled water (item 4). Bring solution to a gentle boil and add agar (item 5) slowly while stirring. When the agar has dissolved fully, distribute the medium into culture vessels, cover, and autoclave.

more suitable substrates or by treatment with substances which improve the growth of plants and their resistance to attacks.

Acknowledgments

My experiments were carried out at the Institut für Bodenkunde und Pflanzenernährung der Fachhochschule Weihenstephan in Freising, West Germany. I am very grateful to the director, F. Penningsfeld, who has supported and encouraged my investigations. Thanks are also due to Joseph Arditti and Nancy Thym-Hochrein for commenting on and assisting with the translation.

Table A-38. NB-1 medium for seed germination and seedling culture of European orchids (Vöth, 1976)

Item number	Component	Amount per liter of culture medium (final concentration in culture medium)	Stock solution (a concentrate prepared for repeated and convenient use)	Volume of stock solution per liter of culture medium	Remarks
	Complex substances				
1	Medical yeast[a]	5 g	no stock	no stock	weigh
2	Peptone[b]	3 g	no stock	no stock	weigh
3	Activated charcoal (veg.)[c]	2 g	no stock	no stock	weigh
	Sugar				
4	Sucrose	10 g	no stock	no stock	weigh
5	Raw sugar[d]	10 g	no stock	no stock	weigh
6	Water, distilled[e]	to 1000 ml			
	Solidifier[e]				
7	Agar	10 g	no stock	no stock	weigh

[a] Medical yeast (*Faex medicinalis*) is beer yeast, washed, bitter substances removed, dried at 40°C, and pulverized. Address of source is not given in the original paper.
[b] Peptone: source and address of supply are not given in the original paper.
[c] Activated charcoal: address of supply is not given in the original paper.
[d] Raw sugar: uncleaned sugar. Source and address of supply are not given in the original paper.
[e] Mix items 1–3 with 800 ml of distilled water (item 6), adjust the pH to 5.5, add the sugar (items 4, 5), and adjust the volume to 1000 ml with more distilled water (item 6). Bring the solution to a gentle boil and add the agar (item 7) slowly while stirring. When the agar has fully dissolved, pour the medium into culture vessels, cover, and autoclave.

Table A-39. Zak medium for seed germination of European orchids (Borriss, 1969)

Item number	Component	Amount per liter of culture medium (final concentration in culture medium)	Stock solution (a concentrate prepared for repeated and convenient use)	Volume of stock solution per liter of culture medium	Remarks
	Complex substance				
1	Coconut water[a]	20 ml	no stock	no stock	measure volume
	Sugar[b]				
2	Sucrose	10 g	no stock	no stock	weigh
3	Water, distilled[b]	to 1000 ml			
	Solidifier[b]				
4	Agar	15 g	no stock	no stock	weigh

[a] Coconut milk from ripe fruits, autoclaved.
[b] Mix item 1 with 900 ml of distilled water (item 3), add sugar (item 2), and adjust the volume to 1000 ml with more distilled water (item 3). Bring the solution to a gentle boil and add agar (item 4) slowly while stirring. When the agar has dissolved completely, distribute the medium into culture vessels, cover, and autoclave.

Literature Cited

Borriss, H. 1969. Samenvermehrung und Anzucht europäischer Erdorchideen. Ber. 2. Europ. Orchideenkongress, Paris, p. 74–78.
Burgeff, H. 1911. Die Anzucht tropischer Orchideen aus Samen. G. Fischer Verlag, Jena.
——. 1936. Samenkeimung der Orchideen und Entwicklung ihrer Keimpflanzen. G. Fischer Verlag, Jena.
——. 1954. Samenkeimung und Kultur europäischer Erdorchideen nebst Versuchen zu ihrer Verbreitung. Gustav Fischer Verlag, Stuttgart.
Champagnat, M. 1971. Recherches sur la multiplication végétative de Neottia nidus-avis Rich. Annales des Sciences Naturelles, Botanique, Ser. 12, 12:209–248.
Downie, D. G. 1940. On the germination and growth of Goodyera repens. Trans. and Proc. Bot. Soc. Edin. 33:36–51.
——. 1943. Notes on the germination of Corallorhiza innata. Trans. and Proc. Bot. Soc. Edin. 33:380–382.
——. 1949. The germination of Goodyera repens (L.) R. Br. in fungal extract. Trans. and Proc. Bot. Soc. Edin. 35:120–125.
Eiberg, H. 1969. Keimung europäischer Erdorchideen. Die Orchidee 20:266–270.
Fast, G. 1974. Über eine Methode der kombinierten generativen-vegetativen Vermehrung von Cypripedium calceolus L. Die Orchidee 25:125–129.
——. 1976. Möglichkeiten zur Massenvermehrung von Cypripedium calceolus und anderen europäischen Wildorchideen. Proc. 8th World Orchid Conf., Frankfurt (1975), p. 359–363.
——. 1978. Über das Keimverhalten europäischer Erdorchideen bei asymbiotischer Aussaat. Die Orchidee 29:270–274.
Haas, N. 1977. Asymbiotische Vermehrung europäischer Erdorchideen. I and II. Die Orchidee 28:27–31, 69–73.
Hadley, G. 1970. The interaction of kinetin, auxin and other factors in the development of north temperate orchids. New Phytol. 69:549–555.
Harbeck, M. 1961. Erfahrungen mit der Aussaat von Orchis maculata auf sterilem Nährboden. Die Orchidee 12:67–70.
——. 1963. Einige Beobachtungen bei der Aussaat europäischer Erdorchideen auf sterilen Nährböden. Die Orchidee 14:58–65.
——. 1964. Die Anzucht von Orchis maculata vom Samen bis zur Blüte. Die Orchidee 15:57–60.
Harvais, G. 1972. The development and growth requirements of Dactylorhiza purpurella in asymbiotic cultures. Can. J. Bot. 50:1223–1229.
Lucke, E. 1971. Zur Samenkeimung mediterraner Ophrys. Die Orchidee 22:62–65.
——. 1976. Erste Ergebnisse zur asymbiotischen Samenkeimung von Himantoglossum hircinum. Die Orchidee 27:60–61.

Seeman, G. 1953. Uber eine neué Wasserungsmethode fur schwerkeimende Orchideesnsamen. Die Orchidee 4(2):56.
Seitz, H. M. 1976. Symbiotische Samenkeimung bei europäischen Orchideen. Proc. 8th World Orchid Conf., Frankfurt (1975), p. 343–350.
Sprau, F. 1939. Über die Anzucht heimischer Orchideen zwecks Salepgewinnung. Orchis 17:41–44.
Stoutamire, W. 1974. Terrestrial orchid seedlings, p. 101–128. *In* C. L. Withner (ed.), The orchids: Scientific studies. Wiley-Interscience, New York.
Thomale, H. 1954. Die Orchideen. Verlag Eugen Ulmer, Stuttgart.
Veyret, Y. 1974. Development of the embryo and the young seedling stages of orchids, p. 223–265. *In* C. L. Withner (ed.), The orchids: Scientific studies. Wiley-Interscience, New York.
Vöth, W. 1975. Asymbiotische Aussaat und Jungpflanzenentwicklung von Orchis sancta, O. coriophora ssp. fragrans und O. x kallithea. Gärtner.-bot. Brief 45:1492–1499.
——. 1976. Aussaat und Kultur von Serapias parviflora und S. orientalis. Proc. 8th World Orchid Conf., Frankfurt (1975), p. 351–358.

Bibliography

Allenberg, H. 1976. Notizen zur Keimung, Meristemkultur und Regeneration von Erdorchideen. Die Orchidee 27:28–31.
Arditti, J. 1967. Factors affecting the germination of orchid seeds. Bot. Rev. 33:1–97.
Borriss, H. and L. Albrecht. 1969. Rationelle Samenvermehrung und Anzucht europäischer Erdorchideen. Gartenwelt 69:511–514.
Curtis, J. T. 1943. Germination and seedling development in five species of Cypripedium L. Amer. J. Bot. 30:199–206.
Hadley, G., and G. Harvais. 1968. The effect of certain growth substances on asymbiotic germination and development of Orchis purpurella. New Phytol. 67:441–445.
Harvais, G. 1974. Notes on the biology of some native orchids of Thunder Bay, their endophytes and symbionts. Can. J. Bot. 52:451–460.

European Terrestrial Orchids

Geoffrey Hadley

North temperate orchids are almost exclusively terrestrial and in general less conspicuous and less frequent than their tropical relatives. It is not surprising that they have not been exploited for horticultural purposes. Nevertheless, much interest has been shown in the germination of north temperate species and they have been used quite extensively for scientific research. Bernard (1899) based his symbiosis hypothesis on observations with *Neottia nidus-avis,* although his subsequent studies and the later development of the nonsymbiotic method of germination by Knudson were largely carried out with subtropical, or at least non-European, species.

In her investigations into the germination of European species Downie (1941) followed Knudson's methods but used "the mineral nutrient solution . . . of Pfeffer" without any explanation of the basis for his choice. The main difference from Knudson C is that Pfeffer medium (Table A-39a) lacks ammonium ions. Downie's methods were followed by Hadley (1970), but there has been little critical work on north temperate species to compare with the investigations of nutritional and environmental factors for other groups of orchids.

The methods described here have been used for several species of the genus *Dactylorhiza* and for representatives of other genera (*Anacamptis, Calypso, Coeloglossum, Corallorhiza, Cypripedium, Epipactis, Goodyera, Gymnadenia, Habenaria, Liparis, Listera, Neottia,*

Nigritella, Ophrys, Orchis, and *Platanthera*). However, in some cases the available information relates to only one or two tests or experiments and may not necessarily be conclusive.

Plant Material. Mature seeds of north temperate species can easily be collected during late summer or autumn when the capsules become mature and start to change color from green to brown. Single intact capsules (not all the ones from a plant, since the stage of development of capsules varies from top to bottom of the flowering spike) should be collected, allowed to dry, and then opened to disperse the seeds over a sheet of white paper. Near-mature capsules will often open naturally if covered and left to dry. Seeds can be liberated by crushing dry capsules if necessary, then filtered through wire gauze.

Immature seeds from intact green capsules can be used following the methods described for tropical orchids (p. 273). Surface sterilization of the capsule is advisable since north temperate species growing in humid conditions carry a surface microflora that may lead to contamination unless suitable precautions are taken.

Seeds can be stored in tightly capped vials at low temperature (0–5° C) and should remain viable for some months, although percentage germination may fall dramatically within one year.

Surface Sterilization. Mature seeds must be sterilized; the usual agent is calcium hypochlorite (pp. 254, 261, 275). The use of readily available commercial (household bleach) preparations that include a wetting agent simplifies the procedure, and shaking for 15 min in a 10% (v/v) solution is sufficient. The techniques described elsewhere (p. 275) apply equally to north temperate terrestrial orchids, but those species which have resistant seed coats can be sterilized for longer periods, often up to 60 min, by which time the brown seed coat may be bleached, without damaging the embryos.

Culture Conditions. Temperatures of 20–25° C are suitable, but germination (leading to slower development) will occur equally at 15–20° C. Illumination is not necessary and inhibits germination of *Coeloglossum viride, Dactylorhiza purpurella,* and possibly other species. Therefore, cultures should be incubated in darkness at least until germination has commenced. Since the protocorms are heterotrophic, transfer to diffuse daylight or artificial illumination becomes necessary only when development of the green shoot occurs, i.e., after some weeks or months, dependent on the species. Procedures for transplanting to fresh containers have been described (p. 277).

Culture Media. Pfeffer medium (Table A-39a) has been used for most work with north temperate terrestrial species. The addition of 0.1% yeast extract or potato extract has been found to be beneficial, especially for symbiotic germination of orchids growing in dual culture with nutritionally exacting fungi. Media containing undefined materials such as coconut milk have been used more extensively in recent studies by Linden (1980).

The sugar normally employed with Pfeffer medium is 1–2% dextrose (glucose). Sucrose is equally suitable. Sugar concentrations as low as 0.1% have been found satisfac-

Table A-39a. Composition (mg/liter) of two media used for germination of north temperate orchids

Item number	Component	Medium	
		Pfeffer	Eg1 (Linden, 1980)
	Macroelements		
1	Calcium nitrate, $Ca(NO_3)_2 \cdot 4H_2O$	800	1000
2	Potassium phosphate, KH_2PO_4	200	250
3	Dipotassium phosphate, K_2HPO_4	–	250
4	Magnesium sulfate, $MgSO_4 \cdot 7H_2O$	200	250
5	Ammonium sulfate, $(NH_4)_2SO_4$	–	250
6	Potassium nitrate, KNO_3	200	–
7	Potassium chloride, KCl	100	–
8	Calcium phosphate, $Ca_3(PO_4)_2$	–	–
9	*Trace elements*[a]		
a	Iron, $FeSO_4 \cdot 7H_2O$	0.97	20
b	Copper, $CuSO_4 \cdot 5H_2O$	0.16	–
c	Zinc, $ZnSO_4 \cdot 7H_2O$	0.74	–
d	Manganese, $MnSO_4 \cdot 4H_2O$	0.08	–
e	Molybdenum, $(NH_4)_6MO_7O_{24} \cdot 4H_2O$	0.26	–
	Other items[b]		
10	Glucose	20 g	–
11	Sucrose	–	10 g
12	Coconut milk (Gibco, USA)		20 ml
13	Agar	10–15 g	10 g

[a] Usually kept as combined stock solution.
[b] 0.1% potato extract, 0.1% yeast extract, or 10% (v/v) coconut milk may be added to Pfeffer medium (Hadley, 1970) if required.

tory for germination of *Dactylorhiza purpurella* and some other species and are used for symbiotic germination tests.

Symbiotic Germination. The use of symbiotic germination methods is outside the scope of most growers unless facilities are available for culturing the fungi involved. There are also difficulties in isolating, recognizing, and identifying the fungi. However, if actively growing cultures are available the technique differs from that of asymbiotic germination only in certain respects. The growth medium should not include the usual 1–2% sugar since this will encourage the fungus to smother the seeds. Not more than 0.1% glucose (100 mg per liter) can be used. Alternatively, very finely divided (ball-milled) cellulose, prepared from chromatography-grade cellulose powder, is suitable, but has the disadvantage of making the growth medium opaque.

The seed is sown in the usual way and the fungus can be added at the same time, or a few days later, as a small section of agar (ca. 5 mm diam.) cut from a culture growing on any suitable medium. From this source the fungus will grow over the orchid seed and, if it is compatible, will infect the cells and give rise to symbiotic (mycorrhizal) protocorms. Those protocorms which show enhanced rates of growth can be transplanted to fresh tubes or flasks as required.

Individual Nutrient Requirements. Pfeffer glucose medium without any supplement is suitable for germination of *Coeloglossum viride, Dactylorhiza purpurella* (and probably most other species of that genus), *Habenaria saccata, Ophrys apifera, Platanthera bifolia,* and *P.*

chlorantha. Species that germinate on Pfeffer with potato extract added include *Corallorhiza innata* (Downie, 1943), *Epipactis gigantea,* and *Goodyera repens;* also, germination of *Platanthera* species is enhanced on this medium. A yeast-extract supplement appears necessary for *Anacamptis pyramidalis* and *Leuchorchis albida,* and probably for *Gymnadenia conopsea* and *Listera ovata.*

Some other species will germinate on media containing yeast extract or coconut milk (Linden, 1980) viz. *Calypso bulbosa, Cypripedium calceolus* and *C. reginae, Dactylorhiza* spp. and *Orchis* spp.; whether these supplements are essential is not known.

General Comments. The fact that undefined supplements are required for germination of some species suggests that north temperate orchids may be more exacting than many tropical forms. It must also be remembered that temperate species usually grow slowly compared with tropical and epiphytic orchids, and it may be many weeks or months before they can be transplanted. The protocorms are usually colorless until they are large enough to produce a shoot initial. This will turn green and the cultures should be kept in daylight or under artificial illumination at this stage.

Literature Cited

Bernard, N. 1899. L'evolution dans la symbiose. Ann. Sci. Nat. Bot. 9:1–196.

Downie, D. G. 1941. Notes on the germination of some British orchids. Trans. and Proc. Bot. Soc. Edin. 33:94–103.

——. 1943. Notes on the germination of Corallorhiza innata. Trans. and Proc. Bot. Soc. Edin. 33:380–390.

Hadley, G. 1970. The interaction of kinetin, auxin and other factors in the development of north temperate orchids. New Phytol. 69:549–555.

Linden, B. 1980. Aseptic germination of seeds of northern terrestrial orchids. Ann. Bot. Fennici 17:174–182.

Native Orchids of France

Joseph Arditti

With the exception of *Cypripedium* and three genera in the tribe Kerosphaeroideae (*Liparis, Malaxis,* and *Corallorhiza*), the orchids in the French flora belong to the tribes Ophryoideae and Polychondroideae. These orchids have, on the whole, proven difficult to germinate asymbiotically, but some of them have been germinated without the aid of mycorrhizae (Veyret, 1969).

Plant Material. Mature seeds of several species seem to have been used in the original research.

Surface Sterilization. No details are given in the original paper. The procedures recommended for tropical species (p. 275) can be used.

Culture Vessels. The culture vessels used for tropical species (p. 275) can be employed.

Culture Conditions. No information is given in the original paper. The conditions recommended for North American (p. 278), Australian (p. 296), and German (p. 309) terrestrial orchids could be used.

Culture Medium. The medium used includes components of Knudson C and Heller's media, two sugars, and a relatively low concentration of agar (Table A-40).

Procedure. Details are not given and therefore the procedures used for other orchids (pp. 276, 278, 296, 309) should be employed.

Developmental Sequence. Some species failed to germinate, others formed protocorms, and several developed seedlings (Table A-41).

General Comments. The problems encountered in the asymbiotic germination of north temperate terrestrial orchids appear to be as formidable in France as they are in the United States.

Table A-40. A medium used for germination and seedling culture of orchids native to France (Veyret, 1969)

Item number	Component	Amount per liter of culture medium (final concentration in culture medium)	Stock solution (a concentrate prepared for repeated and convenient use)	Volume of stock solution per liter of culture medium	Remarks
	Macroelements				
1	Calcium nitrate, $Ca(NO_3)_2 \cdot 4H_2O$[a]	1 g	100 g/liter[a]	10 ml[a]	or weigh[a]
2	Potassium phosphate, KH_2PO_4	250 mg	25 g/liter	10 ml	
3	Magnesium sulfate, $MgSO_4 \cdot 7H_2O$	250 mg	25 g/liter	10 ml	
4	Ammonium sulfate, $(NH_4)_2SO_4$[a]	500 mg	50 g/liter[a]	10 ml[a]	or weigh[a]
	Iron				
5	Ferric chloride, $FeCl_3 \cdot 6H_2O$[b]	1 mg	100 g/liter[b]	10 ml[b]	
6	*Microelements*[c]				
a	Manganese sulfate, $MnSO_4 \cdot 4H_2O$	0.01 mg	10 mg/liter	1 ml[c]	one solution[c]
b	Boric acid, H_3BO_3	1 mg	1 g/liter		
c	Molybdic acid, MoO_3	0.02 mg	20 mg/liter		
d	Aluminum chloride, $AlCl_3$	0.03 mg	30 mg/liter		
e	Cupric sulfate (anh.), $CuSO_4$	0.03 mg	30 mg/liter		
f	Nickel chloride, $NiCl_2 \cdot 6H_2O$	0.03 mg	30 mg/liter		
g	Zinc sulfate, $ZnSO_4 \cdot 7H_2O$	1 mg	1 g/liter		
h	Potassium iodide, KI	0.01 mg	10 mg/liter		
	Sugar[d]				
7	Sucrose	10 g	no stock		weigh
8	Glucose	15 g	no stock		
9	Water, distilled[d]	to 1000 ml			
	Solidifier[d]				
10	Agar	8 g	no stock		weigh

[a] Solutions containing ammonium and/or nitrate may become contaminated on standing. Therefore, stock solutions should not be prepared. If made, they must be kept frozen between uses.

[b] Shake well before use.

[c] Add all microelements to the same one liter of distilled water; stir and/or heat until they are dissolved, and add 1 ml per liter of culture medium.

[d] Add items 1–6 to 650 ml distilled water (item 9); adjust the pH to 5.2; add sugar (items 7, 8) and adjust volume to 1000 ml with more distilled water (item 9). Bring solution to a gentle boil and add agar (item 10) slowly while stirring. When the agar has dissolved completely (and the solution turns a clear amber color), distribute the medium into culture vessels, cover, autoclave, cool, and use.

Table A-41. Results of attempts to germinate French orchids on asymbiotic media (Veyret, 1969)

Failed to germinate	Protocorm formation only	Good protocorm formation or plant development
Anacamptis pyramidalis	*Aceras anthropophorum*	*Coeloglossum viride*
Corallorhiza innata	*Goodyera repens*	*Dactylorchis durandii*
Cypripedium calceolus	*Gymnadenia conopea*	*D. maculata*
Epipactis atrorubens	*Listera ovata*	*D. majalis*
E. latifolia	*Orchis maculata*	*D. munbyana*
E. palustris	*O. maculata* var. *elodes*	*D. praetermissa*
Goodyera repens	*O. militaris*	*Liparis loeselii*[a]
Gymnadenia albida	*O. morio*	*Listera cordata*[a]
Liparis loeselii[a]	*Platanthera bifolia*	
Listera cordata[a]		
Neottia nidus-avis		
Ophrys apifera		
O. arachnites		
O. aranifera		
Orchis latifolia		

[a] These species appear on both lists in the original paper (Veyret, 1969).

Literature Cited

Veyret, Y. 1969. La structure des semences des Orchidaceae et leur aptitude a la germination in vitro en cultures pures. Mus. Natl. d'Hist. Nat., Trav. Lab "La Jaysinia" 3:89–97.

Japanese Orchids

Goro Nishimura

Procedures have been developed for the germination of several orchid species which are native to Japan.

Calanthe

Calanthe is primarily a terrestrial genus and it is therefore not surprising that seed germination has posed difficulties (Hasegawa *et al.*, 1976; Hibino *et al.*, 1978; Ichihashi and Yamashita, 1977; Ito, 1976; Kano, 1965, 1967, 1968; Mii, 1978; Mii and Kako, 1974, 1976; Nagashima, 1977, 1979a,b; Suzuki and Abe, 1977).

Plant Material. Depending on the species, both mature and immature seeds can be germinated (Table A-42).

Table A-42. Age and germination in *Calanthe* seeds

Species	Age
Calanthe discolor	Mature
	Immature, starting 50 days following pollination; best germination after 60–80 days.
C. discolor × *C. sieboldii*	Mature
C. furcata	Immature, starting 60 days following pollination; best germination on 70th and 120th days
C. izu-insularis	Mature
C. sieboldii	Mature
	Immature, starting 70 days following pollination; best germination after 80–120 days.

Table A-43. Surface sterilization of *Calanthe* seeds

Species	Procedure
Mature seeds:	
Calanthe discolor	Soak in calcium hypochlorite (5 g/250 ml) for 10 min, calcium hypochlorite (10 g/140 ml) for 15–20 min, potassium hypochlorite (1 g/100 ml) for 12 min, or potassium hypochlorite (10 g/100 ml) for 8 min.
C. discolor × *C. sieboldii*	See *C. discolor.*
C. furcata	Not given in the literature; try procedure used for *C. discolor.*
C. izu-insularis	Soak in 1/10 dilution of commercial sodium hypochlorite (Antiformin; available chlorine, 5%) for 8 min.
C. sieboldii	See *C. discolor*
Immature seeds:	
All species	Surface-sterilize capsules as indicated for tropical orchids (p. 275).

Surface Sterilization. Sterilization procedures vary with the species and maturity of the seeds (Table A-43).

Culture Vessels. Culture vessels recommended for tropical orchids can also be used for *Calanthe.* The seeds germinate equally well in 50-, 100-, 200-, or 300-ml flasks, but the plantlets grow better in larger vessels.

Culture Conditions. The best conditions for seed germination and seedling growth vary with the species (Table A-44).

Culture Medium. The selection of culture medium depends on the species being germinated (Table A-45).

Procedure. Pretreat and sterilize seeds as required (Table A-49) and place on medium using the procedures outlined for tropical orchids (p. 273). Then place the flasks under the appropriate conditions (Table A-44). When the flasks become crowded or the seedlings grow large they should be transflasked.

Table A-44. Culture conditions for *Calanthe* species

	Conditions	
Species	Seed germination	Seedling growth
Calanthe discolor	25°C and 16-hr photoperiods of 400–500 lux; or 27°C and 16-hr photoperiods of 2300 lux; or 23°C and 24 hr of 4000 lux for 7 months; or 25°C in the dark followed by light 5 months later	Same as for seed germination
C. discolor × *C. sieboldii*	See *C. discolor*	Same as for seed germination
C. furcata	27°C and 12 hr light	Same as for seed germination
C. izu-insularis	23–25°C in the dark for 130 days	Move gradually to 23–25°C and 16-hr photoperiods of 400–500 or 1200 lux
C. sieboldii	See *C. furcata*; 30°C enhances germination	Same as for seed germination

Table A-45. Culture media for the germination of *Calanthe*

Species	Medium
Calanthe discolor	
Mature seeds	Hyponex[a] (Table A-47) or Ichihashi and Yamashita (Table A-46), pH 5.0–5.4
Immature seeds	Hyponex[a] (Table A-47) or Murashige and Skoog (inorganic elements only)[b] plus 2% peptone, pH 5.2–5.4
C. discolor × *C. sieboldii*	
Mature seeds	Hyponex[a] (Table A-47), pH 5.0
Immature seeds	Hyponex[a] (Table A-47), or Murashige and Skoog (inorganic elements only) plus 2% peptone, pH 5.2–5.4
C. izu-insularis	Nitsch (Table A-48)

[a] Hydroponic Chemical Co., Inc., Copley, OH, USA.
[b] Arditti, 1977.

Developmental Sequence. Embryos will swell after approximately 2 to 3 months then burst through the testae, form protocorms, and develop into plantlets.

General Comments. Like most terrestrial orchids *Calanthe* species are not easy to germinate but the procedures described here can be used successfully.

Table A-46. Ichihashi and Yamashita (1977) medium used for the germination of ripe seeds of *Calanthe discolor*

Item number	Component	Amount per liter of culture medium (final concentration in culture medium)	Stock solution (a concentrate prepared for repeated and convenient use)	Volume of stock solution per liter of culture medium	Remarks
	Macroelements				
1	Calcium nitrate, $C(NO_3)_2 \cdot 4H_2O$[a]	826 mg	82.6 g/liter[a]	10 ml[a]	or weigh[a]
2	Potassium nitrate, KNO_3[a]	747 mg	74.7 g/liter[a]	10 ml[a]	or weigh[a]
3	Ammonium phosphate, $NH_4H_2PO_4$[a]	391 mg	39.1 g/liter[a]	10 ml[a]	or weigh[a]
4	Magnesium sulfate $MgSO_4 \cdot 7H_2O$	172 mg	17.2 g/liter	10 ml	
5	*Chelated iron and microelements*[b]				
a	Na_2EDTA	37.3 mg	3.73 g/liter	10 ml[b]	one solution[b]
b	Ferrous sulfate, $FeSO_4 \cdot 7H_2O$	27.8 mg	2.78 g/liter		
c	Boric acid, H_3BO_3	6.2 mg	620 mg/liter		
d	Manganese sulfate, $MnSO_4 \cdot 4H_2O$	22.3 mg	2.23 g/liter		
e	Zinc sulfate, $ZnSO_4 \cdot 4H_2O$	8.6 mg	860 mg/liter		
f	Potassium iodide, KI	0.83 mg	83 mg/liter		
g	Sodium molybdate, $Na_2MoO_2 \cdot 2H_2O$	0.25 mg	25 mg/liter		
h	Copper sulfate, $CuSO_4 \cdot 5H_2O$	0.025 mg	2.5 mg/liter		
i	Cobaltous chloride, $CoCl_2 \cdot 6H_2O$	0.025 mg	2.5 mg/liter		
	Sugar[c]				
6	Sucrose	20 g	no stock	no stock	weigh
	Solidifier[c]				
7	Agar	10 g	no stock	no stock	weigh
8	Water, distilled[c]	to 1000 ml			

[a] Solutions containing ammonium or nitrate may become contaminated on standing. Therefore, stock solutions should not be prepared. If made, they must be kept frozen between uses.
[b] Add the chelated iron (items 5a–5b) and all microelements (items 5c–5i) to the same one liter of distilled water; stir and/or heat until dissolved.
[c] Add items 1–5 to 800 ml of distilled water (item 8), adjust pH to 5.0–5.4, and bring volume to 1000 ml with distilled water (item 8); add sugar (item 6); heat to gentle boil and add agar (item 7) with stirring; when agar has dissolved completely distribute into culture vessels, autoclave, cool, and use.

Table A-47. Hyponex medium for the germination of *Calanthe* species

Item number	Component	Amount per liter of culture medium (final concentration in culture medium)
	Macroelements	
1	Hyponex[a]	3 g
	Sugar	
2	Sucrose	20, 30, or 35 g[b]
	Complex additive	
3	Peptone or tryptone	2 g[c]
	Darkening agent	
4	Activated charcoal[d]	2 g
	Solidifier[e,f]	
5	Agar	8, 10, 12, or 15 g[e]
6	Water, distilled[f]	to 1000 ml

[a] Available from most nurseries and orchid dealers or Hydroponic Chemical Co., Inc., Copley, OH, USA.

[b] Use 30 g/liter for immature seeds of *Calanthe sieboldii, C. furcata, Cymbidium goeringii,* and *Cymbidium gyokuchin;* 20 g/liter for immature seeds of *Calanthe discolor;* and 35 g/liter for all others, including mature seeds of the two *Cymbidium* species.

[c] Use 2 g/liter for all species; tryptone can be used for *C. discolor.*

[d] There are no reports of charcoal incorporation into media used for *C. furcata* germination, but there is no reason to believe that it could not be used for this species also.

[e] Use 10 g/liter for immature seeds of *C. discolor* and *C. sieboldii,* 12 g/liter for *C. furcata* (8 and 15 g/liter have been used for mature seeds), and 15 g/liter for all others.

[f] Dissolve items 1–3 in 800 ml distilled water (item 6); adjust pH as indicated in Table A-45; bring volume to 1000 ml with more distilled water (item 6); heat to a boil and add agar slowly while stirring; when agar has dissolved completely, add charcoal and stir in completely, distribute into culture vessels, autoclave, cool, and use.

Table A-48. Nitsch medium as used for the germination of *Calanthe*

Item number	Component	Amount per liter of culture medium (final concentration in culture medium)	Stock solution (a concentrate prepared for repeated and convenient use)	Volume of stock solution per liter of culture medium	Remarks
	Macroelements				
1	Potassium nitrate, KNO_3[a]	950 mg	95 g/liter[a]	10 ml[a]	or weigh[a]
2	Ammonium nitrate, NH_4NO_3[a]	720 mg	72 g/liter[a]	10 ml[a]	or weigh[a]
3	Magnesium sulfate, $MgSO_4 \cdot 7H_2O$	185 mg	18.5 g/liter	10 ml	
4	Calcium chloride, $CaCl_2 \cdot 2H_2O$	166 mg	16.6 g/liter	10 ml	
5	Potassium phosphate, KH_2PO_4	68 mg	6.8 g/liter	10 ml	
6	Iron[b]				
a	Na_2EDTA	37.25 mg	7.45 g/liter	5 ml[b]	one solution[b]
b	Ferrous sulfate, $FeSO_4 \cdot 7H_2O$	27.9 mg	5.57 g/liter		
7	*Microelements*[c]				
a	Manganese sulfate, $MnSO_4 \cdot 4H_2O$	25 mg	2.5 g/liter	10 ml[c]	one solution[c]
b	Boric acid, H_3BO_3	10 mg	1 g/liter		
c	Zinc sulfate, $ZnSO_4 \cdot 7H_2O$	10 mg	1 g/liter		
d	Sodium molybdate, $Na_2MoO_4 \cdot 2H_2O$	0.25 mg	2.5 mg/liter		
e	Copper sulfate, $CuSO_4 \cdot 5H_2O$	0.25 mg	2.5 mg/liter		

(*continued*)

Table A-48.—Continued

Item number	Component	Amount per liter of culture medium (final concentration in culture medium)	Stock solution (a concentrate prepared for repeated and convenient use)	Volume of stock solution per liter of culture medium	Remarks
	Polyol				
8	*myo*-Inositol	100 mg	no stock	no stock	weigh
	Amino acid				
9	Glycine[d]	2 mg	400 mg/100 ml 70% ethanol[e]	0.5 ml	
	Vitamins				
10	Nicotinic acid	5 mg	1 g/100 ml 70% ethanol[e]	0.5 ml	
11	Pyridoxine · HCl	0.5 mg	100 mg/100 ml 70% ethanol[e]	0.5 ml	
12	Thiamine · HCl	0.5 mg	100 mg/100 ml 70% ethanol[e]	0.5 ml	
13	Folic acid	0.5 mg	100 mg/100 ml 70% ethanol[e]	0.5 ml	
14	Biotin	0.05 mg	10 mg/100 ml 70% ethanol[e]	0.5 ml	
	Hormone				
15	Indoleacetic acid[f]	0.1 mg	20 mg/100 ml 70% ethanol[e]	0.5 ml	
	Sugar[g]				
16	Sucrose	20 g	no stock	no stock	weigh
	Solidifiers[g]				
17	Agar	8 g	no stock	no stock	weigh
18	Water, distilled[g]	to 1000 ml			

[a] Solutions containing nitrate and ammonium tend to become contaminated on standing. Therefore, it is better to weigh these components each time. If a stock solution is prepared it must be kept frozen.
[b] Dissolve both components in one solution.
[c] Add all microelements to the same one liter of distilled water, stir and/or heat until dissolved.
[d] Keep in refrigerator or freezer.
[e] Pour 50 ml ethanol in a 100-ml volumetric flask, add substance to be dissolved, and shake well. If substance does not dissolve, add 2–3 drops of 1 M HCl and shake again. Add another 20 ml of ethanol and then bring total volume to 100 ml with distilled water. Keep stock solution in freezer or refrigerator.
[f] Mii and Kako (1974) did not add this hormone.
[g] Add items 1–15 to 800 ml distilled water (item 18); adjust pH to 5.5; add sugar (item 16) and adjust volume to 1000 ml with distilled water (item 18); bring to a slow boil and add agar (item 17) with stirring. Distribute into culture vessels, autoclave, cool, and use.

Table A-49. Pretreatment of *Calanthe* seeds

Species	Treatment
Calanthe discolor	Wash with water before sterilization, or soak in methyl ethyl ketone, benzene, n-hexane, ether, toluene, or acetic ethanol for 23 days at 5°C, or in ethyl acetate for 30 days, or in 1 M KOH for 5 min, or in 1 M HCl for 5 min before sterilization, or soak in sterile distilled water for 5 hr after sterilization.
C. discolor × *C. sieboldii*	Wash with water before sterilization.
C. furcata	None; seeds are removed under sterile conditions from immature capsules.
C. izu-insularis	Wash with water before sterilization, or place on filter paper in a funnel and leach with water drops (100/min) for 14 days before sterilization and wash 5–6 times with sterile distilled water after seeds have been sterilized.
C. sieboldii	Remove seeds under sterile conditions from immature capsules.

Literature Cited

Arditti, J. 1977. Clonal propagation of orchids by means of tissue culture—a manual, p. 203–293. *In* J. Arditti (ed.), Orchid biology and perspectives, I. Cornell University Press, Ithaca, New York.
Hasegawa, K., M. Sato, and K. Goi. 1976. Studies on germination of Calanthe species. Proc. Japan. Soc. Hort. Sci. Conf., p. 302–303, Autumn 1976.
Hibino, K., N. Mizuno, and S. Kako. 1978. Studies on germination of Calanthe discolor. I. Effect of pretreatment on germination. Proc. Jap. Soc. Hort. Sci. Conf., p. 356–357, Autumn 1978.
Ichihashi, S., and M. Yamashita. 1977. Studies on the media for orchid seed germination. I. J. Japan. Soc. Hort. Sci. 45:407–413.
Ito, I. 1976. Effect of growth regulators, which were supplied to the immature seeds by putting the flower stalk in regulator containing water, on seed germination of Calanthe discolor, p. 317–320. *In* H. Torikata (ed.), Seed formation and sterile culture of the orchids: Supplement. Seibundo-Shinkosha, Tokyo.
Kano, K. 1965. Studies on the media for orchid seed germination. Mem. Fac. Agr. Kagawa Univ. No. 20:1–68.
——. 1967. Studies on germination of Calanthe discolor. Shinkaki 55:52–55. Takii Nursery, Inc., Kyoto.
——. 1968. Studies on the media for orchid seed germination, p. 95–152. *In* H. Torikata (ed.), Seed formation and sterile culture of the orchids. Seibundo-Shinkosha, Tokyo.
Mii, M. 1978. Effect of organic solvent pretreatment on germination of Calanthe discolor. Proc. Japan. Soc. Hort. Sci. Conf., p. 330–331, Spring 1978.
Mii, M., and S. Kako. 1974. Studies on germination of Calanthe izu-insularis. I. Effect of water treatment and light condition on germination. Proc. Japan. Soc. Hort. Sci. Conf., p. 326–327, Autumn 1974.
——. 1976. Effect of water treatment and light condition on germination of *Calanthe izu-insularis*, p. 320–324. *In* H. Torikata (ed.), Seed formation and sterile culture of the orchids: Supplement. Siebundo-Shinkosha, Tokyo.
Nagashima, T. 1977. Studies on seed formation and germination of Calanthe discolor. Proc. Japan. Soc. Hort. Sci. Conf., p. 380–381, Spring 1977.
——. 1979a. Studies on seed formation and germination of Calanthe sieboldii. Proc. Japan. Soc. Hort. Sci. Conf., p. 276–277, Spring 1979.
——. 1979b. Studies on the seed formation and germination of Calanthe furcata. Proc. Japan. Soc. Hort. Sci. Conf., p. 304–305, Autumn 1979.
Suzuki, S., and S. Abe. 1977. Studies on germination of Calanthe discolor and C. sieboldii. Proc. Japan. Soc. Hort. Sci. Conf., p. 382–383, Spring 1977.

Cymbidium

The procedures presented here are based on a number of methods (Hagiya and Fujita, 1968; Kako 1968; Kano, 1965, 1967, 1968a,b, 1971a,b; Kano and Goi, 1964; Nagashima, 1978; Sawa and Namba, 1974, 1976; Sawa and Torikata, 1968; Ueda and Torikata, 1976).

Plant Material. Both mature and immature seeds can be germinated (Table A-50).

Surface Sterilization. Calcium hypochlorite has been used to sterilize mature seeds of Japanese *Cymbidium* species (Table A-51) before or after other treatments (Table A-52).

Culture Vessels. A variety of culture vessels can be used (p. 275). Germination of *Cymbidium goeringii* is accelerated if the vessels are sealed for the first 300 days. After that they should be covered with a stopper fitted with a cotton-filled glass tube. Sealing with the latter improves the germination of *Cymbidium kanran.*

Table A-50. Age at which seeds of Japanese *Cymbidium* species were germinated

Species	Age
Cymbidium goeringii	Starting 100 days after pollination, or 48–50 days after fertilization (Nagashima, 1978). Starting 4 months after fruit set, 6 months being optimal (Hagiya and Fujita, 1968). Starting on the 96th day, best germination from the 156th to 218th day (Kano, 1965). Starting at the 4th month, which is also optimal (Sawa and Torikata, 1968).
C. gyokuchin × *C. kanran*	Mature
C. kanran	Seeds from 6-month-old capsules germinate best.
C. sinense	Mature

Table A-51. Surface sterilization procedures for seeds of Japanese *Cymbidium* species

Species	Procedure
Cymbidium goeringii (*C. virescens*)	Soak the seeds in calcium hypochlorite (7 g/100 ml) for 20 min or 1.5 hr.
C. gyokuchin	Soak the seeds in calcium hypochlorite (7 g/100 ml) for 10 min.
C. gyokuchin × *C. kanran*	Soak the seeds in calcium hypochlorite (7 g/100 ml) for 10 min.
C. kanran	Soak the seeds in calcium hypochlorite (7 g/100 ml) for 10 min.

Table A-52. Pre- and poststerilization treatments that enhance seed germination of Japanese *Cymbidium* species

	Treatment	
Species	Presterilization	Poststerilization
Cymbidium goeringii (*C. virescens*)	Soak ripe seeds in 0.1 M KOH for 10 min or in 1 M KOH for 3 min.	Soak in sterile distilled water.
C. gyokuchin	Soak ripe seeds in 0.1 M KOH for 5 min.	Soak in sterile distilled water for 5 hr.
C. gyokuchin × *C. kanran*	Soak ripe seeds in 0.1 M KOH for 5–7 min.	
C. kanran	Soak ripe seeds in 0.1 M KOH for 10 min or 1 M KOH for 5 min.	
C. sinense		Soak in sterile distilled water for 5 hr.

Table A-53. Culture conditions for seed germination and seedling growth of Japanese *Cymbidium* species

	Conditions			
		Illumination		
Species	Temperature	Duration (hr)	Intensity (lux)	Remarks
Cymbidium goeringii (*C. virescens*)	25–26°C	12–16	2000–3000	Cap flask tightly with a rubber stopper which is impervious to air until seeds germinate (about 300 days). Then replace the stopper with one which allows for air movement.
	25°C	Dark		
	25–30°C	16	5000–7000 (daylight plus fluorescent)	
C. gyokuchin	20–38°C	Continuous		
C. gyokuchin × *C. kanran*	No data	No data		
C. kanran	24–32°C	16	4000–6000 (daylight plus fluorescent)	
C. sinense	Room temp. spring to autumn; min. 20°C in winter	Natural daylength	1000–2000, rainy days; 10,000–20,000, sunny days	

Culture Conditions. Optimal conditions for germination and seedling growth vary (Table A-53).

Culture Media. Cymbidium goeringii and *C. gyokuchin* germinate best on Hyponex medium (Table A-47). Modified Knudson C (Table A-55) is best for *C. kanran* and Hyponex medium (Table A-47) for *C. sinense* (Table A-54).

Procedure. The best procedure for the germination of *Cymbidium goeringii* seeds is: (1) soak in saturated KOH for 10 min (Table A-52); (2) sterilize in calcium hypochlorite (Table A-52); (3) soak in sterile distilled water (Table A-52); and (4) seal the vessels until the seeds germinate (Table A-52).

The seeds of most temperate-zone *Cymbidium* species (*C. goeringii, C. kanran,* etc.) form a rhizome on germination. Shoots and roots differentiate from the rhizome only after a year or more. To accelerate differentiation, rhizome tips should be excised and cultured on a modified Knudson C (Table A-56) in the dark. After shoot and root formation, the cultures should be gradually moved to light.

Table A-54. Media for the germination of seeds of Japanese *Cymbidium* species

Species	Medium
Cymbidium goeringii (*C. virescens*)	Knudson C (Table A-56).
Immature seeds (100 days after pollination)	Hyponex medium (Table A-47), pH 5.2–5.4, plus 2 g peptone and 30 g sucrose per liter, solidified with 12 g agar.
Mature seeds	Hyponex medium (Table A-47), pH 4.8–5.3, plus 2 g peptone and 35 g sucrose per liter, solidified with 15 g agar.
C. gyokuchin	Same as mature seeds of *C. goeringii*
C. kanran	Modified Knudson C (Table A-55).
C. sinense	Hyponex medium (Table A-47), pH 5.0, plus 2 g tryptone and 35 g sucrose per liter, solidified with 15 g agar.

Table A-55. Knudson C medium (Knudson, 1946) modified for the germination of *Cymbidium kanran* seeds (Sawa and Namba, 1974)

Item number	Component	Amount per liter of culture medium (final concentration in culture medium)	Stock solution (a concentrate prepared for repeated and convenient use)	Volume of stock solution per liter of culture medium	Remarks
	Macroelements				
1	Calcium nitrate, $Ca(NO_3)_2 \cdot 4H_2O$[a]	1 g	100 g/liter[a]	10 ml[a]	or weigh[a]
2	Potassium phosphate, KH_2PO_4[b]	250 mg	25 g/liter[b]	10 ml[b]	
3	Magnesium sulfate, $MgSO_4 \cdot 7H_2O$	250 mg	25 g/liter	10 ml	
4	Ammonium sulfate, $(NH_4)_2SO_4$[a]	500 mg	50 g/liter[a]	10 ml[a]	or weigh[a]
5	Ferrous sulfate, $FeSO_4 \cdot 7H_2O$[c]	25 mg	2.5 g/liter[c]	10 ml[c]	
6	Manganese sulfate, $MnSO_4 \cdot 4H_2O$	7.5 mg	750 mg/liter	10 ml	
	Microelement[d]				
7	Boric acid, H_3BO_3	1.0 mg	100 mg/liter	10 ml	
	Vitamins				
8	Nicotinic acid	0.1 mg	100 mg/100 ml 70% ethanol[e]	0.1 ml	
9	Riboflavin	0.1 mg	100 mg/100 ml 70% ethanol[e]	0.1 ml	
10	Pyridoxine · HCl	0.1 mg	100 mg/100 ml 70% ethanol[e]	0.1 ml	
11	Thiamine · HCl	0.1 mg	100 mg/100 ml 70% ethanol[e]	0.1 ml	
12	Ascorbic acid	0.1 mg	100 mg/100 ml 70% ethanol[e]	0.1 ml	
	Amino acids[f]				
13	Arginine	5 mg	100 mg/100 ml distilled water	5 ml	
14	Asparagine	5 mg	100 mg/100 ml distilled water	5 ml	
	Sugar[g]				
15	Sucrose	30 g	no stock	no stock	weigh
	Solidifier[g]				
16	Agar	17 g	no stock	no stock	weigh
17	Water, distilled[g]	to 100 ml			

[a] Solutions containing ammonium and nitrate may become contaminated on standing. Therefore, stock solutions should not be prepared. If made, they must be kept frozen between uses.

[b] A phosphate buffer which will keep the pH constant may be substituted. Prepare buffer by mixing 975 ml of 0.1 M KH_2PO_4 (monopotassium phosphate) solution (13.6 g/liter) with 25 ml of a 0.1 M K_2HPO_4 (dipotassium phosphate) solution (17.4 g/liter); measure the pH to be certain it is correct (pH 5.1–5.4), adjust if necessary, and use 18 ml/liter of culture medium.

[c] A rust-colored precipitate may form on standing. This solution must, therefore, be shaken vigorously before use.

[d] Keep refrigerated between uses.

[e] Pour 50 ml ethanol in a 100-ml volumetric flask, add substance to be dissolved, and shake well. If substance does not dissolve, add 2–3 drops of 1 M HCl and shake again. Add another 20 ml of ethanol and then bring total volume to 100 ml with distilled water. Keep stock solution in freezer or refrigerator.

[f] Keep frozen between uses. If the amino acid fails to dissolve in the stock solution, add a pellet or two of NaOH.

[g] Add items 1–14 to 600 ml distilled water (item 17), stir well; adjust the pH to 5.3; add sugar (item 15) and adjust volume to 1000 ml with more distilled water (item 17). Bring solution to a gentle boil and add agar (item 16) slowly while stirring. When the agar has dissolved completely (and the solution turns a clear amber color), distribute the medium into culture vessels, cover, and autoclave.

Developmental Sequence. A rhizome is formed following germination. Shoots and roots are formed after one year or more.

General Comments. As with other temperate-zone orchids, seed germination of Japanese *Cymbidium* species may pose problems, but the procedures outlined here can overcome the difficulties.

Table A-56. Modified Knudson C medium (Knudson, 1946) as used for plantlet differentiation of *Cymbidium goeringii* (Ueda and Torikata, 1976)

Item number	Component	Amount per liter of culture medium (final concentration in culture medium)	Stock solution (a concentrate prepared for repeated and convenient use)	Volume of stock solution per liter of culture medium	Remarks
	Macroelements				
1	Calcium nitrate, $Ca(NO_3)_2 \cdot 4H_2O$[a]	1 g	100 g/liter[a]	10 ml[a]	or weigh[a]
2	Monopotassium phosphate, KH_2PO_4[b]	250 mg	25 g/liter[b]	10 ml[b]	
3	Magnesium sulfate, $MgSO_4 \cdot 7H_2O$	250 mg	25 g/liter	10 ml	
4	Ammonium sulfate, $(NH_4)_2SO_4$[a]	500 mg	50 g/liter[a]	10 ml[a]	or weigh[a]
5	Ferrous sulfate, $FeSO_4 \cdot 7H_2O$[c]	25 mg	2.5 g/liter[c]	10 ml[c]	
6	Manganese sulfate, $MnSO_4 \cdot 4H_2O$	7.5 mg	750 mg/liter	10 ml	
7	*Microelements*[d]				
a	Manganese sulfate, $MnSO_4 \cdot 4H_2O$	25 mg	2.5 g/liter		
b	Boric acid, H_3BO_3	10 mg	1 g/liter		
c	Zinc sulfate, $ZnSO_4 \cdot 7H_2O$	10 mg	1 g/liter	10 ml[d]	one solution[d]
d	Sodium molybdate, $Na_2MoO_4 \cdot 2H_2O$	0.25 mg	25 mg/liter		
e	Copper sulfate, $CuSO_4 \cdot 5H_2O$	0.025 mg	2.5 mg/liter		
	Hormone				
8	Kinetine[e]	10 mg	100 mg/10 ml 70% ethanol plus a few drops of KOH	1 ml	
	Amino acid				
9	L-Arginine	174 mg[f]	no stock	no stock	weigh
	Sugar[g]				
10	Sucrose	20 g	no stock	no stock	weigh
	Solidifier[g]				
11	Agar	12–15 g	no stock	no stock	weigh
12	Water, distilled[g]	to 1000 ml			

[a] Solutions containing ammonium and nitrate may become contaminated on standing. Therefore, stock solutions should not be prepared. If made, they must be kept frozen between uses.
[b] A phosphate buffer which will keep the pH constant may be substituted. Prepare buffer by mixing 975 ml of 0.1 M KH_2PO_4 (monopotassium phosphate) solution (13.6 g/liter) with 25 ml of a 0.1 M K_2HPO_4 (dipotassium phosphate) solution (17.4 g/liter); measure the pH to be certain it is correct (pH 5.1–5.4), adjust if necessary, and use 18 ml/liter of culture medium.
[c] A rust-colored precipitate may form on standing. This solution must, therefore, be shaken vigorously before use.
[d] Add all microelements to the same one liter of distilled water; stir and/or heat until they are dissolved, and add 1 ml per liter of culture medium.
[e] Keep stock solution refrigerated or frozen.
[f] This is 10^{-3} moles.
[g] Add items 1–9 to 800 ml distilled water (item 12) and mix well; adjust the pH to 5.3; add sugar (item 10) and adjust volume to 1000 ml with more distilled water (item 12). Bring solution to a gentle boil and add agar (item 11) slowly while stirring. When the agar has dissolved completely (and the solution turns a clear amber color), distribute the medium into culture vessels, cover, and autoclave.

Literature Cited

Hagiya, K., and T. Fujita. 1968. Effect of light and temperature on germination of Cymbidium goeringii, p. 238–244. *In* H. Torikata (ed.), Seed formation and sterile culture of the orchids. Seibundo-Shinkosha, Tokyo.

Kako, S. 1968. Studies on germination of Cymbidium goeringii, p. 174–237. *In* H. Torikata (ed.), Seed formation and sterile culture of the orchids. Seibundo-Shinkosha, Tokyo.

Kano, K. 1965. Studies on the media for orchid seed germination. Mem. Fac. Agr. Kagawa Univ. No. 20:1–68.

——. 1967. Studies on the orchid germination in an aseptic condition. X. Acceleration of the germination of "hard-to-germinate" orchid seeds. Proc. Japan. Soc. Hort. Sci. Conf., p. 264–265, Autumn 1967.

——. 1968a. Studies on the media for orchid seed germination, p. 95–152. *In* H. Torikata (ed.), Seed formation and sterile culture of the orchids. Seibundo-Shinkosha, Tokyo.
——. 1968b. Acceleration of the germination of so-called "hard-to-germinate" orchid seeds. Amer. Orch. Soc. Bull. 37:690–698.
——. 1971a. Studies on acceleration of the germination in Cymbidium goeringii. Proc. Japan. Soc. Hort. Sci. Conf., p. 220–221, Spring 1971.
——. 1971b. Seed germination of oriental Cymbidium and their shoot-tip culture. Proc. 6th World Orchid Conf., p. 133–142.
Kano, K., and M. Goi. 1964. Studies on the orchid germination in an aseptic condition. VIII. Proc. Japan. Soc. Hort. Sci. Conf., p. 36, Autumn 1964.
Nagashima, T. 1978. Studies on seed formation of Cymbidium goeringii. Proc. Japan. Soc. Hort. Sci. Conf., p. 332–333, Spring 1978.
Sawa, K., and M. Namba. 1974. Studies about the germination of Cymbidium kanran. Proc. Japan. Soc. Hort. Sci. Conf., p. 296–301, Autumn 1974.
——. 1976. Apical meristem culture and seed germination in *Cymbidium kanran*, p. 296–301. *In* H. Torikata (ed.), Seed formation and sterile culture of the orchids: Supplement. Seibundo-Shinkosha, Tokyo.
Sawa, K., and H. Torikata. 1968. Studies on germination of *Cymbidium*, p. 153–173. *In* H. Torikata (ed.), Seed formation and sterile culture of the orchids. Seibundo-Shinkosha, Tokyo.
Ueda, H., and H. Torikata. 1976. Organogenesis in Cymbidium, p. 309–313. *In* H. Torikata (ed.), Seed formation and sterile culture of the orchids: Supplement. Seibundo-Shinkosha, Tokyo.

Galeola septentrionalis

Galeola septentrionalis is an achlorophyllous orchid that can be germinated aseptically to some extent (Nakamura, 1962, 1976, 1978; Nakamura *et al.*, 1975).

Plant Material. The fruit was chopped in water, filtered through a sieve (pore diameter 0.75 mm), dried and, until used, the seeds were stored at 4–5° C in an airtight vessel with some silica gel. They retained full germinating ability for at least 5 years (Nakamura *et al.*, 1975).

Surface Sterilization. Seeds are sterilized by soaking them in freshly prepared calcium hypochlorite (7 g/100 ml) for 10–20 min or potassium hypochlorite solution (active chlorine 1.8%) for 15 min.

Culture Vessels. Test tubes (17 × 170 mm or 18 × 180 mm) were employed by the original workers, but other vials could be used.

Culture Conditions. Germination of seeds and the subsequent growth require specific atmospheric conditions (Table A-57) which can be provided in airtight containers.

Culture Media. Seeds are germinated and maintained for about 40 days on a relatively simple medium (Table A-58). After 40–50 days the protocorms should be transferred to another solution (Table A-59).

Procedure. Introduce the seeds into test tubes containing the first medium (Table A-58) and place the cultures under appropriate conditions (Table A-57). After 40–50

Table A-57. Conditions for seed germination, protocorm formation and enlargement, and organ differentiation and development of *Galeola septentrionalis* (Numbers in parentheses indicate minimum and maximum levels)

Factor	Germination (0–7 weeks after sowing)	Protocorm formation (8–14 weeks)	Protocorm enlargement (15–19 weeks)	Organ differentiation (20–24 weeks)	Organ development (25 weeks or more)
Atmospheric pressure[a]	1.8 kg/cm^2 (1.0–2.0)[b]	1.4 kg/cm^2 (1.0–2.0)[b]	1.2 kg/cm^2 (1.0–1.4)[b]	1 kg/cm^2 (1.0–1.4)[b]	1 kg/cm^2 (1.0–1.4)[b]
Oxygen[a]	5% (3–21)	10% (5–12)	15% (5–12)	17.5% (15–21)	21% (15–21)
Carbon dioxide[a]	8% (0.03–10)	6% (0.03–6)	2% (0.03–4)	0.03% (0.03–2)	0.03% (0.03–2)
Nitrogen[a]	87%	84%	83%	82.5%	79% (normal air)
Temperature	30°C (28–32)	24°C (20–26)	24°C (20–26)	24°C (20–26)	24°C (20–26)
Illumination	Dark	Dark	Dark	Dark	Dark
Ethylene $\mu l/l$	2–8	None	None	None	None

[a] One atmosphere equals 1 kg/cm^2 and is the normal pressure at sea level; air is composed of approximately 78% nitrogen, 21% oxygen, 0.03% cabon dioxide, and other gases.

days transfer the protocorms to the second medium (Table A-59) and the required culture conditions, which should be changed following an additional 40–50 days. When organs start to form, move the cultures to normal air at normal pressure in the dark.

Developmental Sequence. Following seed-coat rupture (7 weeks after sowing), the spherical embryos grow into protocorms and differentiate within 14 weeks. Organs (adventitious roots and scaly leaves) differentiate in 20–24 weeks.

General Comments. No information is available on physiological effects of the unusual atmospheric conditions required by this achlorophyllous orchid.

Table A-58. Medium for the germination of *Galeola septentrionalis* seeds (Nakamura, 1962, 1976; Nakamura *et al.*, 1975)

Item number	Component	Amount per liter of culture medium (final concentration in culture medium)
	Macroelement	
1	Potassium chloride, KCl	7.5 g
	Complex additive	
2	Diffusate from 10 g beer-yeast powder (Ebios[a]) or malt extract broth (Difco)	10 g
	Sugar	
3	Sucrose	10 g
	Solidifier	
4	Agar[b]	10 g
5	Water, distilled[c]	to 1000 ml

[a] Available from Ebios Yakuhin Kogyo Co., Ltd., Toyko, Japan.

[b] May be omitted.

[c] Mix all components, adjust pH to 4.8–5.0, distribute into culture vessels, and autoclave for 15 min.

Table A-59. Medium for protocorm culture of *Galeola septentrionalis* (Nakamura *et al.*, 1975)

Item number	Component	Amount per liter of culture medium (final concentration in culture medium)	Stock solution (a concentrate prepared for repeated and convenient use)	Volume of stock solution per liter of culture medium	Remarks
	Macroelements[a]				
1	Calcium nitrate, $Ca(NO_3)_2 \cdot 4H_2O$[a]	283.32 mg	28.33 g/liter[a]	10 ml[a]	or weigh[a]
2	Potassium nitrate, KNO_3[a]	80 mg	8 g/liter[a]	10 ml[a]	or weigh[a]
3	Sodium phosphate, $NaH_2PO_4 \cdot H_2O$	82.5 mg	8.25 g/liter	10 ml	
4	Potassium chloride, KCl	65 mg	6.5 g/liter	10 ml	
5	Magnesium sulfate, $MgSO_4 \cdot 7H_2O$	5.7 mg	570 mg/liter	10 ml	
6	*Iron*				
a	Ferric citrate	4 mg	400 mg/liter	10 ml (a and b)	one solution (a and b)
b	Na_2EDTA	18.6 mg	1.86 g/liter		
7	*Microelements*[b]				
a	Zinc sulfate, $ZnSO_4 \cdot H_2O$	2.7 mg	2.7 g/liter	1 ml[b] (a–c)	one solution[b] (a–c)
b	Potassium iodide, KI	0.75	750 mg/liter		
c	Boric acid, H_3BO_3	1.5 mg	1.5 g/liter		
	Vitamins				
8	Thiamine · HCl[c]	0.1 mg	100 mg/100 ml 70% ethanol	0.1 ml	
9	Nicotinic acid[c]	0.1 mg	100 mg/100 ml 70% ethanol	0.5 ml	
10	Pyridoxine HCl[c]	0.1 mg	100 mg/100 ml 70% ethanol	0.1 ml	
11	Ascorbic acid[c]	1.8 mg	100 mg/100 ml 70% ethanol	1.8 ml	
	Hormones				
12	Indoleacetic acid (IAA)[c]	0.28	100 mg/100 ml 70% ethanol	0.28 ml	
13	Kinetin[c]	0.22 mg	100 mg/100 ml 70% ethanol	0.22 ml	
	Complex additive				
14	Casein hydrolysate (Casamino acids, vitamin-free)	5 g	no stock		weigh
	Sugar[d]				
15	Glucose	36 g	no stock		weigh
	Solidifier[d]				
16	Agar	6 g	no stock		weigh
17	Water, distilled[d]	to 1000 ml			

[a] Solutions containing ammonium and nitrate may become contaminated on standing. Therefore, stock solutions should not be prepared. If made, they must be kept frozen between uses.
[b] Add all microelements to the same one liter of distilled water, stir and/or heat until dissolved. Add 1 ml per liter of culture medium.
[c] Keep refrigerated or frozen.
[d] Dissolve items 1–14 in 350 ml of distilled water (item 17). Adjust the pH to 5.0, bring volume to 400 ml, and sterilize by filtering through Millipore GSWP filters (Millipore Filter Corporation) and dispense 2 ml at a time into preautoclaved 18 × 180 mm test tubes. Add the agar (item 16) and sugar (item 15) to 600 ml distilled water (item 17), autoclave for 15 min and pour 3 ml at a time into the 2 ml of medium in the 18 × 180 mm tubes; shake well to mix, wait for solidification, and use.

Acknowledgment

I thank Dr. Shin Ichi Nakamura for commenting on the manuscript of this section.

Literature Cited

Nakamura, S. I. 1962. Zur Samenkeimung einer chlorophyllfreien Erdorchidee Galeola septentrionalis Reichb. f. Z. Bot. 50:487–497.

——. 1976. Atmospheric condition required for the growth of Galeola septentrionalis seedlings. Bot. Mag. Tokyo 89:211–218.
——. 1978. Can the holomycotrophyte be cultured without fungal symbionts? A gnotobiological view. Trans. Mycol. Soc. Japan 19:325–331 (in Japanese).
Nakamura, S. I., T. Uchida, and M. Hamada. 1975. Atmospheric conditions controlling the seed germination of an achlorophyllous orchid, Galeola septentrionalis. Bot. Mag. Tokyo 88:103–109.

Liparis nervosa

Liparis nervosa is a terrestrial species with small reddish-brown and green flowers and bright green, relatively broad leaves. The germination of its seeds was investigated together with those of *Cymbidium sinense* (Kano and Goi, 1964).

Plant Material. The germination of mature seeds has been investigated.

Surface Sterilization. Seeds should be sterilized by soaking them in calcium hypochlorite (7 g/100 ml) for 10 min.

Culture Vessels. The same vessels as those employed for Japanese *Cymbidium* species can be used (p. 336).

Culture Conditions. Cultures should be maintained under the same conditions as *Cymbidium sinense* (p. 338, Table A-53).

Culture Medium. Hyponex medium (3 g Hyponex, 2 g tryptone, 35 g sucrose per liter, solidified with 15 g agar as in Table A-47) is appropriate.

Procedure. Following sterilization, the seeds should be soaked for 5 hours in sterile distilled water and then placed on the culture medium. Seedlings should be transferred when they become large and crowded.

General Comments. This is an effective method for the germination of a north temperate terrestrial orchid.

Literature Cited

Kano, K., and M. Goi. 1964. Studies on the orchid germination in an aseptic condition. VIII. Proc. Japan. Soc. Hort. Sci. Conf., p. 36 Autumn 1964.

Neofinetia falcata

The flowers of this angrecoid orchid are white and intensely fragrant. It is an epiphyte and relatively easy to germinate.

Plant Material. Immature seeds can germinate well 50–90 days after pollination. Surface sterilization of the immature fruit, culture vessels, conditions, medium and procedures are the same as for tropical orchids.

Surface Sterilization. The fruits should be sterilized by soaking them in calcium hypochlorite (7 g/100 ml), for 20–30 min.

Culture Vessels. No description is given, but probably test tubes (24 × 200 mm) were used.

Culture Conditions. Test tubes were kept at 25° C, and 16-hr photoperiods of 3000 lux provided by fluorescent lamps.

Culture Media. Knudson C (Table A-5) and Karasawa (Table A-61) media are appropriate.

Procedure. Immature seeds were removed from sterilized capsules under sterile conditions and sown on both media at about 10-day intervals from the 15th to 21st day after pollination.

Developmental Sequence. No description beyond: "Good germination is obtained" (Sawa, Taneda, and Fujimori, 1979).

General Comments. Immature seeds germinate well between 40 days and 100 days after pollination.

Literature Cited

Sawa, K., M. Taneda, and R. Fujimori. 1979. Studies on germination of orchids native to Japan. I. Germination of orchids native to Shikoku Island. Proc. Japan. Soc. Hort. Sci. Conf., p. 278–279, Spring 1979.

Phajus tankervilliae

Essentially a tropical terrestrial species, *Phaius tankervilliae* should be germinated like tropical species (Uesato, 1974) on unmodified Knudson C medium (Table A-5) or modified by the addition of the Nitsch's microelements (Table A-48).

Literature Cited

Uesato, K. 1974. Concerning the germination of Phajus tankervilliae Bl. Proc. Japan. Soc. Hort. Sci. Conf., p. 330–331, Autumn 1974.

Other Japanese Orchids

A number of Japanese orchids (Table A-60) have been germinated on Karasawa medium (Table A-61) or on Knudson C (Table A-5) with the following additions per liter: arginine, 1 mg; thiamine (vitamin B_1), 0.1 mg; pyridoxine (vitamin B_6), 0.1 mg; and niacin (nicotinic acid), 0.1 mg (but *Gastrochilus matsuran* does not germinate well on

Table A-60. Japanese orchids which have been germinated on Knudson C and Karasawa media (Sawa *et al.*, 1979)

Aerides japonicum	*Galeola septentrionalis*	*L. odorata*
Amitostigma gracile[a]	*Galeorchis cyclochila*[a]	
A. keiskei	*Gastrochilus japonicus*	*Malaxis monophyllos*
Aplectrum unguiculatum	*G. matsuran*	
	Gastrodia confusa[a]	*Neofinetia falcata*
Bletilla striata	*G. nipponica*	
Bulbophyllum inconspicuum	*Goodyera macrantha*	*Oberonia japonica*
	G. maximowicziana	*Oreorchis patens*
Cremastra appendiculata	*G. schlechtendaliana*	
Cymbidium goeringii[a]	*G. velutina*	*Pogonia japonica*
C. kanran		*Ponerorchis chidori*
C. lancifolium[a]	*Habenaria dentata*	*P. graminifolia*
C. nipponicum	*H. cyoensis*	
Cypripedium debile[a]	*H. radiata*	*Sarcanthus scolopendrifolius*
C. japonicum[a]	*H. sagittifera*	*Sarcochilus japonicus*
		Spiranthes sinensis
Dendrobium	*Liparis kumokiro*	
D. tosaense	*L. nervosa*	*Tulotis ussuriensis*

[a] Soaked in 0.1 M KOH for 5 min.

this combination). Since all species germinate on the former (Sawa *et al.*, 1979), it is the only one given in detail here (Table A-61).

The cultures should be maintained at 25° C, 16-hr photoperiods, and 3000 lux.

Literature Cited

Karasawa, K. 1966. On the media with banana and honey added for seed germination and subsequent growth of orchids. Orchid Rev. 74: 313–318.

Sawa, K., M. Taneda, and R. Fujimori. 1979. Studies on germination of orchids native to Japan. I. Germination of orchids native to Shikoku Island. Proc. Japan. Soc. Hort. Sci. Conf., p. 278–279, Spring 1979.

Table A-61. Karasawa (1966) medium as used for the germination of Shikoku Island (Japan) orchids (Sawa *et al.*, 1979)

Item number	Component	Amount per liter of culture medium (final concentration in culture medium)
	Macroelements	
1	Hyponex	3 g
	Complex additives	
2	Peptone	2 g
3	Banana	15 g
4	Bee honey	30 g
	Sugar	
5	Sucrose	20 g
	Solidifier[a]	
6	Agar	15 g
7	Water, distilled[a]	to 1000 ml

[a] Place items 1–5 in a homogenizer and add 700 ml distilled water (item 7). Homogenize thoroughly, adjust pH to 5.2, and bring volume to 1000 ml with more distilled water (item 7). Heat solution to a slow boil, add agar (item 6) with stirring, distribute into culture vessels, autoclave, cool, and use.

Ophrys sphegodes

Joseph Arditti

Assessments of the role of endophytic fungi in the germination of north temperate terrestrial orchids can be made by culturing these species on asymbiotic media enriched with certain additives. Such experiments have led to the formulation of a medium for seed germination and seedling culture of *Ophrys sphegodes* (Mead and Bulard, 1975).

Plant Material. Seeds used in the original research were harvested from ripe capsules produced by cross-pollination on plants collected in the French Alps and cultivated in pots. These capsules contained 10,000–16,000 seeds of which 20–25% had no visible embryos. The seeds were stored at 4° C (39° F).

Surface Sterilization. The procedures used for *Orchis laxiflora* (p. 350) and tropical epiphytic orchids are suitable.

Culture Vessels. Pyrex test tubes, 25 mm × 250 mm containing 10 ml (for seed germination) or 15 ml (for seedling transflasking) sloped medium were used in the original research. Vessels of the kind recommended for tropical epiphytes can also be employed.

Culture Conditions. Germination is almost completely inhibited by light. Therefore, cultures must be placed in the dark at 23° C. Following transflasking, the cultures should be moved to 16-hr days (in the original research the illumination was provided by white fluorescent tubes, but other light sources should also prove suitable) under the same temperature.

Culture Medium. A medium containing casein hydrolysate and four vitamins is suitable (Table A-62).

Procedure. Cultures must be maintained under conditions suitable for germination (i.e., in the dark) until protocorms are formed (usually at least 2 months). After that they may be transferred to light. The seedlings are moved to fresh medium 7–12 months after sowing.

Developmental Sequence. Germination occurs within 2 months and rarely after 30 days. The protocorms eventually become covered with thick hairs, 3 mm in length. Some protocorms may die or turn black.

After transfer, mortality may be high and development heterogeneous. Some protocorms develop into individual plants. Others may produce one to several normal plantlets or a number of new protocorms. A few protocorms form callus that may give rise to protocorms and plants. Eventually plants with many roots and green leaves 4–7 cm long can be produced. These plants form tubers. On transfer to fresh medium the tubers develop into new plantlets which eventually form tubers again. Planting some of these tubers enabled the original researchers "to reach the third generation."

Table A-62. Mead and Bulard (1975) medium for seed germination and seedling culture of *Ophrys sphegodes*

Item number	Component	Amount per liter of culture medium (final concentration in culture medium)	Stock solution (a concentrate prepared for repeated and convenient use)	Volume of stock solution per liter of culture medium	Remarks
	Macroelements[a,b]				
1	Calcium sulfate, $CaSO_4 \cdot 2H_2O$	80 g	8 g/liter	10 ml	
2	Potassium phosphate, KH_2PO_4	250 mg	25 g/liter	10 ml	
3	Magnesium sulfate, $MgSO_4 \cdot 7H_2O$	240 mg	24 g/liter	10 ml	
4	Ammonium nitrate, NH_4NO_3[a]	80 mg	8 g/liter[a]	10 ml[a]	or weigh[a]
5	Calcium phosphate, $CaH_4(PO_4)_2 \cdot H_2O$[b]	100 mg	10 g/liter	10 ml[b]	
	Iron				
6	Ferric salt of EDTA, $FeNA_2$ EDTA[c]	35 mg	3.5 g/liter	10 ml	
7	*Microelements*[d]				
a	Manganese sulfate, $MnSO_4 \cdot 4H_2O$	0.1 mg	100 mg/liter	1 ml[d]	one solution[d]
b	Potassium iodide, KI	0.01 mg	10 mg/liter		
c	Boric acid, H_3BO_3	1 mg	1 g/liter		
d	Copper sulfate, $CuSO_4 \cdot 5H_2O$	0.03 mg	30 mg/liter		
e	Zinc sulfate, $ZnSO_4 \cdot 7H_2O$	1 mg	1 g/liter		
f	Nickel chloride, $NiCl_2 \cdot 6H_2O$	0.03 mg	30 mg/liter		
g	Aluminum chloride, $AlCl_3$	0.03 mg	30 mg/liter		
	Vitamins				
8	Thiamine	0.1 mg	100 mg/100 ml 70% ethanol[e]	0.1 ml	
9	Pyridoxine	0.1 mg	100 mg/100 ml 70% ethanol[e]	0.1 ml	
10	Niacin	0.1 mg	100 mg/100 ml 70% ethanol[e]	0.1 ml	
11	Biotin	0.005 mg	0.5 mg/100 ml 70% ethanol[e]	1 ml	
	Complex additive				
12	Vitamin-free casamino acids[f]	5 g	no stock	no stock	weigh
	Sugar[g]				
13	Sucrose	10 to 20 g	no stock	no stock	weigh
14	Water, distilled[g]	to 1000 ml			
	Solidifier[g]				
15	Agar	6–8 g	no stock	no stock	weigh

[a] Solutions containing ammonium and/or nitrate may become contaminated on standing. Therefore, stock solutions should not be prepared. If made, they must be kept frozen between uses.
[b] Shake well before dispensing.
[c] Available from Sigma Chemical Co.
[d] Add all microelements to the same one liter of distilled water; stir and/or heat until they are dissolved, and add 1 ml per liter of culture medium.
[e] Store in a refrigerator between uses.
[f] Available from Difco Laboratories.
[g] Add items 1–12 to 650 ml distilled water (item 14); adjust the pH to 5.6; add sugar (item 13) and adjust volume to 1000 ml with more distilled water (item 14). Bring solution to a gentle boil and add agar (item 15) slowly while stirring. When the agar has dissolved completely (and the solution turns a clear amber color), distribute the medium into culture vessels, cover, and autoclave.

General Comments. This method should prove to be of value to those who wish to germinate *Ophrys sphegodes* and/or study the requirements of related species.

Literature Cited

Mead, J. W., and C. Bulard. 1975. Effects of vitamins and nitrogen sources on asymbiotic germination and development of Orchis laxiflora and Ophrys sphegodes. New Phytol. 74:33–40.

Orchis laxiflora
Joseph Arditti

The available evidence seems to suggest that seed and seedlings of terrestrial orchids may have very specific requirements for germination and growth. Careful experiments are necessary for the elucidation of these requirements, as has been done with *Orchis laxiflora* (Mead and Bulard, 1979).

Plant Material. In the original research, seeds collected from cross-pollinated plants in the French Maritime Alps were stored at 4° C until used.

Surface Sterilization. Calcium hypochlorite, 3%, was used in the original research. The seeds were placed in small nylon bags and immersed in the solution for 15 min. The procedures used for tropical epiphytic orchids should also prove suitable.

Culture Vessels. Test tubes containing slanted medium were used in the initial research. However, vessels of the kind employed for tropical epiphytes should also prove suitable.

Culture Conditions. Cultures must be maintained in the dark at 24° C until the protocorms are "large enough to be transplanted." After that the protocorms and developing plants should be transferred to "a daily regime of 16 hr white fluorescent lights [or other light source] at a temperature of 23 ± 1° C" (Mead and Bulard, 1979).

Culture Media. Two media should be used, one for germination (Table A-63) and a second for transflasking (Table A-64) and seedling culture. The sterilization method is not described in the original paper and therefore it is possible to assume that it was autoclaving.

Procedure. After sowing, the cultures are maintained in the dark until the seedlings are ready for transfer. At that time they are transferred to the second medium (Table A-64) and placed under illumination until they develop large leaves and roots and turn into complete plantlets.

Developmental Sequence. The embryos enlarge, burst through the seed coat, and form small translucent protocorms which are eventually covered with "epidermal hairs."

General Comments. It is not clear whether seedlings produced by this method can be successfully moved to soil. Nevertheless, this method constitutes a significant advance in the germination of north temperate terrestrial orchids because it points to specific requirements of one species and to the procedures which should be employed in research with others.

Literature Cited

Mead, J. W., and C. Bulard, 1979. Vitamins and nitrogen requirements of Orchis laxiflora Lamk. New Phytol. 83:129–136.

Table A-63. Mead and Bulard medium (1979) for germination of *Orchis laxiflora* seeds

Item number	Component	Amount per liter of culture medium (final concentration in culture medium)	Stock solution (a concentrate prepared for repeated and convenient use)	Volume of stock solution per liter of culture medium	Remarks
	Macroelements				
1	Calcium sulfate, $CaSO_4 \cdot 2H_2O$	80 mg	8 g/liter	10 ml	
2	Potassium phosphate, KH_2PO_4	270 mg	27 g/liter	10 ml	
3	Magnesium sulfate, $MgSO_4 \cdot 7H_2O$	240 mg	24 g/liter	10 ml	
4	Ammonium nitrate, NH_4NO_3[a]	80 mg	8 g/liter[a]	10 ml[a]	or weigh[a]
5	Calcium phosphate, $CaH_4(PO_4)_2 \cdot H_2O$[b]	100 mg	10 g/liter[b]	10 ml[b]	
	Iron				
6	Ferric salt of EDTA, $FeNa_2$ EDTA[c]	35 mg	3.5 g/liter	10 ml	
7	*Microelements*[d]				
a	Manganese sulfate, $MnSO_4 \cdot H_2O$	0.1 mg	100 mg/liter		
b	Potassium iodide, KI	0.01 mg	10 mg/liter		
c	Boric acid, H_3BO_3	1 mg	1 g/liter		
d	Copper sulfate, $CuSO_4 \cdot 5H_2O$	0.03 mg	30 mg/liter	1 ml[d]	one solution[d]
e	Zinc sulfate, $ZnSO_4 \cdot 7H_2O$	1 mg	1 g/liter		
f	Nickel chloride, $NiCl_2 \cdot 6H_2O$	0.03 mg	30 mg/liter		
g	Aluminum chloride, $AlCl_3$	0.03 mg	30 mg/liter		
	Vitamins				
8	Thiamine · HCl	0.1 mg	100 mg/100 ml 70% ethanol[e]	0.1 ml	
9	Pyridoxine · HCl	0.1 mg	100 mg/100 ml 70% ethanol[e]	0.1 ml	
10	Niacin · HCl	0.1 mg	100 mg/100 ml 70% ethanol[e]	0.1 ml	
11	Biotin	0.005 mg	0.5 mg/100 ml 70% ethanol[e]	1 ml	
	Complex additive				
12	Vitamin-free casamino acids[f]	5 g	no stock	no stock	weigh
	Sugar[g]				
13	Sucrose	10 g	no stock	no stock	weigh
14	Water, distilled[g]	to 1000 ml			
	Solidifier[g]				
15	Agar	8 g	no stock	no stock	weigh

[a] Solutions containing ammonium and/or nitrate may become contaminated on standing. Therefore, stock solutions should not be prepared. If made, they must be kept frozen between uses.
[b] Shake stock solution well before dispensing.
[c] Available from Sigma Chemical Co.
[d] Add all microelements to the same one liter of distilled water; stir and/or heat until they are dissolved, and add 1 ml per liter of culture medium.
[e] Keep refrigerated between uses.
[f] Available from Difco Laboratories.
[g] Add items 1–12 to 650 ml distilled water (item 14); adjust the pH to 5.6; add sugar (item 13) and adjust volume to 1000 ml with more distilled water (item 14). Bring solution to a gentle boil and add agar (item 15) slowly while stirring. When the agar has dissolved completely (and the solution turns a clear amber color), distribute the medium into culture vessels, cover, and autoclave.

Paphiopedilum

Robert Ernst

Paphiopedilum species and hybrids are popular with hobbyists and have considerable commercial value. However, their seeds are more difficult to germinate than those of tropical epiphytes. Efforts have therefore been made over the years to develop suitable

Table A-64. Mead and Bulard medium (1979) for the culture of *Orchis laxiflora* seedlings

Item number	Component	Amount per liter of culture medium (final concentration in culture medium)	Stock solution (a concentrate prepared for repeated and convenient use)	Volume of stock solution per liter of culture medium	Remarks
	Macroelements				
1	Calcium sulfate, $CaSO_4 \cdot 2H_2O$	8 mg	8 g/liter	10 ml	
2	Potassium phosphate, KH_2PO_4	270 mg	27 g/liter	10 ml	
3	Magnesium sulfate, $MgSO_4 \cdot 7H_2O$	240 mg	24 g/liter	10 ml	
4	Ammonium phosphate, $(NH_4)_2HPO_4$[a]	132 mg[b]	13.2 g/liter[a]	10 ml[a]	or weigh[a]
	Iron				
5	Ferric salt of EDTA, $FeNa_2EDTA$[c]	35 mg	3.5 g/liter	10 ml	
6	*Microelements*[d]				
a	Manganese sulfate, $MnSO_4 \cdot H_2O$	0.1 mg	100 mg/liter	1 ml[d]	one solution[d]
b	Potassium iodide, KI	0.01 mg	10 mg/liter		
c	Boric acid, H_3BO_3	1 mg	1 g/liter		
d	Copper sulfate, $CuSO_4 \cdot 5H_2O$	0.03 mg	30 mg/liter		
e	Zinc sulfate, $ZnSO_4 \cdot 7H_2O$	1 mg	1 g/liter		
f	Nickel chloride, $NiCl_2 \cdot 6H_2O$	0.03 mg	30 mg/liter		
g	Aluminum chloride, $AlCl_3$	0.03 mg	30 mg/liter		
	Vitamins				
7	Thiamine · HCl	0.1 mg	100 mg/100 ml 70% ethanol[e]	0.1 ml	
8	Pyridoxine · HCl	0.1 mg	100 mg/100 ml 70% ethanol[e]	0.1 ml	
9	Niacin	0.1 mg	100 mg/100 ml 70% ethanol	0.1 ml	
10	Biotin	0.005 mg	0.5 mg/100 ml 70% ethanol	1 ml	
	Complex additive				
11	Vitamin-free casamino acids[f]	5 g	no stock	no stock	weigh
	Sugar[g]				
12	Sucrose	20 g	no stock	no stock	weigh
13	Water, distilled[g]	to 1000 ml			
	Solidifier[g]				
14	Agar	6 g	no stock	no stock	weigh

[a] Solutions containing ammonium and/or nitrate may become contaminated on standing. Therefore, stock solutions should not be prepared. If made, they must be kept frozen between uses.
[b] This is an estimate based on the amount of NH_4NO_3 being replaced, which is 80 mg (or 1 mmole) per liter.
[c] Available from Sigma Chemical Co.
[d] Add all microelements to the same one liter of distilled water; stir and/or heat until they are dissolved, and add 1 ml per liter of culture medium.
[e] Store in a refrigerator between uses.
[f] Available from Difco Laboratories.
[g] Add items 1–11 to 650 ml distilled water (item 13); adjust the pH to 5.6; add sugar (item 12) and adjust volume to 1000 ml with more distilled water (item 13). Bring solution to a gentle boil and add agar (item 14) slowly while stirring. When the agar has dissolved completely (and the solution turns a clear amber color), distribute the medium into culture vessels, cover, and autoclave.

asymbiotic seed-germination media (Burgeff, 1936; Liddell, 1953; Thomale, 1954; Yamada, 1952). The procedure outlined here (Ernst, unpublished) has proved to be particularly successful.

Plant Material. It is preferable to culture immature seeds. The seed capsule should be harvested about 10 months after pollination and in any event prior to dehiscing (capsules generally split open one year or longer after setting). Mature seeds from ripe capsules can also be used.

Surface Sterilization. Capsules and seeds must be surface-sterilized by the procedures outlined for tropical orchids (p. 275).

Culture Vessels. The type of vessels used for tropical orchids (p. 275) are also suitable for *Paphiopedilum.*

Culture Conditions. Flasks should be placed under Gro Lux or lights with an intensity of 100–300 ft-candles. The photoperiods and temperatures used for tropical orchids (p. 275) are also suitable for *Paphiopedilum.*

Culture Medium. A medium designated as RE is used for the germination of *Paphiopedilum* (Table A-65).

Procedure. The procedures recommended for planting the seeds of tropical orchids (p. 276) on culture media can be employed for *Paphiopedilum.* When the seedlings have reached the protocorm stage, with many of them having one or more leaves, they should be thinned by transflasking onto medium containing banana and charcoal (Table A-65). Within 8–10 months of transflasking, the seedlings will generally reach a height of about

Table A-65. Medium RE for *Paphiopedilum* seed germination and seedling culture

Item number	Component	Amount per liter of culture medium (final concentration in culture medium)	Stock solution (a concentrate prepared for repeated and convenient use)	Volume of stock solution per liter of culture medium	Remarks
	Macroelements				
1	Ammonium sulfate, $(NH_4)_2SO_4$[a]	150 mg	15 g/liter[a]	10 ml[a]	or weigh[a]
2	Ammonium nitrate, NH_4NO_3[a]	400 mg	40 g/liter[a]	10 ml[a]	or weigh[a]
3	Potassium nitrate, KNO_3[a]	400 mg	40 g/liter[a]	10 ml[a]	or weigh[a]
4	Potassium phosphate, KH_2PO_4	300 mg	30 g/liter	10 ml	
5	Magnesium nitrate, $Mg(NO_3)_2 \cdot 6H_2O$[a]	100 mg	10 g/liter[a]	10 ml[a]	or weigh[a]
6	Calcium nitrate, $Ca(NO_3)_2 \cdot 4H_2O$[a]	150 mg	15 g/liter[a]	10 ml[a]	or weigh[a]
7	Ferrous sulfate, $FeSO_4 \cdot 7H_2O$[b]	25 mg	2.5 g/liter[b]	10 ml[b]	
	Sugar				
8	Fructose (levulose)	20 g	no stock		weigh
	Darkening agent[c]				
9	Vegetable charcoal[d]	2 g	no stock		weigh
	Complex additive[c]				
10	Ripe banana	100 g	no stock		weigh
11	Water, distilled[e]	to 1000 ml			
	Solidifier[e]				
12	Agar	16 g	no stock		weigh

[a] Stock solutions containing nitrogen tend to become contaminated. Therefore, it is preferable not to make stock solutions. If prepared, they must be kept frozen.

[b] A rust-colored precipitate may form on standing. The solution must, therefore, be shaken vigorously before use.

[c] Use only in medium to be employed for the culture of seedlings. Do not use in seed-germination medium.

[d] Obtainable from J. T. Baker Chemical Co.

[e] For seed-germination medium add items 1–7 to 700 ml of distilled water (item 11); adjust pH to 5.2 ± 0.2; add sugar (item 8) and dissolve; adjust volume to 1000 ml with distilled water (item 11); heat to 80–90°C (about 180°F) and add agar (item 12) with constant stirring; when the agar has dissolved dispense medium into culture vessels, autoclave, cool, and use. For seedling medium dissolve items 1–7 in 400 ml of distilled water in a homogenizer, add banana (item 10) and charcoal (item 9), and homogenize for about 3 min; transfer to larger vessel, stir well, add sugar (item 8), and adjust volume to 1000 ml with distilled water (item 11). Adjust pH to 5.2 ± 0.2 and heat to 80–90°C (about 180°F). Add agar while stirring. When the agar is dissolved distribute the medium into culture flasks, autoclave, cool, and use.

5 cm (2 in.) and will be ready for removal to flats. Crowding will substantially retard development.

Developmental Sequence. Embryos will swell within 1–3 months and turn green 4–6 weeks later. Protocorm formation usually occurs 4–8 weeks after greening. The seedlings develop leaves within 1–2 months after protocorm formation.

General Comment. This procedure has proved very successful for the germination of a genus which is otherwise difficult to germinate.

Literature Cited

Burgeff, H. 1936. Samenkeimung der Orchideen und Entwicklung ihrer Keimpflanzen, p. 200. G. Fischer Verlag, Jena.
Liddell, R. W. 1953. Notes on germinating *Cypripedium* seed. Amer. Orchid Soc. Bull. 22:195–197.
Thomale, H. 1954. Die Orchideen. Eugen Ulmer Verlag, Stuttgart.
Yamada, M. 1952. Progress report on germination work. Hawaii Orchid Soc. Bull. 4:24–26.

Rhynchostylis gigantea

Joseph Arditti

Transflasking of seedlings is a common practice among orchid growers, and suitable media for the purpose are important. In the case of *Rhynchostylis gigantea* the use of two media (one for seed germination and another for seedling culture) coupled with frequent transflasking can accelerate growth (Kaewbamrung, 1967).

Plant Material. The procedure is suitable for mature seed and three-month-old seedlings.

Surface Sterilization. Seeds can be sterilized by the procedures recommended for tropical orchids (p. 275). There is no need to surface-sterilize the seedlings.

Culture Vessels. The vessels recommended for tropical orchids (p. 275) can be used.

Culture Conditions. Cultures can be maintained under conditions which are suitable for tropical orchids (p. 275).

Culture Media. One medium is used for seed germination (Table A-66) and another (Table A-67) for seedling culture.

Procedure. Seeds are germinated on the first medium (Table A-66) in the same manner as those of tropical orchids (p. 276). Three months later the seedlings should be transferred to the second medium (Table A-67). After that the seedling should be transferred every 6 months using the same medium (Table A-67). When the seedlings become large enough they should be moved into community pots.

Table A-66. Burgeff's Eg1 medium as used for the germination of *Rhynchostylis gigantea* seeds (Kaewbamrung, 1967)

Item number	Component	Amount per liter of culture medium (final concentration in culture medium)	Stock solution (a concentrate prepared for repeated and convenient use)	Volume of stock solution per liter of culture medium	Remarks
	Macroelements				
1	Calcium nitrate, $Ca(NO_3)_2 \cdot 4H_2O$[a]	1 g	100 g/liter[a]	10 ml[a]	or weigh[a]
2	Magnesium sulfate, $MgSO_4 \cdot 7H_2O$	250 mg	25 g/liter	10 ml	
3	Ammonium sulfate, $(NH_4)_2SO_4$[a]	250 mg	25 g/liter[a]	10 ml[a]	or weigh[a]
4	Ferrous sulfate, $FeSO_4 \cdot 7H_2O$[b]	20 mg	2 g/liter[b]	10 ml[b]	
5	Potassium phosphate, KH_2PO_4	250 mg	25 g/liter	10 ml	
6	Dipotassium phosphate, K_2HPO_4	250 mg	25 g/liter	10 ml	
	Sugar[c]				
7	Sucrose	20 g	no stock	no stock	weigh
	Complex additive[c]				
8	Coconut water from semi-mature fruits	200 ml	no stock	no stock	measure
9	Water, distilled[c]	to 1000 ml			
	Solidifier[c]				
10	Agar	12 g	no stock	no stock	weigh

[a] Solutions containing ammonium and/or nitrate may become contaminated on standing. Therefore, stock solutions should not be prepared. If made, they must be kept frozen between uses.
[b] A rust-colored precipitate may form on standing. This solution must, therefore, be shaken vigorously before use.
[c] Add items 1–6 and 8 to 500 ml distilled water (item 9); adjust the pH to 5.0; add sugar (item 7) and adjust volume to 1000 ml with more distilled water (item 9). Bring solution to a gentle boil and add agar (item 10) slowly while stirring. When the agar has dissolved completely (and the solution turns a clear amber color), distribute the medium into culture vessels, cover, and autoclave.

Developmental Sequence. Seedlings develop additional leaves and roots.

General Comments. The reason for the growth-promoting effect of complex additives (Tables A-66, A-67) is not clear, but their use has obvious practical advantages.

Literature Cited

Kaewbamrung, M. 1967. Transflasking media for Rhynchostylis seedlings. Orchid Soc. Thailand Bull. 1:18–22.

Spathoglottis plicata

Joseph Arditti

A terrestrial orchid, *Spathoglottis plicata* is widespread and can be found from Sumatra (Indonesia) to the Philippines. *Spathoglottis* species found in New Guinea and certain Pacific islands were at one time referred to as *S. plicata* but are now regarded as distinct (Holttum, 1957). The procedure described here was developed in India (Chennaveeraiah and Patil, 1971).

Plant Material. Seeds are obtained from young green (i.e., immature) capsules of *Spathoglottis plicata.*

Surface Sterilization. Capsules should be surface-sterilized in saturated calcium hypochlorite (7 g/100 ml) solution (pp. 254, 261, 275) for 10 min.

Table A-67. Vacin and Went medium as used for seedling culture of *Rhynchostylis gigantea* (Kaewbamrung, 1967)

Item number	Component	Amount per liter of culture medium (final concentration in culture medium)	Stock solution (a concentrate prepared for repeated and convenient use)	Volume of stock solution per liter of culture medium	Remarks
	Macroelements				
1	Potassium nitrate, $KNO_3 \cdot 4H_2O$[a]	525 mg	52.5 g/liter[a]	10 ml[a]	or weigh[a]
2	Potassium phosphate, KH_2PO_4	250 mg	25 g/liter	10 ml	
3	Magnesium sulfate, $MgSO_4 \cdot 7H_2O$	250 mg	25 g/liter	10 ml	
4	Ammonium sulfate, $(NH_4)_2SO_4$[a]	500 mg	50 g/liter[a]	10 ml[a]	or weigh[a]
5	Manganese sulfate, $MnSO_4 \cdot 4H_2O$	7.5 mg	750 mg/liter	10 ml	
6	Tricalcium phosphate, $Ca_3(PO_4)_2$	200 mg	no stock	no stock	weigh
	Iron				
7	Ferric tartrate, $Fe_4(C_4H_2O_5) \cdot 4H_2O$[b]	28 mg	no stock	no stock	weigh
	Sugar				
8	Sucrose	20 g	no stock	no stock	weigh
	Complex additives				
9	Ripe banana[c]	2 half slices/ culture vessel			
10	Potato extract[d]	250 ml			
11	Coconut water from semi-mature fruits	200 ml			
12	Water, distilled[e]	to 1000 ml			
	Solidifier[e]				
13	Agar	8 g	no stock	no stock	weigh

[a] Solutions containing ammonium and nitrate may become contaminated on standing. Therefore, stock solutions should not be prepared. If made, they must be kept frozen between uses.
[b] An equimolar amount of iron as Fe EDTA can be substituted. The formula weight of the ferric tartrate used is 685.4, which means that 28 mg is equivalent to 0.04 mmoles of compound or 0.16 mmoles of Fe^{++}.
[c] Cut a banana into 1 cm slices; cut each slice in half and put both halves into one culture vessel.
[d] Slice 100 g peeled potatoes into 1 cm (0.4 in.) cubes, boil in 500 ml distilled water until the volume is reduced to 250 ml, pass through a sieve, and add to the culture medium. Excess amounts may be stored in a freezer.
[e] Add items 1–7 and 10–11 to 350 ml distilled water (item 12) in a blender, and homogenize for 2 min; adjust the pH to 5.0; add sugar (item 8) and adjust volume to 1000 ml with more distilled water (item 12). Bring solution to a gentle boil and add agar (item 13) slowly while stirring. When the agar has dissolved completely (and the solution turns a clear amber color), distribute the medium into culture vessels, add the banana slices (item 9), cover, autoclave, cool, and use.

Culture Vessels. Culture vessels of the type recommended for tropical orchids (p. 275) can be used.

Culture Conditions. The cultures can be maintained under diffuse daylight, natural day length, and 50–60% relative humidity at 25° ± 2° C or under the conditions recommended for tropical orchids (p. 275).

Culture Medium. A modified White's medium (Table A-68) should be used.

Procedure. Seeds are removed from surface-sterilized immature capsules under sterile conditions and placed on the culture medium. Cultures are then moved to appropriate conditions.

Developmental Sequence. The embryos swell in the 3rd or 4th week and emerge from the testae, and protocorms covered with numerous hairs (rhizoids) are formed. First leaf

Table A-68. Modified White's medium (Mellor and Stace-Smith, 1977; Yeoman and MacLeod, 1977) as used for seed germination and seedling culture of *Spathoglottis plicata* (Chennaveeraiah and Patil, 1973)

Item number	Component	Amount per liter of culture medium (final concentration in culture medium)	Stock solution (a concentrate prepared for repeated and convenient use)	Volume of stock solution per liter of culture medium	Remarks
	Macroelements				
1	Calcium nitrate, $Ca(NO_3)_2 \cdot 4H_2O$[a]	300 mg	30 g/liter[a]	10 ml[a]	or weigh[a]
2	Monosodium phosphate, $NaH_2PO_4 \cdot H_2O$	16.5 mg	16.5 g/liter	10 ml	
3	Potassium chloride, KCl	65 mg	6.5 g/liter	10 ml	
4	Magnesium sulfate, $MgSO_4 \cdot 7H_2O$	720 mg	72 g/liter	10 ml	
5	Potassium nitrate, KNO_3[a]	80 mg	8 g/liter[a]	10 ml[a]	or weigh[a]
6	Sodium sulfate, Na_2SO_4	200 mg	20 g/liter	10 ml	
	Iron				
7	Ferric sulfate, $Fe_2(SO_4)_3$	2.5 mg	250 mg/liter[b]	10 ml[b]	
8	*Microelements*[c]				
a	Manganese sulfate, $MnSO_4 \cdot 4H_2O$	7 mg	7 g/liter	1 ml[c]	one solution[c]
b	Boric acid, H_3BO_3	1.5 mg	1.5 g/liter		
c	Potassium iodide, KI	0.75 mg	750 mg/liter		
d	Zinc sulfate, $ZnSO_4 \cdot 7H_2O$	3 mg	3 g/liter		
	Amino acids				
9	Glycine	3 mg	100 mg/100 ml 70% ethanol[d]	3 ml	
10	Cysteine · HCl	1 mg	100 mg/100 ml 70% ethanol adjusted to pH 8.5[d]	1 ml	
	Vitamins				
11	Niacin (nicotinic acid)	0.5 mg	100 mg/100 ml 70% ethanol[d]	0.5 ml	
12	Pyridoxine · HCl	0.1 mg	100 mg/100 ml 70% ethanol[d]	0.1 ml	
13	Thiamine · HCl	0.1 mg	100 mg/100 ml 70% ethanol[d]	0.1 ml	
	Hormones				
14	Kinetin	1 mg	100 mg/100 ml 70% ethanol[d]	1 ml	
15	2,4-dichlorophenoxyacetic acid (2,4-D)	1 mg	100 mg/100 ml 70% ethanol[d]	1 ml	
	Complex additives[e]				
16	Coconut water	100 ml	no stock	no stock	
17	Casein hydrolysate	1 g	no stock	no stock	weigh
	Sugar[e]				
18	Sucrose	20 g	no stock	no stock	weigh
19	Water, distilled[e]	to 1000 ml			
	Solidifier[e]				
20	Agar	12–15 g	no stock	no stock	weigh

[a] Solutions containing ammonium and/or nitrate may become contaminated on standing. Therefore, stock solutions should not be prepared. If made, they must be kept frozen between uses.
[b] Shake well before use.
[c] Add all microelements to the same one liter of distilled water; stir and/or heat until they are dissolved, and add 1 ml per liter of culture medium.
[d] Store in a refrigerator or a freezer.
[e] Add items 1–8 and 16–17 to 650 ml distilled water (item 19); mix well; adjust the pH to 5.2; add sugar (item 18) and adjust volume to 1000 ml with more distilled water (item 19). Add agar (item 20) and autoclave. While the solution is being autoclaved combine items 9–15 in a small vial and mix well. On removal of the agar solution from the autoclave add to it the mixture, and swirl to mix. Then dispense the complete medium into preautoclaved culture vessels, cool, and use.

primordia appear on the upper surfaces of the protocorms in approximately 8 weeks, and then a pointed vegetative apex appears. The first small leaves appear during the subsequent 2–3 weeks. Roots are formed in the 13th week. Complete plantlets develop after that and must be transflasked when they become crowded.

General Comments. When this procedure is used 40% of the seeds germinate. More than one plant (and up to 10) may grow from a single protocorm in 5% of the seedlings; 15% develop a callus and 10% form adventitious buds.

Literature Cited

Chennaveeraiah, M. S., and S. J. Patil. 1973. In vitro morphogenesis in seed cultures of an orchid Spathoglottis plicata. Proc. 60th Indian Sci. Congr., Part 3:410–411.

Holttum, R. E. 1957. A revised flora of Malaya. Vol. 1. Orchids of Malaya, 2d ed. Government Printing Office, Singapore.

Mellor, F. C., and R. Stace-Smith. 1977. Virus free potatoes by tissue culture, p. 616–646. *In* J. Reinert and Y. P. S. Bajaj (eds.), Applied and fundamental aspects of plant cell tissue and organ culture. Springer Verlag, New York.

Yeoman, M. M., and A. J. MacLeod. 1977. Tissue (callus) culture techniques, p. 31–59. *In* N. E. Street (ed.), Plant tissue and cell structure. University of California Press.

Thunia marshalliana

Joseph Arditti

A native of Burma, *Thunia marshalliana* has been described as both "epiphytal" (Grant, 1895) and terrestrial (Warner *et al.*, 1884). The germination of *T. marshalliana* seeds and seedling development can be enhanced by vitamins (Henriksson, 1951).

Plant Material. This procedure was developed with mature seeds.

Surface Sterilization. The seeds are surface-sterilized by soaking them in saturated calcium hypochlorite (7 g/100 ml water) solution (pp. 254, 261, 275) for 10 min.

Culture Vessels. Erlenmeyer flasks and other vessels of the kind suitable for tropical orchids (p. 275) can be used.

Culture Conditions. Cultures can be maintained under 20° C (68° F), 10-hr days, and a light intensity of 2000 lux or the conditions recommended for tropical orchids.

Culture Media. Vitamin B_6 (pyridoxine) stimulates germination but retards the growth of seedlings, whereas Vitamin B_1 (thiamine) and niacin (nicotinic acid) enhance their development. Therefore, seedlings should be germinated on a medium containing all three vitamins (Table A-69). After germination the seedlings should be transferred to a medium which contains only thiamine and niacin (Table A-69).

Procedure. The procedure outlined for tropical orchids (p. 276) should be followed.

Table A-69. Modified Burgeff Eg1 medium as used for seed germination and seedling culture of *Thunia marshalliana* (Henriksson, 1951)

Item number	Component	Amount per liter of culture medium (final concentration in culture medium)	Stock solution (a concentrate prepared for repeated and convenient use)	Volume of stock solution per liter of culture medium	Remarks
	Macroelements				
1	Calcium nitrate, $Ca(NO_3)_2 \cdot 4H_2O$[a]	1 g	100 g/liter[a]	10 ml[a]	or weigh[a]
2	Monopotassium phosphate, KH_2PO_4	250 mg	25 g/liter	10 ml	
3	Dipotassium phosphate, K_2HPO_4	250 mg	25 g/liter	10 ml	
4	Magnesium sulfate, $MgSO_4 \cdot 7H_2O$	250 mg	25 g/liter	10 ml	
5	Ammonium sulfate, $(NH_4)_2SO_4$[a]	250 mg	25 g/liter[a]	10 ml[a]	or weigh[a]
6	Ferrous phosphate, $Fe(PO_4) \cdot 7H_2O$[b]	50 mg	5 g/liter[b]	10 ml[b]	
	Vitamins				
7	Thiamine · HCl (Vitamin B_1)[c]	0.2 mg	100 mg/100 ml 70% ethanol[d]	0.2 ml	
8	Pyridoxine[e]	0.1 mg	100 mg/100 ml 70% ethanol[d]	0.1 ml	
9	Niacin (nicotinic acid)[c]	0.2 mg	100 mg/100 ml 70% ethanol[d]	0.2 ml	
	Sugar[f]				
10	Glucose	10 g	no stock	no stock	weigh
11	Water, distilled[f]	to 1000 ml			
	Solidifier[f]				
12	Agar	15 g	no stock	no stock	weigh

[a] Solutions containing ammonium and nitrate may become contaminated on standing. Therefore, stock solutions should not be prepared. If made, they must be kept frozen between uses.
[b] Shake well before use.
[c] Use in seed-germination and seedling culture media.
[d] Store in refrigerator and freezer.
[e] Use only in seed-germination medium.
[f] Add items 1–9 to 650 ml distilled water (item 11); adjust the pH to 5.1; add sugar (item 10) and adjust volume to 1000 ml with more distilled water (item 11). Bring solution to a gentle boil and add agar (item 12) slowly while stirring. When the agar has dissolved completely (and the solution turns a clear amber color), distribute the medium into culture vessels, cover, autoclave, cool, and use.

Developmental Sequence. The seeds swell after 14 days and by the end of a month the seedlings appear. Green leaves and roots form following 150 days in culture.

General Comments. Information obtained from basic research has made possible the formulation of separate media for seed germination and seedling growth.

Literature Cited

Grant, B. 1895. The orchids of Burma. The Hanthawaddy Press, Rangoon.
Henriksson, L. E. 1951. Asymbiotic germination of orchids and some effects of vitamins on Thunia marshalliana. Svensk Bot. Tdskr. 45:447–459.
Warner, R., B. S. Williams, and T. Moore. 1884. The orchid album. Vol. 3, pl. 130. B. S. Williams, Publisher, Victoria and Paradise Nurseries, Upper Holloway, England.

Vanilla

Joseph Arditti

Germination of *Vanilla* is of considerable economic importance because this orchid is the source of the spice vanilla (Bouriquet, 1954). It is not surprising therefore that many investigators have worked to develop practical methods which can be used in propagation and breeding programs (Bouriquet, 1947a,b, 1948, 1949, 1950, 1954; Bouriquet and Boiteau, 1937; Knudson, 1950; Lugo Lugo, 1955a,b; Withner, 1955). The culture of immature seeds was first reported in 1950 (Knudson, 1950). Improved methods for the culture of mature (Lugo Lugo, 1955a,b) and immature (Withner, 1955) seeds were published 5 years later.

Plant Material. Mature (Lugo Lugo, 1955a,b) and immature (Withner, 1955) seeds can be cultured.

Surface Sterilization. Mature seeds can be sterilized by soaking them in filtered saturated calcium hypochlorite (7 g/100 ml water) solution (pp. 254, 261, 275). Immature fruits should first be rinsed with 95% ethyl alcohol (ethanol), then immersed in saturated calcium hypochlorite for five min, and after that dipped twice in 95% ethanol and flamed.

Culture Vessels. Test tubes 25 × 200 mm with tightly fitting cotton stoppers and containing 25–30 ml of medium were used in the original research with mature seeds. Screw-cap vials filled with 10–15 ml medium, or Erlenmeyer flasks stoppered with "one-hole rubber stoppers through which a short length of glass tubing was inserted can also be used. The tubing was stuffed with a small wad of cotton." Since the cultures are maintained under relatively high temperatures and low humidities screw-cap tubes (with the caps screwed on tightly) would seem preferable as a means to prevent drying of the medium.

Culture Conditions. Cultures of mature seeds (Lugo-Lugo, 1955a,b) should be maintained in the dark at 32° C (90° F). In the original research cultures of immature seeds

Table A-70. Knudson B medium modified for seed germination and seedling culture of *Vanilla* (Lugo Lugo, 1955)

Item number	Component	Amount per liter of culture medium (final concentration in culture medium)	Stock solution (a concentrate prepared for repeated and convenient use)	Volume of stock solution per liter of culture medium	Remarks
	Macroelements				
1	Calcium chloride, $CaCl_2 \cdot 2H_2O$	623 mg	62.3 g/liter	10 ml	
2	Monopotassium phosphate, KH_2PO_4	250 mg	25 g/liter	10 ml	
3	Magnesium sulfate, $MgSO_4 \cdot 7H_2O$	250 mg	25 g/liter	10 ml	
4	Sodium sulfate, Na_2SO_4	537 mg	53.7 g/liter	10 ml	
5	Ferrous sulfate, $FePO_4 \cdot 4H_2O$[a]	25 mg	2.5 g/liter[a]	10 ml[a]	
6	Urea, NH_2CONH_2[b]	250 mg	25 g/liter[b]	10 ml[b]	or weigh[b]
	Sugar[c]				
7	Sucrose	20 g	no stock	no stock	weigh
8	Water, distilled[c]	to 1000 ml			
	Solidifier[c]				
9	Agar	15 g	no stock	no stock	weigh

[a] Shake well before use.
[b] Solutions containing organic substances may become contaminated on standing. Therefore, stock solutions should not be prepared. If made, they must be kept frozen between uses.
[c] Add items 1–6 to 650 ml distilled water (item 8); adjust the pH to 6.25; add sugar (item 7) and adjust volume to 1000 ml with more distilled water (item 8). Bring solution to a gentle boil and add agar (item 9) slowly while stirring. When the agar has dissolved completely (and the solution turns a clear amber color), distribute the medium into culture vessels, cover, autoclave, cool, and use.

were maintained at temperatures which "varied anywhere from 20° C to approximately 34° C and were usually in the neighborhood of 25–27° C" (68°, 93°, 77°, and 81° F, respectively). Illumination was constant, "provided by banks of red, blue and white 40 watt fluorescent lights about 10 inches from the cultures." It seems safe to assume that such cultures can be maintained under the conditions recommended for tropical orchids (p. 275).

Culture Media. A modified Knudson B medium (Table A-70) is suitable for mature seeds, and Burgeff's N_3f with additives (Table A-71) should be used for immature ones.

Procedure. Mature seeds should be sterilized and placed on the medium in the manner described for tropical orchids. Their germination can be improved by a 24-hr immersion in 3% hydrogen peroxide. Surface-sterilized immature capsules must be split open with a sterile scalpel, and the seeds scraped "with as little placental or pod tissue as possible" and "introduced into the flasks or tubes all spread out" (Withner, 1955). Further care for *Vanilla* seedlings is the same as for those of tropical orchids.

Table A-71. Burgeff N_3f medium as modified for immature seeds and seedling culture of *Vanilla* (Withner, 1955)

Item number	Component	Amount per liter of culture medium (final concentration in culture medium)	Stock solution (a concentrate prepared for repeated and convenient use)	Volume of stock solution per liter of culture medium	Remarks
	Macroelements				
	Solution A				
1	Calcium nitrate, $Ca(NO_3)_2 \cdot 4H_2O$[a]	1 g	100 g/liter[a]	10 ml[a]	or weigh[a]
2	Potassium chloride, KCl[b]	250 mg	25 g/liter[b]	10 ml[b]	
3	Magnesium sulfate, $MgSO_4 \cdot 7H_2O$	250 mg	25 g/liter	10 ml	
4	Ammonium sulfate, $(NH_4)_2SO_4$[a]	250 mg	25 g/liter[a]	10 ml[a]	or weigh[a]
5	Ferrous sulfate, $FeSO_4 \cdot 7H_2O$[b]	20 mg	2 g/liter[b]	10 ml[b]	
	Solution B				
6	Citric acid[a]	90 mg	9 g/liter[a]	10 ml[a]	or weigh[a]
7	Potassium phosphate, K_2HPO_4	250 mg	25 g/liter	10 ml	
	Amino acids				
8	Arginine	12 mg	no stock	no stock	weigh
9	Lysine	18 mg	no stock	no stock	weigh
	Auxin				
10	Indolebutyric acid[c]	1 mg	100 mg/100 ml 70% ethanol[c]	1 ml	
	Sugar[d]				
11	Glucose	10 g	no stock	no stock	weigh
12	Fructose	10 g	no stock	no stock	weigh
13	Water, distilled[d]	to 1000 ml			
	Solidifier[d]				
14	Agar	12–15 g	no stock	no stock	weigh

[a] Solutions containing ammonium, nitrate, or organic compounds may become contaminated on standing. Therefore, stock solutions should not be prepared. If made, they must be kept frozen between uses.

[b] A rust-colored precipitate may form on standing. This solution must, therefore, be shaken vigorously before use.

[c] Store in refrigerator or freezer between uses.

[d] Add items 1–5 to 350 ml distilled water (item 13) and items 5–7 to another 350 ml of distilled water (item 13); mix the two solutions, add the amino acids and auxin; adjust the pH to 5.0; add the sugars (items 11, 12) and adjust volume to 1000 ml with more distilled water (item 13). Bring solution to a gentle boil and add agar (item 14) slowly while stirring. When the agar has dissolved completely (and the solution turns a clear amber color), distribute the medium into culture vessels, cover, autoclave, cool, and use.

Developmental Sequence. When germination starts the testae of mature seeds split and the embryos protrude. The black seed coats of immature seeds split and small white protocorms are formed. Seedlings develop within a year.

General Comments. The development of germination methods for mature and immature seeds made possible the production of *Vanilla* hybrids. It is important to note, however, that incorrect terminology has been and is still being used in connection with these procedures. The fruits of *Vanilla* and other orchids are capsules because they are dehiscent and formed from a compound ovary. They are not pods (more appropriately termed legumes), which are simple fruits formed from a single carpel and typical of the Fabaceae (Leguminosae). Thus it is incorrect to refer to *Vanilla* fruit as pods (green or otherwise). It is even worse to refer to culture of seeds from immature capsules as "green podding."

Ovules consist of a female gametophyte (megagametophyte) surrounded by two integuments (coats). After fertilization a zygote is formed by the fusion of the male and female gametes and develops into a seed. In other words, the ovules as such cease to exist after fertilization has taken place. Therefore, immature seeds should not be called ovules even if they were erroneously referred to by this term in the early literature.

Literature Cited

Bouriquet, G. 1947a. Sur la germination des graines de vanillier (Vanilla planifolia And.). Bull L'Acad. Malagache, n.s. 25, 1:11.

——. 1947b. Sur la germination des graines de vanillier (Vanilla planifolia, And.). L'Agron. Trop. 2:150–164.

——. 1948. La germination des graines de vanillier. L'Agron. Trop. 3:498–499.

——. 1949. La transplantation des jeunes vanilliers issus de semis asymbiotiques. L'Agron. Trop. 4:614–618.

——. 1950. La transplantation des jeunes vanilliers issus de semis asymbiotiques. Bull. Agric. Madagascar No. 23, Tananarive.

——. 1954. Le vanillier et la vanille dans le monde. Editions Paul Lechevalier, Paris.

Bouriquet, G., and P. Boiteau. 1937. Germination asymbiotique de graines de vanillier (Vanilla planifolia And.) Bull. Acad. Malagache, n.s. 20:415–17.

Knudson, L. 1950. Germination of seeds of Vanilla. Amer. J. Bot. 37:241–247.

Lugo Lugo, M. 1955a. Effects of nitrogen on the germination of Vanilla planifolia seeds. Amer. Orchid Soc. Bull. 24:309–312.

——. 1955b. The effect of nitrogen on the germination of Vanilla planifolia. Amer. J. Bot. 42:673–684.

Withner, C. L. 1955. Ovule culture and growth of Vanilla seedlings. Amer. Orchid. Soc. Bull. 24:380–392.

Discussion

Joseph Arditti

The requirements of germinating orchid seeds and developing seedlings have been the subject of several reviews (Arditti, 1967, 1979; Stoutamire, 1974; Warcup, 1975; Withner, 1959, 1974). Therefore, only a short summary will be presented here.

Mineral Nutrition. All germinating orchid seeds and developing seedlings require a well-balanced mixture of minerals (even if some species germinate better following a period in distilled water). The total concentration of these mixtures varies (Table A-72).

Table A-72. Major element composition (in millimoles) of several media used for orchid-seed germination and seedling culture and of orchid-nurturing tree trunk effluate (numbers in parentheses refer to tables)

	Burgeff				Curtis		Fast			Pfeffer	Harvais		
Ion	Eg1 (A-30)	N_3f (A-31)	MN + N (A-26)	Of/N-free (A-27)	1936 (A-10)	modif. (A-7)	1971	F, 1976 (A-34)	FN, 1978 (A-35)	1770 (A-32)	1973 (A-11)	1974 (A-9)	1972 (A-25)
Nitrate	8.40	8.40			5.71	5.71	3.96	2.77	1.36	8.76	8.76	10.37	8.76
Ammonium	3.80	3.80	7.56		2.75	2.75	7.57	2.07	0.81		0.005[a]	5.13[a]	
Nitrate : ammonium ratio	2.2	2.2			2.08	2.08	0.52	1.34	1.68		1.752[a]	2.02	
Phosphate	3.20	1.40	6.86	3.11	0.90	0.88	1.84	0.61	0.24	1.47	1.47	1.47	1.47
Sulfate	2.90	2.90	5.04	1.71	1.06	1.08	4.80	0.34	0.13	0.81	0.81	0.81	0.81
Chloride		3.40	3.07	1.36				2.22	0.44	1.34	1.34	1.34	1.34
Potassium	4.60	6.20	8.58	5.12	0.88	0.88	5.79	2.83	0.68	4.79	4.79	4.79	4.79
Magnesium	1.00	1.00	1.22	1.01	1.06	1.06	1.01	0.34	0.13	0.81	0.81	0.81	0.81
Calcium	4.20	4.20	0.68	0.68	1.48	1.48	0.13	0.35	0.28	3.39	3.39	1.69	3.39
Citrate		0.43									0.003[a]	0.08[a]	0.003[a]
Iron	0.07	0.07	0.04	0.7	0.02	0.02	10 mg chelate	0.014	0.012		0.004[a]	0.1[a]	0.004[a]
Manganese											0.002	0.002	0.002
Sodium			1.71										
Urea													
Ammonium : urea ratio													
Total concentration	28.17	31.8	34.76	13.69	13.86	13.86	6.91	11.54	4.08	21.37	21.38	26.59	21.38
Ranking[c]	20	27	31	10	11	11	5	8	2	14	15	19	15

[a] Estimated.
[b] 5.0 plus an estimated 0.13.
[c] Lowest = 1.

Table A-72—Continued

Ion	Henriksson 1951 (A-69)	Ichihashi and Yamashita (A-24,A-46)	Kaewbamrung 1967 (A-66)	Knudson B (A-6)	Knudson C (A-5)	Lucke CT-1 1976 (A-36)	Lugo Lugo 1955 (A-70)	Mead and Bulard 1975 (A-62)	Mead and Bulard 1979 (A-63)	Mead and Bulard 1979 (A-64)	Murashige and Skoog	Pfeffer (A-32)
Nitrate	12.72	14.38	8.48	8.48	8.40			0.99	0.99		39.4	8.73
Ammonium	3.78	3.39	3.78	7.66	7.60	1.52		0.99	0.99	1	20.61	
Nitrate : ammonium ratio	3.37	4.2	2.24	1.10	1.10			1	1		1.9	
Phosphate	3.56	3.39	3.28	2.16	1.80	1.5	1.95	2.76	2.76	2.98	1.24	1.46
Sulfate	2.90	0.70	2.97	4.79	4.80	0.20	4.79	1.43	1.43	1.43	1.50	0.81
Chloride							8.48				5.98	1.34
Potassium	3.28	7.38	4.72	1.83	1.80	0.74	1.84	1.98	1.98	1.98	20.03	4.77
Magnesium	1.01	0.70	1.01	1.01	1.00	0.20	1.01	0.97	0.97	0.97	1.50	0.81
Calcium	4.24	3.49	4.24	4.24	4.20		4.24	0.85	0.85	0.46	2.99	3.38
Citrate												
Iron	0.42		0.07	0.33	0.09		0.11	0.009	0.009	0.009		
Manganese					0.034							
Sodium							7.56					
Urea							4.16					
Ammonium : urea ratio												
Total concentration	31.91	33.43	28.55	30.5	29.72	4.16	34.14	10.06	10.06	8.91	93.25	21.3
Ranking[c]	28	29	21	25	22	3	30	7	7	6	34	13

[a] Estimated.
[b] 5.0 plus an estimated 0.13.
[c] Lowest = 1.

(continued)

Table A-72—Continued

Ion	Polypodium nutrient (A-28)	Raghavan and Torrey (A-15)	RE (A-65)	Sladden modif. Burgeff	Thomale GD (A-33)	Thompson	Tree trunk effluate	Vacin and Went (A-15)	Veyret 1969 (A-40)	Modif. White 1973 (A-68)	Wynd	
Nitrate		2.00	11.57	8.40	10.06		0.0025	5.19	8.48	3.33	9.8	7.8
Ammonium		2.00	7.27	10.60	5.50	2.99	0.0880	7.56	7.56			
Nitrate : ammonium ratio		1	1.59	0.8	1.82		0.02	0.69	1.12			
Phosphate	1.72	2.76	2.20	2.94	2.20	2.99	0.0105	3.14	1.84	0.12	2.5	3.8
Sulfate	0.41	1.43	1.22	6.50	2.16	1.49	0.0052	4.83	4.79	4.38	1.23	9.7
Chloride							0.1430		0.01			
Potassium	2.88	1.98	6.16	2.94	6.16	3.99	0.0770	7.03	1.84	0.79	2.5	19.4
Magnesium	0.41	0.97	0.67	1.20	0.74	1.49	0.1770	1.01	1.01	2.92	1.23	3.9
Calcium		0.85	0.63	4.80		0.49	0.0250	1.95	4.24	1.27	4.9	1.9
Citrate				1.89								
Iron	0.42		0.09	0.67	0.07		0.0073	0.19	0.004	0.13[b]		
Manganese								0.04		0.03		
Sodium							0.1310			2.94		
Urea						8.99						
Ammonium : urea ratio						0.33						
Total concentration	5.84	12.99	30.46	39.94	24.3	22.43	0.686	30.94	29.77	15.91	22.16	46.5
Ranking[c]	4	9	24	26	18	17	1	26	23	12	16	33

[a] Estimated.
[b] 5.0 plus an estimated 0.13.
[c] Lowest = 1.

Terrestrial species seem to germinate better on more dilute media than those used for epiphytes, but there are exceptions to this generalization. Microelements may or may not be added to media used for orchid-seed germination and seedling culture. This does not mean that some orchids require microelements and others do not. Rather, it means that enough microelements are present as impurities in most media to satisfy requirements.

Sugars. Germinating orchid seeds and young seedlings require an exogenous source of energy. For most species sucrose is satisfactory. However, some orchids germinate better on media which contain glucose and/or fructose (Table A-73) or honey.

Hormones. The seeds of most orchid species do not require an exogenous supply of hormones (Table A-74). Some do, however, or at least they benefit from the addition of hormones to culture media.

Vitamins. As with hormones, the seeds of most orchids do not require the addition of vitamins to culture media, but in some instances vitamins may improve germination and enhance seedling growth (Table A-74).

Complex Additives. A number of complex additives are used as components of some culture media (Table A-74). The reasons for their variable effects are not clear due to the numerous and varied components they contain (Tables A-3, A-75).

Illumination. Epiphytic orchids and some terrestrial species germinate in the light. However, a number of terrestrials germinate best in the dark or under reduced illumination.

Temperature. The optimal temperature for seed germination of most orchids is 20–25° C, and the range extends from 6° to 40° C. Some species may require chilling for varying periods during, before, or after germination, but the most suitable temperature even for these species is 20–25° C.

pH. The seeds of most species germinate best at pH 4.8–5.2 but can tolerate pH as low as 3.3–3.7.

Atmosphere. Normal atmospheric pressure and composition meet the requirements of most orchids, but some species germinate better under different conditions.

Practical Implications. The practical implications of the information presented in this appendix are that there is no single "best" medium or set of conditions for orchid-seed germination. Knudson C medium, sucrose, pH 5.2, 20–25° C, reasonable illumination, and a 12-hr day are optimal or nearly so for most tropical epiphytes. However, the requirements of terrestrial species are different. Tropical terrestrials such as *Pa-*

Table A-73. Sugars used in media employed for seed germination and seedling culture of some orchids[a]

Species	Sugar (g/liter)		
	Sucrose	Glucose	Fructose
Aceras anthropophorum	10	15	
Arundina bambusifolia	20		
Bletilla striata		15	
Calopogon puchellus	20		
Calypso bulbosa		10	
Coeloglossum viride	10	15	
Corallorhiza maculata		10	
Cypripedium acaule		10	
C. parviflorum		10	
C. reginae	10		
Dactylorchis durandii	10	15	
D. maculata	10	15	
D. majalis	10	15	
D. munbyana	10	15	
D. praetermissa	10	15	
D. purpurella	10 or	10	
Dactylorhiza purpurella	10 or 20		
Epipactis gigantea		10	
Epiphytic species, tropical	20		
European terrestrial species	7, 10, 12, 20	10, 12	5, 10
Goodyera oblongifolia		10	
G. pubescens		20	
Gymnadenia conopea	10	15	
Habenaria (see *Platanthera*)			
Liparis loeselii	10	15	
Listera cordata	10	15	
L. ovata	10	15	
Ophrys sphegodes	10, 20	10	
Orchis laxiflora	10, 20		
O. maculata	10	15	
O. maculata var. *elodes*	10	15	
O. militaris	10	15	
O. morio	10	15	
Paphiopedilum			20
Platanthera bifolia	10	15	
P. dilatata		10	
P. hyperborea		10	
P. obtusata		10	
Pogonia ophioglossoides		10	
Rhynchostylis gigantea	20		
Serapias parviflora	20		
Spathoglottis plicata	20		
Spiranthes cernua		10	
S. gracilis		10	
Thunia marshalliana		10	
Vanilla (immature seeds)		10	10
Vanilla (mature seeds)	20		

[a] The nomenclature used in this table is that of the original papers. Therefore, it is possible that synonyms may appear as separate entries.

phiopedilum can now be germinated easily on the RE medium (Table A-65). It is possible, therefore, that this medium may also be suitable for other tropical terrestrials. North temperate terrestrial species, on the other hand, are still difficult to germinate, but there is every reason to believe that with time appropriate methods will be developed for these orchids too.

Table A-74. Organic additives in media used for seed germination and seedling culture of some orchids (amount per liter of culture medium)

Species	Complex additives	Vitamins	Darkening agents	Hormones: Cytokinin	Hormones: Auxins	Amino acids and other additives
Calypso bulbosa	50 ml coconut water, 0.1 ml growth substance $W66_f$ 100 ml potato extract					
Cypripedium reginae	100 ml potato extract	5 mg niacin, 0.5 mg pantothenic acid, 0.5 mg pyridoxine, 5 mg thiamine	2 g graphite			
Dactylorhiza (*Orchis*) *purpurella*	1 g yeast extract, 5 g casamino acids					
Epipactis gingantea	50 ml coconut water, 0.1 ml growth substance $W66_f$					
Epiphytic species, tropical	100–150 g ripe banana		2 g vegetable charcoal			
European terrestrial species	20 ml coconut water, 1.3–3 g peptone, 2 g grass extract, 6 ml fish emulsion, 8.0 ml autolyzed beer yeast, 5 g medical yeast, 3 g potato starch	0.1 mg thiamine, 0.2 mg biotin	2 g vegetable charcoal			130 mg Na nucleinate
Goodyera oblongifolia	100 ml potato extract					
Habenaria (see *Platanthera*)						
Orchis laxiflora	5 g casamino acids					
Paphiopedilum	100 g ripe banana		2 g vegetable charcoal			
Platanthera hyperborea	5 g casamino acids					
Rhynchostylis gigantea	200 ml coconut water, 1 slice ripe banana, 250 ml potato extract					
Spathoglottis plicata	100 ml coconut water, 1 g casein hydrolysate	0.5 mg niacin, 0.1 mg pyridoxine, 0.1 mg thiamine		1 mg kinetin	1 mg 2,4-D, 1 mg indolebutyric acid	3 mg glycine, 1 mg cysteine HCl
Thunia marshalliana		0.2 mg niacin, 0.1 mg pyridoxine, 0.2 mg thiamine				
Vanilla (immature seeds)						12 mg arginine, 18 mg lysine, 90 mg citric acid
Vanilla (mature seeds)						250 mg urea

Table A-75. Approximate partial analysis of banana fruit pulp

Component	Content
Minerals	
Aluminum	small amount
Boric acid	some
Calcium (CaO)	0.028–0.37% (DW, pulp), 8 mg/100 g edible portion
Chlorides (Cl)	0.171–0.308% (DW, pulp)
Copper	0.09 mg/100 g pulp
Iodine	5–200 ppb (fresh fruit)
Iron (Fe_2O_3)	0.0064–0.0079% (DW, pulp), 0.7 mg/100 g edible portion
Magnesium (MgO)	0.18% (DW, pulp)
Phosphates (P_2O_5)	0.179–0.304% (DW, pulp), 26 mg/100 g edible portion
Potassium (K_2O)	1.21–1.68% (DW, pulp), 370 mg/100 g edible portion
Silica (SiO_2)	0.058–0.096% (DW, pulp)
Sodium (Na_2O)	0.201–0.273% (DW, pulp), 1 mg/100 g edible portion
Sulfur (SO_3)	0.046–0.053% (DW, pulp)
Zinc	2.8 mg/kg edible portion of plantain
Ash	
Ash (CO_2 deducted)	3.78% (fruit, just ripe)
Ash	1.33% (in bits of peel which stick to pulp)
Ash	0.70–0.76% (FW, pulp), 0.8 g/100 g edible portion
Nitrogenous substances	
Nitrogenous matter	8.91% (DW, fruit, just ripe)
Protein	0.81–1.49% (FW, pulp), 1.1 g/100 g edible portion
Amide nitrogen	10.13% (N fraction of albumin-globulin protein)
Humin nitrogen	2.92% (N fraction of albumin-globulin protein)
Monoamino nitrogen	64–37% (N fraction of albumin-globulin protein)
Nonamino nitrogen	7.65% (N fraction of albumin-globulin protein)
Amino acids	
Alanine	0.22 (g/g nitrogen or g/100 g edible portion)[a]
Arginine	0.21 (g/g nitrogen or g/100 g edible portion)[a] 11.08% (nitrogen fraction of albumin-globulin protein)
Aspartic acid	0.80 (g/g nitrogen or g/100 g edible portion)[a]
Cystine	0.042 (g/g nitrogen or g/100 g edible portion)[a]
Glutamic acid	1.00 (g/g nitrogen or g/100 g edible portion)[a]
Glycine	0.20 (g/g nitrogen or g/100 g edible portion)[a]
Histidine	0.42 (g/g nitrogen or g/100 g edible portion)[a] 0.39% (nitrogen fraction of albumin-globulin protein)
Iso leucine	0.11 (g/g nitrogen or g/100 g edible portion)[a]
Leucine	0.29 (g/g nitrogen or g/100 g edible portion)[a]
Lysine	0.23 (g/g nitrogen or g/100 g edible portion)[a] 3.26% (nitrogen fraction of albumin-globulin protein)
Methionine	0.036 (g/g nitrogen or g/100 g edible portion)[a]
Phenylalanine	0.14 (g/g nitrogen or g/100 g edible portion)[a]
Proline	0.19 (g/g nitrogen or g/100 g edible portion)[a]
Serine	0.20 (g/g nitrogen or g/100 g edible portion)[a]
Threonine	0.16 (g/g nitrogen or g/100 g edible portion)[a]
Tryptophan	0.072 (g/g nitrogen or g/100 g edible portion)[a]
Tyrosine	0.072 (g/g nitrogen or g/100 g edible portion)[a]
Valine	0.17 (g/g nitrogen or g/100 g edible portion)[a]
Carbohydrates	
Glucose	11.81% (DW, fruit, just ripe)
Nonreducing sugars	6.12–13.38% (FW, pulp)
Reducing sugars	5.44% (in bits of peel which stick to pulp)
Reducing sugars	6.19–10.73% (FW, pulp)
Starch	2.93–6.54% (DW, pulp)
Starch and related material	1.54% (DW, fruit, just ripe)
Sucrose	4.50% (DW, fruit, just ripe)
Sucrose	1.05% (in bits of peel which stick to pulp)
Total carbohydrates	23.7–24.7% (FW, pulp), 22.2 g/100 g edible portion

(*continued*)

Table A-75.—Continued

Component	Content
Cellulose, lignin, fiber, and related substances	
Cellulose	0.15% (FW, fruit)
Crude fiber	22.82% (DW, fruit, just ripe)
Hemicelluloses	1–2% (DW, fruit)
Hemicellulose	0.16% (FW, fruit)
Lignin	0.50% (FW, fruit)
Total fiber	0.80% (FW, fruit), 0.5 g/100 g edible portion
Vitamins	
Vitamin A	131.43 international units/100 mg or 50–332 IU/100 g pulp, 190 IU/100 g edible portion
Thiamine (Vitamin B_1)	0.03 mg/100 g fruit, 0.05 mg/100 g edible portion
Riboflavin (Vitamin B_2)	0.04 mg/100 g fruit
Pyridoxine (Vitamin B_6)	0.34 mg/100 g fruit
Ascorbic acid (Vitamin C)	6.86 mg/100 g fruit or 0.1 mg/g, 10 mg/100 g edible portion
Niacin	0.4 mg/100 g fruit, 0.7 mg/100 g edible portion
Folic acid	0.095 mg/100 g fruit
Biotin	0.0044 mg/100 g fruit
Organic acids	
Acetic acid	some
Butyric acid	some
Citric acid	0.15–0.32% (fruit)
Malic acid	0.053–0.5% (fruit)
Oxalic acid	0.0064% (fruit)
Tartaric acid	some
Pectic substances	
Pectin	0.34–0.57% (FW, pulp)
Protopectin	0.29–0.35% (FW, pulp)
Pectin (calcium pectate)	0.93% (in bits of peel which stick to pulp)
Protopectin (calcium pectate)	0.21% (in bits of peel which stick to pulp)
Fats and fatty acids	
Crude fat	0.53% (in bits of peel which stick to pulp)
Crude fat	0.30–0.47 (FW, pulp)
Linoleic acid	some (in hydrolysate of fat extracted from starch)
Linolenic acid	some (in hydrolysate of fat extracted from starch)
Oleic acid	some (in hydrolysate of fat extracted from starch)
Palmitic acid	some (in hydrolysate of fat extracted from starch)
Fat	0.2 g/100 g edible portion
Sterols	
Phytosterol	some (in hydrolysate of fat extracted from starch)
Sterol	some (in fatty material from fruits)
Polyols and related substances	
Inositol	34 mg/100 g fruit
Phytin [CaMg salt of phytic acid (inositol hexaphosphoric acid)]	0.41–5.11 mg (oven-dried fruit)
Acidity	
pH	4.2–4.75 (ripe fruit)
Titration	4.06–4.46 ml of 1N NaOH needed to neutralize 100 g of pulp
Tanins	
Tanin	2.57–4.35 standard units/100 g fruit
Moisture	
Moisture	88.28% (in bits of peel which stick to pulp)
Moisture	70.6–75.9% (FW, pulp)
Water	75.7% in 100 g edible portion

Sources: Anon., n.d.; Tamura, 1970; Von Loesecke, 1950; Watt and Merrill, 1963.

DW, dry weight; FW, fresh weight.

[a] The original paper (Tamura, 1970) states that amino acids, grams per "gram nitrogen edible portion" or "in 100 gram edible portion," were used to indicate the values but does not state which.

Literature Cited

Anonymous. No date. Nutritive values for bananas. United Brands Co.

Arditti, J. 1967. Factors affecting the germination of orchid seeds. Bot. Rev. 33:1–97.

——. 1979. Aspects of the physiology of orchids. Adv. Bot. Res. 7:421–655. Academic Press, London.

Stoutamire, W. 1974. Terrestrial orchid seedlings, p. 101–128. *In* C. L. Withner (ed.), The orchids: Scientific studies. Wiley-Interscience, New York.

Tamura, S. 1970. Amino acid composition of food in Japan. Japan Agric. Res. Quart. 5:56–60.

Von Loesecke, H. W. 1950. Bananas, chemistry, physiology, technology. Interscience Publishers, New York.

Warcup, J. H. 1975. Factors affecting symbiotic germination of orchid seed, p. 87–104. *In* F. E. Sanders, B. Mosse, and P. B. Tinker (eds.), Endomycorrhizas. Academic Press, London.

Watt, B. K., and A. L. Merrill. 1963. Composition of food. U.S. Dept. of Agric. Handbook No. 8.

Withner, C. L. 1959. Orchid physiology, p. 315–360. *In* C. L. Withner (ed.), The orchids: A scientific survey. Ronald Press, New York.

——. 1974. Developments in orchid physiology, p. 129–168. *In* C. L. Withner (ed.), The orchids: Scientific studies. Wiley-Interscience, New York.

INDEX OF PERSONS

Where only one name is given in the text or in footnotes or literature cited, initials or names have been added by the indexer for identification. Initials were not added for individuals usually referred to in the botanical literature by their last names only (e.g., Linnaeus).

INDEX OF PLANT NAMES

This index includes common, scientific, and regional names as well as other taxonomic designations. Scientific names (*Cattleya labiata,* for example) are italicized. All others appear in Roman type. Taxonomic groupings above the generic level are in capital letters. Boldface numerals denote illustrations.

SUBJECT INDEX

Boldface numerals indicate an illustration.

Library of Congress Cataloging in Publication Data (Revised)

Main entry under title:
Orchid biology.

Includes bibliographies and indexes.
1. Orchids. I. Arditti, Joseph.
QK495.064053 584'.15 76-25648
ISBN 0-8014-1040-1 (v. 1) AACR1
ISBN 0-8014-1276-5 (v. 2)